Lecture Notes in Computer Science 8136

Commenced Publication in 1973
Founding and Former Series Editors:
Gerhard Goos, Juris Hartmanis, and Jan van Leeuwen

Editorial Board

David Hutchison
  Lancaster University, UK
Takeo Kanade
  Carnegie Mellon University, Pittsburgh, PA, USA
Josef Kittler
  University of Surrey, Guildford, UK
Jon M. Kleinberg
  Cornell University, Ithaca, NY, USA
Alfred Kobsa
  University of California, Irvine, CA, USA
Friedemann Mattern
  ETH Zurich, Switzerland
John C. Mitchell
  Stanford University, CA, USA
Moni Naor
  Weizmann Institute of Science, Rehovot, Israel
Oscar Nierstrasz
  University of Bern, Switzerland
C. Pandu Rangan
  Indian Institute of Technology, Madras, India
Bernhard Steffen
  TU Dortmund University, Germany
Madhu Sudan
  Microsoft Research, Cambridge, MA, USA
Demetri Terzopoulos
  University of California, Los Angeles, CA, USA
Doug Tygar
  University of California, Berkeley, CA, USA
Gerhard Weikum
  Max Planck Institute for Informatics, Saarbruecken, Germany

T0213934

Vladimir P. Gerdt   Wolfram Koepf
Ernst W. Mayr   Evgenii V. Vorozhtsov (Eds.)

Computer Algebra in Scientific Computing

15th International Workshop, CASC 2013
Berlin, Germany, September 9-13, 2013
Proceedings

 Springer

Volume Editors

Vladimir P. Gerdt
Joint Institute for Nuclear Research (JINR)
Laboratory of Information Technologies (LIT)
141980 Dubna, Russia
E-mail: gerdt@jinr.ru

Wolfram Koepf
Universität Kassel, Institut für Mathematik
Heinrich-Plett-Straße 40, 34132 Kassel, Germany
E-mail: koepf@mathematik.uni-kassel.de

Ernst W. Mayr
Technische Universität München
Institut für Informatik
Boltzmannstraße 3, 85748 Garching, Germany
E-mail: mayr@in.tum.de

Evgenii V. Vorozhtsov
Institute of Theoretical and Applied Mechanics
Russian Academy of Sciences, 630090 Novosibirsk, Russia
E-mail: vorozh@itam.nsc.ru

ISSN 0302-9743   e-ISSN 1611-3349
ISBN 978-3-319-02296-3   e-ISBN 978-3-319-02297-0
DOI 10.1007/978-3-319-02297-0
Springer Cham Heidelberg New York Dordrecht London

Library of Congress Control Number: 2013947183

CR Subject Classification (1998): F.2, G.2, E.1, I.1, I.3.5, G.1, F.1

LNCS Sublibrary: SL 1 – Theoretical Computer Science and General Issues

© Springer International Publishing Switzerland 2013
This work is subject to copyright. All rights are reserved by the Publisher, whether the whole or part of the material is concerned, specifically the rights of translation, reprinting, reuse of illustrations, recitation, broadcasting, reproduction on microfilms or in any other physical way, and transmission or information storage and retrieval, electronic adaptation, computer software, or by similar or dissimilar methodology now known or hereafter developed. Exempted from this legal reservation are brief excerpts in connection with reviews or scholarly analysis or material supplied specifically for the purpose of being entered and executed on a computer system, for exclusive use by the purchaser of the work. Duplication of this publication or parts thereof is permitted only under the provisions of the Copyright Law of the Publisher's location, in ist current version, and permission for use must always be obtained from Springer. Permissions for use may be obtained through RightsLink at the Copyright Clearance Center. Violations are liable to prosecution under the respective Copyright Law.
The use of general descriptive names, registered names, trademarks, service marks, etc. in this publication does not imply, even in the absence of a specific statement, that such names are exempt from the relevant protective laws and regulations and therefore free for general use.
While the advice and information in this book are believed to be true and accurate at the date of publication, neither the authors nor the editors nor the publisher can accept any legal responsibility for any errors or omissions that may be made. The publisher makes no warranty, express or implied, with respect to the material contained herein.

Typesetting: Camera-ready by author, data conversion by Scientific Publishing Services, Chennai, India
Printed on acid-free paper

Springer is part of Springer Science+Business Media (www.springer.com)

Preface

Since the inception of the object and the subject area of computer algebra, from the 1960s, German scientists have played an important, and in some areas decisive, role in the development of this area of mathematics and computer science. For example, Rüdiger Loos (University of Karlsruhe and then University of Tübingen) is one of the pioneers of the development of algorithmic and software methods of computer algebra; together with Bruno Buchberger and George Collins he produced the first monograph in the world that covered the main areas of computer algebra [Buchberger, B., Collins, G.E., Loos, R. (eds.), Computer Algebra. Symbolic and Algebraic Computation. Springer-Verlag 1982]. This book (translated into Russian in 1986, driven by the rapid growth of interest in computer algebra in the USSR) included four chapters written by R. Loos, two of which were co-authored with B. Buchberger and G. Collins.

In addition to developing algorithms for computer algebra, their software implementation and application in scientific and technical computing, on the initiative of, and (at least) under the initial guidance of experts from Germany, a whole series of computer algebra systems (CASs) for special and general mathematical destination were developed:

— Simath (Horst Zimmer, University Saarbrücken) and Kash / Kant (Michael Pohst, `http://page.math.tu-berlin.de/~kant/`) for computing in algebraic number theory
— MuPAD (Benno Fuchssteiner, University of Paderborn, `http://www.mathworks.com/discovery/mupad.html`) – general-purpose mathematical system
— GAP (Joachim Neubüser, RWTH-Aachen, `http://www.gap-system.org/`) – for calculations in the theory of groups
— Singular (Gert-Martin Greuel and Gerhard Pfister, University of Kaiserslautern, `http://www.singular.uni-kl.de/`) – for computing in polynomial algebra and algebraic geometry
— Felix (Joachim Apel and Uwe Klaus, University of Leipzig, `http://felix.hgb-leipzig.de/`) for computing in polynomial algebra
— Molgen (Adalbert Kerber, University of Bayreuth, `http://www.molgen.de/`) for the generation of molecular structures

It is also to be noted that the CAS Reduce – one of the oldest systems — was co-developed at Konrad-Zuse-Zentrum Berlin (ZIB), our host, for a long time, and Winfried Neun, our local organizer, was the responsible person for Reduce at ZIB.

At present, research on the development and application of methods, algorithms, and programs of computer algebra is performed at universities of Aachen,

Bayreuth, Berlin, Bochum, Bonn, Hannover, Kaiserslautern, Karlsruhe, Kassel, Leipzig, Munich, Oldenburg, Paderborn, Passau, Saarbrücken, Tübingen, and others as well as at research centers (ZIB, MPI, DESY, Fraunhofer Institute).

In connection with the above, it was decided to hold the 15th CASC Workshop in Berlin. The 14 earlier CASC conferences, CASC 1998, CASC 1999, CASC 2000, CASC 2001, CASC 2002, CASC 2003, CASC 2004, CASC 2005, CASC 2006, CASC 2007, CASC 2009, CASC 2010, CASC 2011, and CASC 2012 were held, respectively, in St. Petersburg (Russia), in Munich (Germany), in Samarkand (Uzbekistan), in Konstanz (Germany), in Yalta (Ukraine), in Passau (Germany), in St. Petersburg (Russia), in Kalamata (Greece), in Chişinău (Moldova), in Bonn (Germany), in Kobe (Japan), in Tsakhkadzor (Armenia), in Kassel (Germany), and in Maribor (Slovenia), and they all proved to be very successful.

This volume contains 33 full papers submitted to the workshop by the participants and accepted by the Program Committee after a thorough reviewing process. Additionally, the volume includes the abstracts of the three invited talks.

Polynomial algebra, which is at the core of computer algebra, is represented by contributions devoted to the complexity of solving systems of polynomial equations with small degrees, highly scalable multiplication of distributed sparse multivariate polynomials on many-core systems, fast approximate polynomial evaluation and interpolation, application of Groebner bases for mechanical theorem proving in geometry, application of quantifier elimination for determining whether a univariate polynomial satisfies the sign definite condition, solution of polynomial systems with approximate complex-number coefficients with the aid of a polyhedral algorithm, the solution of a problem of interpolating a sparse, univariate polynomial with the aid of a recursive algorithm using probes of smaller degree than in previously known methods, computation of limit points of the quasi-component of a regular chain with the aid of Puiseux series expansion, solution of a system of polynomial equations as part of algebraic cryptoanalysis by reducing to a mixed integer linear programming problem, an improved QRGCD algorithm for computing the greatest common divisor of two univariate polynomials, construction of classes of irreducible bivariate polynomials.

The invited talk by D. Grigoriev surveys complexity results concerning the solution of tropical linear systems and tropical polynomial systems.

One paper deals with the theory of matrices: deterministic recursive algorithms for the computation of generalized Bruhat decomposition of the matrix are presented therein. It is to be noted that the matrix computations are widely used in many papers in the area of polynomial algebra, which were summarized above.

A number of papers included in the proceedings are devoted to using computer algebra for the investigation of various mathematical and applied topics related to ordinary differential equations (ODEs): computing divisors and common multiples of quasi-linear ordinary differential equations, investigation of local integrability of the ODE systems near a degenerate stationary point, the computation of the dimension of the solution space of a given full-rank system of linear ODEs, application of symbolic calculations and polynomial invariants to the classification of singularities of planar polynomial systems of ODEs, the use of Vessiot's vector field based approach for an analysis of geometric singularities of ODEs.

Several papers deal with applications of symbolic computations for solving partial differential equations (PDEs) in mathematical physics. In one of them, a general symbolic framework is described for boundary problems for linear PDEs. The methods of computer algebra are used intensively in the other two papers for deriving new methods for the numerical solution of two- and three-dimensional viscous incompressible Navier–Stokes equations.

The invited talk by T. Wolf is devoted to the problems arising at the application of computer algebra methods for finding infinitesimal symmetries, first integrals or conservation laws, Lax-pairs, etc. when investigating the integrability of PDEs or ODEs.

Several papers deal with applications of symbolic and symbolic-numeric algorithms in mechanics and physics: the investigation of gyrostat satellite dynamics, modeling of identical particles with pair oscillator interactions, tunneling of clusters through repulsive barriers, application of CAS Maple for investigating a quantum measurements model of hydrogen-like atoms, development of efficient methods to compute the Hopf bifurcations in chemical networks with the aid of the package REDLOG, which is an integral part of CAS Reduce, determination of stationary points for the family of Fermat–Torricelli–Coulomb-like potential functions, the determination of stationary sets of Euler's equations on the Lie algebra with the aid of CASs Maple and *Mathematica.*

The invited talk by A. Griewank is concerned with methods, algorithms, software for, and some history about, the field of automatic differentiation, highlighting original developments and the use of adjoints.

The other topics include the application of the CAS *Mathematica* for the simulation of quantum error correction in quantum computing, the application of the CAS GAP for the enumeration of Schur rings over the group A_5, constructive computation of zero separation bounds for arithmetic expressions, the parallel implementation of fast Fourier transforms with the aid of the SPIRAL library generation system, the use of object-oriented languages such as Java or Scala for implementation of categories as type classes, a survey of industrial applications of approximate computer algebra, i.e., algebraic computation of expressions with inaccurate coefficients represented by floating-point numbers.

The CASC 2013 workshop was supported financially by a generous grant from the Deutsche Forschungsgemeinschaft (DFG). Our particular thanks are due to the members of the CASC 2013 local Organizing Committee in Berlin, i.e., Winfried Neun and Uwe Pöhle (Zuse Institute Berlin), who ably handled all the local arrangements in Berlin. Furthermore, we want to thank all the members of the Program Committee for their thorough work. Finally, we are grateful to W. Meixner for his technical help in the preparation of the camera-ready manuscript for this volume and the design of the conference poster.

July 2013

V.P. Gerdt
W. Koepf
E.W. Mayr
E.V. Vorozhtsov

Organization

CASC 2013 was organized jointly by the Department of Informatics at the Technische Universität München, Germany, and the Konrad Zuse-Zentrum für Informationstechnik Berlin (ZIB), Germany.

Workshop General Chairs

Vladimir P. Gerdt (JINR, Dubna)　　　Ernst W. Mayr (TU München)

Program Committee Chairs

Wolfram Koepf (Kassel)　　　Evgenii V. Vorozhtsov (Novosibirsk)

Program Committee

Sergei Abramov (Moscow)
François Boulier (Lille)
Hans-Joachim Bungart (München)
Victor F. Edneral (Moscow)
Ioannis Z. Emiris (Athens)
Jaime Gutierrez (Santander)
Victor Levandovskyy (Aachen)
Marc Moreno Maza (London, CAN)
Alexander Prokopenya (Warsaw)
Eugenio Roanes-Lozano (Madrid)

Valery Romanovski (Maribor)
Markus Rosenkranz (Canterbury)
Werner M. Seiler (Kassel)
Doru Stefanescu (Bucharest)
Thomas Sturm (Saarbrücken)
Agnes Szanto (Raleigh)
Stephen M. Watt (W. Ontario, CAN)
Andreas Weber (Bonn)
Kazuhiro Yokoyama (Tokyo)

External Reviewers

Ainhoa Aparicio Monforte
Atanas Atanasov
Benjamin Batistic
Carlos Beltran
Francisco Botana
Juergen Braeckle
Alexander Bruno
Morgan Deters
Jean-Guillaume Dumas
Wolfgang Eckhardt
Mark Giesbrecht
Domingo Gomez

Hans-Gert Graebe
Dima Grigoryev
Andy Hone
Max Horn
Martin Horvat
Denis Khmelnov
Kinji Kimura
Alexander Kobel
Christos Konaxis
Christoph Koutschan
Istvan Kovacs
Ryszard Kozera

Heinz Kredel
Wen-Shin Lee
Franois Lemaire
Michael Lieb
Gennadi Malaschonok
Hirokazu Murao
Philipp Neumann
Ulrich Oberst
Dmitrii Pasechnik
Pavel Pech
Ludovic Perret
Eckhard Pfluegel
Nalina Phisanbut
Adrien Poteaux
Andreas Ruffing
Tateaki Sasaki

Yosuke Sato
Raimund Seidel
Takeshi Shimoyama
Ashish Tiwari
Elias Tsigaridas
Benjamin Uekermann
Raimundas Vidunas
Sergue Vinitsky
Dingkang Wang
Yonghui Xia
Josephine Yu
Zafeirakis Zafeirakopoulos
Christoph Zengler
Eugene Zima
Miloslav Znojil

Local Organization

Winfried Neun (Berlin)

Website

http://wwwmayr.in.tum.de/CASC2013/

The Many Faces of Integrability from an Algebraic Computation Point of View (Abstract; *Invited Talk*)

Thomas Wolf

Department of Mathematics, Brock University,
500 Glenridge Avenue,
St. Catharines,
Ontario, Canada L2S 3A1
twolf@brocku.ca

Abstract. Integrability investigations can be performed on smooth objects, like partial differential equations (PDEs), ordinary differential equations (ODEs) and such systems formulated for scalar-, vector-, matrix- or supersymmetric functions, or on discrete objects, for example, in discrete differential geometry. The aim can be to find infinitesimal symmetries, first integrals or conservation laws, Lax-pairs, pre-Hamiltonian operators, recursion operators or consistent face relations. What all the resulting conditions for the existence of these structures have in common is that they are overdetermined systems of equations.

In the talk a number of integrability investigations are discussed and the resulting conditions are characterized. In each case the computer algebra challenges are explained and ways are shown how these challenges can be answered.

Table of Contents

On the Dimension of Solution Spaces of Full Rank Linear Differential Systems

S.A. Abramov[1,*] and M.A. Barkatou[2]

[1] Computing Centre of the Russian Academy of Sciences, Vavilova, 40, Moscow
119333, Russia
`sergeyabramov@mail.ru`
[2] Institut XLIM, Département Mathématiques et Informatique,
Université de Limoges, CNRS, 123, Av. A. Thomas,
87060 Limoges Cedex, France
`moulay.barkatou@unilim.fr`

Abstract. Systems of linear ordinary differential equations of arbitrary orders of full rank are considered. We study the change in the dimension of the solution space that occurs while differentiating one of the equations. Basing on this, we show a way to compute the dimension of the solution space of a given full rank system. In addition, we show how the change in the dimension can be used to estimate the number of steps of some algorithms to convert a given full rank system into an appropriate form.

1 Introduction

Given a system of linear homogeneous differential equations with the coefficients from some "functional" field. Suppose we differentiate one of its equations. What would then happen to the solution space of the system? Would it remain unchanged or would we always get some extra solutions?

In the scalar case, when we differentiate equation $L(y) = 0$, the resulting equation $(L(y))' = 0$ has a larger order than the original one. Let the coefficients of the equations belong to some differential field $\mathbb{K}$, and the solutions be in some "functional" space Λ. If $\mathbb{K}$ and Λ are such that every equation of order m has a solution space of dimension m then equation $(L(y))' = 0$ has more solutions than equation $L(y) = 0$.

For the systems of linear ordinary differential equations the problem is not as simple as for scalar equations. Indeed, the solution space of a system of equations is the intersection of the solution spaces of all the equations of the system. Thus, the fact the solution spaces of the individual equations becomes larger does not imply that their intersection becomes larger too.

---

* Supported in part by the Russian Foundation for Basic Research, project no. 13-01-00182-a. The first author thanks also Department of Mathematics and Informatics of XLIM Institute of Limoges University for the hospitality during his visits.

V.P. Gerdt et al. (Eds.): CASC 2013, LNCS 8136, pp. 1–9, 2013.
© Springer International Publishing Switzerland 2013

In the present paper we prove that differentiating one of the equations in a full rank system, one increases the dimension of the solution space by one (Section 3). In Section 4 some applications of this are discussed.

In Appendix the difference case is briefly considered.

2   Preliminaries

The ring of $m \times m$ matrices with entries in a ring R is denoted by $\mathrm{Mat}_m(R)$. By I_m we denote the identity matrix of order m. The notation M^T is used for the transpose of a matrix (vector) M.

Let $(\mathbb{K}, \partial)$, $\partial =\,'$, be a differential field of characteristic 0 with an algebraically closed constant field $Const(\mathbb{K}) = \{c \in \mathbb{K} \mid \partial c = 0\}$. We denote by Λ a fixed *universal differential extension field* of $\mathbb{K}$ (see [9, Sect. 3.2]). This is a differential extension Λ of $\mathbb{K}$ with $Const(\Lambda) = Const(\mathbb{K})$ such that any differential system

$$\partial y = Ay, \tag{1}$$

with $A \in \mathrm{Mat}_m(\mathbb{K})$ has a solution space of dimension m over the constants.

If, e.g., $\mathbb{K}$ is a subfield of the field $\mathbb{C}((x))$ of formal Laurent series with complex coefficients with $\partial = \frac{d}{dx}$ then we can consider Λ as the quotient field of the ring generated by expressions of form $e^{P(x)}x^\gamma(\psi_0 + \psi_1 \log x + \cdots + \psi_s(\log x)^s)$, where in any such expression

- $P(x)$ is a polynomial in $x^{-1/p}$, where p is a positive integer,
- $\gamma \in \mathbb{C}$,
- s is a non-negative integer and $\psi_i \in \mathbb{C}[[x^{1/p}]]$, $i = 0, 1, \ldots, s$.

Besides first-order systems of form (1) we will consider differential systems of order $r \geqslant 1$ which have the form

$$A_r y^{(r)} + A_{r-1}y^{(r-1)} + \cdots + A_0 y = 0. \tag{2}$$

The coefficient matrices

$$A_0, A_1, \ldots, A_r \tag{3}$$

belong to $\mathrm{Mat}_m(\mathbb{K})$, and A_r (the *leading matrix* of the system) is non-zero.

Remark 1. *If A_r is invertible in $\mathrm{Mat}_m(\mathbb{K})$ then the system (2) is equivalent to the first order system having mr equations: $Y' = AY$, with*

$$A = \begin{pmatrix} 0 & I_m & \cdots & 0 \\ \vdots & \vdots & \ddots & \vdots \\ 0 & 0 & \cdots & I_m \\ \hat{A}_0 & \hat{A}_1 & \cdots & \hat{A}_{r-1} \end{pmatrix}, \tag{4}$$

where $\hat{A}_k = -A_r^{-1}A_k$, $k = 0, 1, \ldots, r - 1$, and

$$Y = \left(y_1 \ldots, y_m, y_1' \ldots, y_m', \ldots, y_1^{(r-1)}, \ldots, y_m^{(r-1)}\right)^T. \tag{5}$$

Therefore if the leading matrix of the system (2) is invertible then the dimension of the solution space of this system is equal to mr.

Denote the ring $\mathrm{Mat}_m(\mathbb{K}[\partial])$ by $\mathcal{D}_m$. System (2) can be written as $L(y) = 0$ where

$$L = A_r \partial^r + A_{r-1} \partial^{r-1} + \cdots + A_0 \in \mathcal{D}_m. \tag{6}$$

System (2) can be also written as a system of m scalar linear equations

$$L_1(y_1, \ldots, y_m) = 0, \quad \ldots, \quad L_m(y_1, \ldots, y_m) = 0, \tag{7}$$

with

$$L_i(y_1, \ldots, y_m) = \sum_{j=1}^{m} l_{ij}(y_j), \quad l_{ij} \in \mathbb{K}[\partial], \quad i, j = 1, \ldots, m, \quad \max_{i,j} \mathrm{ord}\, l_{ij} = r. \tag{8}$$

When a system is represented in form (7) we can rewrite it in form (2) and vice versa. The matrix A_r is the leading matrix of the system regardless of representation form. We suppose also that the system is of *full rank*, i.e., that equations (7) are independent over $\mathbb{K}[\partial]$, in other words the rows

$$\ell_i = (l_{i1}, \ldots, l_{im}), \tag{9}$$

$i = 1, \ldots, m$, are linearly independent over $\mathbb{K}[\partial]$. We say that an operator $L \in \mathcal{D}_m$ is of full rank if the system $L(y) = 0$ is. The leading matrix of L is the leading matrix of the system $L(y) = 0$.

3 Differentiating of an Equation of a Full Rank System

3.1 Formulation of the Main Theorem

Our nearest purpose is to prove the following theorem:

Theorem 1. *Let a system of the form (7) be of full rank. Let the system*

$$L_1(y_1, \ldots, y_m) = 0, \quad \ldots, \quad L_{m-1}(y_1, \ldots, y_m) = 0, \quad \tilde{L}_m(y_1, \ldots, y_m) = 0, \tag{10}$$

be such that its first $m - 1$ equations are as in the system (7) while the m-th equation is the result of differenting of the m-th equation of (7), thus the equation $\tilde{L}_m(y_1, \ldots, y_m) = 0$ is equivalent to the equation $(L_m(y_1, \ldots, y_m))' = 0$. Then the dimension of the solution space of (10) exceeds by 1 the dimension of the solution space of (7).

To prove this theorem we consider first the case when a given system has an invertible leading matrix (in this case the system is certainly of full rank). After this we consider the general case of a system of full rank.

The set of solutions of (10) coincides with the union of the set of solutions of all the systems

$$L_1(y_1, \ldots, y_m) = 0, \quad \ldots, \quad L_{m-1}(y_1, \ldots, y_m) = 0, \quad L_m(y_1, \ldots, y_m) = c, \tag{11}$$

when c runs through the set of constants of Λ (any constant c specifies a system). Note that the fact that the dimension of the solution space of (10) does not exceed the dimension of the solution space of (7) more than by 1 is trivial: if c_1, c_2 are constants and $\varphi, \psi \in \Lambda^m$ are solutions of the system (11) with $c = c_1$, resp. $c = c_2$, then $c_2\varphi - c_1\psi$ is a solution of (7). Thus it is sufficient to prove simply that the differentiation increases the dimension of the solution space of a full rank system.

It is also trivial that if (7) is of full rank then (10) is also of full rank. Going back to (8), (9), let $\tilde{\ell}_m = \partial l_m = (\tilde{l}_{m1}, \ldots, \tilde{l}_{mm})$. If $u_1, \ldots, u_m \in \mathbb{K}[\partial]$ are such that $u_1\ell_1 + \cdots + u_{m-1}\ell_{m-1} + u_m\tilde{\ell}_m = 0$ then $v_1\ell_1 + \cdots + v_{m-1}\ell_{m-1} + v_m\ell_m = 0$ where $v_1 = u_1, \ldots, v_{m-1} = u_{m-1}$, $v_m = u_m\partial$, and if $u_i \neq 0$, $0 \leqslant i \leqslant m$, then $v_i \neq 0$.

3.2   Invertible Leading Matrix Case

Lemma 1. *Let the leading matrix of (7) be invertible. Then the dimension of the solution space of (10) is larger than the dimension of the solution space of (7).*

Proof. Together with the union of the set of solutions of all the systems (11) when c runs through the set of constants of Λ, we consider the system

$$L_1(y_1, \ldots, y_m) = 0, \quad \ldots, \quad L_{m-1}(y_1, \ldots, y_m) = 0,$$
$$L_m(y_1, \ldots, y_m) = y_{m+1}, \quad y'_{m+1} = 0. \quad (12)$$

Observe that the system (12) is equivalent to the system $\tilde{Y}' = \tilde{A}\tilde{Y}$ where the matrix $\tilde{A} \in \mathrm{Mat}_{rm+1}(\mathbb{K})$ is obtained from the matrix (4) by adding the last row of zeros and the last column $(0, \ldots, 0, 1, 0)^T$. The column vector $\tilde{Y}$ is obtained from Y (see (5)) by adding y_{m+1} as the last component. The dimension of the solution space of $\tilde{Y}' = \tilde{A}\tilde{Y}$ is equal to $mr + 1$, while the dimension of the solution space of the original system is equal to mr (Remark 1).

Therefore the system (12) has a solution $(\tilde{y}_1, \ldots, \tilde{y}_m, \tilde{y}_{m+1})$ with $\tilde{y}_{m+1} \neq 0$. Evidently $(\tilde{y}_1, \ldots, \tilde{y}_m)$ is a solution of (11) with $c = \tilde{y}_{m+1} \neq 0$, but $(\tilde{y}_1, \ldots, \tilde{y}_m)$ is not a solution of (7) since $L_m(\tilde{y}_1, \ldots, \tilde{y}_m) \neq 0$. The claim follows.

Thus the dimension of the solution space of (10) is equal to $mr + 1$. This proves the Theorem 1 in the case when the given system has an invertible leading matrix.

3.3   General Case of a System of Full Rank

In [2] the following proposition has been proved:

Proposition 1. *Let L be a full rank operator of the form (6). Then there exists $N \in \mathcal{D}_m$ such that the leading matrix of LN is invertible. (In addition, N can be taken such that LN is of order r).*

Using Lemma 1 and Proposition 1 we can complete the proof of Theorem 1. Let a given system of the form (2) be represented as $L(y) = 0$ where L is as in (6). If the leading matrix A_r of L is invertible then the statement of the theorem follows from Lemma 1. Otherwise let N be an operator such that the leading matrix of LN is invertible (Proposition 1). Set

$$
D = \begin{pmatrix} 0 & \cdots & 0 & 0 \\ \vdots & \ddots & \vdots & \vdots \\ 0 & \cdots & 0 & 0 \\ 0 & \cdots & 0 & 1 \end{pmatrix} \partial + \begin{pmatrix} 1 & \cdots & 0 & 0 \\ \vdots & \ddots & \vdots & \vdots \\ 0 & \cdots & 1 & 0 \\ 0 & \cdots & 0 & 0 \end{pmatrix},
\tag{13}
$$

$D \in \mathcal{D}_m$.

By Lemma 1 the dimension of the solution space of the system $DLN(y) = 0$ is larger than the dimension of the solution space of the system $LN(y) = 0$. This implies that there exists $\varphi \in \Lambda^m$ such that $N(\varphi)$ is a solution of the system $DL(y) = 0$ but is not a solution of $L(y) = 0$. In turn this implies that the dimension of the solution space of the system $DL(y) = 0$ is larger than the dimension of the solution space of the system $L(y) = 0$. Theorem 1 is proved.

Remark 2. *Theorem 1 is valid for the case of a full rank inhomogeneous system as well. That is a system of the form $L(y) = b$, with $L \in \mathcal{D}_m$ of full rank and $b \in \mathbb{K}^m$. First of all note that this system has at least one solution in Λ^m since by adding to y an $(m+1)$-st component with value 1, one can transform the given system into a homogeneous system with a matrix belonging to $\mathrm{Mat}_{m+1}(\mathbb{K})$. The set of solutions in Λ^m of $L(y) = b$ is an affine space over the $\mathrm{Const}(\Lambda)$ and is given by $V_L + f$ where $V_L \subset \Lambda^m$ is the solution space of the homogeneous system $L(y) = 0$ and $f \in \Lambda^m$ is a particular solution of $L(y) = b$. When we differentiate the m-th equation of the system $L(y) = b$ we get a new system $\tilde{L}(y) = \tilde{b}$ where the operator $\tilde{L}$ corresponds to system (10). By Theorem 1 $\dim V_{\tilde{L}} = \dim V_L + 1$.*

4   Some Applications

4.1   The Dimension of the Solution Space of a Given Full Rank System

By Remark 1, if the leading matrix of the system (2) is invertible then the dimension of the solution space of this system is equal to mr. How to find the dimension of the solution space in the general case?

We use the notation

$$[M]_{i,*}, \quad 1 \leqslant i \leqslant m,$$

for the $(1 \times m)$-matrix which is the i-th row of an $(m \times m)$-matrix M. Let a full rank operator $L \in \mathcal{D}_m$ be of the form (6). If $1 \leqslant i \leqslant m$ then define $\alpha_i(L)$ as the maximal integer k, $1 \leqslant k \leqslant r$, such that $[A_k]_{i,*}$ is a nonzero row. The matrix $M \in \mathrm{Mat}_m(\mathbb{K})$ such that $[M]_{i,*} = [A_{\alpha_i(L)}]_{i,*}$, $i = 1, 2, \ldots, m$, is the *row frontal matrix* of L.

Theorem 2. *Let the row frontal matrix of a full rank system $L(y) = 0$, $L \in \mathcal{D}_m$, be invertible. Then the dimension of the solution space of this system is $\sum_{i=1}^{m} \alpha_i(L)$.*

Proof. It follows directly from Theorem 1: when we differentiate $r - \alpha_i(L)$ times the i-th equation of the given system, $i = 1, 2, \ldots, m$, we increase the dimension of the solution space by $mr - \sum_{i=1}^{m} \alpha_i(L)$, and the received full rank system has the leading matrix which coincides with the row frontal matrix of the original system, therefore the obtained system has an invertible leading matrix and the dimension of its solution space is equal to mr.

In [6,7] algorithms to convert a given full rank system into an equivalent system having an invertible row frontal matrix were proposed. It is supposed that the field $\mathbb{K}$ is constructive, in particular that there exists a procedure for recognizing whether a given element of $\mathbb{K}$ is equal to 0. Therefore in such situations we are able to compute the dimension of the solution space of a given full rank system.

4.2   Faster Computation of l-Embracing Systems

Suppose that $\mathbb{K} = K(x)$, $\partial = \frac{d}{dx}$, where K is a field of characteristic zero such that each of its elements is a constant. For any system S of the form (2) the algorithm EG_δ ([4,5]) constructs an l-*embracing* system $\bar{S}$:

$$\bar{A}_r(x)y^{(r)}(x) + \cdots + \bar{A}_1(x)y'(x) + \bar{A}_0(x)y(x) = 0,$$

of the same form, but with the leading matrix $\bar{A}_r(x)$ being invertible, and with the solution space containing all the solutions of S. EG_δ is used for finding a finite super-set of the set of singular points of solutions of the given system.

First we describe briefly the algorithm EG_δ, and then discuss its improvement which is due to Theorem 1.

Let the i-th row of the matrix $A_s(x)$, $0 \leqslant s \leqslant r$, be nonzero and the i-th rows of the matrices $A_{s-1}(x), A_{s-2}(x), \ldots, A_0(x)$ be zero. Let the t-th entry, $1 \leqslant t \leqslant m$, be the last nonzero entry of the i-th row of $A_s(x)$. Then, the number $(r - s) \cdot m + t$ is called the *length* of the i-th equation of the system, and the entry of matrix $A_s(x)$ having indices i, t is called the *last nonzero coefficient* of the i-th equation of the system.

Algorithm EG_δ is based on alternation of reductions and differential shifts. Let us explain how the reduction works. It is checked whether the rows of the leading matrix are linearly dependent over $K(x)$. If they are, coefficients of the dependence $v_1(x), v_2(x), \ldots, v_m(x) \in K[x]$ are found. From the equations of the system corresponding to nonzero coefficients, we select the equation of the greatest length. Let it be the i-th equation. This equation is replaced by the linear combination of the equations with the coefficients $v_1(x), v_2(x), \ldots, v_m(x)$. As a result, the i-th row of the leading matrix vanishes. This step is called *reduction* (the reduction does not increase lengths of the equations).

Let the i-th row of the leading matrix be zero, and let $a(x)$ be the last nonzero coefficient of the i-th equation. Let us divide this equation by $a(x)$, differentiate it, and clear the denominators. This operation is called *differential shift* of the i-th equation of the system. Due to the performed division by the last nonzero coefficient, this operation decreases the length of the i-th equation in the system (2).

The algorithm EG_δ is as follows. If the rows of the leading matrix are linearly dependent over $K(x)$, then the reduction is performed. Suppose that this makes the i-th row of the leading matrix zero. Then, we perform the differential shift of the i-th equation and continue the process of alternated reductions and differential shifts until the leading matrix becomes nonsingular. (We never get the equation $0 = 0$ since the equations of the original system are independent over $K(x)[\partial]$.)

As we have mentioned no single equation increases its length due to the reduction. The differential shift decreases the length of the corresponding equation. Thus the sum of all the lengths is decreased by a "reduction + differential shift" step. This implies that algorithm EG_δ always terminates and the number of "reduction + differential shift" steps does not exceed $(r+1)m^2$.

Note that the division by the last coefficient of an equation before differentiating the equation is produced to ensure decreasing of the length of the equation. This division and clearing the denominators after the differentiation are quite expensive. If we exclude this division then the cost of a step "reduction + differentiation" will be in general significantly less than the cost of a "reduction + differential shift" step. By Theorem 1 the corresponding sequence of "reduction + differentiation" steps will be finite (thus the new version of EG_δ terminates for any system of the form (2)) and the number k of the "reduction + differentiation" steps does not exceed mr. (By Theorem 1 the dimension of the solution space of the original system is equal to $mr - k$; thus we have one more way to compute the dimension of the solution space besides the one given by Theorem 2). Note that Theorem 1 is applicable since $K(x) \subset \bar{K}(x)$ where $\bar{K}$ is the algebraic closure of K.

This improvement trick works also in the case of an inhomogeneous system when the corresponding homogeneous system is of full rank. The corresponding homogeneous system is transformed independently on the right-hand side when we produce "reduction + differentiation" steps. Therefore the dimension of the solution space of the corresponding homogeneous system increases due to every "reduction + differentiation" step. The upper bound mr keeps valid in the inhomogeneous case (one can also use Remark 2 for proving this).

In addition, we note that due to Theorem 1 it is not necessary to select an equation of maximal length in the reduction substep; therefore various strategies of a row selection on the reduction substep of each "reduction + differentiation" step are possible. Such strategies make it possible to slow down the growth of degrees of system coefficients when applying EG_δ (due to Appendix this works also in the case of difference systems, i.e. gives an improvement of EG_σ [1,3,5]).

Appendix: The Difference Case

A statement similar to Theorem 1 is valid in the difference case, when $(\mathbb{K}, \sigma)$ is a difference field (σ is an automorphism of $\mathbb{K}$) of characteristic 0 with an algebraically closed constant field $Const(\mathbb{K}) = \{c \in \mathbb{K} \mid \sigma c = c\}$. Let Λ the universal Picard-Vessiot ring extension of $\mathbb{K}$ (see [8, Sect. 1.4]). A system is of full rank if its equations are independent over the ring $\mathbb{K}[\sigma]$. The application of the operator $\Delta = \sigma - 1$ is used instead of the differentiation of an equation of a given full rank system.

The proof is a little more complicated since the invertibility of A is needed to guarantee that the dimension space of a system $\sigma y = Ay$, $A \in \mathrm{Mat}_m(\mathbb{K})$, is equal to m . However there is no problem with proving the analog of Lemma 1. After applying Δ to the last equation of the original system we get the system $\sigma \tilde{Y} = \tilde{A}\tilde{Y}$ where the matrix $\tilde{A} \in \mathrm{Mat}_{rm+1}(\mathbb{K})$ is obtained from the matrix (4) by adding the last row $(0, \ldots, 0, 1)$ and the last column $(0, \ldots, 0, 1, 1)^T$. If A is invertible then $\tilde{A}$ is invertible too. Similarly to the differential case the column vector $\tilde{Y}$ is obtained from Y (see (5)) by adding y_{m+1} as the last component.

Consider a system of order $r \geqslant 1$ which has the form

$$A_r \sigma^r y + A_{r-1} \sigma^{r-1} y + \cdots + A_0 y = 0. \tag{14}$$

The coefficient matrices $A_0, A_1, \ldots, A_r$ belong to $\mathrm{Mat}_m(\mathbb{K})$, and if A_r, A_0 (the leading and *trailing* matrices of the system) are invertible then the system (14) is equivalent to the first order system having mr equations: $\sigma Y = AY$, with A as in (4), and A is invertible since $\det A = -\det \hat{A}_0 = \det A_r^{-1} \det A_0 \neq 0$. Therefore if both the leading and trailing matrices of the system (14) are invertible then the dimension of the solution space of this system is equal to mr.

Denote the ring $\mathrm{Mat}_m(\mathbb{K}[\sigma])$ by $\mathcal{E}_m$. System (14) can be written as $L(y) = 0$ where

$$L = A_r \sigma^r + A_{r-1}\sigma^{r-1} + \cdots + A_0 \in \mathcal{E}_m. \tag{15}$$

Similarly to the differential case, we say that the operator $L \in \mathcal{E}_m$ is of full rank if the system $L(y) = 0$ is of full rank.

It can be shown (see [5, Sect. 3.5]) that for any full rank operator L of the form (15) there exists $F \in \mathcal{E}_m$ such that the product FL is an operator of order $r+1$ with both the leading and trailing matrices are invertible. Using adjoint difference operators we can analogously to the differential case prove that there exists $N \in \mathcal{E}_m$ such that the operator LN has invertible both the leading and trailing matrices (N can be taken such that LN is of order $r+1$). We can consider the operator D which is obtained from (13) by replacing ∂ by $\Delta = \sigma - 1$ and repeat the reasoning given in the last paragraph above Remark 2.

References

1. Abramov, S.A.: EG–eliminations. J. of Difference Equations and Applications 5(4-5), 393–433 (1999)
2. Abramov, S.A., Barkatou, M.A., Khmelnov, D.E.: On full rank differential systems with power series coefficients. J. of Symbolic Computation (submitted)
3. Abramov, S.A., Bronstein, M.: On solutions of linear functional systems. In: Proc. ISSAC 2001, pp. 1–6 (2001)
4. Abramov, S.A., Khmelnov, D.E.: On singular points of solutions of linear differential systems with polynomial coefficients. Journal of Mathematical Sciences 185(3), 347–359 (2012)
5. Abramov, S.A., Khmelnov, D.E.: Linear differential and difference Systems: EG_δ- and EG_σ-eliminations. Programming and Computer Software 39(2), 91–109 (2013)
6. Barkatou, M.A., El Bacha, C., Pflügel, E.: Simultaneously row- and column-reduced higher-order linear differential systems. In: Proc. of ISSAC 2010, pp. 45–52 (2010)
7. Barkatou, M.A., El Bacha, C., Labahn, G., Pflügel, E.: On simultaneously row and column reduction of higher-order linear differential systems. J. of Symbolic Comput. 49(1), 45–64 (2013)
8. van der Put, M., Singer, M.F.: Galois Theory of Difference Equations. Lectures Notes in Mathematics, vol. 1666. Springer, Heidelberg (1997)
9. van der Put, M., Singer, M.F.: Galois Theory of Linear Differential Equations. Grundlehren der mathematischen Wissenschaften, vol. 328. Springer, Heidelberg (2003)

Polyhedral Methods for Space Curves Exploiting Symmetry Applied to the Cyclic n-roots Problem[*]

Danko Adrovic and Jan Verschelde

Department of Mathematics, Statistics, and Computer Science
University of Illinois at Chicago
851 South Morgan (M/C 249)
Chicago, IL 60607-7045, USA
{jan,adrovic}@math.uic.edu
www.math.uic.edu/~jan, www.math.uic.edu/~adrovic

Abstract. We present a polyhedral algorithm to manipulate positive dimensional solution sets. Using facet normals to Newton polytopes as pretropisms, we focus on the first two terms of a Puiseux series expansion. The leading powers of the series are computed via the tropical prevariety. This polyhedral algorithm is well suited for exploitation of symmetry, when it arises in systems of polynomials. Initial form systems with pretropisms in the same group orbit are solved only once, allowing for a systematic filtration of redundant data. Computations with cddlib, Gfan, PHCpack, and Sage are illustrated on cyclic n-roots polynomial systems.

Keywords: Algebraic set, Backelin's Lemma, cyclic n-roots, initial form, Newton polytope, polyhedral method, polynomial system, Puiseux series, symmetry, tropism, tropical prevariety.

1   Introduction

We consider a polynomial system $\mathbf{f}(\mathbf{x}) = \mathbf{0}$, $\mathbf{x} = (x_0, x_1, \ldots, x_{n-1})$, $\mathbf{f} = (f_1, f_2, \ldots, f_N)$, $f_i \in \mathbb{C}[\mathbf{x}]$, $i = 1, 2, \ldots, N$. Although in many applications the coefficients of the polynomials are rational numbers, we allow the input system to have approximate complex numbers as coefficients. For $N = n$ (as many equations as unknowns), we expect in general to find only isolated solutions. In this paper we focus on cases $N \geq n$ where the coefficients are so special that $\mathbf{f}(\mathbf{x}) = \mathbf{0}$ has an algebraic set as a solution.

Our approach is based on the following observation: if the solution set of $\mathbf{f}(\mathbf{x}) = \mathbf{0}$ has a space curve, then this space curve extends from $\mathbb{C}^* = \mathbb{C} \setminus \{0\}$ to infinity. In particular, the space curve intersects hyperplanes at infinity at isolated points. We start our series development of the space curve at these

[*] This material is based upon work supported by the National Science Foundation under Grant No. 0713018 and Grant No. 1115777.

© Springer International Publishing Switzerland 2013

isolated points. Computing series developments for solutions of polynomial systems is a hybrid symbolic-numeric method, appropriate for inputs which consist of approximate numbers (the coefficients) and exact data (the exponents).

In this paper we will make various significant assumptions. First we assume that the algebraic sets we consider are reduced, that is: free of multiplicities. Moreover, an algebraic set of dimension d is in general position with respect to the first d coordinate planes. For example, we assume that a space curve is not contained in a plane perpendicular to the first coordinate axis. Thirdly, we assume the algebraic set of dimension d to intersect the first d coordinate planes at regular solutions.

Our approach consists of two stages. The computation of the candidates for the leading powers of the Puiseux series is followed by the computation of the leading coefficients and the second term of the Puiseux series, if the leading term of the series does not already entirely satisfy the system. Following our assumptions, the second term of the Puiseux series indicates the existence of a space curve. If the system is invariant to permutation of the variables, then it suffices to compute only the generators of the solution orbits. We then develop the Puiseux series only at the generators. Although our approach is directed at general algebraic sets, our approach of exploiting symmetry applies also to the computation of all isolated solutions. Our main example is one family of polynomial systems, the cyclic n-roots system.

Related Work. Our approach is inspired by the constructive proof of the fundamental theorem of tropical algebraic geometry in [32] (an alternative proof is in [39]) and related to finiteness proofs in celestial mechanics [27], [30]. The initial form systems allow the elimination of variables with the application of coordinate transformations, an approach presented in [29] and related to the application of the Smith normal form in [25]. The complexity of polyhedral homotopies is studied in [33] and generalized to affine solutions in [28]. Generalizations of the Newton-Puiseux theorem [43], [58], can be found in [5], [7], [37], [38], [45], and [47]. A symbolic-numeric computation of Puiseux series is described in [40], [41], and [42]. Algebraic approaches to exploit symmetry are [13], [20], [23], and [50]. The cyclic n-roots problem is a benchmark for polynomial system solvers, see e.g: [9], [13], [14], [16], [17], [18], [20], [35], [50], and relevant to operator algebras [10], [26], [54]. Our results on cyclic 12-roots correspond to [46].

Our Contributions. This paper is a thorough revision of the unpublished preprint [2], originating in the dissertation of the first author [1], which extended [3] from the plane to space curves. In [4] we gave a tropical version of Backelin's Lemma in case $n = m^2$, in this paper we generalize to the case $n = \ell m^2$. Our approach improves homotopies to find all isolated solutions. Exploiting symmetry we compute only the generating cyclic n-roots, more efficiently than the symmetric polyhedral homotopies of [57].

2 Initial Forms, Cyclic n-roots, and Backelin's Lemma

In this section we introduce our approach on the cyclic 4-roots problem. For this problem we can compute an explicit representation for the solution curves. This explicit representation as monomials in the independent parameters for positive dimensional solution sets generalizes into the tropical version of Backelin's Lemma.

2.1 Newton Polytopes, Initial Forms, and Tropisms

In this section we first define Newton polytopes, initial forms, pretropisms, and tropisms. The sparse structure of a polynomial system is captured by the sets of exponents and their convex hulls.

Definition 1. Formally we denote a polynomial $f \in \mathbb{C}[\mathbf{x}]$ as

$$f(\mathbf{x}) = \sum_{\mathbf{a} \in A} c_{\mathbf{a}} \mathbf{x}^{\mathbf{a}}, \quad c_{\mathbf{a}} \in \mathbb{C}^*, \quad \mathbf{x}^{\mathbf{a}} = x_0^{a_0} x_1^{a_1} \cdots x_{n-1}^{a_{n-1}}, \tag{1}$$

and we call the set A of exponents *the support of* f. The convex hull of A is *the Newton polytope of* f. The tuple of supports $\mathbf{A} = (A_1, A_2, \ldots, A_N)$ span the Newton polytopes $\mathbf{P} = (P_1, P_2, \ldots, P_N)$ of the polynomials $\mathbf{f} = (f_1, f_2, \ldots, f_N)$ of the system $\mathbf{f}(\mathbf{x}) = \mathbf{0}$.

The development of a series starts at a solution of an initial form of the system $\mathbf{f}(\mathbf{x}) = \mathbf{0}$, with supports that span faces of the Newton polytopes of $\mathbf{f}$.

Definition 2. Let $\mathbf{v} \neq \mathbf{0}$, denote $\langle \mathbf{a}, \mathbf{v} \rangle = a_0 v_0 + a_1 v_1 + \cdots + a_{n-1} v_{n-1}$, and let f be a polynomial supported on A. Then, *the initial form of f in the direction of* $\mathbf{v}$ is

$$\mathrm{in}_{\mathbf{v}}(f) = \sum_{\mathbf{a} \,\in\, \mathrm{in}_{\mathbf{v}}(A)} c_{\mathbf{a}} \mathbf{x}^{\mathbf{a}}, \quad \text{where} \quad \mathrm{in}_{\mathbf{v}}(A) = \{ \, \mathbf{a} \in A \mid \langle \mathbf{a}, \mathbf{v} \rangle = \min_{\mathbf{b} \in A} \langle \mathbf{b}, \mathbf{v} \rangle \, \}. \tag{2}$$

The initial form of a system $\mathbf{f}(\mathbf{x}) = \mathbf{0}$ with polynomials in $\mathbf{f} = (f_1, f_2, \ldots, f_N)$ in the direction of $\mathbf{v}$ is denoted by $\mathrm{in}_{\mathbf{v}}(\mathbf{f}) = (\mathrm{in}_{\mathbf{v}}(f_1), \mathrm{in}_{\mathbf{v}}(f_2), \ldots, \mathrm{in}_{\mathbf{v}}(f_N))$. If the number of monomials with nonzero coefficient in each $\mathrm{in}_{\mathbf{v}}(f_k)$, for all $k = 1, 2, \ldots, N$, is at least two, then $\mathbf{v}$ is a *pretropism*.

The notation $\mathrm{in}_{\mathbf{v}}(f)$ follows [53], where $\mathbf{v}$ represents a weight vector to order monomials. The polynomial $\mathrm{in}_{\mathbf{v}}(f)$ is homogeneous with respect to $\mathbf{v}$. Therefore, in solutions of $\mathrm{in}_{\mathbf{v}}(f)(\mathbf{x}) = \mathbf{0}$ we can set x_0 to the free parameter t. In [12] and [34], initial form systems are called truncated systems.

Faces of Newton polytopes P spanned by two points are edges and all vectors $\mathbf{v}$ that lead to the same $\mathrm{in}_{\mathbf{v}}(P)$ (the convex hull of $\mathrm{in}_{\mathbf{v}}(A)$) define a polyhedral cone (see e.g. [59] for an introduction to polytopes).

Definition 3. Given a tuple of Newton polytopes $\mathbf{P}$ of a system $\mathbf{f}(\mathbf{x}) = \mathbf{0}$, *the tropical prevariety* of $\mathbf{f}$ is the common refinement of the normal cones to the edges of the Newton polytopes in $\mathbf{P}$.

Our definition of a tropical prevariety is based on the algorithmic characterization in [11, Algorithm 2], originating in [44]. Consider for example the special case of two polytopes P_1 and P_2 and take the intersection of two cones, normal to two edges of the two polytopes. If the intersection is not empty, then the intersection contains a vector $\mathbf{v}$ that defines a tuple of two edges $(\text{in}_\mathbf{v}(P_1), \text{in}_\mathbf{v}(P_2))$.

Definition 4. For space curves, the special role of x_0 is reflected in the normal form of the Puiseux series:

$$\begin{cases} x_0 = t^{v_0} \\ x_i = t^{v_i}(y_i + z_i t^{w_i}(1 + O(t))), & i = 1, 2, \ldots, n-1, \end{cases} \tag{3}$$

where the leading powers $\mathbf{v} = (v_0, v_1, \ldots, v_{n-1})$ define a *tropism*.

In the definition above, it is important to observe that the tropism $\mathbf{v}$ defines as well the initial form system $\text{in}_\mathbf{v}(\mathbf{f})(\mathbf{x}) = \mathbf{0}$ that has as solution the initial coefficients of the Puiseux series.

Every tropism is a pretropism, but not every pretropism is a tropism, because pretropisms depend only on the Newton polytopes. For a d-dimensional algebraic set, a d-dimensional polyhedral cone of tropisms defines the exponents of Puiseux series depending on d free parameters.

2.2 The Cyclic n-roots Problem

For $n = 3$, the cyclic n-roots system originates naturally from the elementary symmetric functions in the roots of a cubic polynomial. For $n = 4$, the system is

$$\mathbf{f}(\mathbf{x}) = \begin{cases} x_0 + x_1 + x_2 + x_3 = 0 \\ x_0 x_1 + x_1 x_2 + x_2 x_3 + x_3 x_0 = 0 \\ x_0 x_1 x_2 + x_1 x_2 x_3 + x_2 x_3 x_0 + x_3 x_0 x_1 = 0 \\ x_0 x_1 x_2 x_3 - 1 = 0. \end{cases} \tag{4}$$

The permutation group which leaves the equations invariant is generated by $(x_0, x_1, x_2, x_3) \to (x_1, x_2, x_3, x_0)$ and $(x_0, x_1, x_2, x_3) \to (x_3, x_2, x_1, x_0)$. In addition, the system is equi-invariant with respect to the action $(x_0, x_1, x_2, x_3) \to (x_0^{-1}, x_1^{-1}, x_2^{-1}, x_3^{-1})$.

With $\mathbf{v} = (+1, -1, +1, -1)$, there is a unimodular coordinate transformation M, denoted by $\mathbf{x} = \mathbf{z}^M$:

$$\text{in}_\mathbf{v}(\mathbf{f})(\mathbf{x}) = \begin{cases} x_1 + x_3 = 0 \\ x_0 x_1 + x_1 x_2 + x_2 x_3 + x_3 x_0 = 0 \\ x_1 x_2 x_3 + x_3 x_0 x_1 = 0 \\ x_0 x_1 x_2 x_3 - 1 = 0 \end{cases}$$

$$M = \begin{bmatrix} +1 & -1 & +1 & -1 \\ 0 & 1 & 0 & 0 \\ 0 & 0 & 1 & 0 \\ 0 & 0 & 0 & 1 \end{bmatrix} \qquad \mathbf{x} = \mathbf{z}^M : \begin{cases} x_0 = z_0^{+1} \\ x_1 = z_0^{-1} z_1 \\ x_2 = z_0^{+1} z_2 \\ x_3 = z_0^{-1} z_3 \end{cases} \qquad (5)$$

The system in$_\mathbf{v}(\mathbf{f})(\mathbf{z}) = \mathbf{0}$ has two solutions. These two solutions are the leading coefficients in the Puiseux series. In this case, the leading term of the series vanishes entirely at the system so we write two solution curves as $\left(t, -t^{-1}, -t, t^{-1}\right)$ and $\left(t, t^{-1}, -t, -t^{-1}\right)$. To compute the degree of the two solution curves, we take a random hyperplane in $\mathbb{C}^4$: $c_0 x_0 + c_1 x_1 + c_2 x_2 + c_3 x_3 + c_5 = 0$, $c_i \in \mathbb{C}^*$. Then the number of points on the curve and on the random hyperplane equals the degree of the curve. Substituting the representations we obtained for the curves into the random hyperplanes gives a quadratic polynomial in t (after clearing the denominator t^{-1}), so there are two quadric curves of cyclic 4-roots.

2.3   A Tropical Version of Backelin's Lemma

In [4], we gave an explicit representation for the solution sets of cyclic n-roots, in case $n = m^2$, for any natural number $m \geq 2$. Below we state Backelin's Lemma [6], in its tropical form.

Lemma 1 (Tropical Version of Backelin's Lemma). *For $n = m^2 \ell$, where $\ell \in \mathbb{N} \setminus \{0\}$ and ℓ is no multiple of k^2, for $k \geq 2$, there is an $(m-1)$-dimensional set of cyclic n-roots, represented exactly as*

$$\begin{aligned} x_{km+0} &= u^k t_0 \\ x_{km+1} &= u^k t_0 t_1 \\ x_{km+2} &= u^k t_0 t_1 t_2 \\ &\vdots \\ x_{km+m-2} &= u^k t_0 t_1 t_2 \cdots t_{m-2} \\ x_{km+m-1} &= \gamma u^k t_0^{-m+1} t_1^{-m+2} \cdots t_{m-3}^{-2} t_{m-2}^{-1} \end{aligned} \qquad (6)$$

for $k = 0, 1, 2, \ldots, m-1$, free parameters $t_0, t_1, \ldots, t_{m-2}$, constants $u = e^{\frac{i 2\pi}{m\ell}}$, $\gamma = e^{\frac{i\pi\beta}{m\ell}}$, with $\beta = (\alpha \mod 2)$, and $\alpha = m(m\ell - 1)$.

Proof. By performing the change of variables $y_0 = t_0$, $y_1 = t_0 t_1$, $y_2 = t_0 t_1 t_2$, $\ldots, y_{m-2} = t_0 t_1 t_2 \cdots t_{m-2}$, $y_{m-1} = \gamma t_0^{-m+1} t_1^{-m+2} \cdots t_{m-3}^{-2} t_{m-2}^{-1}$, the solution (6) can be rewritten as

$$x_{km+j} = u^k y_j, \quad j = 0, 1, \ldots, m-1. \qquad (7)$$

The solution (7) satisfies the cyclic n-roots system by plain substitution as in the proof of [19, Lemma 1.1], whenever the last equation $x_0 x_1 x_2 \cdots x_{n-1} - 1 = 0$ of the cyclic n-roots problem can also be satisfied.

We next show that we can always satisfy the equation $x_0 x_1 x_2 \cdots x_{n-1} - 1 = 0$ with our solution. First, we perform an additional change of coordinates to separate the γ coefficient. We let $y_0 = Y_0$, $y_1 = Y_1$, ..., $y_{m-2} = Y_{m-2}$, $y_{m-1} = \gamma Y_{m-1}$. Then on substitution of (7) into $x_0 x_1 x_2 \cdots x_{n-1} - 1 = 0$, we get

$$(\gamma^{m\ell} \underbrace{u^0 u^0 \cdots u^0}_{m} \underbrace{u^1 u^1 \cdots u^1}_{m} \cdots$$

$$\cdots \underbrace{u^{m\ell-1} u^{m\ell-1} \cdots u^{m\ell-1}}_{m} Y_0^{m\ell} Y_1^{m\ell} Y_2^{m\ell} \cdots Y_{m-2}^{m\ell} Y_{m-1}^{m\ell}) - 1 = 0$$

$$(\gamma^{m\ell} u^{m(0+1+2+\cdots+(m\ell-1))} Y_0^{m\ell} Y_1^{m\ell} Y_2^{m\ell} \cdots Y_{m-2}^{m\ell} Y_{m-1}^{m\ell}) - 1 = 0 \tag{8}$$

$$(\gamma u^{\frac{m(m\ell-1)}{2}} Y_0 Y_1 Y_2 \cdots Y_{m-2} Y_{m-1})^{m\ell} - 1 = 0.$$

The last equation in (8) has now the same form as in [19, Lemma 1.1]. We are done if we can satisfy it. We next show that it can always be satisfied with our solution.

Since all the tropisms in the cone add up to zero, the product $(Y_0 Y_1 Y_2 \cdots \cdots Y_{m-2} Y_{m-1})$, which consists of free parameter combinations, equals to 1. Since $(Y_0 Y_1 Y_2 \cdots Y_{m-2} Y_{m-1}) = 1$, we are left with

$$(\gamma u^{\frac{m(m\ell-1)}{2}})^{m\ell} - 1 = 0. \tag{9}$$

We distinguish two cases:

1. $\gamma = 1$, implied by (m is even, ℓ is odd) or (m is odd, ℓ is odd) or (m is even, ℓ is even).
   To show that (9) is satisfied, we rewrite (9):

$$(u^{\frac{m(m\ell-1)}{2}})^{m\ell} - 1 = 0 \Leftrightarrow (u^{\frac{m^2\ell(m\ell-1)}{2}}) - 1 = 0 \Leftrightarrow ((u^{m\ell})^{\frac{m(m\ell-1)}{2}}) - 1 = 0, \tag{10}$$

   which is satisfied by $u = e^{\frac{i2\pi}{m\ell}}$ and $m(m\ell - 1)$ being even.
2. $\gamma \neq 1$, implied by (m is odd, ℓ is even).
   To show that our solution satisfies (9), we rewrite (9):

$$(\gamma u^{\frac{m(m\ell-1)}{2}})^{m\ell} - 1 = 0 \Leftrightarrow (\gamma u^{\frac{m^2\ell}{2}} u^{\frac{-m}{2}})^{m\ell} - 1 = 0 \Leftrightarrow (\gamma (u^{m\ell})^{\frac{m}{2}} u^{\frac{-m}{2}})^{m\ell} - 1 = 0. \tag{11}$$

Since $u = e^{\frac{i2\pi}{m\ell}}$, $u^{m\ell} = 1$, we can simplify (11) further

$$(\gamma u^{\frac{-m}{2}})^{m\ell} - 1 = 0$$
$$(e^{\frac{i\pi}{m\ell}} (e^{\frac{i2\pi}{m\ell}})^{\frac{-m}{2}})^{m\ell} - 1 = 0$$
$$(e^{\frac{i\pi}{m\ell}} (e^{\frac{-i\pi}{\ell}}))^{m\ell} - 1 = 0 \tag{12}$$
$$(e^{i\pi} e^{-i\pi m}) - 1 = 0$$
$$(e^{(1-m)i\pi}) - 1 = 0.$$

Since m is odd, we can write $m = 2j + 1$, for some j. The last equation of (12) has the form

$$\left(e^{(1-m)i\pi}\right) - 1 = 0 \;\Leftrightarrow\; \left(e^{(1-(2j+1))i\pi}\right) - 1 = 0 \;\Leftrightarrow\; \left(e^{(-2j)i\pi}\right) - 1 = 0. \quad (13)$$

Since $\left(e^{(-2j)i\pi}\right) = 1$, for any j, the equation $\left(e^{(-2j)i\pi}\right) - 1 = 0$ is satisfied, implying (9). □

Backelin's Lemma comes to aid when applying a homotopy to find all isolated cyclic n-roots as follows. We must decide at the end of a solution path whether we have reached an isolated solution or a positive dimension solution set. This problem is especially difficult in the presence of isolated singular solutions (such as 4-fold isolated cyclic 9-roots [36]). With the form of the solution set as in Backelin's Lemma, we solve a triangular binomial system in the parameters t and with as x values the solution found at the end of a path. If we find values for the parameters for an end point, then this solution lies on the solution set.

3   Exploiting Symmetry

We illustrate the exploitation of permutation symmetry on the cyclic 5-roots system. Adjusting polyhedral homotopies to exploit the permutation symmetry for this system was presented in [57].

3.1   The Cyclic 5-roots Problem

The mixed volume for the cyclic 5-roots system is 70, which equals the exact number of roots. The first four equations of the cyclic 5-roots system $C_5(\mathbf{x}) = \mathbf{0}$, define solution curves:

$$\mathbf{f}(\mathbf{x}) = \begin{cases} x_0 + x_1 + x_2 + x_3 + x_4 = 0 \\ x_0 x_1 + x_0 x_4 + x_1 x_2 + x_2 x_3 + x_3 x_4 = 0 \\ x_0 x_1 x_2 + x_0 x_1 x_4 + x_0 x_3 x_4 + x_1 x_2 x_3 + x_2 x_3 x_4 = 0 \\ x_0 x_1 x_2 x_3 + x_0 x_1 x_2 x_4 + x_0 x_1 x_3 x_4 + x_0 x_2 x_3 x_4 + x_1 x_2 x_3 x_4 = 0. \end{cases} \quad (14)$$

where $\mathbf{v} = (1, 1, 1, 1, 1)$. As the first four equations of C_5 are homogeneous, the first four equations of C_5 coincide with the first four equations of $\mathrm{in}_{\mathbf{v}}(C_5)(\mathbf{x}) = \mathbf{0}$. Because these four equations are homogeneous, we have lines of solutions. After computing representations for the solution lines, we find the solutions to the original cyclic 5-roots problem intersecting the solution lines with the hypersurface defined by the last equation. In this intersection, the exploitation of the symmetry is straightforward.

The unimodular matrix with $\mathbf{v} = (1, 1, 1, 1, 1)$ and its corresponding coordinate transformation are

$$M = \begin{bmatrix} 1 & 1 & 1 & 1 & 1 \\ 0 & 1 & 0 & 0 & 0 \\ 0 & 0 & 1 & 0 & 0 \\ 0 & 0 & 0 & 1 & 0 \\ 0 & 0 & 0 & 0 & 1 \end{bmatrix} \quad \mathbf{x} = \mathbf{z}^M : \begin{cases} x_0 = z_0 \\ x_1 = z_0 z_1 \\ x_2 = z_0 z_2 \\ x_3 = z_0 z_3 \\ x_4 = z_0 z_4. \end{cases} \tag{15}$$

Applying $\mathbf{x} = \mathbf{z}^M$ to the initial form system (14) gives

$$\mathrm{in}_\mathbf{v}(\mathbf{f})(\mathbf{x} = \mathbf{z}^M) = \begin{cases} z_1 + z_2 + z_3 + z_4 + 1 = 0 \\ z_1 z_2 + z_2 z_3 + z_3 z_4 + z_1 + z_4 = 0 \\ z_1 z_2 z_3 + z_2 z_3 z_4 + z_1 z_2 + z_1 z_4 + z_3 z_4 = 0 \\ z_1 z_2 z_3 z_4 + z_1 z_2 z_3 + z_1 z_2 z_4 + z_1 z_3 z_4 + z_2 z_3 z_4 = 0. \end{cases} \tag{16}$$

The system (16) has 14 isolated solutions of the form $z_1 = c_1$, $z_2 = c_2$, $z_3 = c_3$, $z_4 = c_4$. If we let $z_0 = t$, in the original coordinates we have

$$x_0 = t, \ x_1 = tc_1, \ x_2 = tc_2, \ x_3 = tc_3, \ x_4 = tc_4 \tag{17}$$

as representations for the 14 solution lines.

Substituting (17) into the omitted equation $x_0 x_1 x_2 x_3 x_4 - 1 = 0$, yields a univariate polynomial in t of the form $kt^5 - 1 = 0$, where k is a constant. Among the 14 solutions, 10 are of the form $t^5 - 1$. They account for $10 \times 5 = 50$ solutions. There are two solutions of the form $(-122.99186938124345)t^5 - 1$, accounting for $2 \times 5 = 10$ solutions and an additional two solutions are of the form $(-0.0081306187557833118)t^5 - 1$ accounting for $2 \times 5 = 10$ remaining solutions. The total number of solutions is 70, as indicated by the mixed volume computation. Existence of additional symmetry, which can be exploited, can be seen in the relationship between the coefficients of the quintic polynomial, i.e. $\frac{1}{(-122.99186938124345)} \approx -0.0081306187557833118$.

3.2 A General Approach

That the first $n - 1$ equations of cyclic n-roots system give explicit solution lines is exceptional. For general polynomial systems we can use the leading term of the Puiseux series to compute witness sets [49] for the space curves defined by the first $n - 1$ equations. Then via the diagonal homotopy [48] we can intersect the space curves with the rest of the system. While the direct exploitation of symmetry with witness sets is not possible, with the Puiseux series we can pick out the generating space curves.

4 Computing Pretropisms

Following from the second theorem of Bernshteĭn [8], the Newton polytopes may be in general position and no normals to at least one edge of every Newton polytope exists. In that case, there does not exist a positive dimensional solution set either. We look for $\mathbf{v}$ so that $\mathrm{in}_{\mathbf{v}}(\mathbf{f})(\mathbf{x}) = \mathbf{0}$ has solutions in $(\mathbb{C}^*)^n$ and therefore we look for pretropisms. In this section we describe two approaches to compute pretropisms. The first approach applies cddlib [21] on the Cayley embedding. Algorithms to compute tropical varieties are described in [11] and implemented in Gfan [31]. The second approach is the application of tropical_intersection of Gfan.

4.1   Using the Cayley Embedding

The Cayley trick formulates a resultant as a discriminant as in [24, Proposition 1.7, page 274]. We follow the geometric description of [52], see also [15, §9.2]. The *Cayley embedding* $E_{\mathbf{A}}$ of $\mathbf{A} = (A_1, A_2, \ldots, A_N)$ is

$$E_{\mathbf{A}} = (A_1 \times \{\mathbf{0}\}) \cup (A_2 \times \{\mathbf{e}_1\}) \cup \cdots \cup (A_N \times \{\mathbf{e}_{N-1}\}) \tag{18}$$

where $\mathbf{e}_k$ is the kth $(N-1)$-dimensional unit vector. Consider the convex hull of the Cayley embedding, the so-called Cayley polytope, denoted by $\mathrm{conv}(E_{\mathbf{A}})$. If $\dim(E_{\mathbf{A}}) = k < 2n - 1$, then a facet of $\mathrm{conv}(E_{\mathbf{A}})$ is a face of dimension $k - 1$.

Proposition 1. *Let $E_{\mathbf{A}}$ be the Cayley embedding of the supports $\mathbf{A}$ of the system $\mathbf{f}(\mathbf{x}) = \mathbf{0}$. The normals of those facets of $\mathrm{conv}(E_{\mathbf{A}})$ that are spanned by at least two points of each support in $\mathbf{A}$ form the tropical prevariety of $\mathbf{f}$.*

Proof. Denote the Minkowski sum of the supports in $\mathbf{A}$ as $\Sigma_{\mathbf{A}} = A_1 + A_2 + \cdots + A_N$. Facets of $\Sigma_{\mathbf{A}}$ spanned by at least two points of each support define the generators of the cones of the tropical prevariety. The relation between $E_{\mathbf{A}}$ and $\Sigma_{\mathbf{A}}$ is stated explicitly in [15, Observation 9.2.2]. In particular, cells in a polyhedral subdivision of $E_{\mathbf{A}}$ are in one-to-one correspondence with cells in a polyhedral subdivision of the Minkowski sum $\Sigma_{\mathbf{A}}$. The correspondence with cells in a polyhedral subdivision implies that facet normals of $\Sigma_{\mathbf{A}}$ occur as facet normals of $\mathrm{conv}(E_{\mathbf{A}})$. Thus the set of all facets of $\mathrm{conv}(E_{\mathbf{A}})$ gives the tropical prevariety of $\mathbf{f}$. $\square$

Note that $\Sigma_{\mathbf{A}}$ can be computed as the Newton polytope of the product of all polynomials in $\mathbf{f}$. As a practical matter, applying the Cayley embedding is better than just plainly computing the convex hull of the Minkowski sum because the Cayley embedding maintains the sparsity of the input, at the expense of increasing the dimension. Running cddlib [21] to compute the H-representation of the Cayley polytope of the cyclic 8-roots problem yields 94 pretropisms. With symmetry we have 11 generators, displayed in Table 1.

For the cyclic 9-roots problem, the computation of the facets of the Cayley polytope yield 276 pretropisms, with 17 generators: $(-2, 1, 1, -2, 1, 1, -2, 1, 1)$, $(-1, -1, 2, -1, -1, 2, -1, -1, 2)$, $(-1, 0, 0, 0, 0, 1, -1, 1, 0)$, $(-1, 0, 0, 0, 0, 1, 0, -1, 1)$, $(-1, 0, 0, 0, 1, -1, 0, 1, 0)$, $(-1, 0, 0, 0, 1, -1, 1, 0, 0)$, $(-1, 0, 0, 0, 1, 0, -1, 0, 1)$, $(-1, 0, 0, 0, 1, 0, -1, 1, 0)$, $(-1, 0, 0, 0, 1, 0, 0, -1, 1)$, $(-1, 0, 0, 1, -1, 0, 1, -1, 1)$, $(-1, 0, 0, 1, -1, 0, 1, 0, 0)$, $(-1, 0, 0, 1, -1, 1, -1, 0, 1)$, $(-1, 0, 0, 1, -1, 1, -1, 1, 0)$, $(-1, 0, 0, 1, -1, 1, 0, -1, 1)$, $(-1, 0, 0, 1, 0, -1, 1,

Table 1. Eleven pretropism generators of the cyclic 8-root problem, the number of solutions of the corresponding initial form systems, and the multidimensional cones they generate, as computed by Gfan

generating pretropisms and initial forms		higher dimensional cones of pretropisms			
pretropism $\mathbf{v}$	#solutions of $\text{in}_{\mathbf{v}}(C_8)(\mathbf{z})$	1D	2D	3D	4D
1. $(-3,1,1,1,-3,1,1,1)$	94	$\{1\}$	$\{1,3\}$	$\{1,6,11\}$	$\{1,2,3,11\}$
2. $(-1,-1,-1,3,-1,-1,-1,3)$	115	$\{2\}$	$\{1,6\}$	$\{1,10,11\}$	
3. $(-1,-1,1,1,-1,-1,1,1)$	112	$\{3\}$	$\{1,10\}$	$\{2,8,11\}$	
4. $(-1,0,0,0,1,-1,1,0)$	30	$\{4\}$	$\{1,11\}$		
5. $(-1,0,0,0,1,0,-1,1)$	23	$\{5\}$	$\{2,3\}$		
6. $(-1,0,0,1,-1,1,0,0)$	32	$\{6\}$	$\{2,8\}$		
7. $(-1,0,0,1,0,-1,1,0)$	40	$\{7\}$	$\{3,7\}$		
8. $(-1,0,0,1,0,0,-1,1)$	16	$\{8\}$	$\{2,11\}$		
9. $(-1,0,1,-1,1,-1,1,0)$	39	$\{9\}$	$\{6,11\}$		
10. $(-1,0,1,0,-1,1,-1,1)$	23	$\{10\}$	$\{8,11\}$		
11. $(-1,1,-1,1,-1,1,-1,1)$	509	$\{11\}$	$\{10,11\}$		

$-1,1)$, $(-1,0,0,1,0,0,-1,0,1)$, and $(-1,0,1,-1,1,-1,0,1,0)$. To get the structure of the two dimensional cones, a second run of the Cayley embedding is needed on the smaller initial form systems defined by the pretropisms.

The computations for $n = 8$ and $n = 9$ finished in less than a second on one core of a 3.07Ghz Linux computer with 4Gb RAM. For the cyclic 12-roots problem, `cddlib` needed about a week to compute the 907,923 facets normals of the Cayley polytope. Although effective, the Cayley embedding becomes too inefficient for larger problems.

4.2   Using `Tropical_intersection` of Gfan

The solution set of the cyclic 8-roots polynomial system consists of space curves. Therefore, all tropisms cones were generated by a single tropism. The computation of the tropical prevariety however, did not lead only to single pretropisms but also to cones of pretropisms. The cyclic 8-roots cones of pretropisms and their dimension are listed in Table 1. Since the one dimensional rays of pretropisms yielded initial form systems with isolated solutions and since all higher dimensional cones are spanned by those one dimensional rays, we can conclude that there are no higher dimensional algebraic sets, as any two dimensional surface degenerates to a curve if we consider only one tropism.

For the computation of the tropical prevariety, the Sage 5.7/Gfan function `tropical_intersection()` ran (with default settings without exploitation of symmetry) on an AMD Phenom II X4 820 processor with 6 GB of RAM, running GNU/Linux, see Table 2. As the dimension n increases so does the running time, but the relative cost factors are bounded by n.

Table 2. Time to compute the tropical prevarieties for cyclic n-roots with Sage 5.7/Gfan and the relative cost factors: for $n = 12$, it takes 9.6 times longer than for $n = 11$

n	seconds	hms format	factor
8	16.37	16 s	1.0
9	79.36	1 m 19 s	4.8
10	503.53	8 m 23 s	6.3
11	3898.49	1 h 4 m 58 s	7.7
12	37490.93	10 h 24 m 50 s	9.6

5   The Second Term of a Puiseux Series

In exceptional cases like the cyclic 4-roots problem where the first term of the series gives an exact solution or when we encounter solution lines like with the first four equations of cyclic 5-roots, we do not have to look for a second term of a series. In general, a pretropism $\mathbf{v}$ becomes a tropism if there is a Puiseux series with leading powers equal to $\mathbf{v}$. The leading coefficients of the series is a solution in $\mathbb{C}^*$ of the initial form system $\mathrm{in}_\mathbf{v}(\mathbf{f})(\mathbf{x}) = \mathbf{0}$. We solve the initial form systems with PHCpack [55] (its blackbox solver incorporates MixedVol [22]). For the computations of the series we use Sage [51].

5.1   Computing the Second Term

In our approach, the calculation of the second term in the Puiseux series is critical to decide whether a solution of an initial form system corresponds to an isolated solution at infinity of the original system, or whether it constitutes the beginning of a space curve. For sparse systems, we may not assume that the second term of the series is linear in t. Trying consecutive powers of t will be wasteful for high degree second terms of particular systems. In this section we explain our algorithm to compute the second term in the Puiseux series.

A unimodular coordinate transformation $\mathbf{x} = \mathbf{z}^M$ with M having as first row the vector $\mathbf{v}$ turns the initial form system $\mathrm{in}_\mathbf{v}(\mathbf{f})(\mathbf{x}) = \mathbf{0}$ into $\mathrm{in}_{\mathbf{e}_1}(\mathbf{f})(\mathbf{z}) = \mathbf{0}$ where $\mathbf{e}_1 = (1, 0, \dots, 0)$ equals the first standard basis vector. When $\mathbf{v}$ has negative components, solutions of $\mathrm{in}_\mathbf{v}(\mathbf{f})(\mathbf{x}) = \mathbf{0}$ that are at infinity (in the ordinary sense of having components equal to ∞) are turned into solutions in $(\mathbb{C}^*)^n$ of $\mathrm{in}_{\mathbf{e}_1}(\mathbf{f})(\mathbf{z}) = \mathbf{0}$.

The following proposition states the existence of the exponent of the second term in the series. After the proof of the proposition we describe how to compute this second term.

Proposition 2. *Let $\mathbf{v}$ denote the pretropism and $\mathbf{x} = \mathbf{z}^M$ denote the unimodular coordinate transformation, generated by $\mathbf{v}$. Let $\mathrm{in}_\mathbf{v}(\mathbf{f})(\mathbf{x} = \mathbf{z}^M)$ denote the transformed initial form system with regular isolated solutions, forming the isolated solutions at infinity of the transformed polynomial system $\mathbf{f}(\mathbf{x} = \mathbf{z}^M)$. If the substitution of the regular isolated solutions at infinity into the transformed* *polynomial system* $\mathbf{f}(\mathbf{x} = \mathbf{z}^M)$ *does not satisfy the system entirely, then the constant terms of* $\mathbf{f}(\mathbf{x} = \mathbf{z}^M)$ *have disappeared, leaving at least one monomial* $c_\ell t^{w_\ell}$ *for some* f_ℓ *in* $\mathbf{f}(\mathbf{x} = \mathbf{z}^M)$ *with minimal value* w_ℓ. *The minimal exponent* w_ℓ *is the candidate for the exponent of the second term in the Puiseux series.*

Proof. Let $\mathbf{z} = (z_0, z_1, \ldots, z_{n-1})$ and $\bar{\mathbf{z}} = (z_1, z_2, \ldots, z_{n-1})$ denote variables after the unimodular transformation. Let $(z_0 = t, z_1 = r_1, \ldots, z_{n-1} = r_{n-1})$ be a regular solution at infinity and t the free variable.

The ith equation of the original system after the unimodular coordinate transformation has the form

$$f_i = z_0^{m_i}(P_i(\bar{\mathbf{z}}) + O(z_0)Q_i(\mathbf{z})), \quad i = 1, 2, \ldots, N, \tag{19}$$

where the polynomial $P_i(\bar{\mathbf{z}})$ consists of all monomials which form the initial form component of f_i and $Q_i(\mathbf{z})$ is a polynomial consisting of all remaining monomials of f_i. After the coordinate transformation, we denote the series expansion as

$$\begin{cases} z_0 = t \\ z_j = r_j + k_j t^{w_\ell}(1 + O(t)), \quad j = 1, 2, \ldots, n - 1. \end{cases} \tag{20}$$

for some ℓ and where at least one k_j is nonzero.

We first show that, for all i, the polynomial $z_0^{m_i} P_i(\bar{\mathbf{z}})$ cannot contain a monomial of the form $c_\ell t^{w_\ell}$ on substitution of (20). The polynomial $z_0^{m_i} P_i(\bar{\mathbf{z}})$ is the initial form of f_i, hence solution at infinity ($z_0 = t, z_1 = r_1, z_2 = r_2, \ldots, z_{n-1} = r_{n-1}$) satisfies $z_0^{m_i} P_i(\bar{\mathbf{z}})$ entirely. Substituting (20) into $z_0^{m_i} P_i(\bar{\mathbf{z}})$ eliminates all constants in $t^{m_i} P_i(\bar{\mathbf{z}})$. Hence, the polynomial $P_i(t) = R_i(t^w)$ and, therefore, $t^{m_i} P_i(t) = R_i(t^{w+m_i})$.

We next show that for some $i = \ell$, the polynomial $Q_i(\mathbf{z})$ contains a monomial $c_\ell t^{w_\ell}$. The polynomial $Q_i(\mathbf{z})$ is rewritten:

$$z_0^{w_\ell} Q_i(\bar{\mathbf{z}}) = z_0^{w_\ell} T_{i0}(\bar{\mathbf{z}}) + z_0^{w_\ell+1} T_{i1}(\bar{\mathbf{z}}) + \cdots. \tag{21}$$

The polynomial $Q_i(\mathbf{z}) = z_0^{w_\ell} Q_i(\bar{\mathbf{z}})$ consists of monomials which are not part of the initial form of f_i. Hence, on substitution of solution at infinity (20), $z_0^{w_\ell} Q_i(\bar{\mathbf{z}})$ $= t^{w_\ell} Q_i(t)$ does not vanish entirely and there must be at least one $i = \ell$ for which constants remain after substitution. Since $Q_\ell(t)$ contains monomials which are constants, $t^{w_\ell} Q_\ell(t)$ must contain a monomial of the form $c_\ell t^{w_\ell}$. □

Now we describe the computation of the second term, in case the initial root does not satisfy the entire original system. Assume the following general form of the series:

$$\begin{cases} z_0 = t \\ z_i = c_i^{(0)} + k_i t^{w_\ell}(1 + O(t)), \quad i = 1, 2, \ldots, n - 1, \end{cases} \tag{22}$$

for some ℓ and where $c_i^{(0)} \in \mathbb{C}^*$ are the coordinates of the initial root, k_i is the unknown coefficient of the second term t^{w_ℓ}, $w_\ell > 0$. Note that only for some k_i nonzero values may exist, but not all k_i may be zero. We are looking for the smallest w_ℓ for which the linear system in the k_i's admits a solution with at least one nonzero coordinate. Substituting (22) gives equations of the form

$$\widehat{c_i^{(0)}} t^{w_\ell} (1 + O(t)) + t^{w_\ell + b_i} \sum_{j=1}^{n} \gamma_{ij} k_j (1 + O(t)) = 0, \quad i = 1, 2, \ldots, n, \qquad (23)$$

for constant exponents w_ℓ, b_i and constant coefficients $\widehat{c_i^{(0)}}$ and γ_{ij}.

In the equations of (23) we truncate the $O(t)$ terms and retain those equations with the smallest value of the exponents w_ℓ, because with the second term of the series solution we want to eliminate the lowest powers of t when we plug in the first two terms of the series into the system. This gives a condition on the value w_ℓ of the unknown exponent of t in the second term. If there is no value for w_ℓ so that we can match with $w_\ell + b_i$ the minimal value of w_ℓ for all equations where the same minimal value of w_ℓ occurs, then there does not exist a second term and hence no space curve. Otherwise, with the matching value for w_ℓ we obtain a linear system in the unknown k variables. If a solution to this linear system exists with at least one nonzero coordinate, then we have found a second term, otherwise, there is no space curve.

For an algebraic set of dimension d, we have a polyhedral cone of d tropisms and we take any general vector $\mathbf{v}$ in this cone. Then we apply the method outlined above to compute the second term in the series in one parameter, in the direction of $\mathbf{v}$.

5.2   Series Developments for Cyclic 8-roots

We illustrate our approach on the cyclic 8-roots problem, denoted by $C_8(\mathbf{x}) = \mathbf{0}$ and take as pretropism $\mathbf{v} = (1, -1, 0, 1, 0, 0, -1, 0)$. Replacing the first row of the 8-dimensional identity matrix by $\mathbf{v}$ yields a unimodular coordinate transformation, denoted as $\mathbf{x} = \mathbf{z}^M$, explicitly defined as

$$x_0 = z_0, \ x_1 = z_1/z_0, \ x_2 = z_2, \ x_3 = z_0 z_3, \ x_4 = z_4, \ x_5 = z_5, \ x_6 = z_6/z_0, \ x_7 = z_7. \qquad (24)$$

Applying $\mathbf{x} = \mathbf{z}^M$ to the initial form system $\mathrm{in}_{\mathbf{v}}(C_8)(\mathbf{x}) = \mathbf{0}$ gives

$$\mathrm{in}_{\mathbf{v}}(C_8)(\mathbf{x} = \mathbf{z}^M) = \begin{cases} z_1 + z_6 = 0 \\ z_1 z_2 + z_5 z_6 + z_6 z_7 = 0 \\ z_4 z_5 z_6 + z_5 z_6 z_7 = 0 \\ z_4 z_5 z_6 z_7 + z_1 z_6 z_7 = 0 \\ z_1 z_2 z_6 z_7 + z_1 z_5 z_6 z_7 = 0 \\ z_1 z_2 z_3 z_4 z_5 z_6 + z_1 z_2 z_5 z_6 z_7 + z_1 z_4 z_5 z_6 z_7 = 0 \\ z_1 z_2 z_3 z_4 z_5 z_6 z_7 + z_1 z_2 z_4 z_5 z_6 z_7 = 0 \\ z_1 z_2 z_3 z_4 z_5 z_6 z_7 - 1 = 0. \end{cases} \qquad (25)$$

By construction of M, observe that all polynomials have the same power of z_0, so z_0 can be factored out. Removing z_0 from the initial form system, we find a solution

$$z_0 = t, \ z_1 = -I, \ z_2 = \frac{-1}{2} - \frac{I}{2}, \ z_3 = -1, \ z_4 = 1 + I, \ z_5 = \frac{1}{2} + \frac{I}{2}, \ z_6 = I, \ z_7 = -1 - I \qquad (26)$$

where $I = \sqrt{-1}$. This solution is a regular solution. We set $z_0 = t$, where t is the variable for the Puiseux series. In the computation of the second term, we assume the Puiseux series of the form at the left of (27). We first transform the cyclic 8-roots system $C_8(\mathbf{x}) = \mathbf{0}$ using the coordinate transformation given by (24) and then substitute the assumed series form into this new system. Since the next term in the series is of the form $k_j t^1$, we collect all the coefficients of t^1 and solve the linear system of equations. The second term in the Puiseux series expansion for the cyclic 8-root system, has the form as at the right of (27).

$$
\begin{cases}
z_0 = t \\
z_1 = -I + c_1 t \\
z_2 = \frac{-1}{2} - \frac{I}{2} + c_2 t \\
z_3 = -1 + c_3 t \\
z_4 = 1 + I + c_4 t \\
z_5 = \frac{1}{2} + \frac{I}{2} + c_5 t \\
z_6 = I + c_6 t \\
z_7 = (-1 - I) + c_7 t
\end{cases}
\qquad
\begin{cases}
z_0 = t \\
z_1 = -I + (-1 - I)t \\
z_2 = \frac{-1}{2} - \frac{I}{2} + \frac{1}{2}t \\
z_3 = -1 \\
z_4 = 1 + I - t \\
z_5 = \frac{1}{2} + \frac{I}{2} - \frac{1}{2}t \\
z_6 = I + (1 + I)t \\
z_7 = (-1 - I) + t
\end{cases}
\tag{27}
$$

Because of the regularity of the solution of the initial form system and the second term of the Puiseux series, we have a symbolic-numeric representation of a quadratic solution curve.

If we place the same pretropism in another row in the unimodular matrix, then we can develop the same curve starting at a different coordinate plane. This move is useful if the solution curve would not be in general position with respect to the first coordinate plane. For symmetric polynomial systems, we apply the permutations to the pretropism, the initial form systems, and its solutions to find Puiseux series for different solution curves, related to the generating pretropism by symmetry.

Also for the pretropism $\mathbf{v} = (1, -1, 1, -1, 1, -1, 1, -1)$, the coordinate transformation is given by the unimodular matrix M equal to the identity matrix, except for its first row $\mathbf{v}$. The coordinate transformation $\mathbf{x} = \mathbf{z}^M$ yields $x_0 = z_0$, $x_1 = z_1/z_0$, $x_2 = z_0 z_2$, $x_3 = z_3/z_0$, $x_4 = z_0 z_4$, $x_5 = z_5/z_0$, $x_6 = z_0 z_6$, $x_7 = z_7/z_0$. Applying the coordinate transformation to $\mathrm{in}_\mathbf{v}(C_8)(\mathbf{x})$ gives

$$
\mathrm{in}_\mathbf{v}(C_8)(\mathbf{x} = \mathbf{z}^M) =
\begin{cases}
z_1 + z_3 + z_5 + z_7 = 0 \\
z_1 z_2 + z_2 z_3 + z_3 z_4 + z_4 z_5 + z_5 z_6 + z_6 z_7 + z_1 + z_7 = 0 \\
z_1 z_2 z_3 + z_3 z_4 z_5 + z_5 z_6 z_7 + z_1 z_7 = 0 \\
z_1 z_2 z_3 z_4 + z_2 z_3 z_4 z_5 + z_3 z_4 z_5 z_6 + z_4 z_5 z_6 z_7 \\
\quad + z_1 z_2 z_3 + z_1 z_2 z_7 + z_1 z_6 z_7 + z_5 z_6 z_7 = 0 \\
z_1 z_2 z_3 z_4 z_5 + z_3 z_4 z_5 z_6 z_7 + z_1 z_2 z_3 z_7 + z_1 z_5 z_6 z_7 = 0 \\
z_1 z_2 z_3 z_4 z_5 z_6 + z_2 z_3 z_4 z_5 z_6 z_7 + z_1 z_2 z_3 z_4 z_5 \\
\quad + z_1 z_2 z_3 z_4 z_7 + z_1 z_2 z_3 z_6 z_7 + z_1 z_2 z_5 z_6 z_7 \\
\quad + z_1 z_4 z_5 z_6 z_7 + z_3 z_4 z_5 z_6 z_7 = 0 \\
z_1 z_2 z_3 z_4 z_5 z_6 z_7 + z_1 z_2 z_3 z_4 z_5 z_7 + z_1 z_2 z_3 z_5 z_6 z_7 \\
\quad + z_1 z_3 z_4 z_5 z_6 z_7 = 0 \\
z_1 z_2 z_3 z_4 z_5 z_6 z_7 - 1 = 0
\end{cases}
\tag{28}
$$

The initial form system (28) has 72 solutions. Among the 72 solutions, a solution of the form

$$z_0 = t, \ z_1 = -1, \ z_2 = I, \ z_3 = -I, \ z_4 = -1, \ z_5 = 1, \ z_6 = -I, \ z_7 = I, \quad (29)$$

here expressed in the original coordinates,

$$x_0 = t, \ x_1 = -1/t, \ x_2 = It, \ x_3 = -I/t, \ x_4 = -t, \ x_5 = 1/t, \ x_6 = -It, \ x_7 = I/t \quad (30)$$

satisfies the cyclic 8-roots entirely. Applying the cyclic permutation of this solution set we can obtain the remaining 7 solution sets, which also satisfy the cyclic 8-roots system.

In [56], a formula for the degree of the curve was derived, based on the coordinates of the tropism and the number of initial roots for the same tropism. We apply this formula and obtain 144 as the known degree of the space curve of the one dimensional solution set, see Table 3.

Table 3. Tropisms, cyclic permutations, and degrees for the cyclic 8 solution curve

$$(1, -1, 1, -1, 1, -1, 1, -1) \qquad\qquad 8 \times 2 = 16$$
$$(1, -1, 0, 1, 0, 0, -1, 0) \to (1, 0, 0, -1, 0, 1, -1, 0) \ 8 \times 2 + 8 \times 2 = 32$$
$$(1, 0, -1, 0, 0, 1, 0, -1) \to (1, 0, -1, 1, 0, -1, 0, 0) \ 8 \times 2 + 8 \times 2 = 32$$
$$(1, 0, -1, 1, 0, -1, 0, 0) \to (1, 0, -1, 0, 0, 1, 0, -1) \ 8 \times 2 + 8 \times 2 = 32$$
$$(1, 0, 0, -1, 0, 1, -1, 0) \to (1, -1, 0, 1, 0, 0, -1, 0) \ 8 \times 2 + 8 \times 2 = 32$$
$$\text{TOTAL} = 144$$

Using the same polyhedral method we can find all the isolated solutions of the cyclic 8-roots system. We conclude this subsection with some empirical observations on the time complexity. In the direction $(1, -1, 0, 1, 0, 0, -1, 0)$, there is a second term in the Puiseux series as for the 40 solutions of the initial form system, there is no first term that satisfies the entire cyclic 8-roots system. Continuing to construct the second term, the total time required is 35.5 seconds, which includes 28 milliseconds that PHCpack needed to solve the initial form system. For $(1, -1, 1, -1, 1, -1, 1, -1)$ there is no second term in the Puiseux series as the first term satisfies the entire system. Hence, the procedure for construction and computation of the second term does not run. It takes PHCpack 12 seconds to solve the initial form system, whose solution set consists of 509 solutions. Determining that there is no second term for the 509 solutions, takes 199 seconds. Given their numbers of solutions, the ratio for time comparison is given by $\frac{509}{40} = 12.725$. However, given that for tropisms $(1, -1, 0, 1, 0, 0, -1, 0)$ the procedure for construction and computation of the second term does run, unlike for tropism $(1, -1, 1, -1, 1, -1, 1, -1)$, the ratio for time comparison is not precise enough. A more accurate ratio for comparison is $\frac{199}{35} \approx 5.686$.

Table 4. Generators of the roots of the initial form system $in_\mathbf{v}(C_{12})(\mathbf{x}) = \mathbf{0}$ with the tropism $\mathbf{v} = (+1, -1, +1, -1, +1, +1, -1, -1, +1, -1, +1, -1)$ in the transformed $\mathbf{z}$ coordinates. Every solution defines a solution curve of the cyclic 12-roots system.

	z_1	z_2	z_3	z_4	z_5	z_6	z_7	z_8	z_9	z_{10}	z_{11}
	$-\frac{1}{2}+\frac{\sqrt{3}}{2}I$	$-\frac{1}{2}-\frac{\sqrt{3}}{2}I$	1	$-\frac{1}{2}+\frac{\sqrt{3}}{2}I$	$-\frac{1}{2}-\frac{\sqrt{3}}{2}I$	-1	$-\frac{1}{2}-\frac{\sqrt{3}}{2}I$	$-\frac{1}{2}+\frac{\sqrt{3}}{2}I$	-1	$-\frac{1}{2}+\frac{\sqrt{3}}{2}I$	$-\frac{1}{2}+\frac{\sqrt{3}}{2}I$

5.3 Cyclic 12-roots

The generating solutions to the quadratic space curve solutions of the cyclic 12-roots problem are in Table 4. As the result in the Table 4 is given in the transformed coordinates, we return the solutions to the original coordinates. For any solution generator $(r_1, r_2, \ldots, r_{11})$ in Table 4:

$$z_0 = t, \; z_1 = r_1, \; z_2 = r_2, \; z_3 = r_3, \; z_4 = r_4, \; z_5 = r_5,$$
$$z_6 = r_6, \; z_7 = r_7, \; z_8 = r_8, \; z_9 = r_9, \; z_{10} = r_{10}, \; z_{11} = r_{11} \tag{31}$$

and turning to the original coordinates we obtain

$$x_0 = t, \; x_1 = r_1/t, \; x_2 = r_2 t, \; x_3 = r_3/t, \; x_4 = r_4 t, \; x_5 = r_5/t$$
$$x_6 = r_6 t, \; x_7 = r_7/t, \; x_8 = r_8 t, \; x_9 = r_9/t, \; x_{10} = r_{10} t, \; x_{11} = r_{11}/t \tag{32}$$

Application of the degree formula of [56] shows that all space curves are quadrics. Compared to [46], we arrive at this result without the application of any factorization methods.

6   Concluding Remarks

Inspired by an effective proof of the fundamental theorem of tropical algebraic geometry, we outlined in this paper a polyhedral method to compute Puiseux series expansions for solution curves of polynomial systems. The main advantage of the new approach is the capability to exploit permutation symmetry. For our experiments, we relied on cddlib and Gfan for the pretropisms, the blackbox solver of PHCpack for solving the initial form systems, and Sage for the manipulations of the Puiseux series.

References

1. Adrovic, D.: Solving Polynomial Systems with Tropical Methods. PhD thesis, University of Illinois at Chicago, Chicago (2012)
2. Adrovic, D., Verschelde, J.: Polyhedral methods for space curves exploiting symmetry. arXiv:1109.0241v1
3. Adrovic, D., Verschelde, J.: Tropical algebraic geometry in Maple: A preprocessing algorithm for finding common factors to multivariate polynomials with approximate coefficients. Journal of Symbolic Computation 46(7), 755–772 (2011); Special Issue in Honour of Keith Geddes on his 60th Birthday, edited by Giesbrecht, M.W., Watt, S.M.
4. Adrovic, D., Verschelde, J.: Computing Puiseux series for algebraic surfaces. In: van der Hoeven, J., van Hoeij, M. (eds.) Proceedings of the 37th International Symposium on Symbolic and Algebraic Computation (ISSAC 2012), pp. 20–27. ACM (2012)
5. Aroca, F., Ilardi, G., López de Medrano, L.: Puiseux power series solutions for systems of equations. International Journal of Mathematics 21(11), 1439–1459 (2011)

6. Backelin, J.: Square multiples n give infinitely many cyclic n-roots. Reports, Matematiska Institutionen 8, Stockholms universitet (1989)
7. Beringer, F., Richard-Jung, F.: Multi-variate polynomials and Newton-Puiseux expansions. In: Winkler, F., Langer, U. (eds.) SNSC 2001. LNCS, vol. 2630, pp. 240–254. Springer, Heidelberg (2003)
8. Bernshteĭn, D.N.: The number of roots of a system of equations. Functional Anal. Appl., 9(3):183–185 (1975); Translated from Funktsional. Anal. i Prilozhen 9(3), 1–4 (1975)
9. Björck, G., Fröberg, R.: Methods to "divide out" certain solutions from systems of algebraic equations, applied to find all cyclic 8-roots. In: Herman, I. (ed.) The Use of Projective Geometry in Computer Graphics. LNCS, vol. 564, pp. 57–70. Springer, Heidelberg (1992)
10. Björck, G., Haagerup, U.: All cyclic p-roots of index 3, found by symmetry-preserving calculations. Preprint available at http://www.math.ku.dk/~haagerup
11. Bogart, T., Jensen, A.N., Speyer, D., Sturmfels, B., Thomas, R.R.: Computing tropical varieties. Journal of Symbolic Computation 42(1), 54–73 (2007)
12. Bruno, A.D.: Power Geometry in Algebraic and Differential Equations. North-Holland Mathematical Library, vol. 57. Elsevier (2000)
13. Colin, A.: Solving a system of algebraic equations with symmetries. Journal of Pure and Applied Algebra 177-118, 195–215 (1997)
14. Dai, Y., Kim, S., Kojima, M.: Computing all nonsingular solutions of cyclic-n polynomial using polyhedral homotopy continuation methods. J. Comput. Appl. Math. 152(1-2), 83–97 (2003)
15. De Loera, J.A., Rambau, J., Santos, F.: *Triangulations. Structures for Algorithms and Applications*. Algorithms and Computation in Mathematics, vol. 25. Springer (2010)
16. Emiris, I.Z.: Sparse Elimination and Applications in Kinematics. PhD thesis, University of California, Berkeley (1994)
17. Emiris, I.Z., Canny, J.F.: Efficient incremental algorithms for the sparse resultant and the mixed volume. Journal of Symbolic Computation 20(2), 117–149 (1995)
18. Faugère, J.C.: A new efficient algorithm for computing Gröbner bases (f_4). Journal of Pure and Applied Algebra 139(1-3), 61–88 (1999); Proceedings of MEGA 1998, Saint-Malo, France, June 22-27 (1998)
19. Faugère, J.C.: Finding all the solutions of Cyclic 9 using Gröbner basis techniques. In: Computer Mathematics - Proceedings of the Fifth Asian Symposium (ASCM 2001). Lecture Notes Series on Computing, vol. 9, pp. 1–12. World Scientific (2001)
20. Faugère, J.C., Rahmany, S.: Solving systems of polynomial equations with symmetries using SAGBI-Gröbner bases. In: Johnson, J., Park, H. (eds.) Proceedings of the 2009 International Symposium on Symbolic and Algebraic Computation (ISSAC 2009), pp. 151–158. ACM (2009)
21. Fukuda, K., Prodon, A.: Double description method revisited. In: Deza, M., Manoussakis, I., Euler, R. (eds.) CCS 1995. LNCS, vol. 1120, pp. 91–111. Springer, Heidelberg (1996)
22. Gao, T., Li, T.Y., Wu, M.: Algorithm 846: MixedVol: a software package for mixed-volume computation. ACM Trans. Math. Softw. 31(4), 555–560 (2005)
23. Gatermann, K.: *Computer Algebra Methods for Equivariant Dynamical Systems*. Lecture Notes in Mathematics, vol. 1728. Springer (2000)
24. Gel'fand, I.M., Kapranov, M.M., Zelevinsky, A.V.: Discriminants, Resultants and Multi-dimensional Determinants. Birkhäuser (1994)

25. Grigoriev, D., Weber, A.: Complexity of solving systems with few independent monomials and applications to mass-action kinetics. In: Gerdt, V.P., Koepf, W., Mayr, E.W., Vorozhtsov, E.V. (eds.) CASC 2012. LNCS, vol. 7442, pp. 143–154. Springer, Heidelberg (2012)
26. Haagerup, U.: Cyclic p-roots of prime length p and related complex Hadamard matrices. Preprint available at http://www.math.ku.dk/~haagerup
27. Hampton, M., Moeckel, R.: Finiteness of relative equilibria of the four-body problem. Invent. math. 163, 289–312 (2006)
28. Herrero, M.I., Jeronimo, G., Sabia, J.: Affine solution sets of sparse polynomial systems. Journal of Symbolic Computation 51(1), 34–54 (2013); Dickenstein, A., Di Rocco, S., Hubert, E., Schicho, J. (eds.): Collected papers of MEGA 2011. Effective Methods in Algebraic Geometry, Stockholm, Sweden, May 30-June 3 (2011)
29. Hubert, E., Labahn, G.: Rational invariants of scalings from Hermite normal forms. In: van der Hoeven, J., van Hoeij, M. (eds.) Proceedings of the 37th International Symposium on Symbolic and Algebraic Computation (ISSAC 2012), pp. 219–226. ACM (2012)
30. Jensen, A., Hampton, M.: Finiteness of spatial central configurations in the five-body problem. Celestial Mechanics and Dynamical Astronomy 109, 321–332 (2011)
31. Jensen, A.N.: Computing Gröbner fans and tropical varieties in Gfan. In: Stillman, M.E., Takayama, N., Verschelde, J. (eds.) Software for Algebraic Geometry. The IMA Volumes in Mathematics and its Applications, vol. 148, pp. 33–46. Springer (2008)
32. Jensen, A.N., Markwig, H., Markwig, T.: An algorithm for lifting points in a tropical variety. Collectanea Mathematica 59(2), 129–165 (2008)
33. Jeronimo, G., Matera, G., Solernó, P., Waissbein, A.: Deformation techniques for sparse systems. Foundations of Computational Mathematics 9(1), 1–50 (2009)
34. Ya, B.: Kazarnovskii. Truncation of systems of polynomial equations, ideals and varieties. Izvestiya: Mathematics 63(3), 535–547 (1999)
35. Lee, T.L., Li, T.Y., Tsai, C.H.: HOM4PS-2.0: a software package for solving polynomial systems by the polyhedral homotopy continuation method. Computing 83(2-3), 109–133 (2008)
36. Leykin, A., Verschelde, J., Zhao, A.: Newton's method with deflation for isolated singularities of polynomial systems. Theoret. Comput. Sci. 359(1-3), 111–122 (2006)
37. Maurer, J.: Puiseux expansion for space curves. Manuscripta Math. 32, 91–100 (1980)
38. McDonald, J.: Fractional power series solutions for systems of equations. Discrete Comput. Geom. 27(4), 501–529 (2002)
39. Payne, S.: Fibers of tropicalization. Mathematische Zeitschrift 262(2), 301–311 (2009)
40. Poteaux, A.: Calcul de développements de Puiseux et application au calcul du groupe de monodromie d'une courbe algébrique plane. PhD thesis, University of Limoges, Limoges (2008)
41. Poteaux, A., Rybowicz, M.: Good reduction of Puiseux series and complexity of the Newton-Puiseux algorithm over finite fields. In: Jeffrey, D. (ed.) Proceedings of the 2008 International Symposium on Symbolic and Algebraic Computation (ISSAC 2008), pp. 239–246. ACM (2008)
42. Poteaux, A., Rybowicz, M.: Good reduction of Puiseux series and applications. Journal of Symbolic Computation 47(1), 32–63 (2012)
43. Puiseux, V.: Recherches sur les fonctions algébriques. J. de Math. Pures et Appl. 15, 365–380 (1850)

44. Richter-Gebert, J., Sturmfels, B., Theobald, T.: First steps in tropical geometry. In: Litvinov, G.L., Maslov, V.P. (eds.) Idempotent Mathematics and Mathematical Physics. Contemporary Mathematics, vol. 377, pp. 289–317. AMS (2005)
45. Rond, G.: About the algebraic closure of the field of power series in several variables in characteristic zero. arXiv:1303.1921v2
46. Sabeti, R.: Numerical-symbolic exact irreducible decomposition of cyclic-12. LMS Journal of Computation and Mathematics 14, 155–172 (2011)
47. Servi, T.: Multivariable Newton-Puiseux theorem for convergent generalised power series. arXiv:1304.0108v3
48. Sommese, A.J., Verschelde, J., Wampler, C.W.: Homotopies for intersecting solution components of polynomial systems. SIAM J. Numer. Anal. 42(4), 552–1571 (2004)
49. Sommese, A.J., Verschelde, J., Wampler, C.W.: Introduction to numerical algebraic geometry. In: Dickenstein, A., Emiris, I.Z. (eds.) Solving Polynomial Equations. Foundations, Algorithms and Applications. Algorithms and Computation in Mathematics, vol. 14, pp. 301–337. Springer (2005)
50. Steidel, S.: Gröbner bases of symmetric ideals. Journal of Symbolic Computation 54(1), 72–86 (2013)
51. Stein, W.A., et al.: Sage Mathematics Software (Version 4.5.2). The Sage Development Team (2010), http://www.sagemath.org
52. Sturmfels, B.: On the Newton polytope of the resultant. Journal of Algebraic Combinatorics 3, 207–236 (1994)
53. Sturmfels, B.: Gröbner Bases and Convex Polytopes. University Lecture Series, vol. 8. AMS (1996)
54. Szöllősi, F.: Construction, classification and parametrization of complex Hadamard matrices. PhD thesis, Central European University, Budapest, arXiv:1110.5590v1 (2011)
55. Verschelde, J.: Algorithm 795: PHCpack: A general-purpose solver for polynomial systems by homotopy continuation. ACM Trans. Math. Softw. 25(2), 251–276 (1999) Software available at http://www.math.uic.edu/~jan/download.html
56. Verschelde, J.: Polyhedral methods in numerical algebraic geometry. In: Bates, D.J., Besana, G., Di Rocco, S., Wampler, C.W. (eds.) Interactions of Classical and Numerical Algebraic Geometry. Contemporary Mathematics, vol. 496, pp. 243–263. AMS (2009)
57. Verschelde, J., Gatermann, K.: Symmetric Newton polytopes for solving sparse polynomial systems. Adv. Appl. Math. 16(1), 95–127 (1995)
58. Walker, R.J.: Algebraic Curves. Princeton University Press (1950)
59. Ziegler, G.M.: Lectures on Polytopes. Graduate Texts in Mathematics, vol. 152. Springer (1995)

Computing the Limit Points
of the Quasi-component of a Regular Chain
in Dimension One

Parisa Alvandi, Changbo Chen, and Marc Moreno Maza

ORCCA, University of Western Ontario (UWO), London, Ontario, Canada
palvandi@uwo.ca, changbo.chen@gmail.com, moreno@csd.uwo.ca

Abstract. For a regular chain R in dimension one, we propose an algorithm which computes the (non-trivial) limit points of the quasi-component of R, that is, the set $\overline{W(R)} \backslash W(R)$. Our procedure relies on Puiseux series expansions and does not require to compute a system of generators of the saturated ideal of R. We provide experimental results illustrating the benefits of our algorithms.

1 Introduction

The theory of regular chains, since its introduction by J.F. Ritt [22], has been applied successfully in many areas including differential systems [8,2,13], difference systems [12], unmixed decompositions [14] and primary decomposition [23] of polynomial ideals, intersection multiplicity calculations [17], cylindrical algebraic decomposition [7], parametric [28] and non-parametric [4] semi-algebraic systems. Today, regular chains are at the core of algorithms computing triangular decomposition of polynomial systems and which are available in several software packages [16,26,27]. Moreover, those algorithms provide back-engines for computer algebra system front-end solvers, such as MAPLE's `solve` command.

This paper deals with a notorious issue raised in all types of triangular decompositions, the *Ritt problem*, stated as follows. Given two regular chains (algebraic or differential) R and S, whose saturated ideals $\mathrm{sat}(R)$ and $\mathrm{sat}(S)$ are radical, check whether the inclusion $\mathrm{sat}(R) \subseteq \mathrm{sat}(S)$ holds or not. In the algebraic case, the challenge is to test such inclusion without computing a system of generators of $\mathrm{sat}(R)$. This question would be answered if one would have a procedure with the following specification: for the regular chain R compute regular chains $R_1, \ldots, R_e$ such that $\overline{W(R)} = W(R_1) \cup \cdots \cup W(R_e)$ holds, where $W(R)$ is the quasi-component of R and $\overline{W(R)}$ is the Zariski closure of $W(R)$.

We propose a solution to this algorithmic quest, in the algebraic case. To be precise, our procedure computes the *non-trivial limit points* of the quasi-component $W(R)$, that is, the set $\lim(W(R)) := \overline{W(R)} \setminus W(R)$ as a finite union of quasi-components of some other regular chains, see Theorem 7 in Section 7. We focus on the case where $\mathrm{sat}(R)$ has dimension one.

When the regular chain R consists of a single polynomial r, primitive w.r.t. its main variable, one can easily check that $\lim(W(R)) = V(r, h_r)$ holds, where

V.P. Gerdt et al. (Eds.): CASC 2013, LNCS 8136, pp. 30–45, 2013.
© Springer International Publishing Switzerland 2013

h_r is the initial of r. Unfortunately, there is no generalization of this result when R consists of several polynomials, unless R enjoys remarkable properties, such as being a *primitive regular chain* [15]. To overcome this difficulty, it becomes necessary to view R as a "parametric representation" of the quasi-component $W(R)$. In this setting, the points of $\lim(W(R))$ can be computed as limits (in the usual sense of the Euclidean topology [1]) of sequences of points along "branches" (in the sense of the theory of algebraic curves) of $W(R)$. It turns out that these limits can be obtained as constant terms of convergent Puiseux series defining the "branches" of $W(R)$ in the neighborhood of the points of interest.

Here comes the main technical difficulty of this approach. When computing a particular point of $\lim(W(R))$, one needs to follow one branch per defining equation of R. Following a branch means computing a truncated Puiseux expansion about a point. Since the equation of R defining a given variable, say X_j, depends on the equations of R defining the variables $X_{j-1}, X_{j-2}, \ldots$, the truncated Puiseux expansion for X_j is defined by an equation whose coefficients involve the truncated Puiseux expansions for $X_{j-1}, X_{j-2}, \ldots$.

From Sections 3 to 7, we show that this principle indeed computes the desired limit points. In particular, we introduce the notion of a *system of Puiseux parametrizations of a regular chain*, see Section 3. This allows us to state in Theorem 3 a concise formula for $\lim(W(R))$ in terms of this latter notion. Then, we estimate to which accuracy one needs to effectively compute such Puiseux parametrizations in order to deduce $\lim(W(R))$, see Theorem 6 in Section 6.

In Section 8, we report on a preliminary implementation of the algorithms presented in this paper. We evaluate our code by applying it to the question of removing redundant components in Kalkbrener's decompositions and observe the benefits of this strategy. Section 2 briefly reviews notions from the theories of regular chains and algebraic curves. We conclude this section with an example.

Consider the regular chain $R = \{r_1, r_2\} \subset \mathbf{k}[X_1, X_2, X_3]$ with $r_1 = X_1 X_2^2 + X_2 + 1, r_2 = (X_1 + 2) X_1 X_3^2 + (X_2 + 1)(X_3 + 1)$. We have $W(R) = V(R) \setminus V(h_R)$ with $h_R = X_1^2(X_1 + 2)$. To determine $\lim(W(R))$, we compute Puiseux series expansions of r_1 at $X_1 = 0$ and $X_1 = -2$. For such calculation, we use MAPLE's command algcurves[puiseux] [24]. The Puiseux expansions of r_1 at $X_1 = 0$ are:

$$[X_1 = T, X_2 = -1 - T + O(T^2)], [X_1 = T, X_2 = -1/T + 1 + T + O(T^2)].$$

Clearly, the second expansion does not yield a limit point. After substituting the first expansion into r_2, we have:

$$r_2' = r_2(X_1 = T, X_2 = -1 - T + O(T^2), X_3) = (T+2)TX_3^2 + (-T + O(T^2))(X_3 + 1).$$

Now, we compute Puiseux series expansions of r_2' which are

$$[T = T, X_3 = 1 - 1/3\, T + O(T^2)], [T = T, X_3 = -1/2 + 1/12\, T + O(T^2)].$$

So the regular chains $\{X_1, X_2 + 1, X_3 - 1\}$ and $\{X_1, X_2 + 1, X_3 + 1/2\}$ give the limit points of $W(R)$ at $X_1 = 0$. Similarly, $\{X_1 + 2, X_2 - 1, X_3 + 1\}$ and $\{X_1 + 2, X_2 + 1/2, X_3 + 1\}$ give the limit points of $W(R)$ at $X_1 = -2$.

---

[1] The closures of $W(R)$ in Zariski and the Euclidean topologies are equal when $\mathbf{k} = \mathbb{C}$.

2   Preliminaries

This section is a review of various notions from the theories of regular chains, algebraic curves and topology. For these latter subjects, our references are the textbooks of R.J. Walker [25], G. Fischer [11] and J. R. Munkres [20]. Notations and hypotheses introduced in this section are used throughout the paper.

Multivariate Polynomials. Let $\mathbf{k}$ be a field which is algebraically closed. Let $X_1 < \cdots < X_s$ be $s \geq 1$ ordered variables. We denote by $\mathbf{k}[X_1, \ldots, X_s]$ the ring of polynomials in the variables $X_1, \ldots, X_s$ and with coefficients in $\mathbf{k}$. For a non-constant polynomial $p \in \mathbf{k}[X_1, \ldots, X_s]$, the greatest variable in p is called *main variable* of p, denoted by $\mathrm{mvar}(p)$, and the leading coefficient of p w.r.t. $\mathrm{mvar}(p)$ is called *initial* of p, denoted by $\mathrm{init}(p)$.

Zariski Topology. We denote by $\mathbb{A}^s$ the *affine s-space* over $\mathbf{k}$. An *affine variety* of $\mathbb{A}^s$ is the set of common zeroes of a collection $F \subseteq \mathbf{k}[X_1, \ldots, X_s]$ of polynomials. The *Zariski topology* on $\mathbb{A}^s$ is the topology whose closed sets are the affine varieties of $\mathbb{A}^s$. The *Zariski closure* of a subset $W \subseteq \mathbb{A}^s$ is the intersection of all affine varieties containing W.

Relation between Zariski Topology and the Euclidean Topology. When $\mathbf{k} = \mathbb{C}$, the affine space $\mathbb{A}^s$ is endowed with both Zariski topology and the Euclidean topology. While Zariski topology is coarser than the Euclidean topology, we have the following (Corollary 1 in I.10 of [19]) key result. Let $V \subseteq \mathbb{A}^s$ be an irreducible affine variety and $U \subseteq V$ be open in the Zariski topology induced on V. Then, the closure of U in Zariski topology and the closure of U in the Euclidean topology are both equal to V.

Limit Points. Let (X, τ) be a topological space. Let $S \subseteq X$ be a subset. A point $p \in X$ is a *limit point* of S if every neighborhood of p contains at least one point of S different from p itself. If the space X is a metric space, the point p is a limit point of S if and only if there exists a sequence $(x_n, n \in \mathbb{N})$ of points of $S \setminus \{p\}$ with p as limit. In practice, the "interesting" limit points of S are those which do not belong to S. For this reason, we call such limit points *non-trivial* and we denote by $\lim(S)$ the set of non-trivial limit points of S.

Regular Chain. A set R of non-constant polynomials in $\mathbf{k}[X_1, \ldots, X_s]$ is called a *triangular set*, if for all $p, q \in R$ with $p \neq q$ we have $\mathrm{mvar}(p) \neq \mathrm{mvar}(q)$. A variable X_i is said *free* w.r.t. R if there exists no $p \in R$ such that $\mathrm{mvar}(p) = X_i$. For a nonempty triangular set R, we define the *saturated ideal* $\mathrm{sat}(R)$ of R to be the ideal $\langle R \rangle : h_R^\infty$, where h_R is the product of the initials of the polynomials in R. The saturated ideal of the empty triangular set is defined as the trivial ideal $\langle 0 \rangle$. The ideal $\mathrm{sat}(R)$ has several properties, in particular it is unmixed [3]. We denote its height, that is the number of polynomials in R, by e, thus $\mathrm{sat}(R)$ has dimension $s - e$. Let $X_{i_1} < \cdots < X_{i_e}$ be the main variables of the polynomials in R. We denote by r_j the polynomial of R whose main variable is X_{i_j} and by h_j the initial of r_j. We say that R is a *regular chain* whenever R is empty or $\{r_1, \ldots, r_{e-1}\}$ is a regular chain and h_e is regular modulo the saturated ideal $\mathrm{sat}(\{r_1, \ldots, r_{e-1}\})$. The regular chain R is said *strongly normalized* whenever each of the main variables of the polynomials of R (that is, $X_{i_1} < \cdots < X_{i_e}$) does not appear in h_R.

Limit Points of the Quasi-component of a Regular Chain. We denote by $W(R) := V(R) \setminus V(h_R)$ the *quasi-component* of R, that is, the common zeros of R that do not cancel h_R. The above discussion implies that the closure of $W(R)$ in Zariski topology and the closure of $W(R)$ in the Euclidean topology are both equal to $V(\mathrm{sat}(R))$, that is, the affine variety of $\mathrm{sat}(R)$. We denote by $\overline{W(R)}$ this common closure and $\lim(W(R))$ this common set of limit points.

Rings of Formal Power Series. Recall that $\mathbf{k}$ is an algebraically closed field. We denote by $\mathbf{k}[[X_1, \ldots, X_s]]$ and $\mathbf{k}\langle X_1, \ldots, X_s \rangle$ the rings of formal and convergent power series in $X_1, \ldots, X_s$ with coefficients in $\mathbf{k}$. When $s = 1$, we write T instead of X_1. For $f \in \mathbf{k}[[X_1, \ldots, X_s]]$, its *order* is defined by $\min\{d \mid f_{(d)} \neq 0\}$ if $f \neq 0$ and by ∞ otherwise, where $f_{(d)}$ is the *homogeneous part* of f in degree d. We denote by $\mathcal{M}_s$ the only maximal ideal of $\mathbf{k}[[X_1, \ldots, X_s]]$, that is, $\mathcal{M}_s = \{f \in \mathbf{k}[[X_1, \ldots, X_s]] \mid \mathrm{ord}(f) \geq 1\}$. Let $f \in \mathbf{k}[[X_1, \ldots, X_s]]$ with $f \neq 0$. Let $k \in \mathbb{N}$. We say that f is (1) *general* in X_s if $f \not\equiv 0 \mod \mathcal{M}_{s-1}$, (2) *general* in X_s of order k if we have $\mathrm{ord}(f \mod \mathcal{M}_{s-1}) = k$.

Formal Puiseux Series. We denote by $\mathbf{k}[[T^*]] = \bigcup_{n=1}^{\infty} \mathbf{k}[[T^{\frac{1}{n}}]]$ the ring of *formal Puiseux series*. For a fixed $\varphi \in \mathbf{k}[[T^*]]$, there is an $n \in \mathbb{N}_{>0}$ such that $\varphi \in \mathbf{k}[[T^{\frac{1}{n}}]]$. Hence $\varphi = \sum_{m=0}^{\infty} a_m T^{\frac{m}{n}}$, where $a_m \in \mathbf{k}$. We call *order of* φ the rational number defined by $\mathrm{ord}(\varphi) = \min\{\frac{m}{n} \mid a_m \neq 0\} \geq 0$. We denote by $\mathbf{k}((T^*))$ the quotient field of $\mathbf{k}[[T^*]]$.

Convergent Puiseux series. Let $\varphi \in \mathbb{C}[[T^*]]$ and $n \in \mathbb{N}$ such that $\varphi = f(T^{\frac{1}{n}})$ holds for some $f \in \mathbb{C}[[T]]$. We say that the Puiseux series φ is *convergent* if we have $f \in \mathbb{C}\langle T \rangle$. Convergent Puiseux series form an integral domain denoted by $\mathbb{C}\langle T^* \rangle$; its quotient field is denoted by $\mathbb{C}(\langle T^* \rangle)$. For every $\varphi \in \mathbb{C}((T^*))$, there exist $n \in \mathbb{Z}$, $r \in \mathbb{N}_{>0}$ and a sequence of complex numbers $a_n, a_{n+1}, a_{n+2}, \ldots$ such that we have $\varphi = \sum_{m=n}^{\infty} a_m T^{\frac{m}{r}}$ and $a_n \neq 0$. Then, we define $\mathrm{ord}(\varphi) = \frac{n}{r}$.

Puiseux Theorem. If $\mathbf{k}$ has characteristic zero, the field $\mathbf{k}((T^*))$ is the algebraic closure of the field of formal Laurent series over $\mathbf{k}$. Moreover, if $\mathbf{k} = \mathbb{C}$, the field $\mathbb{C}(\langle T^* \rangle)$ is algebraically closed as well. From now on, we assume $\mathbf{k} = \mathbb{C}$.

Puiseux Expansion. Let $\mathbb{B} = \mathbb{C}((X^*))$ or $\mathbb{C}(\langle X^* \rangle)$. Let $f \in \mathbb{B}[Y]$, where $d := \deg(f, Y) > 0$. Let $h := \mathrm{lc}(f, Y)$. According to Puiseux Theorem, there exists $\varphi_i \in \mathbb{B}$, $i = 1, \ldots, d$, such that $\frac{f}{h} = (Y - \varphi_1) \cdots (Y - \varphi_d)$. We call $\varphi_1, \ldots, \varphi_d$ the *Puiseux expansions* of f at the origin.

Puiseux Parametrization. Let $f \in \mathbb{C}\langle X \rangle[Y]$. A *Puiseux parametrization* of f is a pair $(\psi(T), \varphi(T))$ of elements of $\mathbb{C}\langle T \rangle$ for some new variable T, such that (1) $\psi(T) = T^\varsigma$, for some $\varsigma \in \mathbb{N}_{>0}$; (2) $f(X = \psi(T), Y = \varphi(T)) = 0$ holds in $\mathbb{C}\langle T \rangle$, and (3) there is no integer $k > 1$ such that both $\psi(T)$ and $\varphi(T)$ are in $\mathbb{C}\langle T^k \rangle$. The index ς is called the *ramification index* of the parametrization $(T^\varsigma, \varphi(T))$. It is intrinsic to f and $\varsigma \leq \deg(f, Y)$. Let $z_1, \ldots, z_\varsigma$ denote the distinct roots of unity of order ς in $\mathbb{C}$. Then $\varphi(z_i X^{1/\varsigma})$, for $i = 1, \ldots, \varsigma$, are ς Puiseux expansions of f. For a Puiseux expansion φ of f, let c minimum such that both $\varphi = g(T^{1/c})$ and $g \in \mathbb{C}\langle T \rangle$ holds. Then $(T^c, g(T))$ is a Puiseux parametrization of f.

3   Puiseux Expansions of a Regular Chain

In this section, we introduce the notion of Puiseux expansions of a regular chain, motivated by the work of [18,1] on Puiseux expansions of space curves. Throughout this section, let $R = \{r_1, \ldots, r_{s-1}\} \subset \mathbb{C}[X_1 < \cdots < X_s]$ be a strongly normalized regular chain whose saturated ideal has dimension one and assume that X_1 is free w.r.t. R.

Lemma 1. *Let $h_R(X_1)$ be the product of the initials of the polynomials in R. Let $\rho > 0$ be small enough such that the set $U_\rho := \{x = (x_1, \ldots, x_s) \in \mathbb{C}^s \mid 0 < |x_1| < \rho\}$ does not contain any zeros of h_R. Define $V_\rho(R) := V(R) \cap U_\rho$. Then, we have $W(R) \cap U_\rho = V_\rho(R)$.*

Proof. It follows from the observation that $U_\rho \cap V(h_R) = \emptyset$.

Notation 1. *Let $W \subseteq \mathbb{C}^s$. Denote $\lim_0(W) := \{x = (x_1, \ldots, x_s) \in \mathbb{C}^s \mid x \in \lim(W)$ and $x_1 = 0\}$.*

Lemma 2. *We have $\lim_0(W(R)) = \lim_0(V_\rho(R))$.*

Proof. By Lemma 1, we have $W(R) \cap U_\rho = V_\rho(R)$. Meanwhile, $\lim_0(W(R)) = \lim_0(W(R) \cap U_\rho)$ holds. Thus $\lim_0(W(R)) = \lim_0(V_\rho(R))$ holds.

Lemma 3. *For $1 \leq i \leq s - 1$, let $d_i := \deg(r_i, X_{i+1})$. Then R generates a zero-dimensional ideal in $\mathbb{C}(\langle X_1^* \rangle)[X_2, \ldots, X_s]$. Let $V^*(R)$ be the zero set of R in $\mathbb{C}(\langle X_1^* \rangle)^{s-1}$. Then $V^*(R)$ has exactly $\prod_{i=1}^{s-1} d_i$ points, counting multiplicities.*

Proof. It follows directly from the definition of regular chain, and the fact that $\mathbb{C}(\langle X_1^* \rangle)$ is an algebraically closed field.

Definition 1. *We use the notations of Lemma 3. Each point in $V^*(R)$ is called a Puiseux expansion of R.*

Notation 2. *Let $m = |V^*(R)|$. Write $V^*(R) = \{\Phi_1, \ldots, \Phi_m\}$. For $i = 1, \ldots, m$, write $\Phi_i = (\Phi_i^1(X_1), \ldots, \Phi_i^{s-1}(X_1))$. Let $\rho > 0$ be small enough such that for $1 \leq i \leq m$, $1 \leq j \leq s - 1$, each $\Phi_i^j(X_1)$ converges in $0 < |X_1| < \rho$. We define $V_\rho^*(R) := \cup_{i=1}^m \{x \in \mathbb{C}^s \mid 0 < |x_1| < \rho, x_{j+1} = \Phi_i^j(x_1), j = 1, \ldots, s - 1\}$.*

Theorem 1. *We have $V_\rho^*(R) = V_\rho(R)$.*

Proof. We prove this by induction on s. For $i = 1, \ldots, s - 1$, recall that h_i is the initial of r_i. If $s = 2$, we have $r_1(X_1, X_2) = h_1(X_1) \prod_{i=1}^{d_1}(X_2 - \Phi_i^1(X_1))$. So $V_\rho^*(R) = V_\rho(R)$ clearly holds.

Now we consider $s > 2$. Write $R' = \{r_1, \ldots, r_{s-2}\}$, $R = R' \cup \{r_{s-1}\}$, $X' = X_2, \ldots, X_{s-1}$, $X = (X_1, X', X_s)$, $x' = x_2, \ldots, x_{s-1}$, $x = (x_1, x', x_s)$, and $m' = |V^*(R')|$. For $i = 1, \ldots, m$, let $\Phi_i = (\Phi_i', \Phi_i^{s-1})$, where Φ_i' stands for $\Phi_i^1, \ldots, \Phi_i^{s-2}$. Assume the theorem holds for R', that is $V_\rho^*(R') = V_\rho(R')$. For any $i = 1, \ldots, m'$, there exist $i_1, \ldots, i_{d_{s-1}} \in \{1, \ldots, m\}$ such that we have

$$r_{s-1}(X_1, X' = \Phi_i', X_s) = h_{s-1}(X_1) \prod_{k=1}^{d_{s-1}}(X_s - \Phi_{i_k}^{s-1}(X_1)). \tag{1}$$

Note that $V^*(R) = \cup_{i=1}^{m'} \cup_{k=1}^{d_s-1} \{(X' = \Phi_i', X_s = \Phi_{i_k}^{s-1})\}$. Therefore, by induction hypothesis and Equation (1), we have

$$
\begin{aligned}
V_\rho^*(R) &= \cup_{i=1}^{m'} \cup_{k=1}^{d_s-1} \{x \mid x \in U_\rho, x' = \Phi_i'(x_1), x_s = \Phi_{i_k}^{s-1}(x_1)\} \\
&= \cup_{k=1}^{d_s-1} \{x \mid (x_1, x') \in V_\rho^*(R'), x_s = \Phi_{i_k}^{s-1}(x_1)\} \\
&= \{x \mid (x_1, x') \in V_\rho^*(R'), r_{s-1}(x_1, x', x_s) = 0\} \\
&= \{x \mid (x_1, x') \in V_\rho(R'), r_{s-1}(x_1, x', x_s) = 0\} \\
&= V_\rho(R).
\end{aligned}
$$

Theorem 2. *Let $V_{\geq 0}^*(R) := \{\Phi = (\Phi^1, \dots, \Phi^{s-1}) \in V^*(R) \mid \mathrm{ord}(\Phi^j) \geq 0, j = 1, \dots, s - 1\}$. Then we have $\lim_0(W(R)) = \cup_{\Phi \in V_{\geq 0}^*(R)} \{(X_1 = 0, \Phi(X_1 = 0))\}$.*

Proof. By definition of $V_{\geq 0}^*(R)$, we immediately have

$$
\lim_0(V_\rho^*(R)) = \cup_{\Phi \in V_{\geq 0}^*(R)} \{(X_1 = 0, \Phi(X_1 = 0))\}.
$$

Next, by Theorem 1, we have $V_\rho^*(R) = V_\rho(R)$. Thus, we have $\lim_0(V_\rho^*(R)) = \lim_0(V_\rho(R))$. Besides, with Lemma 2, we have $\lim_0(W(R)) = \lim_0(V_\rho(R))$. Thus the theorem holds.

Definition 2. *Let $V_{\geq 0}^*(R)$ be as defined in Theorem 2. Let $M = |V_{\geq 0}^*(R)|$. For each $\Phi_i = (\Phi_i^1, \dots, \Phi_i^{s-1}) \in V_{\geq 0}^*(R)$, $1 \leq i \leq M$, we know that $\Phi_i^j \in \mathbb{C}(\langle X_1^* \rangle)$. Moreover, by Equation (1), we know that for $j = 1, \dots, s - 1$, Φ_i^j is a Puiseux expansion of $r_j(X_1, X_2 = \Phi_i^1, \dots, X_j = \Phi_i^{j-1}, X_{j+1})$. Let $\varsigma_{i,j}$ be the ramification index of Φ_i^j and $(T^{\varsigma_{i,j}}, X_{j+1} = \varphi_i^j(T))$, where $\varphi_i^j \in \mathbb{C}\langle T \rangle$, be the corresponding Puiseux parametrization of Φ_i^j. Let ς_i be the least common multiple of $\{\varsigma_{i,1}, \dots, \varsigma_{i,s-1}\}$. Let $g_i^j = \varphi_i^j(T = T^{\varsigma_i/\varsigma_{i,j}})$. We call the set $\mathfrak{G}_R := \{(X_1 = T^{\varsigma_i}, X_2 = g_i^1(T), \dots, X_s = g_i^{s-1}(T)), i = 1, \dots, M\}$ a system of Puiseux parametrizations of R.*

Theorem 3. *We have $\lim_0(W(R)) = \mathfrak{G}_R(T = 0)$.*

Proof. It follows directly from Theorem 2 and Definition 2.

Remark 1. *The limit points of $W(R)$ at $X_1 = \alpha \neq 0$ can be reduced to the computation of $\lim_0(W(R))$ by a coordinate transformation $X_1 = X_1 + \alpha$. Given an arbitrary one-dimensional regular chain R, the set $\lim(W(R))$ can be computed in the following manner. Compute a regular chain N which is strongly normalized and such that $\mathrm{sat}(R) = \mathrm{sat}(N)$ and $V(h_N) = V(\widehat{h_R})$ both hold, where $\widehat{h_R}$ is the iterated resultant of h_R w.r.t R. See [6]. Let $X_i := \mathrm{mvar}(h_R)$. Note that N is still a regular chain w.r.t. the new order $X_i < \{X_1, \dots, X_n\} \setminus \{X_i\}$. Observe that $\lim(W(R)) \subseteq \lim(W(N))$ holds. Thus we have $\lim(W(R)) = \lim(W(N)) \setminus W(R)$.*

4   Puiseux Parametrization in Finite Accuracy

In this section, we define the Puiseux parametrizations of a polynomial $f \in \mathbb{C}\langle X \rangle[Y]$ in finite accuracy, see Definition 4. For $f \in \mathbb{C}\langle X \rangle[Y]$, we also define the approximation of f for a given accuracy, see Definition 3. This approximation of f is a polynomial in $\mathbb{C}[X, Y]$. In Sections 5 and 6, we prove that to compute a Puiseux parametrization of f of a given accuracy, it suffices to compute a Puiseux parametrization of an approximation of f of some finite accuracy.

In this section, we review and adapt the classical Newton-Puiseux algorithm to compute Puiseux parametrizations of a polynomial $f \in \mathbb{C}[X, Y]$ of a given accuracy. Since we do not need to compute the singular part of Puiseux parametrizations, the usual requirement $\operatorname{discrim}(f, Y) \neq 0$ is dropped.

Definition 3. *Let $f = \sum_{i=0}^{\infty} a_i X^i \in \mathbb{C}[[X]]$. For any $\tau \in \mathbb{N}$, we call $f^{(\tau)} := \sum_{i=0}^{\tau} a_i X^i$ the polynomial part of f of accuracy $\tau + 1$. Let $f = \sum_{i=0}^{d} a_i(X) Y^i \in \mathbb{C}\langle X \rangle [Y]$. For any $\tau \in \mathbb{N}$, we call $\widehat{f}^{(\tau)} := \sum_{i=0}^{d} a_i^{(\tau)} Y^i$ the approximation of f of accuracy $\tau + 1$.*

Definition 4. *Let $f \in \mathbb{C}\langle X \rangle [Y]$, with $\deg(f, Y) > 0$. Let $\sigma, \tau \in \mathbb{N}_{>0}$ and $g(T) = \sum_{k=0}^{\tau-1} b_k T^k$. Let $\{T^{k_1}, \ldots, T^{k_m}\}$ be the support of $g(T)$. The pair $(T^\sigma, g(T))$ is called a Puiseux parametrization of f of accuracy τ if there exists a Puiseux parametrization $(T^\varsigma, \varphi(T))$ of f such that: (i) σ divides ς; (ii) $\gcd(\sigma, k_1, \ldots, k_m) = 1$; and (iii) $g(T^{\varsigma/\sigma})$ is the polynomial part of $\varphi(T)$ of accuracy $(\varsigma/\sigma)(\tau - 1) + 1$. Note that if $\sigma = \varsigma$, then $g(T)$ is the polynomial part of $\varphi(T)$ of accuracy τ.*

Definition 5 ([10]). *A $\mathbb{C}$-term[2] is defined as a triple $t = (q, p, \beta)$, where q and p are coprime integers, $q > 0$ and $\beta \in \mathbb{C}$ is non-zero. A $\mathbb{C}$-expansion is a sequence $\pi = (t_1, t_2, \ldots)$ of $\mathbb{C}$-terms, where $t_i = (q_i, p_i, \beta_i)$. We say that π is finite if there are only finitely many elements in π.*

Definition 6. *Let $\pi = (t_1, \ldots, t_N)$ be a finite $\mathbb{C}$-expansion. We define a pair $(T^\sigma, g(T))$ of polynomials in $\mathbb{C}[T]$ in the following manner: (i) if $N = 1$, set $\sigma = 1$, $g(T) = 0$ and $\delta_N = 0$; (ii) otherwise, let $a := \prod_{i=1}^{N} q_i$, $c_i := \sum_{j=1}^{i} \left(p_j \prod_{k=j+1}^{N} q_k \right)$ $(1 \le i \le N)$, $\delta_i := c_i / \gcd(a, c_1, \ldots, c_N)$ $(1 \le i \le N)$. Set $\sigma := a / \gcd(a, c_1, \ldots, c_N)$ and $g(T) := \sum_{i=1}^{N} \beta_i T^{\delta_i}$. We call the pair $(T^\sigma, g(T))$ the Puiseux parametrization of π of accuracy $\delta_N + 1$. Denote by ConstructParametrization an algorithm to compute $(T^\sigma, g(T))$ from π.*

Definition 7. *Let $f \in \mathbb{C}\langle X \rangle [Y]$ and write f as $f(X, Y) := \sum_{i=0}^{d} \left(\sum_{j=0}^{\infty} a_{i,j} X^j \right) Y^i$. The Newton Polygon of f is defined as the lower part of the convex hull of the set of points (i, j) in the plane such that $a_{i,j} \neq 0$.*

Let $f \in \mathbb{C}\langle X \rangle [Y]$. We denote by NewtonPolygon(f, I) an algorithm to compute the segments in the Newton Polygon of f, where I is a flag controlling the algorithm specification as follows. If $I = 1$, only segments with non-positive slopes are computed. If $I = 2$, only segments with negative slopes are computed. Such an algorithm can be found in [25]. Next we introduce some notations which are necessary to present Algorithm 2.

---

[2] It is a simplified version of Duval's definition.

Let $f \in \mathbb{C}[X, Y]$, $t = (q, p, \beta)$ be a $\mathbb{C}$-term and $\ell \in \mathbb{N}$ s.t. $\mathsf{NewPoly}(f, t, \ell) := X^{-\ell} f(X^q, X^p(\beta + Y)) \in \mathbb{C}[X, Y]$. Let $f = \sum_{i=0}^{d} \sum_{j=0}^{m} a_{i,j} X^j Y^i \in \mathbb{C}[X, Y]$ and let Δ be a segment of the Newton Polygon of f. Denote $\mathsf{SegmentPoly}(f, \Delta) := (q, p, \ell, \phi)$ such that the following holds: (1) $q, p, \ell \in \mathbb{N}$; $\phi \in \mathbb{C}[Z]$; q and p are coprime, $q > 0$; (2) for any $(i, j) \in \Delta$, we have $qj + pi = \ell$; and (3) letting $i_0 := \min(\{i \mid (i, j) \in \Delta\})$, we have $\phi = \sum_{(i,j) \in \Delta} a_{i,j} Z^{(i-i_0)/q}$.

Theorem 4. *Algorithm 2 terminates and is correct.*

Proof. It directly follows from the proof of the Newton-Puiseux algorithm in Walker's book [25], the relation between $\mathbb{C}$-expansion and Puiseux parametrization discussed in Duval's paper [10], and Definitions 6 and 4.

Algorithm 1. $\mathsf{NonzeroTerm}(f, I)$

    Input: $f \in \mathbb{C}[X, Y]$; $I = 1$ or 2.
    Output: A finite set of pairs (t, ℓ), where t is a $\mathbb{C}$-term, and $\ell \in \mathbb{N}$.
1 $S := \emptyset$;
2 **for** *each* $\Delta \in \mathsf{NewtonPolygon}(f, I)$ **do**
3      $(q, p, \ell, \phi) := \mathsf{SegmentPoly}(f, \Delta)$;
4   **for** *each root* ξ *of* ϕ *in* $\mathbb{C}$ **do**
5   **for** *each root* β *of* $U^q - \xi$ *in* $\mathbb{C}$ **do** $\{t := (q, p, \beta); S := S \cup \{(t, \ell)\}\}$;
6 **return** S

Algorithm 2. $\mathsf{NewtonPuiseux}$

    Input: $f \in \mathbb{C}[X, Y]$; a given accuracy $\tau > 0 \in \mathbb{N}$.
    Output: All the Puiseux parametrizations of f of accuracy τ.
1 $\pi := (\ )$; $S := \{(\pi, f)\}$; $P := \emptyset$;
2 **while** $S \neq \emptyset$ **do**
3   let $(\pi^*, f^*) \in S$; $S := S \setminus \{(\pi^*, f^*)\}$; **if** $\pi^* = (\ )$ **then** $I := 1$ **else** $I := 2$;
      $(T^\sigma, g(T)) := \mathsf{ConstructParametrization}(\pi^*)$;
4   **if** $\deg(g(T), T) + 1 < \tau$ **then**
5         $C := \mathsf{NonzeroTerm}(f^*, I)$;
6   **if** $C = \emptyset$ **then**
7            $P := P \cup \{(T^\sigma, g(T))\}$
8   **else**
9   **for** *each* $(t = (p, q, \beta), \ell) \in C$ **do**
10              $\pi^{**} := \pi^* \cup (t)$; $f^{**} := \mathsf{NewPoly}(f^*, t, \ell)$; $S := S \cup \{(\pi^{**}, f^{**})\}$
11   **else**
12         $P := P \cup \{(T^\sigma, g(T))\}$
13 **return** P

---

5 Computing in Finite Accuracy

Let $f \in \mathbb{C}\langle X \rangle[Y]$. In this section, we consider the following problems: (a) Is it possible to use an approximation of f of some finite accuracy m in order to compute a Puiseux parametrization of f of a prescribed finite accuracy τ? (b) If yes, how to calculate m from f and τ? (c) Provide an upper bound on m. Theorem 5 provides the answers to (a) and (b) while Lemma 6 answers (c).

In the rest of this paper, the proof of a lemma is omitted if it is a routine.

Lemma 4. *Let $f \in \mathbb{C}\langle X\rangle[Y]$. Let $d := \deg(f, Y) > 0$. Let $q \in \mathbb{N}_{>0}$, $p, \ell \in \mathbb{N}$ and assume that q and p are coprime. Let $\beta \neq 0 \in \mathbb{C}$. Assume that q, p, ℓ define the segment $qj + pi = \ell$ of the Newton Polygon of f. Let $f_1 := X_1^{-\ell} f(X_1^q, X_1^p(\beta + Y_1))$.*

Then, we have the following results: (i) for any given $m_1 \in \mathbb{N}$, there exists a number $m \in \mathbb{N}$ such that the approximation of f_1 of accuracy m_1 can be computed from the approximation of f of accuracy m; (ii) moreover, it suffices to take $m = \lfloor \frac{m_1 + \ell}{q} \rfloor$.

Theorem 5. *Let $f \in \mathbb{C}\langle X\rangle[Y]$. Let $\tau \in \mathbb{N}_{>0}$. Let $\sigma \in \mathbb{N}_{>0}$ and $g(T) = \sum_{k=0}^{\tau-1} b_k T^k$. Assume that $(T^\sigma, g(T))$ is a Puiseux parametrization of f of accuracy τ. Then one can compute a number $m \in \mathbb{N}$ such that $(T^\sigma, g(T))$ is a Puiseux parametrization of accuracy τ of $\tilde{f}^{m-1}$, where $\tilde{f}^{m-1}$ is the approximation of f of accuracy m. We denote by $\mathsf{AccuracyEstimate}$ an algorithm to compute m from f and τ.*

Proof. By Lemma 4 and the construction of the Newton-Puiseux algorithm, we conclude that there exists a number $m \in \mathbb{N}$ such that $(T^\sigma, g(T))$ is a Puiseux parametrization of accuracy τ of the approximation of f of accuracy m.

Next we show that there is an algorithm to compute m. We initially set $m' := \tau$. Let $f_0 := \sum_{i=0}^d \left(\sum_{j=0}^{m'} a_{i,j} X^j \right) Y^i$. That is, f_0 is the approximation of f of accuracy $m' + 1$. We run the Newton-Puiseux algorithm to check whether the terms $a_{k,m'} X^{m'} Y^k$, $0 \leq k \leq d$, make any contributions in constructing the Newton Polygons of all f_i. If at least one of them make contributions, we increase the value of m' and restart the Newton-Puiseux algorithm until none of the terms $a_{k,m'} X^{m'} Y^k$, $0 \leq k \leq d$, makes any contributions in constructing the Newton Polygons of all f_i. We set $m := m'$.

Lemma 5. *Let $d, \tau \in \mathbb{N}_{>0}$. Let $a_{i,j}$, $0 \leq i \leq d$, $0 \leq j < \tau$, and b_k, $0 \leq k < \tau$ be symbols. Write $\mathbf{a} = (a_{0,0}, \ldots, a_{0,\tau-1}, \ldots, a_{d,0}, \ldots, a_{d,\tau-1})$ and $\mathbf{b} = (b_0, \ldots, b_{\tau-1})$. Let $f(\mathbf{a}, X, Y) = \sum_{i=0}^d \left(\sum_{j=0}^{\tau-1} a_{i,j} X^j \right) Y^i \in \mathbb{C}[\mathbf{a}][X, Y]$ and let $g(\mathbf{b}, X) = \sum_{k=0}^{\tau-1} b_k X^k \in \mathbb{C}[\mathbf{b}][X]$. Let $p := f(\mathbf{a}, X, Y = g(\mathbf{b}, X))$. Let $F_k := \mathrm{coeff}(p, X^k)$, $0 \leq k < \tau - 1$, and $F := \{F_0, \ldots, F_{\tau-1}\}$. Then under the order $\mathbf{a} < \mathbf{b}$ and $b_0 < b_1 < \cdots < b_{\tau-1}$, F forms a zero-dimensional regular chain in $\mathbb{C}(\mathbf{a})[\mathbf{b}]$ with main variables $(b_0, b_1, \ldots, b_{\tau-1})$ and main degrees $(d, 1, \ldots, 1)$. In addition, we have (i) $F_0 = \sum_{i=0}^d a_{i,0} b_0^i$, and (ii) $\mathrm{init}(F_1) = \cdots = \mathrm{init}(F_{\tau-1}) = \mathrm{der}(F_0, b_0) = \sum_{i=1}^d i \cdot a_{i,0} b_0^{i-1}$.*

Proof. Write $p = \sum_{i=0}^d \left(\sum_{j=0}^{\tau-1} a_{i,j} X^j \right) \left(\sum_{k=0}^{\tau-1} b_k X^k \right)^i$ as a univariate polynomial in X. Observe that $F_0 = \sum_{i=0}^d a_{i,0} b_0^i$. Therefore F_0 is irreducible in $\mathbb{C}(\mathbf{a})[\mathbf{b}]$. Moreover, we have $\mathrm{mvar}(F_0) = b_0$ and $\mathrm{mdeg}(F_0) = d$.

Since $d > 0$, we know that $a_{1,0} \left(\sum_{k=0}^{\tau-1} b_k X^k \right)$ appears in p. Thus, for $0 \le k < \tau$, b_k appears in F_k. Moreover, for any $k \ge 1$ and $i < k$, b_k can not appear in F_i since b_k and X^k are always raised to the same power. For the same reason, for any $i > 1$, b_k^i cannot appear in F_k, for $1 \le k < \tau$. Thus $\{F_0, \ldots, F_{\tau-1}\}$ is a triangular set with main variables $(b_0, b_1, \ldots, b_{\tau-1})$ and main degrees $(d, 1, \ldots, 1)$.

Moreover, we have $\mathrm{init}(F_1) = \cdots = \mathrm{init}(F_{\tau-1}) = \sum_{i=1}^{d} i \cdot a_{i,0} b_0^{i-1}$, which is coprime with F_0. Thus $F = \{F_0, \ldots, F_{\tau-1}\}$ is a regular chain.

As a direct corollary, we have the following lemma.

Lemma 6. *Let* $f = \sum_{i=0}^{d} \left(\sum_{j=0}^{\infty} a_{i,j} X^j \right) Y^i \in \mathbb{C}[[X]][Y]$. *Assume that* $d = \deg(f, Y) > 0$ *and* f *is general in* Y. *Let* $\varphi(X) = \sum_{k=0}^{\infty} b_k X^k \in \mathbb{C}[[X]]$ *such that* $f(X, \varphi(X)) = 0$ *holds. Let* $\tau > 0 \in \mathbb{N}$. *Then all coefficients* b_i, *for* $0 \le i < \tau$, *can be completely determined by* $\{a_{i,j} \mid 0 \le i \le d, 0 \le j < \tau\}$ *if and only if* b_0 *is a simple zero of* $f(0, Y)$. *Therefore, "generically", all coefficients* b_i, *for* $0 \le i < \tau$, *can be completely determined by the approximation of* f *of accuracy* τ.

6   Accuracy Estimates

Let $R := \{r_1(X_1, X_2), \ldots, r_{s-1}(X_1, \ldots, X_s)\} \subset \mathbb{C}[X_1 < \cdots < X_s]$ be a strongly normalized regular chain. In this section, we show that to compute the limit points of $W(R)$, it suffices to compute the Puiseux parametrizations of R of some accuracy. Moreover, we provide accuracy estimates in Theorem 6.

Lemma 7. *Let* $f = a_d(X)Y^d + \cdots + a_0(X) \in \mathbb{C}\langle X\rangle[Y]$, *where* $d = \deg(f, Y) > 0$. *For* $0 \le i \le d$, *let* $\delta_i := \mathrm{ord}(a_i)$. *Let* $k := \min(\delta_0, \ldots, \delta_d)$. *Let* $\tilde{f} := X^{-k} f$. *Then we have* $\tilde{f} \in \mathbb{C}\langle X\rangle[Y]$ *and* $\tilde{f}$ *is general in* Y. *This operation of producing* $\tilde{f}$ *from* f *is called "making* f *general" and we denote it by* MakeGeneral.

The following lemma shows that computing limit points reduces to making a polynomial f general.

Lemma 8. *Let* $f \in \mathbb{C}\langle X\rangle[Y]$, *where* $\deg(f, Y) > 0$, *be general in* Y. *Let* $\rho > 0$ *be small enough such that* f *converges in* $|X| < \rho$. *Let* $V_\rho(f) := \{(x, y) \in \mathbb{C}^2 \mid 0 < |x| < \rho, f(x, y) = 0\}$. *Then* $\lim_0(V_\rho(f)) = \{(0, y) \in \mathbb{C}^2 \mid f(0, y) = 0\}$ *holds.*

Proof. With $1 \le i \le c$, for some c such that $1 \le c \le \deg(f, Y)$, let $(X = T^{s_i}, Y = \varphi_i(T))$ be the distinct Puiseux parametrizations of f. By Lemma 1 and Theorem 3, we have $\lim_0(V_\rho(f)) = \cup_{i=1}^{c} \{(0, y) \in \mathbb{C}^2 \mid y = \varphi_i(0)\}$. Let $(X = T^{\sigma_i}, g_i(T))$, $i = 1, \ldots, c$, be the corresponding Puiseux parametrizations of f of accuracy 1. By Theorem 5, there exists an approximation $\tilde{f}$ of f of some finite accuracy such that $(X = T^{\sigma_i}, g_i(T))$, $i = 1, \ldots, c$, are also Puiseux parametrizations of $\tilde{f}$ of accuracy 1. Thus, we have $\varphi_i(0) = g_i(0)$, $i = 1, \ldots, c$. Since $\tilde{f}$ is general in Y, by Theorem 2.3 in [25], we have $\cup_{i=1}^{c} \{(0, y) \in \mathbb{C}^2 \mid y = g_i(0)\} = \{(0, y) \in \mathbb{C}^2 \mid \tilde{f}(0, y) = 0\}$. Since $\tilde{f}(0, y) = f(0, y)$, the lemma holds.

Lemma 9. *Let $a(X_1, \ldots, X_s) \in \mathbb{C}[X_1, \ldots, X_s]$. Let $g_i = \sum_{j=0}^{\infty} c_{ij} T^j \in \mathbb{C}\langle T \rangle$, for $i = 1 \cdots s$. We write $a(g_1, \ldots, g_s)$ as $\sum_{k=0}^{\infty} b_k T^k$. To compute a given coefficient b_k, one only needs to know the coefficients of the polynomial a and the coefficients $c_{i,j}$ for $1 \leq i \leq s, 0 \leq j \leq k$.*

Lemma 10. *Let $f = a_d(X)Y^d + \cdots + a_0(X) \in \mathbb{C}\langle X \rangle[Y]$, where $d = \deg(f, Y) > 0$. Let $\delta := \operatorname{ord}(a_d(X))$. Then "generically", a Puiseux parametrization of f of accuracy τ can be computed from an approximation of f of accuracy $\tau + \delta$.*

Proof. Let $\widetilde{f} := \mathsf{MakeGeneral}(f)$. Observe that f and $\widetilde{f}$ have the same system of Puiseux parametrizations. Then the conclusion follows from Lemma 7 and 6.

Let $R := \{r_1(X_1, X_2), \ldots, r_{s-1}(X_1, \ldots, X_s)\} \subset \mathbb{C}[X_1 < \cdots < X_s]$ be a strongly normalized regular chain. For $1 \leq i \leq s-1$, let $h_i := \operatorname{init}(r_i)$, $d_i := \deg(r_i, X_{i+1})$ and $\delta_i := \operatorname{ord}(h_i)$. We define f_i, ς_i, T_i, $\varphi_i(T_i)$, $1 \leq i \leq s-1$, as follows. Let $f_1 := r_1$. Let $(X_1 = T_1^{\varsigma_1}, X_2 = \varphi_1(T_1))$ be a Puiseux parametrization of f_1. For $i = 2, \ldots, s-1$ do
(i) Let $f_i := r_i(X_1 = T_1^{\varsigma_1}, X_2 = \varphi_1(T_1), \ldots, X_i = \varphi_{i-1}(T_{i-1}), X_{i+1})$.
(ii) Let $(T_{i-1} = T_i^{\varsigma_i}, X_{i+1} = \varphi_i(T_i))$ be a Puiseux parametrization of f_i.
Before stating our main result on the bound, we first present several lemmas.

Lemma 11. *For $0 \leq i \leq s-2$, define $g_i(T_{s-2}) := T_{s-2}^{\prod_{k=i+1}^{s-2} \varsigma_k}$. Let $T_0 := X_1$. Then we have $T_i = g_i(T_{s-2})$, $0 \leq i \leq s-2$.*

Proof. We prove it by induction. Clearly it holds for $i = s-2$. Suppose it holds for i. Then we have $T_{i-1} = T_i^{\varsigma_i} = \left(T_{s-2}^{\prod_{k=i+1}^{s-2} \varsigma_k} \right)^{\varsigma_i} = \left(T_{s-2}^{\prod_{k=i}^{s-2} \varsigma_k} \right)$. Therefore it also holds for $i-1$. So it holds for all $0 \leq i \leq s-2$.

Lemma 12. *There exist numbers $\tau_1, \ldots, \tau_{s-2} \in \mathbb{N}$ such that in order to make f_{s-1} general in X_s, it suffices to compute the polynomial parts of φ_i of accuracy τ_i, $1 \leq i \leq s-2$. Moreover, if we write the algorithm $\mathsf{AccuracyEstimate}$ for short as θ, the accuracies τ_i can be computed in the following manner: $\tau_{s-2} := (\prod_{k=1}^{s-2} \varsigma_k) \delta_{s-1} + 1$, $\tau_{i-1} := \max(\theta(f_i, \tau_i), (\prod_{k=1}^{i-1} \varsigma_k) \delta_{s-1} + 1)$, for $2 \leq i \leq s-2$.*

Proof. By Lemma 11, we have $g_0(T_{s-2}) = T_{s-2}^{\prod_{k=1}^{s-2} \varsigma_k}$. Since $\operatorname{ord}(h_{s-1}(X_1)) = \delta_{s-1}$, we have $\operatorname{ord}(h_{s-1}(X_1 = g_0(T_{s-2}))) = \left(\prod_{k=1}^{s-2} \varsigma_k \right) \delta_{s-1}$. Let $\tau_{s-2} := (\prod_{k=1}^{s-2} \varsigma_k) \delta_{s-1} + 1$. By Lemma 7, to make f_{s-1} general in X_s, it suffices to compute the polynomial parts of the coefficients of f_{s-1} of accuracy τ_{s-2}.

By Lemma 9, we need to compute the polynomial parts of $\varphi_i(g_i(T_{s-2}))$, $1 \leq i \leq s-2$, of accuracy τ_{s-2}. Since $\operatorname{ord}(g_i(T_{s-2})) = \prod_{k=i+1}^{s-2} \varsigma_k$, to achieve this accuracy, it is enough to compute the polynomial parts of φ_i of accuracy $(\prod_{k=1}^{i} \varsigma_k) \delta_{s-1} + 1$, for $1 \leq i \leq s-2$.

Since we have $f_i = r_i(X_1 = T_1^{\varsigma_1}, X_2 = \varphi_1(T_1), \ldots, X_i = \varphi_{i-1}(T_{i-1}), X_{i+1})$ and $(T_{i-1} = T_i^{\varsigma_i}, X_{i+1} = \varphi_i(T_i))$ is a Puiseux parametrization of f_i, by Theorem 5 and Lemma 9, to compute the polynomial part of φ_i of accuracy τ_i, we need the polynomial part of φ_{i-1} of accuracy $\theta(f_i, \tau_i)$. Thus, $\tau_{s-2} := (\prod_{k=1}^{s-2} \varsigma_k) \delta_{s-1} + 1$

and $\tau_{i-1} = \max(\theta(f_i, \tau_i), (\prod_{k=1}^{i-1} \varsigma_k)\delta_{s-1} + 1)$ for $2 \le i \le s - 2$ will guarantee f_{s-1} can be made general in X_s.

Theorem 6. *One can compute positive integer numbers $\tau_1, \ldots, \tau_{s-1}$ such that, in order to compute $\lim_0(W(R))$, it suffices to compute Puiseux parametrizations of f_i of accuracy τ_i, for $i = 1, \ldots, s - 1$. Moreover, generically, one can choose $\tau_{s-1} := 1$, $\tau_{s-2} := (\prod_{k=1}^{s-2} \varsigma_k)\delta_{s-1} + 1$, $\tau_i = (\prod_{k=1}^{s-2} \varsigma_k)(\sum_{k=2}^{s-1} \delta_i) + 1$, for $i = 1, \ldots, s - 3$, and each index ς_k can be set to d_k, for $k = 1, \ldots, s - 2$.*

Proof. By Lemma 12, we know that $\tau_1, \ldots, \tau_{s-1}$ can be computed. By Lemma 11, we have $X_1 = T_{i-1}^{\prod_{k=1}^{i-1} \varsigma_k}$. Since $\mathrm{ord}(h_i(X_1)) = \delta_i$, we have $\mathrm{ord}(h_i(X_1 = T_{i-1}^{\prod_{k=1}^{i-1} \varsigma_k})) = \left(\prod_{k=1}^{i-1} \varsigma_k\right) \delta_i$. By Lemma 10, generically a Puiseux parametrization of f_i of accuracy τ_i can be computed from an approximation of f_i of accuracy $\tau_i + \delta_i$. In Lemma 12, let $\theta(f_i, \tau_i) = \tau_i + (\prod_{k=1}^{i-1} \varsigma_k)\delta_i$, $2 \le i \le s - 2$, which implies the bound in the theorem. Finally we observe that $\varsigma_k \le d_k$ holds, for $1 \le k \le s - 2$.

7 Algorithm

In this section, we provide a complete algorithm for computing the non-trivial limit points of the quasi-component of a one-dimensional strongly normalized regular chain based on the results of the previous sections.

Proposition 1. *Algorithm 4 is correct and terminates.*

Proof. This follows from Theorem 3, Theorem 5, Theorem 6 and Lemma 8.

Theorem 7. *Let $R \subset \mathbb{Q}[X_1, \ldots, X_n]$ be a regular chain such that $\dim(\mathrm{sat}(R)) = 1$. Then there exists an algorithm to compute regular chains $R_i \in \mathbb{Q}[X_1, \ldots, X_n]$, $i = 1, \ldots, e$, such that $\lim(W(R)) = \cup_{i=1}^e W(R_i)$.*

Proof. By Remark 1, we can assume that R is strongly normalized and X_1 is free w.r.t. R. By Proposition 1, there is an algorithm to compute $\lim(W(R))$. Thus, it suffices to prove that $\lim(W(R))$ can be represented by regular chains in $\mathbb{Q}[X_1, \ldots, X_n]$, whenever $R \subset \mathbb{Q}[X_1, \ldots, X_n]$ holds. By examining carefully Algorithms 1, 2, 3, 4, and their subroutines, one observes that only Algorithms 1 and 4 may introduce numbers that are in the algebraic closure $\overline{\mathbb{Q}}$ of $\mathbb{Q}$, and not in $\mathbb{Q}$ itself. In fact, for each $x = (x_1, \ldots, x_n) \in \lim(W(R))$, Algorithms 1 and 4 introduce a field extension $\mathbb{Q}(\xi_1, \ldots, \xi_m)$ such that we have $x_i \in \mathbb{Q}[\xi_1, \ldots, \xi_m]$. Let $Y_1, \ldots, Y_m$ be m new symbols. Let $G := \{g_1(Y_1), g_2(Y_1, Y_2), \ldots, g_m(Y_1, Y_2, \ldots, Y_m)\}$ be an irreducible regular chain (i.e. generating a maximal ideal over $\mathbb{Q}$) such that $G(Y_1 = \xi_1, \ldots, Y_m = \xi_m) = 0$ holds. Since $x_i \in \mathbb{Q}[\xi_1, \ldots, \xi_m]$, there exists $f_i \in \mathbb{Q}[Y_1, \ldots, Y_m]$, $i = 1, \ldots, n$, such that $x_i = f_i(Y_1 = \xi_1, \ldots, Y_m = \xi_m)$. Let $\mathcal{S}_x := \{X_1 = f_1(Y_1, \ldots, Y_m), \ldots, X_n = f_n(Y_1, \ldots, Y_m), G(Y_1, \ldots, Y_m) = 0\}$. The projection of the zero set of $\mathcal{S}_x$ on the $(X_1, \ldots, X_n)$-space is the zero set of an irregular chain $R_x \in \mathbb{Q}[X_1, \ldots, X_m]$ and we have $\lim(W(R)) = \cup_{x \in \lim(W(R))} W(R_x)$.

Algorithm 3. LimitPointsAtZero

Input: A regular chain $R := \{r_1(X_1, X_2), \ldots, r_{s-1}(X_1, \ldots, X_s)\}$.

Output: The non-trivial limit points of $W(R)$ whose X_1-coordinates are 0.

1 let $S := \{(T_0)\}$;

2 compute the accuracy estimates $\tau_1, \ldots, \tau_{s-2}$ by Theorem 6; let $\tau_{s-1} = 1$;

3 **for** i *from* 1 *to* $s - 1$ **do**

4 $\quad$ $S' := \emptyset$;

5 $\quad$ **for** $\Phi \in S$ **do**

6 $\quad\quad$ $f_i := r_i(X_1 = \Phi_1, \ldots, X_i = \Phi_i, X_{i+1})$;

7 $\quad\quad$ **if** $i > 1$ **then**

8 $\quad\quad\quad$ let $\delta := \mathrm{ord}(f_i, T_{i-1})$; let $f_i := f_i/T_{i-1}^\delta$;

9 $\quad\quad$ $E := \mathsf{NewtonPuiseux}(f_i, \tau_i)$;

10 $\quad\quad$ **for** $(T_{i-1} = \phi(T_i), X_{i+1} = \varphi(T_i)) \in E$ **do**

11 $\quad\quad\quad$ $S' := S' \cup \{\Phi(T_{i-1} = \phi(T_i)) \cup (\varphi(T_i))\}$

12 $\quad$ $S := S'$

13 **if** $S = \emptyset$ **then** return $\emptyset$;

14 **else** return eval$(S, T_{s-1} = 0)$;

Algorithm 4. LimitPoints

Input: A regular chain $R := \{r_1(X_1, X_2), \ldots, r_{s-1}(X_1, \ldots, X_s)\}$.

Output: All the non-trivial limit points of $W(R)$.

1 let $h_R := \mathrm{init}(R)$; let L be the set of zeros of h_R in $\mathbb{C}$; $S := \emptyset$;

2 **for** $\alpha \in L$ **do**

3 $\quad$ $R_\alpha := R(X_1 = X_1 + \alpha)$; $S_\alpha := \mathsf{LimitPointsAtZero}(R_\alpha)$;

4 $\quad$ update S_α by replacing the first coordinate of every point in S_α by α;

5 $\quad$ $S := S \cup S_\alpha$

6 return S

---

8   Experimentation

We have implemented Algorithm 4 of Section 7, which computes the limit points of the quasi-component of a one-dimensional strongly normalized regular chain. The implementation is based on the **RegularChains** library and the command algcurves[puiseux] [24] of MAPLE. The code is available at http://www.orcca.on.ca/cchen/ACM13/LimitPoints.mpl. This preliminary implementation relies on algebraic factorization, whereas, as suggested in [10], applying the D5 principle [9], in the spirit of triangular decomposition algorithms [6], would be sufficient when computations need to split into different cases. This would certainly improve performance greatly and this enhancement is work in progress.

As pointed out in the introduction, the computation of the limit points of the quasi-component of a regular chain can be applied to removing redundant components in a Kalkbrener triangular decomposition. In Table 1, we report on experimental results of this application.

The polynomial systems listed in this table are one-dimensional polynomial systems selected from the literature [5,6]. For each system, we first call the Triangularize command of the library RegularChains, with the option "'normalized='strongly', 'radical'='yes'". For the input system, this process computes a Kalkbrener triangular decomposition $\mathcal{R}$ where the regular chains are strongly normalized and their saturated ideals are radical. Next, for each one-dimensional regular chain R in the output, we compute the limit points $\lim(W(R))$, thus deducing a set of regular chains $R_1, \ldots, R_e$ such that the union of their quasi-components equals the Zariski closure $\overline{W(R)}$. The algorithm Difference [5] is then called to test whether or not there exists a pair R, R' of regular chains of $\mathcal{R}$ such that the inclusion $\overline{W(R)} \subseteq \overline{W(R')}$ holds. In Table 1, the columns T and #(T) denote respectively the timings spent by Triangularize and the number of regular chains returned by this command; the columns d-1 and d-0 denote respectively the number of 1-dimensional and 0-dimensional regular chains; the columns R and #(R) denote respectively the timings spent on removing redundant components in the output of Triangularize and the number of regular chains in the output irredundant decomposition. As we can see in the table, most of the decompositions are checked to be irredundant, which we could not do before this work by means of triangular decomposition algorithms. In addition, the three redundant 0-dimensional components in the Kalkbrener triangular decomposition of system f-744 are successfully removed in about 7 minutes, whereas we cannot draw this conclusion in more than one hour by a brute-force method computing the generators of the saturated ideals of regular chains. Therefore, we have verified experimentally the benefits provided by the proposed algorithms.

Table 1. Removing redundant components

Sys	T	#(T)	d-1	d-0	R	#(R)
f-744	14.360	4	1	3	432.567	1
Liu-Lorenz	0.412	3	3	0	216.125	3
MontesS3	0.072	2	2	0	0.064	2
Neural	0.296	5	5	0	1.660	5
Solotareff-4a	0.632	7	7	0	32.362	7
Vermeer	1.172	2	2	0	75.332	2
Wang-1991c	3.084	13	13	0	6.280	13

9 Concluding Remarks

In this paper, we proposed an algorithm for computing the limit points of the quasi-component of a regular chain in dimension one by means of Puiseux series expansions. In the future, we will investigate how to compute the limit points in higher dimension with the help of the Abhyankar-Jung theorem [21].

References

[1] Alonso, M.E., Mora, T., Niesi, G., Raimondo, M.: An algorithm for computing analytic branches of space curves at singular points. In: Proc. of the 1992 International Workshop on Mathematics Mechanization, pp. 135–166 (1992)

[2] Boulier, F., Lazard, D., Ollivier, F., Petitot, M.: Representation for the radical of a finitely generated differential ideal. In: Proc. of ISSAC 1995, pp. 158–166 (1995)

[3] Boulier, F., Lemaire, F., Moreno Maza, M.: Well known theorems on triangular systems and the D5 principle. In: Proc. of Transgressive Computing 2006, Granada, Spain, pp. 79–91 (2006)

[4] Chen, C., Davenport, J.H., May, J.P., Moreno Maza, M., Xia, B., Xiao, R.: Triangular decomposition of semi-algebraic systems. J. Symb. Comput. 49, 3–26 (2013)

[5] Chen, C., Golubitsky, O., Lemaire, F., Maza, M.M., Pan, W.: Comprehensive triangular decomposition. In: Ganzha, V.G., Mayr, E.W., Vorozhtsov, E.V. (eds.) CASC 2007. LNCS, vol. 4770, pp. 73–101. Springer, Heidelberg (2007)

[6] Chen, C., Moreno Maza, M.: Algorithms for computing triangular decomposition of polynomial systems. J. Symb. Comput. 47(6), 610–642 (2012)

[7] Chen, C., Moreno Maza, M., Xia, B., Yang, L.: Computing cylindrical algebraic decomposition via triangular decomposition. In: Proc. of ISSAC 2009, pp. 95–102 (2009)

[8] Chou, S.C., Gao, X.S.: A zero structure theorem for differential parametric systems. J. Symb. Comput. 16(6), 585–595 (1993)

[9] Della Dora, J., Dicrescenzo, C., Duval, D.: About a new method for computing in algebraic number fields. In: Proc. of EUROCAL 1985, pp. 289–290 (1985)

[10] Duval, D.: Rational Puiseux expansions. Compos. Math. 70(2), 119–154 (1989)

[11] Fischer, G.: Plane Algebraic Curves. American Mathematical Society (2001)

[12] Gao, X.S., Van der Hoeven, J., Yuan, C.M., Zhang, G.L.: Characteristic set method for differential-difference polynomial systems. J. Symb. Comput. 44(9), 1137–1163 (2009)

[13] Hubert, E.: Factorization-free decomposition algorithms in differential algebra. J. Symb. Comput. 29(4-5), 641–662 (2000)

[14] Kalkbrener, M.: Algorithmic properties of polynomial rings. J. Symb. Comput. 26(5), 525–581 (1998)

[15] Lemaire, F., Moreno Maza, M., Pan, W., Xie, Y.: When does $\langle t \rangle$ equal sat(t)? J. Symb. Comput. 46(12), 1291–1305 (2011)

[16] Lemaire, F., Moreno Maza, M., Xie, Y.: The RegularChains library. In: Maple 10, Maplesoft, Canada (2005); refereed software

[17] Marcus, S., Maza, M.M., Vrbik, P.: On fulton's algorithm for computing intersection multiplicities. In: Gerdt, V.P., Koepf, W., Mayr, E.W., Vorozhtsov, E.V. (eds.) CASC 2012. LNCS, vol. 7442, pp. 198–211. Springer, Heidelberg (2012)

[18] Maurer, J.: Puiseux expansion for space curves. Manuscripta Math. 32, 91–100 (1980)

[19] Mumford, D.: The Red Book of Varieties and Schemes, 2nd edn. Springer (1999)

[20] Munkres, J.R.: Topology, 2nd edn. Prentice Hall (2000)

[21] Parusiński, A., Rond, G.: The Abhyankar-Jung theorem. J. Algebra 365, 29–41 (2012)

[22] Ritt, J.F.: Differential Equations from an Algebraic Standpoint, vol. 14. American Mathematical Society (1932)

[23] Shimoyama, T., Yokoyama, K.: Localization and primary decomposition of polynomial ideals. J. Symb. Comput. 22(3), 247–277 (1996)

[24] van Hoeij, M.: An algorithm for computing an integral basis in an algebraic function field. J. Symb. Comput. 18(4), 353–363 (1994)
[25] Walker, R.J.: Algebraic Curves. Springer (1978)
[26] Wang, D.K.: The Wsolve package, `http://www.mmrc.iss.ac.cn/~dwang/wsolve.html`
[27] Wang, D.M.: Epsilon 0.618, `http://www-calfor.lip6.fr/~wang/epsilon`
[28] Yang, L., Hou, X.R., Xia, B.: A complete algorithm for automated discovering of a class of inequality-type theorems. Science in China, Series F 44(1), 33–49 (2001)

On Consistency of Finite Difference Approximations to the Navier-Stokes Equations

Pierluigi Amodio[1], Yuri Blinkov[2], Vladimir Gerdt[3], and Roberto La Scala[1]

[1] Department of Mathematics, University of Bari, Bari, Italy
{pierluigi.amodio,roberto.lascala}@uniba.it
[2] Department of Mathematics and Mechanics,
Saratov State University, Saratov, Russia
BlinkovUA@info.sgu.ru
[3] Laboratory of Information Technologies, Joint Institute for Nuclear Research,
Dubna, Russia
gerdt@jinr.ru

Abstract. In the given paper, we confront three finite difference approximations to the Navier–Stokes equations for the two-dimensional viscous incompressible fluid flows. Two of these approximations were generated by the computer algebra assisted method proposed based on the finite volume method, numerical integration, and difference elimination. The third approximation was derived by the standard replacement of the temporal derivatives with the forward differences and the spatial derivatives with the central differences. We prove that only one of these approximations is strongly consistent with the Navier–Stokes equations and present our numerical tests which show that this approximation has a better behavior than the other two.

1 Introduction

By its completion to involution [1], the well-known Navier–Stokes system of equations [2] for unsteady two-dimensional motion of incompressible viscous liquid of constant viscosity may be written in the following dimensionless form [11]

$$
\begin{cases}
f_1 := u_x + v_y = 0\,, \\
f_2 := u_t + uu_x + vu_y + p_x - \frac{1}{\mathrm{Re}}(u_{xx} + u_{yy}) = 0\,, \\
f_3 := v_t + uv_x + vv_y + p_y - \frac{1}{\mathrm{Re}}(v_{xx} + v_{yy}) = 0\,, \\
f_4 := u_x^2 + 2v_xu_y + v_y^2 + p_{xx} + p_{yy} = 0\,.
\end{cases}
\tag{1}
$$

Here (u, v) is the velocity field, f_1 is the continuity equation, f_2 and f_3 are the proper Navier–Stokes equations [2], and f_4 is the pressure Poisson equation [3]. The constant Re denotes the Reynolds number.

For discretization we use the finite difference method [4,5] and consider orthogonal and uniform computational grid. In this method, the finite difference approximation (FDA) to the differential equations combined with appropriate initial or/and boundary conditions in their discrete form constitutes the finite

V.P. Gerdt et al. (Eds.): CASC 2013, LNCS 8136, pp. 46–60, 2013.
© Springer International Publishing Switzerland 2013

difference scheme (FDS) for construction of a numerical solution. The main requirement to the scheme is convergence of its numerical solution to the solution of differential equation(s) when the grid spacings go to zero.

The fundamental problem in numerical solving of partial differential equation (PDE) or a system of PDEs is to construct such FDA that for any initial- or/and boundary-value problem, providing existence and uniqueness of the solution to PDE(s) with a smooth dependence on the initial or/and boundary data, the corresponding FDS is convergent. For polynomially-nonlinear PDEs, e.g., the Navier–Stokes equations, to satisfy this requirement FDA must inherit all algebraic properties of the differential equation(s). The necessary condition for the inheritance is the property of s(strong)-consistency of FDA to PDEs introduced first in [6] for linear equations and extended in [13] to nonlinear ones.

The conventional consistency [5], called in [6,13] by weak-consistency implies reduction of FDA to the original PDE(s) when the grid spacings go to zero. This consistency can be verified by a Taylor expansion of the difference equations in the FDA about a grid point. The strong consistency implies reduction of any element in the perfect difference ideal generated by the FDA to an element in the radical differential ideal generated by the PDE(s). In [13], it was shown that s-consistency can be checked in terms of a difference Gröbner basis of the ideal generated by the FDA. Since difference polynomial ring [7] is non Noetherian, in the nonlinear case, generally, one cannot verify s-consistency of a given FDA through computation of associated difference Gröbner basis. However, if the FDA under consideration is w-consistent, then it is not s-consistent if and only if at some step of the Buchberger-like algorithm (cf. [9,10] and [13]) applied to construction of the Gröbner basis, a difference S-polynomial arises which in not w-consistent with any of the consequences of the original PDE(s). In practice, this may help to detect s-inconsistency.

In [11], the algorithmic approach to generation of FDA suggested in [12] was applied to the Navier–Stokes equations (1). The approach is based on the finite volume method combined with numerical integration and difference elimination. As a result, three different w-consistent FDAs were obtained in [11]. Two of them were analyzed in [13] from the viewpoint of s-consistency. One of these FDAs was qualified as a "good" one, i.e., s-consistent, by the claim that it itself is a Gröbner basis. Another FDA was qualified as s-inconsistent by inspection (observed already in [11]) that one of its differential consequences is not reduced to a differential consequence of the system (1) when the grid spacings go to zero. However, as explicit computation with the Maple-based implementation [9] of the Buchberger-like algorithm [9,10] showed, the "good" FDA is not a Gröbner basis what generates a need for the further investigation of its s-consistency.

In this paper, we prove that the "good" FDA generated in [11] is indeed s-consistent. In doing so we avoid the Gröbner basis computation what is rather cumbersome. In addition, we consider universally adopted standard method to discretization, which consists in the replacement of the temporal derivatives in (1) with the forward differences and the spatial derivatives with the central differences, and show that it yields FDA which is not s-consistent. To see numerical impact of the property of s-consistency we confronted the three FDAs and compared their behavior for the mixed initial-boundary value problem whose data originate from the exact solution [14] of (1). This comparison clearly shows superiority of the s-consistent FDA over the other two which are not s-consistent.

This paper is organized as follows. Section 2 introduces the main objects of difference algebra related to discretization of (1). Section 3 is concerned with definition of FDA to (1) and its s-consistency. In Section 4, we consider three particular FDAs to the Navier–Stokes system (1) and establish their s-consistency properties. Section 5 presents the results of our numerical computer experiments. Some concluding remarks are given in Section 6.

2   Preliminaries

The left-hand sides of the PDEs in the Navier–Stokes system (1) can be considered as elements in the *differential polynomial ring* [8]

$$f_i = 0 \ (1 \leq i \leq 4), \quad F := \{f_1, f_2, f_3, f_4\} \subset R := \mathcal{K}[u, v, p], \tag{2}$$

where $\{u, v, p\}$ is the set of *differential indeterminates* and $\mathcal{K} := \mathbb{Q}(\mathrm{Re})$ is the *differential field of constants*.

Remark 1. It is easy to check that the differential ideal $[F] \subset R$ generated by F is *radical* [8].

To approximate the differential system (2) by a difference one, we use an orthogonal and uniform computational grid (mesh) as the set of points $(jh, kh, n\tau) \in \mathbb{R}^3$. Here $\tau > 0$ and $h > 0$ are the grid spacings (mesh steps), and the triple of integers $(j, k, n) \in \mathbb{Z}^3$ numerates the grid points. In doing so, in a grid node $(jh, kh, n\tau)$ a solution to (1) is approximated by the triple of grid functions

$$\{u^n_{j,k}, v^n_{j,k}, p^n_{j,k}\} := \{u, v, p\} \,|_{x=jh, y=kh, t=\tau n} \,. \tag{3}$$

Now we introduce the set of mutually commuting *differences* $\{\sigma_x, \sigma_y, \sigma_t\}$ acting on a grid function $\phi(x, y, t)$, which is to approximate a solution of (1) on the grid points, as the forward shift operators

$$\begin{cases} \sigma_x \circ \phi(x, y, t) = \phi(x + h, y, t), \\ \sigma_y \circ \phi(x, y, t) = \phi(x, y + h, t), \\ \sigma_t \circ \phi(x, y, t) = \phi(x, y, t + \tau). \end{cases} \tag{4}$$

The *monoid* generated by the differences will be denoted by Σ, i.e.,

$$\Sigma := \{\, \sigma_x^{i_1} \sigma_y^{i_2} \sigma_t^{i_3} \mid i_1, i_2, i_3 \in \mathbb{N}_{\geq 0} \,\}, \quad (\forall \sigma \in \Sigma) \, [\sigma \circ 1 = 1],$$

and the ring of *difference polynomials* over $\mathcal{K}$ will be denoted by $\mathcal{R}$. The elements in $\mathcal{R}$ are polynomials in the *difference indeterminates* u, v, p (dependent variables) defined on the grid points and in their shifted values

$$\{\, \sigma_x^{i_1} \sigma_y^{i_2} \sigma_t^{i_3} \circ w \mid w \in \{u, v, p\}, \ \{i_1, i_2, i_3\} \in \mathbb{N}_{\geq 0}^3 \,\}.$$

Definition 1. [7] *A total order $\prec$ on $\{\, \sigma \circ w \mid \sigma \in \Sigma,\ w \in \{u,v,p\} \,\}$ is* ranking *if for all $\sigma, \sigma_1, \sigma_2, \sigma_3 \in \Sigma$ and $w, w_1, w_2 \in \{u,v,p\}$*

$$\sigma\,\sigma_1 \circ w \succ \sigma_1 \circ w, \quad \sigma_1 \circ w_1 \succ \sigma_2 \circ w_2 \iff \sigma \circ \sigma_1 \circ w_1 \succ \sigma \circ \sigma_2 \circ w_2$$

The set $\mathcal{M}$ of *monomials* in the ring $\mathcal{R}$ reads

$$\mathcal{M} := \{\, (\sigma_1 \circ u)^{i_1} (\sigma_2 \circ v)^{i_2} (\sigma_3 \circ p)^{i_3} \mid \sigma_j \in \Sigma,\ i_j \in \mathbb{N}_{\geq 0},\ 1 \leq j \leq 3 \,\}. \tag{5}$$

Definition 2. [13] *A total order $\succ$ on $\mathcal{M}$ is* admissible *if it extends a ranking and*

$$(\forall\, \mu \in \mathcal{M}\setminus\{1\})\,[\mu \succ 1] \wedge (\forall\, \sigma \in \Sigma)\,(\forall\, \mu, a, b \in \mathcal{M}\,)\,[\,a \succ b \iff \mu \cdot \sigma \circ a \succ \mu \cdot \sigma \circ b\,].$$

Given an admissible monomial order $\succ$, every difference polynomial p has the *leading monomial* $\mathrm{lm}(p) \in \mathcal{M}$ and the *leading term* $\mathrm{lt}(p) := \mathrm{lm}(p)\,\mathrm{lc}(p)$ with the *leading coefficient* $\mathrm{lc}(p)$. Throughout this session, every difference polynomial is to be *normalized (i.e., monic)* by division of the polynomial by its leading coefficient. This provides $(\forall p \in \mathcal{R})\,[\mathrm{lc}(p) = 1]$.

Now we consider the notions of difference ideal [7] and its standard basis. The last notion was introduced in [13] in the full analogy to that in differential algebra [16].

Definition 3. [7] *A set $\mathcal{I} \subset \mathcal{R}$ is* difference polynomial ideal *or* σ-ideal *if*

$$(\forall\, a, b \in \mathcal{I})\,(\forall\, c \in \mathcal{R})\,(\forall\, \sigma \in \Sigma)\,[\,a + b \in \mathcal{I},\ a \cdot c \in \mathcal{I},\ \sigma \circ a \in \mathcal{I}\,].$$

If $F \subset \mathcal{R}$, then the smallest σ-ideal containing F is said to be generated by F and denoted by $[F]$.

Definition 4. [13] *If for $\alpha, \beta \in \mathcal{M}$ the equality $\beta = \mu \cdot \sigma \circ \alpha$ holds with $\sigma \in \Sigma$ and $\mu \in \mathcal{M}$ we shall say that α divides β and write $\alpha \mid \beta$. It is easy to see that this divisibility relation yields a partial order.*

Definition 5. [13] *Given a σ-ideal $\mathcal{I}$ and an admissible monomial ordering $\succ$, a subset $G \subset \mathcal{I}$ is its* (difference) standard basis *if $[G] = \mathcal{I}$ and*

$$(\forall\, p \in \mathcal{I})(\exists\, g \in G)\ [\mathrm{lm}(g) \mid \mathrm{lm}(p)].$$

If the standard basis is finite we shall call it Gröbner basis.

Remark 2. Based on Definition 4, one can introduce (see [13]) in difference algebra the concepts of *polynomial reduction* and *normal form* of a difference polynomial p modulo a set of difference polynomials P (notation: $\mathrm{NF}(p, P)$). A *reduced standard basis* G is such that $(\forall g \in G)\,[g = NF(g, G \setminus \{g\})]$.

The algorithmic characterization of standard bases and their construction in difference polynomial rings is done in terms of difference S-polynomials.

Definition 6. [13] *Given an admissible order, and monic difference polynomials p and q (they not need to be distinct), the polynomial $S(p,q) := m_1 \cdot \sigma_1 \circ p - m_2 \cdot \sigma_2 \circ q$ is called (difference)S-polynomial associated to p and q if $m_1 \cdot \sigma_1 \circ \mathrm{lm}(p) = m_2 \cdot \sigma_2 \circ \mathrm{lm}(q)$ with co-prime $m_1 \cdot \sigma_1$ and $m_2 \cdot \sigma_2$.*

Remark 3. This characterization immediately implies [13,10] a difference version of the Buchberger algorithm (cf. [15,16]). The algorithm always terminates when the input polynomials are linear. If this is not the case, the algorithm may not terminate. Additionally, one can take into account Buchberger's criteria to avoid some useless zero reductions. The difference criteria are similar to the differential ones [16].

Definition 7. [7] *A perfect difference ideal generated by a set $F \subset \mathcal{R}$ and denoted by $[\![F]\!]$ is the smallest difference ideal containing F and such that for any $f \in \mathcal{R}$, $\sigma_1, \ldots, \sigma_r \in \Sigma$ and $k_1, \ldots, k_r \in \mathbb{N}_{\geq 0}$*

$$(\sigma_1 \circ f)^{k_1} \cdots (\sigma_r \circ f)^{k_r} \in [\![F]\!] \Longrightarrow f \in [\![F]\!].$$

Remark 4. In difference algebra, perfect ideals play the same role (cf. [17]) as radical ideals in commutative and differential algebra. Obviously, $[F] \subseteq [\![F]\!]$.

3   Consistency of Difference Approximations

Let a finite set of difference polynomials

$$\tilde{f}_1 = \cdots = \tilde{f}_p = 0, \quad \tilde{F} := \{\tilde{f}_1, \ldots \tilde{f}_p\} \subset \mathcal{R} \tag{6}$$

be a FDA to (1). It should be noted that generally the number p in (6) needs not to be equal to the number of equations in (1).

Definition 8. *A differential (resp. difference) polynomial $f \in R$ (resp. $\tilde{f} \in \mathcal{R}$) is* differential-algebraic *(resp.* difference-algebraic*) consequence of (1) (resp. (6)) if $f \in [\![F]\!]$ (resp. $\tilde{f} \in [\![\tilde{F}]\!]$).*

Definition 9. *We shall say that a difference equation $\tilde{f} = 0$ implies (in the continuous limit) the differential equation $f = 0$ and write $\tilde{f} \triangleright f$ if f does not contain the grid spacings h, τ and the Taylor expansion about a grid point $(u_{j,k}^n, v_{j,k}^n, p_{j,k}^n)$ transforms equation $\tilde{f} = 0$ into $f + O(h, \tau) = 0$ where $O(h, \tau)$ denotes expression which vanishes when h and τ go to zero.*

Definition 10. [13,6] *The difference approximation (6) to (1) is* weakly consistent *or* w-consistent *with (1) if $p = 4$ and*

$$(\forall \tilde{f} \in \tilde{F})\,(\exists f \in F)\,[\tilde{f} \triangleright f]. \tag{7}$$

The requirement of weak consistency which has been universally accepted in the literature, is not satisfactory by the following two reasons:

1. The cardinality of FDA in (13) may be different from that of the original set of differential equations. For example, the systems $\{\, u_{xz}+yu = 0,\ u_{yw}+zu = 0 \,\}$ and $\{\, yu_y - zu_z = 0,\ u_x - u_w = 0,\ u_{xw} + yu_y = 0 \,\}$ in one dependent and four dependent variables are fully equivalent (see [6], Example 3). Thus, to construct a FDA, one can use them interchangeably. Whereas Definition 10 fastens $\tilde{F}$ to F.

2. A w-consistent FDA may not be good in view of inheritance of properties of differential systems at the discrete level. We shall demonstrate this in the next section.

Another concept of consistency was introduced in [6] for linear FDA and then extended in [13] to the nonlinear case. For the Navier–Stokes system, it is specialized as follows.

Definition 11. *An FDA (6) to (1) is* strongly consistent *or s-consistent if*

$$(\forall \tilde{f} \in [\![\tilde{F}]\!])\ (\exists f \in [F])\ [\tilde{f} \triangleright f].\tag{8}$$

The algorithmic approach of paper [13] to verification of s-consistency is based on the following theorem.

Theorem 1. [13] *A difference approximation (6) to (1) is s-consistent if and only if a (reduced) standard basis G of the difference ideal $[\tilde{F}]$ satisfies*

$$(\forall g \in G)\ (\exists f \in [F])\ [g \triangleright f].\tag{9}$$

Irrespective of possible infiniteness of the (nonlinear) difference standard basis G, it may be useful to apply an algorithm for its construction (see, for example, the algorithms in [13,10]) and to verify s-consistency of the intermediate polynomials. In doing so, one should check first the w-consistency of the polynomials in the input FDA. Then, if the normal form $\tilde{p}$ of an S-polynomial modulo the current basis is nonzero, then before insertion of $\tilde{p}$ into the intermediate basis one has to construct p such that $\tilde{p} \triangleright p$ and check the condition $p \in [F]$.

Remark 5. Given a differential polynomial $f \in R$, one can algorithmically check its membership in $[\![F]\!]$ by performing the involutive Janet reduction [13].

4   Three Difference Approximations to the Navier–Stokes Equations

To analyze strong consistency of difference approximations to (1) we shall need the following statements.

Proposition 1. *Let* $\tilde{f} \in R$ *be a difference polynomial. Suppose* $\tilde{f} \rhd f$ *where* f *is a differential-algebraic consequence of the Navier–Stokes system (1),* $f \in [F]$. *Then a finite sum of the form*

$$\tilde{p} := \sum_i \tilde{g}_i \cdot \sigma_i \circ \tilde{f}, \quad \sigma_i \in \Sigma, \quad \tilde{g}_i \in \mathcal{R} \tag{10}$$

also implies a differential-algebraic consequence of (1).

Proof. The shift operators in (4) are expanded in the Taylor series as follows

$$\sigma_x = \sum_{k \geq 0} \frac{h^k}{k!} \partial_x^k, \quad \sigma_y = \sum_{k \geq 0} \frac{h^k}{k!} \partial_y^k, \quad \sigma_t = \sum_{k \geq 0} \frac{\tau^k}{k!} \partial_t^k.$$

By the Taylor expansion over a grid point, in the limit when h and τ go to zero, the right-hand side of (10) becomes differential polynomial of the form

$$p := \sum_\mu b_\mu \partial^\mu \circ f, \quad b_\mu \in R, \quad \partial^\mu \in \{ \partial_x^i \partial_y^j \partial_t^k \mid i, j, k \in \mathbb{N}_{\geq 0} \}.$$

Thus, $\tilde{p} \rhd p \in [F]$. □

Corollary 1. *Let* $\tilde{F}$ *be a FDA (6) to (1) and* $\succ$ *be an admissible order on the monomial set (5). Suppose* $(\forall \tilde{f} \in \tilde{F}) \left[\tilde{f} \rhd f \in [F] \right]$. *Then, every element* $\tilde{p}$ *in the difference ideal* $[\tilde{F}]$ *that admits the representation*

$$\tilde{q} := \sum_{k=1}^p \sum_i \tilde{g}_{i,k} \cdot \sigma_i \circ \tilde{f}_k, \quad \sigma_i \in \Sigma, \quad \tilde{g}_i \in \mathcal{R}, \tag{11}$$

where the leading terms of the polynomials in $\sum_i \tilde{g}_{i,k} \cdot \sigma_i \circ \tilde{f}_k$ *do not cancel out, satisfies* $\tilde{q} \rhd q \in [F]$.

Proof. Denote by p_k the continuous limit of $\sum_i \tilde{g}_{i,k} \cdot \sigma_i \circ \tilde{f}_k$. Since $p_k \in [F]$, the no-cancellation assumption implies $\tilde{p} \rhd \sum_{k=1}^p p_k \in [F]$. □

Now we consider three difference approximations to system (1). The first two of them were constructed in [11] by applying the algorithmic approach to discretization proposed in [12] and based on the finite volume method combined with numerical integration and difference elimination. The third approximation is obtained by the conventional discretization what consists of replacing in (1) the temporal derivatives with the forward differences and the spatial derivatives with the central differences.

Every difference equation in an approximation must be satisfied in every node of the grid. As this takes place, one can apply to every equation a finite number of the forward shift operators (4) as well as of their inverses (the backward shift operators) to transform the approximation into an equivalent form. Because of this, we consider the difference approximations generated in [11] in the form which is commonly used for numerical solving of PDEs.

FDA 1 ([11], Eqs. 13)

$$
\begin{cases}
e1^n_{j,k} := \dfrac{u^n_{j+1,k}-u^n_{j-1,k}}{2h} + \dfrac{v^n_{j,k+1}-v^n_{j,k-1}}{2h} = 0, \\[2mm]
e2^n_{j,k} := \dfrac{u^{n+1}_{jk}-u^n_{jk}}{\tau} + \dfrac{{u^n_{j+1,k}}^2-{u^n_{j-1,k}}^2}{2h} + \dfrac{v^n_{j,k+1}u^n_{j,k+1}-v^n_{j,k-1}u^n_{j,k-1}}{2h} + \dfrac{p^n_{j+1,k}-p^n_{j-1,k}}{2h} \\[2mm]
\qquad - \dfrac{1}{\mathrm{Re}}\left(\dfrac{u^n_{j+2,k}-2u^n_{jk}+u^n_{j-2,k}}{4h^2} + \dfrac{u^n_{j,k+2}-2u^n_{jk}+u^n_{j,k-2}}{4h^2} \right) = 0, \\[2mm]
e3^n_{j,k} := \dfrac{v^{n+1}_{jk}-v^n_{jk}}{\tau} + \dfrac{u^n_{j+1,k}v^n_{j+1,k}-u^n_{j-1,k}v^n_{j-1,k}}{2h} + \dfrac{{v^n_{j,k+1}}^2-{v^n_{j,k-1}}^2}{2h} + \dfrac{p^n_{j,k+1}-p^n_{j,k-1}}{2h} \\[2mm]
\qquad - \dfrac{1}{\mathrm{Re}}\left(\dfrac{v^n_{j+2,k}-2v^n_{jk}+v^n_{j-2,k}}{4h^2} + \dfrac{v^n_{j,k+2}-2v^n_{jk}+v^n_{j,k-2}}{4h^2} \right) = 0, \\[2mm]
e4^n_{j,k} := \dfrac{{u^n_{j+2,k}}^2-2{u^n_{j,k}}^2+{u^n_{j-2,k}}^2}{4h^2} + \dfrac{{v^n_{j,k+2}}^2-2{v^n_{j,k}}^2+{v^n_{j,k-2}}^2}{4h^2} \\[2mm]
\qquad + 2\dfrac{u^n_{j+1,k+1}v^n_{j+1,k+1}-u^n_{j+1,k-1}v^n_{j+1,k-1}-u^n_{j-1,k+1}v^n_{j-1,k+1}+u^n_{j-1,k-1}v^n_{j-1,k-1}}{4h^2} \\[2mm]
\qquad + \dfrac{p^n_{j+2,k}-2p^n_{jk}+p^n_{j-2,k}}{4h^2} + \dfrac{p^n_{j,k+2}-2p^n_{jk}+p^n_{j,k-2}}{4h^2} = 0.
\end{cases}
$$

FDA 2 ([11], Eqs. 18)

$$
\begin{cases}
e1^n_{j,k} := \dfrac{u^n_{j+1,k}-u^n_{j-1,k}}{2h} + \dfrac{v^n_{j,k+1}-v^n_{j,k-1}}{2h} = 0, \\[2mm]
e2^n_{j,k} := \dfrac{u^{n+1}_{jk}-u^n_{jk}}{\tau} + u^n_{jk}\dfrac{u^n_{j+1,k}-u^n_{j-1,k}}{2h} + v^n_{jk}\dfrac{u^n_{j,k+1}-u^n_{j,k-1}}{2h} + \dfrac{p^n_{j+1,k}-p^n_{j-1,k}}{2h} \\[2mm]
\qquad - \dfrac{1}{\mathrm{Re}}\left(\dfrac{u^n_{j+1,k}-2u^n_{jk}+u^n_{j-1,k}}{h^2} + \dfrac{u^n_{j,k+1}-2u^n_{jk}+u^n_{j,k-1}}{h^2} \right) = 0, \\[2mm]
e3^n_{j,k} := \dfrac{v^{n+1}_{jk}-v^n_{jk}}{\tau} + u^n_{jk}\dfrac{v^n_{j+1,k}-v^n_{j-1,k}}{2h} + v^n_{jk}\dfrac{v^n_{j,k+1}-v^n_{j,k-1}}{2h} + \dfrac{p^n_{j,k+1}-p^n_{j,k-1}}{2h} \\[2mm]
\qquad - \dfrac{1}{\mathrm{Re}}\left(\dfrac{v^n_{j+1,k}-2v^n_{jk}+v^n_{j-1,k}}{h^2} + \dfrac{v^n_{j,k+1}-2v^n_{jk}+v^n_{j,k-1}}{h^2} \right) = 0, \\[2mm]
e4^n_{j,k} := \left(\dfrac{u^n_{j+1,k}-u^n_{j-1,k}}{2h} \right)^2 + 2\dfrac{v^n_{j+1,k}-v^n_{j-1,k}}{2h}\dfrac{u^n_{j,k+1}-u^n_{j,k-1}}{2h} + \left(\dfrac{v^n_{j,k+1}-v^n_{j,k-1}}{2h} \right)^2 \\[2mm]
\qquad + \dfrac{p^n_{j+1,k}-2p^n_{jk}+p^n_{j-1,k}}{h^2} + \dfrac{p^n_{j,k+1}-2p^n_{jk}+p^n_{j,k-1}}{h^2} = 0
\end{cases}
$$

FDA 3

$$
\begin{cases}
e1^n_{j,k} := \dfrac{u^n_{j+1,k}-u^n_{j-1,k}}{2h} + \dfrac{v^n_{j,k+1}-v^n_{j,k-1}}{2h} = 0, \\[2mm]
e2^n_{j,k} := \dfrac{u^{n+1}_{jk}-u^n_{jk}}{\tau} + u^n_{jk}\dfrac{u^n_{j+1,k}-u^n_{j-1,k}}{2h} + v^n_{jk}\dfrac{u^n_{j,k+1}-u^n_{j,k-1}}{2h} + \dfrac{p^n_{j+1,k}-p^n_{j-1,k}}{2h} \\[2mm]
\qquad - \dfrac{1}{\mathrm{Re}}\left(\dfrac{u^n_{j+1,k}-2u^n_{jk}+u^n_{j-1,k}}{h^2} + \dfrac{u^n_{j,k+1}-2u^n_{jk}+u^n_{j,k-1}}{h^2} \right) = 0, \\[2mm]
e3^n_{j,k} := \dfrac{v^{n+1}_{jk}-v^n_{jk}}{\tau} + u^n_{jk}\dfrac{v^n_{j+1,k}-v^n_{j-1,k}}{2h} + v^n_{jk}\dfrac{v^n_{j,k+1}-v^n_{j,k-1}}{2h} + \dfrac{p^n_{j,k+1}-p^n_{j,k-1}}{2h} \\[2mm]
\qquad - \dfrac{1}{\mathrm{Re}}\left(\dfrac{v^n_{j+1,k}-2v^n_{jk}+v^n_{j-1,k}}{h^2} + \dfrac{v^n_{j,k+1}-2v^n_{jk}+v^n_{j,k-1}}{h^2} \right) = 0, \\[2mm]
e4^n_{j,k} := \left(\dfrac{u^n_{j+1,k}-u^n_{j-1,k}}{2h} \right)^2 + 2\dfrac{v^n_{j+1,k}-v^n_{j-1,k}}{2h}\dfrac{u^n_{j,k+1}-u^n_{j,k-1}}{2h} + \left(\dfrac{v^n_{j,k+1}-v^n_{j,k-1}}{2h} \right)^2 \\[2mm]
\qquad + \dfrac{p^n_{j+1,k}-2p^n_{jk}+p^n_{j-1,k}}{h^2} + \dfrac{p^n_{j,k+1}-2p^n_{jk}+p^n_{j,k-1}}{h^2} = 0
\end{cases}
$$

The difference approximations in the form (6) constructed in [11] are obtained from FDA 1,2 by applying the forward shift operators (4) as follows.

$$
\tilde{F} := \left\{ \sigma \circ e1^n_{j,k},\, \sigma^2 \circ e2^n_{j,k},\, \sigma^2 \circ e3^n_{j,k},\, \sigma^2 \circ e4^n_{j,k} \right\}, \quad \sigma := \sigma_x \sigma_y. \qquad (12)
$$

All three FDAs are w-consistent. This can be easily verified by the Taylor expansion of the finite differences in the set

$$\tilde{F} := \{e1_{j,k}^n, e2_{j,k}^n, e3_{j,k}^n, e4_{j,k}^n\} \tag{13}$$

about the grid point $\{hj, hk, n\tau\}$ when the grid spacings h and τ go to zero.

To study s-consistency, fix admissible monomial order $\succ$ on (5) such that the leading monomials of difference polynomials in (13) read, respectively, as

$$\{u_{j+1,k}^n, u_{jk}^{n+1}, v_{jk}^{n+1}, p_{j+2,k}^n \text{ for FDA 1 and } p_{j+1,k}^n \text{ for FDA 2,3}\}. \tag{14}$$

Such monomial order can be easily constructed by extension of the block (lexdeg) orderly (cf. [13], Remark 2) ranking $\{\sigma_t\}\{\sigma_x, \sigma_y\}$ with $p \succ u \succ v$.

Proposition 2. *Among weakly consistent FDAs 1,2, and 3 only FDA 1 is strongly consistent.*

Proof. ¿From the leading monomial set (14) and the structure of FDAs it follows that every of the approximations has the only nontrivial S-polynomial

$$S(e1_{j,k}^n, e2_{j,k}^n) := \frac{e1_{j,k}^{n+1}}{\tau} - \frac{e2_{j+1,k}^n}{2h}. \tag{15}$$

In the case of FDA 1, the S-polynomial (15) is expressed in terms of the difference polynomials in (13) as follows

$$S(e1_{j,k}^n, e2_{j,k}^n) = -\frac{e1_{j,k}^n}{\tau} + \frac{e2_{j-1,k}^n}{2h} + \frac{e3_{j,k+1}^n + e3_{j,k-1}^n}{2h}$$
$$+ \frac{1}{\text{Re}} \left(\frac{e1_{j+2,k}^n - 2e1_{jk}^n + e1_{j-2,k}^n}{4h^2} + \frac{e1_{j,k+2}^n - 2e1_{jk}^n + e1_{j,k-2}^n}{4h^2} \right) - e4_{j,k}^n. \tag{16}$$

The summands in the right-hand side of (16) have distinct leading terms, and thus cannot be cancelled out. Furthermore, every summand implies a differential consequence of the corresponding equation in the system (1). Hence, by Corollary 1, the S-polynomial (15) implies, in the continuous limit, an algebraic-differential consequence of (1).

Consider now an element of the form (11) in the ideal $[\tilde{F}] \subset \mathcal{R}$ generated by the difference polynomials appearing in FDA 1. If cancellation occurs in $\tilde{p}$ among the leading terms, then the sum in (11) can be rewritten by means of the right-hand side of (16) so that cancellation of the leading terms cannot occur (cf. [15],Ch.2,§ 6,Th.6). Consequently, $\tilde{p}$ implies an element in $[F]$. This proves s-consistency of FDA 1.

In the case of FDAs 2 and 3, the corresponding S-polynomial in addition to the expression shown in right-hand side of (16) has extra and rather cumbersome polynomial additive which we denote by Δ_2 and Δ_3, respectively. In the continuous limit, $\Delta_2 = 0$ implies

$$2vv_{yyyy} + 8v_y v_{yyy} + 6v_{yy}^2 + 2uu_{xxxx} + 8u_x u_{xxx} + 6u_{xx}^2 + p_{yyyy} + p_{xxxx} = 0.$$

Δ_3 is given by

$$\Delta_3 := -u_{j,k}^n \frac{e1_{j+1,k}^n + e1_{j-1,k}^n}{2h} - v_{j,k}^n \frac{e1_{j,k+1}^n + e1_{j,k-1}^n}{2h} + \Delta_3' .$$

Clearly, the explicitly written terms of Δ_3 in the continuous limit imply an element in the differential ideal $[F]$. Further, $\Delta_3' = 0$ implies PDE

$$2vv_{yyyy} + 8v_y v_{yyy} + 6v_{yy}^2 + 2uu_{xxxx} + 8u_x u_{xxx} + 6u_{xx}^2 + p_{yyyy} + p_{xxxx} = 0 .$$

The both of obtained differential equations do not follow from the Navier–Stokes equations that can easily be verified[1] by using the Janet reduction of their left-hand sides modulo the system of polynomials in (1). Therefore, FDAs 2 and 3 are not strongly consistent. □

¿From Theorem 1, we immediately conclude:

Corollary 2. *A difference standard basis G of the ideal $[\tilde{F}]$ generated by the set (13) for FDA 1 satisfies the condition (9).*

5   Numerical Comparison

In this section, we perform some numerical tests for experimental comparison of the three FDAs of the previous section. To this aim, we suppose that the Navier–Stokes system (1) is defined for $t \geq 0$ in the square domain $\Omega = [0, \pi] \times [0, \pi]$ and provide initial conditions for $t = 0$ and boundary conditions for $t > 0$ and $(x, y) \in \partial\Omega$. Initial and boundary conditions are defined according to (17). Moreover, since we are essentially interested in the behavior of the different space discretizations used by the FDAs, any required additional values near the boundary ones are supposed to be known exactly.

Let $[0, \pi] \times [0, \pi]$ be discretized in the (x, y)-directions by means of the $(m+2)^2$ equispaced points $x_j = jh$ and $y_k = kh$, for $j, k = 0, \ldots m+1$, and $h = \pi/(m+1)$. Considering difference equations (13) we observe that, starting from the initial conditions, the second and the third equations give explicit formulae to compute u_{jk}^{n+1} and v_{jk}^{n+1} for $j, k = 1, \ldots, m$, respectively. Vice versa, the fourth equation may be used to derive a $m^2 \times m^2$ linear system that computes the unknowns p_{jk}^{n+1} for $j, k = 1, \ldots, m$. In doing so, the first equation is unnecessary to evaluate the unknowns but may be used to validate the obtained solution. This procedure may be iterated for $n = 0, 1, \ldots, N$ being $t_f = N\tau$ the end point of the time interval. Since in our experiments we are essentially interested in comparing different discretizations of u, v, and p on the space domain, the value of the time step τ was always chosen in order to provide stability.

---

[1] The Maple library implementing the differential Thomas decomposition [18] is capable of making Janet reduction of a differential polynomial modulo a nonlinear system of PDEs. The library is free download from
http://wwwb.math.rwth-aachen.de/thomasdecomposition/index.php

In the following figures, we compare the error behavior in $t_f = 1$ given by the three methods for different values of the Reynolds number Re. Error is computed by means of the formula

$$e_g = \max_{j,k} \frac{|g_{j,k}^N - g(x_j, y_k, t_f)|}{1 + |g(x_j, y_k, t_f)|}.$$

where $g \in \{u, v, p\}$ and $g(x, y, t)$ belongs to the exact solution [14] to (1)

$$\begin{cases} u := -e^{-2t/\mathrm{Re}} \cos(x) \sin(y), \\ v := e^{-2t/\mathrm{Re}} \sin(x) \cos(y), \\ p := -e^{-4t/\mathrm{Re}}(\cos(2x) + \cos(2y))/4. \end{cases} \tag{17}$$

Figure 1 shows the numerical results obtained for (17) with the Reynolds number set to Re $= 10^5$. Each subplot represents the error of a difference approximation for several values of m and $N = 10$. The three lines in each subplot represent the error in u, v, and p. Even if the behavior of the three schemes is essentially the same, for $m = 50$, the scheme based on FDA 1 is able to obtain the solution with an error less than 10^{-7} while the schemes based on FDAs 2 and 3 do not obtain an approximation of the solution with an error less than 10^{-4}.

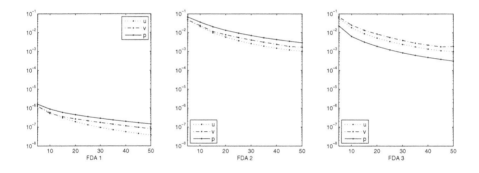

Fig. 1. Relative error in (17) for FDA 1, FDA 2 and FDA 3 with $N = 10$, $t_f = 1$, Re $= 10^5$ and varying m from 5 to 50

Figure 2 shows the value of the first difference polynomial $e_1{}_{j,k}^n$ in (13) for the three FDAs and for growing m obtained by the numerical solution. It is clear that the discretizations FDA 2 and 3 can not get along without the continuity equation f_1 in the Navier-Stokes system (1).

Figure 3 shows the results obtained for problem (17) with the Reynolds number set to Re $= 10^2$. Again each subplot represents the error of a difference scheme and the three lines inside each subplot represent the error in u, v, and p, respectively, for several values of m and $N = 40$. Similar considerations to the

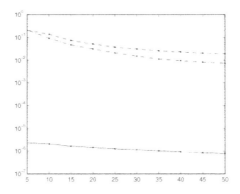

Fig. 2. Computed value of f_1 in (1) for FDA 1 (solid line), FDA 2 (dashed line) and FDA 3 (dash-dotted line) with $N = 10$, $t_f = 1$, $Re = 10^5$ and varying m from 5 to 50

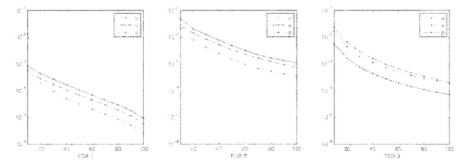

Fig. 3. Computed errors in u, v and p for FDA 1 (left), 2 (middle) and 3 (right): $N = 40$, $t_f = 1$, $Re = 10^2$ and varying m from 10 to 100

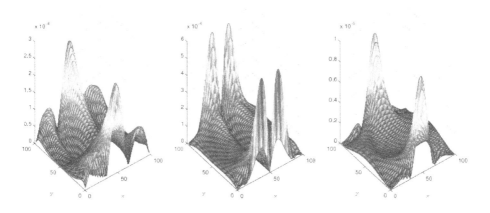

Fig. 4. Computed error with FDA 1 (u, v and p, respectively): $N = 40$, $t_f = 1$, $Re = 10^2$ and $m = 100$

previous example may be done: the scheme based on FDA 1 works much better than the others and the scheme with FDA 2 is the worst.

We conclude showing in Figure 4 the computed error in $t_f = 1$ using the s-consistent FDA 1 applied to the problem (17) (Re = 10^2) with $N = 40$ and $m = 100$. Larger errors are near the boundaries, and u and v seem to be better approximated than p.

6 Conclusion

As it has been already demonstrated in [6] for overdetermined systems of linear PDEs, it may be highly nontrivial to construct strongly consistent difference approximations. In the given paper, we have demonstrated that the demands of s-consistency impose strong limitations on the finite difference approximations to the nonlinear system of Navier–Stokes equations. These limitations proceed from the fact that s-consistent approximations inherit at the discrete level all basic algebraic properties of the initial differential equations.

It turned out that among two distinctive approximations generated in [12] (by applying the same algorithmic technique with different choice of numerical integration method), the one with a 5×5 stencil (FDA 1) is strongly consistent whereas the other one with a 3×3 stencil (FDA 2) is not. This result is at variance with universally accepted opinion that discretization with a more compact stencil is numerically favoured. One more discretization with a 3×3 stencil (FDA 3), obtained from the differential Navier–Stokes equations by the replacement of spatial derivatives with the central differences and of temporal derivatives with the forward differences, also failed to be s-consistent. As this takes place, our computer experimentation revealed much better numerical behavior of the s-consistent approximation in comparison with the considered s-inconsistent ones. The question of existence of s-consistent FDA to (1) with a 3×3 stencil is open.

Unlike the linear case [6], given a difference approximation on a grid with equisized grid spacings, one cannot fully algorithmically check its s-consistency. This is owing to non-noetherianity of difference polynomial rings that may lead to non-existence of a finite difference Gröbner basis for the ideal generated by the approximation. And even with the existence of a Gröbner basis, its construction and algorithmic verification of s-consistency may be very hard. For example, by using experimental implementation in Maple [9] of the algorithm of papers [9,10][2], many finite Gröbner bases have been constructed for the s-consistent approximation FDA-1 and for many different monomial orders. In doing so, the smallest obtained basis consists of 5 different polynomials, and one of the polynomials has 404 terms. In distinction to those rather tedious computations, the verification of s-consistency for FDA 1 and s-inconsistency for the other two was done by analysing the only S-polynomial and required much less symbolic computation.

---

[2] To our knowledge, it is the only software available for computation of difference Gröbner bases for the nonlinear case.

It should be noted that in our paper, we use the collocated arrangement of the dependent variables u, v, and p in the system (1) that often gives rise to oscillations of the variables (cf. [19]) and makes impossible convergence of numerical solutions. Our experiments presented in Section 5 demonstrate no spurious oscillations of the numerical solution. This can be considered as a significant positive property of the obtained FDAs.

Acknowledgements. The authors thank the anonymous referees for constructive comments and recommendations which helped to improve the readability and quality of the paper. The contribution of the second and third authors (Yu.B. and V.G.) was partially supported by grant 13-01-00668 from the Russian Foundation for Basic Research and by grant 3802.2012.2 from the Ministry of Education and Science of the Russian Federation.

References

1. Seiler, W.M.: Involution: The Formal Theory of Differential Equations and its Applications in Computer Algebra. Algorithms and Computation in Mathematics, vol. 24. Springer, Heidelberg (2010)
2. Pozrikidis, C.: Fluid Dynamics: Theory, Computation and Numerical Simulation. Kluwer, Norwell (2001)
3. Gresho, P.M., Sani, R.L.: On pressure boundary conditions for the incompressible Navier–Stokes Equations. Int. J. Numer. Meth. 7, 1111–1145 (1987)
4. Samarskii, A.A.: Theory of Difference Schemes. Marcel Dekker, New York (2001)
5. Strikwerda, J.C.: Finite Difference Schemes and Partial Differential Equations, 2nd edn. SIAM, Philadelphia (2004)
6. Gerdt, V.P., Robertz, D.: Consistency of Finite Difference Approximations for Linear PDE Systems and its Algorithmic Verification. In: Watt, S.M. (ed.) Proc. ISSAC 2010, pp. 53–59. Association for Computing Machinery, New York (2010)
7. Levin, A.: Difference Algebra, vol. 8. Springer (2008)
8. Hubert, E.: Notes on Triangular Sets and Triangulation-Decomposition Algorithms II: Differential Systems. In: Winkler, F., Langer, U. (eds.) SNSC 2001. LNCS, vol. 2630, pp. 40–87. Springer, Heidelberg (2003)
9. La Scala, R.: Gröbner Bases and Gradings for Partial Difference Ideals. arXiv:math.RA/1112.2065
10. Gerdt, V., La Scala, R.: Noetherian Quotient of the Algebra of Partial Difference Polynomials and Gröbner Bases of Symmetric Ideals. arXiv:math.AC/1304.7967
11. Gerdt, V.P., Blinkov, Y.A.: Involution and Difference Schemes for the Navier–Stokes Equations. In: Gerdt, V.P., Mayr, E.W., Vorozhtsov, E.V. (eds.) CASC 2009. LNCS, vol. 5743, pp. 94–105. Springer, Heidelberg (2009)
12. Gerdt, V.P., Blinkov, Y.A., Mozzhilkin, V.V.: Gröbner Bases and Generation of Difference Schemes for Partial Differential Equations. SIGMA 2, 051 (2006); arXiv:math.RA/0605334
13. Gerdt, V.P.: Consistency Analysis of Finite Difference Approximations to PDE Systems. In: Adam, G., Buša, J., Hnatič, M. (eds.) MMCP 2011. LNCS, vol. 7125, pp. 28–42. Springer, Heidelberg (2012)

14. Pearson, C.E.: A computational method for time dependent two dimensional in-compressible viscous flow problems. Sperry-Rand Research Center, Sudbury, Mass., Report No. SRRC-RR-64-17 (1964); Chorin, A.J.: The Numerical Solution of the Navier–Stokes Equations for an Incompressible Fluid. AEC Research and Devel-opment Report, NYO-1480-82, New York University, New York (1967)
15. Cox, D., Little, J., O'Shea, D.: Ideals, Varieties and Algorithms. An Introduction to Computational Algebraic Geometry and Commutative Algebra, 3rd edn. Springer, New York (2007)
16. Ollivier, F.: Standard Bases of Differential Ideals. In: Sakata, S. (ed.) AAECC 1990. LNCS, vol. 508, pp. 304–321. Springer, Heidelberg (1991)
17. Trushin, D.V.: Difference Nullstellensatz. arXiv:math.AC/0908.3865
18. Bächler, T., Gerdt, V., Lange-Hegermann, M., Robertz, D.: Algebraic Thomas Decomposition of Algebraic and Differential Systems. J. Symb. Comput. 47, 1233–1266 (2012); arXiv:math.AC/1108.0817
19. Ferziger, J.H., Peric, M.: Computational Methods for Fluid Dynamics. Springer, Heidelberg (1996)

Faster Sparse Interpolation of Straight-Line Programs

Andrew Arnold[1], Mark Giesbrecht[1], and Daniel S. Roche[2]

[1] Cheriton School of Computer Science, University of Waterloo, Canada
{a4arnold,mwg}@uwaterloo.ca
http://cs.uwaterloo.ca/~{a4arnold,mwg}
[2] United States Naval Academy, USA
roche@usna.edu
http://www.usna.edu/Users/cs/roche/

Abstract. We give a new probabilistic algorithm for interpolating a "sparse" polynomial f given by a straight-line program. Our algorithm constructs an approximation f^* of f, such that $f - f^*$ probably has at most half the number of terms of f, then recurses on the difference $f - f^*$. Our approach builds on previous work by Garg and Schost (2009), and Giesbrecht and Roche (2011), and is asymptotically more efficient in terms of the total cost of the probes required than previous methods, in many cases.

1 Introduction

We consider the problem of interpolating a sparse, univariate polynomial

$$f = c_1 z^{e_1} + c_2 z^{e_2} + \cdots + c_t z^{e_t} \in \mathcal{R}[z]$$

of degree d with t non-zero coefficients $c_1, \ldots, c_t$ (where t is called the *sparsity* of f) over a ring $\mathcal{R}$. More formally, we are given a *straight-line program* that evaluates f at any point, as well as bounds $D \geq d$ and $T \geq t$. The straight-line program is a simple but useful abstraction of a computer program without branches, but our interpolation algorithm will work in more common settings of "black box" sampling of f.

We summarize our final result as follows.

Theorem 1. *Let $f \in \mathcal{R}[z]$, where $\mathcal{R}$ is any ring. Given any straight-line program of length L that computes f, and bounds T and D for the sparsity and degree of f, one can find all coefficients and exponents of f using $\mathcal{O}\tilde{}(LT \log^3 D + LT \log D \log(1/\mu))^\dagger$ ring operations in $\mathcal{R}$, plus a similar number of bit operations. The algorithm is probabilistic of the Monte Carlo type: it can generate random bits at unit cost and on any invocation returns the correct answer with probability greater than $1 - \mu$, for a user-supplied tolerance $\mu > 0$.*

---

† For summary convenience we use soft-Oh notation: for functions $\phi, \psi \in \mathbb{R}_{>0} \to \mathbb{R}_{>0}$ we say $\phi \in \mathcal{O}\tilde{}(\psi)$ if and only if $\phi \in \mathcal{O}(\psi(\log \psi)^c)$ for some constant $c \geq 0$.

V.P. Gerdt et al. (Eds.): CASC 2013, LNCS 8136, pp. 61–74, 2013.
© Springer International Publishing Switzerland 2013

1.1   The Straight-Line Program Model and Interpolation

Straight-line programs are a useful model of computation, both as a theoretical construct and from a more practical point of view; see, e.g., (Bürgisser et al., 1997, Chapter 4). Our interpolation algorithms work more generally for N-variate sparse polynomials $f \in \mathcal{R}[z_1, \ldots, z_N]$ given by a straight-line program $\mathcal{S}_f$ defined as follows. $\mathcal{S}_f$ takes an input $(a_1, \ldots, a_N) \in \mathcal{R}^N$ of length N, and produces a vector $b \in \mathcal{R}^L$ via a series of L instructions $\Gamma_i : 1 \leq i \leq L$ of the form

$$\Gamma_i = \begin{cases} \gamma_i \longleftarrow \alpha_1 \star \alpha_2, & or \\ \gamma_i \longleftarrow \delta \in \mathcal{R} & (i.e., \ a \ constant \ from \ \mathcal{R}), \end{cases}$$

where $\star$ is a ring operation '$+$', '$-$', or '$\times$', and either $\alpha_\ell \in \{a_j\}_{1 \leq j \leq n}$ or $\alpha_\ell \in \{\gamma_k\}_{1 \leq k < i}$ for $\ell = 1, 2$. When we say $\mathcal{S}_f$ computes f, we mean $\mathcal{S}_f$ sets γ_L to $f(a_1, \ldots, a_N) \in \mathcal{R}$.

To interpolate an N-variate polynomial $f \in \mathcal{R}[z_1, \ldots, z_N]$, we apply a Kronecker substitution, and interpolate

$$\hat{f}(z) = f\left(z, z^{(D+1)}, z^{(D+1)^2}, \ldots, z^{(D+1)^{N-1}}\right) \in \mathcal{R}[z].$$

While this certainly increases the degree, f and $\hat{f}$ have the same number of non-zero terms, and f can be easily recovered from $\hat{f}$. This reduces the problem of interpolating the N-variate polynomial f of partial degree at most D to interpolating a univariate polynomial $\hat{f}$ of degree at most $(D+1)^N$. For the remainder of this paper we thus assume f is univariate.

It will also be necessary to evaluate our polynomial $f \in \mathcal{R}[z]$, or rather our straight-line program $\mathcal{S}_f$ for f, in an extension ring of $\mathcal{R}$. Precisely, we want to evaluate f at symbolic ℓth roots of unity for various choices of ℓ, or algebraically, in $\mathcal{R}[z]/(z^\ell - 1)$. This may be regarded as transforming our straight-line program by substituting operations in $\mathcal{R}$ with operations in $\mathcal{R}[z]/(z^\ell - 1)$, where each element is represented by a polynomial in $\mathcal{R}[z]$ of degree less than ℓ. Each instruction Υ_i in the transformed branching program now potentially requires $\mathsf{M}(\ell)$ operations in $\mathcal{R}$, where $\mathsf{M}(\ell)$ is the number of operations in $\mathcal{R}$ and bit operations needed to multiply two degree-ℓ polynomials over the base ring $\mathcal{R}$. By Cantor and Kaltofen (1991), we may assume $\mathsf{M}(\ell) = \mathcal{O}(\ell \log \ell \log \log \ell)$.

Each evaluation of our straight-line program for f in $\mathcal{R}[z]/(z^\ell - 1)$ is called a *probe of degree* ℓ. Thus, the cost of a degree-ℓ probe to $\mathcal{S}_f$ is $\mathcal{O}^\sim(L\ell)$ operations in $\mathcal{R}$, and similarly many bit operations.

This is easily connected to the more "classical" view of sparse interpolation, in which probes are simply evaluations of a "black-box" polynomial at a single point (and we do not have any representation for how f is calculated). Each probe in the straight-line program model can be thought of as evaluating f at *all* ℓth roots of unity in the classical model. Since we charge $\mathsf{M}(\ell) = \mathcal{O}^\sim(\ell)$ operations in $\mathcal{R}$ for a degree ℓ probe in the straight-line program model, i.e., about ℓ times as much as a single black-box probe, this is consistent with the costs in a classical model. We note that algorithms for sparse interpolation presented below could be stated in this classical model, though we find the straight-line program model convenient and will continue with it throughout this paper.

1.2   Previous Work

Straight-line programs, or equivalently algebraic circuits, are important both as a computational model and as a data structure for polynomial computation. Their rich history includes both algorithmic advances and practical implementations (Kaltofen, 1989; Sturtivant and Zhang, 1990; Bruno et al., 2002).

One can naively interpolate a polynomial $f \in \mathcal{R}[z]$ given by a straight-line program using a dense method, with D probes of degree 1. Prony's (1795) interpolation algorithm — see (Ben-Or and Tiwari, 1988; Kaltofen et al., 1990; Giesbrecht et al., 2009) — is a sparse interpolation method that uses evaluations at only $2T$ powers of a root of unity whose order is greater than D. However, in the straight-line program model for a general ring, this would require evaluating at a symbolic Dth root of unity, which would use at least $\Omega(D)$ ring operations and defeat the benefit of sparsity. Problems with Prony's algorithm are also seen in the classical model in that the underlying base ring $\mathcal{R}$ must also support an efficient discrete logarithm algorithm on entries of high multiplicative order (which, for example, is not feasible over large finite fields).

We mention two algorithms specifically intended for straight-line programs.

The Garg-Schost Deterministic Algorithm. Garg and Schost (2009) describe a novel deterministic algorithm for interpolating a multivariate polynomial f given by a straight-line program. Their algorithm entails constructing an integer symmetric polynomial with roots at the exponents of f:

$$\chi = \prod_{i=1}^{t}(y - e_i) \in \mathbb{Z}[y],$$

which is then factored to obtain the exponents e_i.

Their algorithm first finds a *good* prime: a prime p for which the terms of f remain distinct when reduced modulo z^p-1. We call such an image $f \bmod (z^p-1)$ a *good image*. Such an image gives us the values $e_i \bmod p$ and hence $\chi(y) \bmod p$.

Example 2. *For* $f = z^{33} + z^3$, 5 *is not a good prime because* $f \bmod (z^5 - 1) = 2z^3$. *We say* z^{33} *and* z^3 *collide modulo* $z^5 - 1$. 7 *is a good prime, as the image* $f(z) \bmod (z^7 - 1) = z^5 + z^3$ *has as many terms as* $f(z)$ *does.*

In order to guarantee that we have a good prime, the algorithm requires that we construct the images $f \bmod (z^p - 1)$ for the first N primes, where N is roughly $\mathcal{O}^\sim(T^2 \log D)$. A good prime will be a prime p for which the image $f \bmod (z^p - 1)$ has maximally many terms, which will be exactly t. Once we know we have a good image we can discard the images $f \bmod (z^q - 1)$ for *bad* primes q, i.e. images with fewer than t terms. We use the remaining images to construct $\chi(y) = \prod_{i=1}^{t} (y - e_i) \in \mathbb{Z}[y]$ by way of Chinese remaindering on the images $\chi(y) \bmod p$.

We factor $\chi(y)$ to obtain the exponents e_i, after which we directly obtain the corresponding coefficients c_i directly from a good image.

The algorithm of Garg and Schost (2009) can be made faster, albeit Monte Carlo, using the following number-theoretic fact.

Fact 3 (Giesbrecht and Roche, 2011). *Let $f \in \mathcal{R}[z]$ be a polynomial with at most T terms and degree at most D. Let $\lambda = \max(21, \lceil \frac{5}{3} T(T-1) \log D \rceil)$. A prime p chosen at random in the range $[\lambda, 2\lambda]$ is a good prime for $f(z)$ with probability at least $\frac{1}{2}$.*

Thus, in order to find a good image with probability at least $1 - \varepsilon$, we can inspect images $f \bmod (z^p - 1)$ for $\lceil \log 1/\varepsilon \rceil$ primes p chosen at random in $[\lambda, 2\lambda]$. As the height of $\chi(y)$ can be roughly as large as D^T, we still require some $\mathcal{O}^{\sim}(T \log D)$ probes to construct $\chi(y)$.

The "Diversified" Interpolation Algorithm. Giesbrecht and Roche (2011) obtain better performance by way of *diversification*. A polynomial f is said to be *diverse* if its coefficients c_i are pairwise distinct. The authors show that, for f over a finite field or $\mathbb{C}$ and for appropriate random choices of α, $f(\alpha z)$ is diverse with probability at least $\frac{1}{2}$. They then try to interpolate the diversified polynomial $f(\alpha z)$.

Once we have t with high probability, we look at images $f(\alpha z) \bmod (z^p - 1)$ for primes p in $[\lambda, 2\lambda]$, discarding bad images. As $f(\alpha z)$ is diverse, we can recognize which terms in different good images are images of the same term. Thus, as all the e_i are at most D, we can get all the exponents e_i by looking at some $\mathcal{O}^{\sim}(\log D)$ good images of f.

1.3   Deterministic Zero Testing

Both the Monte Carlo algorithms of Garg and Schost (2009) and Giesbrecht and Roche (2011) can be made Las Vegas (i.e., no possibility of erroneous output, but unbounded worst-case running time) by way of deterministic zero-testing. Given a polynomial f represented by a straight-line program, each of these algorithms produces a polynomial f^* that is probably f.

Fact 4 (Bläser et al. (2009); Lemma 13). *Let $\mathcal{R}$ be an integral domain, and suppose $f = f^* \bmod (z^p - 1)$ for $T \log D$ primes. Then $f = f^*$.*

Thus, testing the correctness of the output of a Monte Carlo algorithm requires some $\mathcal{O}^{\sim}(T \log D)$ probes of degree at most $\mathcal{O}^{\sim}(T \log D)$. This cost does not dominate the cost of either Monte Carlo algorithm. We note that this deterministic zero test can dominate the cost of the interpolation algorithm presented in this paper if T is asymptotically dominated by $\log D$.

1.4 Summary of Results

We state as a theorem the number and degree of probes required by our new algorithm presented in this paper.

Theorem 5. *Let $f \in \mathcal{R}[z]$, where $\mathcal{R}$ is a ring. Given a straight-line program for f, one can find all coefficients and exponents of f with probability at least $1 - \mu$ using $\tilde{\mathcal{O}}\left(\log T(\log D + \log \frac{1}{\mu})\right)$ probes of degree at most $\mathcal{O}(T \log^2 D)$.*

Table 1. A "soft-Oh" comparison of interpolation algorithms for straight-line programs

	Probes	Probe degree	Cost of probes	Type
Dense	D	1	LD	deterministic
Garg & Schost	$T^2 \log D$	$T^2 \log D$	$LT^3 \log^2 D$	deterministic
*Las Vegas G & S	$T \log D$	$T^2 \log D$	$LT^3 \log^2 D$	Las Vegas
*Diversified	$\log D$	$T^2 \log D$	$LT^2 \log^2 D$	Las Vegas
†Recursive	$\log T \log D$	$T \log^2 D$	$LT \log^3 D$	Monte Carlo

*Average # of probes given; † for a fixed probability of failure μ

Table 1 gives a rough comparison of known algorithms. Our recursive algorithm improves by a factor of $T/\log D$ over the Giesbrecht-Roche diversification algorithm — ignoring "soft" multiplicative factors of $(\log(T/\log D))^{O(1)}$ — and as such is better suited for moderate values of T. Our algorithm recursively interpolates a series of polynomials of decreasing sparsity. An advantage of this method is that, when we cross a threshold where $\log D$ begins to dominate T, we can merely call the Monte Carlo diversification algorithm instead.

2 A Recursive Algorithm for Interpolating f

Entering each recursive step in our algorithm we have our polynomial f represented by a straight-line program, and an explicit sparse polynomial f^* "approximating" f, that is, whose terms mostly appear in the sparse representation of f. At each recursive step we try to interpolate the difference $g = f - f^*$. To begin with, f^* is initialized to zero.

We first find an "ok" prime p which separates most of the terms of g. We then use that prime p to build a approximation f^{**}, containing most of the terms of g, plus possibly some additional "deceptive" terms. The polynomial f^{**} is constructed such that $g = f - f^*$ has, with high probability, at most $T/2$ terms. We then recursively interpolate the difference $g - f^{**}$.

Producing images $f^* \mod (z^\ell - 1)$ is straightforward, we merely reduce the exponents of terms of f^* modulo ℓ. We assume g has a sparsity bound $T_g \leq T$.

2.1   A Weaker Notion of "Good" Primes

To interpolate a polynomial g, the sparse interpolation algorithm described by Giesbrecht and Roche (2011) requires a *good* prime p which keeps the exponents of g distinct modulo p. That is, $g \bmod (z^p - 1)$ has the same number of terms as g. We define a weaker notion of a good prime, an *ok prime*, which separates most of the terms of g. To that end we measure, for fixed g and prime p, how well p separates the terms of g.

Definition 6. *Fix a polynomial* $g = \sum_{i=1}^{t} c_i z^{e_i} \in \mathcal{R}[z]$ *with non-zero* $c_1, \ldots, c_t \in \mathcal{R}$, *where* $e_i < e_j$ *for* $i < j$, *we say* $c_i z^{e_i}$ *and* $c_j z^{e_j}$, $i \neq j$, *collide modulo* $z^p - 1$ *if* $e_i \equiv e_j \bmod p$. *We call any term* $c_i z^{e_i}$ *of* f *which collides with any other term of* f *a* colliding term *of* f *modulo* $z^p - 1$. *We let* $\mathcal{C}_g(p) \in [0, t]$ *denote the number of colliding terms of* g *modulo* $z^p - 1$.

Example 7. *For the polynomial* $g = 1 + z^5 + z^7 + z^{10}$, $\mathcal{C}_g(2) = 4$, *since* 1 *collides with* z^{10} *and* z^5 *collides with* z^7 *modulo* $z^2 - 1$. *Similarly,* $\mathcal{C}_g(5) = 2$, *since* z^5 *collides with* z^{10} *modulo* $z^5 - 1$.

We say $c_i z^{e_i}$ and $c_j z^{e_j}$ collide modulo $z^p - 1$ because both terms have the same exponent once reduced modulo $z^p - 1$. All other terms of g we will call *non-colliding* terms modulo $z^p - 1$.

In the sparse interpolation algorithm of Giesbrecht and Roche (2011), one chooses a $\lambda \in \mathbb{Z}_{>0}$ such that the probability of a prime $p \in [\lambda, 2\lambda]$, chosen at random and having $\mathcal{C}_g(p) = 0$, is at least $\frac{1}{2}$. However, in order to guarantee that we find such a prime with high probability, we need to choose $\lambda \in \mathcal{O}(T^2 \log D)$.

In this paper we will search over a range of smaller primes, while allowing for a reasonable number of collisions. We try to pick λ such that

$$\Pr\left(\mathcal{C}_g(p) \geq \gamma \text{ for a random prime } p \in [\lambda, 2\lambda]\right) < 1/2,$$

for a parameter γ to be determined.

Lemma 8. *Let* $g \in \mathcal{R}[z]$ *be a polynomial with* $t \leq T$ *terms and degree at most* $d \leq D$. *Suppose we are given* T *and* D, *and let* $\lambda = \max\left(21, \left\lceil \frac{10T(T-1)\ln(D)}{3\gamma} \right\rceil\right)$. *Let* p *be a prime chosen at random in the range* $\lambda, \ldots, 2\lambda$. *Then* $\mathcal{C}_g(p) \geq \gamma$ *with probability less than* $\frac{1}{2}$.

Proof. The proof follows similarly to the proof of Lemma 2.1 in (Giesbrecht and Roche, 2011).

Let B be the set of unfavourable primes for which $\mathcal{C}_g(p) \geq \gamma$ terms collide modulo $z^p - 1$, and denote the size of B by $\#B$. As every colliding term collides with at least one other term modulo $z^p - 1$, we know $p^{\mathcal{C}_g(p)}$ divides $\prod_{1 \leq i \neq j \leq t} (e_i - e_j)$. Thus, as $\mathcal{C}_g(p) \geq \gamma$ for $p \in B$,

$$\lambda^{\#B\gamma} \leq \prod_{p \in B} p^\gamma \leq \prod_{1 \leq i \neq j < t} (e_i - e_j) \leq d^{t(t-1)} \leq D^{T(T-1)}.$$

Solving the inequality for $\#B$ gives us

$$\#B \leq \frac{T(T-1)\ln(D)}{\ln(\lambda)\gamma}.$$

The total number of primes in $[\lambda, 2\lambda]$ is greater than $3\lambda/(5\ln(\lambda))$ for $\lambda \geq 21$ by Corollary 3 to Theorem 2 of (Rosser and Schoenfeld, 1962). From our definition of λ we have

$$\frac{3\lambda}{5\ln(\lambda)} > \frac{2T(T-1)\ln(D)}{\ln(\lambda)\gamma} \geq 2\#B,$$

completing the proof. $\square$

Relating the Sparsity of g mod $(z^p - 1)$ with $\mathcal{C}_g(p)$

Suppose we choose λ according to Lemma 8, and make k probes to compute g mod $(z^{p_1} - 1), \ldots, g$ mod $(z^{p_k} - 1)$. One of the primes p_i will yield an image with fewer than γ colliding terms (i.e. $\mathcal{C}_g(p_i) < \gamma$) with probability at least $1 - 2^{-k}$. Unfortunately, we do not know which prime p maximizes $\mathcal{C}_g(p)$. A good heuristic might be to select the prime p for which g mod $(z^p - 1)$ has maximally many terms. However, this does not necessarily minimize $\mathcal{C}_g(p)$. Consider the following example.

Example 9. *Let*

$$g = 1 + z + z^4 - 2z^{13}.$$

We have

$$g \text{ mod } (z^2 - 1) = 2 - z, \quad and \quad g \text{ mod } (z^3 - 1) = 1.$$

While g mod $(z^2 - 1)$ has more terms than g mod $(z^3 - 1)$, we see that $\mathcal{C}_g(2) = 4$ is larger than $\mathcal{C}_g(3) = 3$.

While we cannot determine the prime p for which g mod $(z^p - 1)$ has maximally many non-colliding terms, we show that choosing the prime p which maximizes the number of terms in g mod $(z^p - 1)$ is, in fact, a reasonable strategy.

We would like to find a precise relationship between $\mathcal{C}_g(p)$, the number of terms of g that collide in the image g mod $(z^p - 1)$, and the sparsity s of g mod $(z^p - 1)$.

Lemma 10. *Suppose that g has t terms, and g mod $(z^p - 1)$ has $s \leq t$ terms. Then $t - s \leq \mathcal{C}_g(p) \leq 2(t - s)$.*

Proof. To prove the lower bound, note that $t - \mathcal{C}_g(p)$ terms of g will not collide modulo $z^p - 1$, and so g mod $(z^p - 1)$ has sparsity s at least $t - \mathcal{C}_g(p)$.

We now prove the upper bound. Towards a contradiction, suppose that $\mathcal{C}_g(p) > 2(t - s)$. There are $\mathcal{C}_g(p)$ terms of g that collide modulo $z^p - 1$. Let h be the $\mathcal{C}_g(p)$-sparse polynomial comprised of those terms of g. As each term of h collides with at least one other term of h, h mod $(z^p - 1)$ has sparsity at most $\mathcal{C}_g(p)/2$. Since none of the terms of $g - h$ collide modulo $z^p - 1$, $(g - h)$ mod $(z^p - 1)$ has sparsity exactly $t - \mathcal{C}_g(p)$. It follows that g mod $(z^p - 1)$ has sparsity at most $t - \mathcal{C}_g(p) + \mathcal{C}_g(p)/2 = t - \mathcal{C}_g(p)/2$. That is, $s \leq t - \mathcal{C}_g(p)/2$, and so $\mathcal{C}_g(p) \leq 2(t - s)$. $\square$

Corollary 11. *Suppose g has sparsity t, $g \bmod (z^q - 1)$ has sparsity s_q, and $g \bmod (z^p - 1)$ has sparsity $s_p \geq s_q$. Then $\mathcal{C}_g(p) \leq 2\mathcal{C}_g(q)$.*

Proof.
$$
\begin{aligned}
\mathcal{C}_g(p) &\leq 2(t - s_p) && \text{by the second inequality of Lemma 10,} \\
&\leq 2(t - s_q) && \text{since } s_p \geq s_q, \\
&\leq 2\mathcal{C}_g(q) && \text{by the first inequality of Lemma 10.}
\end{aligned}
$$
$\square$

Suppose then that we have computed $g \bmod (z^p - 1)$, for p belonging to some set of primes S, and the minimum value of $\mathcal{C}_g(p)$, $p \in S$, is less than γ. Then a prime $p^* \in S$ for which $g \bmod (z^{p^*} - 1)$ has maximally many terms satisfies $\mathcal{C}_g(p^*) < 2\gamma$. We will call such a prime p^* an *ok prime*.

We then choose $\gamma = wT$ for an appropriate proportion $w \in (0, 1)$. Note in general we will use the better bound $T_g \leq T$ in place of T. We show that setting $w = 3/16$ allows that each recursive call reduces the sparsity of the subsequent polynomial by at least half. This would make $\lambda = \lceil \frac{10}{3w}(T-1)\ln(D) \rceil = \lceil \frac{160}{9}(T - 1)\ln(D) \rceil$. As per Lemma 8, in order to guarantee with probability $1 - \varepsilon$ that we have come across a prime p such that $\mathcal{C}_g(p) \leq \gamma$, we will need to perform $\lceil \log 1/\varepsilon \rceil$ probes of degree $\mathcal{O}(T \log D)$. Procedure 1 summarizes how we find an ok prime.

Procedure `FindOkPrime`$(\mathcal{S}_f, f^*, T_g, D, \varepsilon)$

Input:
- $\mathcal{S}_f$, a straight-line program that computes a polynomial f
- f^*, a current approximation to f
- T_g and D, bounds on the sparsity and degree of $g = f - f^*$ respectively
- ε, a bound on the probability of failure

Output: With probability at least $1 - \varepsilon$, we return an "ok prime" for $g = f - f^*$

$\lambda \longleftarrow \max\left(21, \lceil \frac{160}{9}(T_g - 1)\ln D \rceil\right)$
$(\texttt{max\_sparsity}, p) \longleftarrow (0, 0)$
for $i \longleftarrow 1$ **to** $\lceil \log 1/\varepsilon \rceil$ **do**
    $p' \longleftarrow$ a random prime in $[\lambda, 2\lambda]$
    if *# of terms of* $(f - f^*) \bmod (z^{p'} - 1) \geq \texttt{max\_sparsity}$ **then**
        $\texttt{max\_sparsity} \longleftarrow$ # of terms of $(f - f^*) \bmod (z^{p'} - 1)$
        $p \longleftarrow p'$
return p

A practical application would probably choose random primes by selecting random integer values in $[\lambda, 2\lambda]$ and then applying probabilistic primality testing. In order to ensure deterministic worst-case run-time, we could pick random primes in the range $[\lambda, 2\lambda]$ by using a sieve method to pre-compute all the primes up to 2λ.

2.2   Generating an Approximation f^{**} of g

We suppose now that we have, with probability at least $1 - \varepsilon$, an ok prime p; i.e., a prime p such that $\mathcal{C}_g(p) \leq 2wT$ for a suitable proportion w. We now use this ok prime p to construct a polynomial f^{**} containing *most* of the terms of $g = f - f^*$.

For a set of coprime moduli $\mathcal{Q} = \{q_1, \ldots, q_k\}$ satisfying $\prod_{i=1}^{k} q_i > D$, we will compute $g \bmod (z^{pq_i} - 1)$ for $1 \leq i \leq k$. Here we make no requirement that the q_i be prime. We merely require that the q_i are pairwise co-prime.

We choose the q_i as follows: denoting the ith prime by p_i, we set $q_i = p_i^{\lfloor \log_{p_i} x \rfloor}$, for an appropriate choice of x. That is, we let q_i be the greatest power of the ith prime that is no more than x. For $p_i \leq x$, we have $q_i \geq x/p_i$ and $q_i \geq p_i$. Either x/p_i or p_i is at least $\sqrt{x}$, and so $q_i \geq \sqrt{x}$ as well.

By Corollary 1 of Theorem 2 in Rosser and Schoenfeld (1962), there are more than $x/\ln x$ primes less than or equal to x for $x \geq 17$. Therefore

$$\prod_{p_i \leq x} q_i \geq \left(\sqrt{x}\right)^{x/\ln x}.$$

As we want this product to exceed D, it suffices that

$$\ln D < \ln\left(\left(\sqrt{x}\right)^{x/\ln x}\right) = x/2.$$

Thus, if we choose $x \geq \max(2\ln(D), 17)$ and $k = \lceil x/\ln x \rceil$, then $\prod_{i=1}^{k} q_i$ will exceed D. This means $q_i \in \mathcal{O}(\log D)$ and $pq_i \in \mathcal{O}(T \log^2 D)$. The number of probes in this step is $k \in \mathcal{O}(\log(D)/\log\log(D))$. Since we will use the same set of moduli $\mathcal{Q} = \{q_1, \ldots, q_k\}$ in every recursive call, we can pre-compute $\mathcal{Q}$ prior to the first recursive call.

We now describe how to use the images $g \bmod (z^{pq_i} - 1)$ to construct a polynomial f^{**} such that $g - f^{**}$ is at most $T/2$-sparse.

If cz^e is a term of g that does not collide with any other terms modulo $z^p - 1$, then it certainly will not collide with other terms modulo $z^{pq} - 1$ for any natural number q. Similarly, if $c^* z^{e^* \bmod p}$ appears in $g \bmod (z^p - 1)$ and there exists a unique term $c^* z^{e^* \bmod pq_i}$ appearing in $g \bmod (z^{pq_i} - 1)$ for $i = 1, 2, \ldots, k$, then $c^* z^{e^*}$ is potentially a term of g. Note that $c^* z^{e^*}$ is not *necessarily* a term of g: consider the following example.

Example 12. *Let*

$$g(z) = 1 + z + z^2 + z^3 + z^{11+4} - z^{14 \cdot 11 + 4} - z^{15 \cdot 11 + 4},$$

with hard sparsity bound $T_g = 7$ and degree bound $D = 170$ and let $p = 11$. We have

$$g(z) \bmod (z^{11} - 1) = 1 + z + z^2 + z^3 - z^4.$$

As $\deg(g) = 170 < 2 \cdot 3 \cdot 5 \cdot 7 = 210$, it suffices to make the probes $g \bmod z^{11q} - 1$ for $q = 2, 3, 5, 7$. Probing our remainder black-box polynomial, we have

$$g \bmod (z^{22} - 1) = 1 + z + z^2 + z^3 - z^{15},$$
$$g \bmod (z^{33} - 1) = 1 + z + z^2 + z^3 - z^{26},$$
$$g \bmod (z^{55} - 1) = 1 + z + z^2 + z^3 - z^{48},$$
$$g \bmod (z^{77} - 1) = 1 + z + z^2 + z^3 - z^{15}.$$

In each of the images $g \bmod z^{pq} - 1$, there is a unique term whose degree is congruent to one of $e = 0, 1, 2, 3, 4$ modulo p. Four of these terms correspond to the terms $1, z, z^2, z^3$ appearing in g. Whereas the remaining term has degree e satisfying $e = 1 \bmod 2$, $e = 2 \bmod 3$, $e = 3 \bmod 5$, and $e = 1 \bmod 7$. By Chinese remaindering on the exponents, this gives a term $-z^{113}$ not appearing in g.

Procedure ConstructApproximation$(\mathcal{S}_f, f^*, D, p, \mathcal{Q})$

Input:
  - $\mathcal{S}_f$, a straight-line program that computes a polynomial f
  - f^*, a current approximation to f
  - D a bound on the degree of $g = f - f^*$
  - p, an ok prime for g (with high probability)
  - $\mathcal{Q}$, a set of co-prime moduli whose product exceeds D

Output: A polynomial f^{**} such that, if p is an ok prime, $g - f^{**}$ has sparsity
       at most $\lfloor T_g/2 \rfloor$, where g has at most T_g terms.

```
// Collect images of g
E ⟵ set of exponents of terms in (f − f*) mod (z^p − 1)
for q ∈ Q do
    h ⟵ (f − f*) mod (z^{pq} − 1)
    for each term cz^e in h do
        if E_(e mod p),q is already initialized then   E ⟵ E/{e mod p}
        else E_(e mod p),q ⟵ e mod q

// Construct terms of new approximation of g, f**
f** ⟵ 0
for e_p ∈ E do
    e ⟵ least nonnegative solution to {e = E_{e_p},q mod q | q ∈ Q}
    c ⟵ coefficient of z^{e_p} term in (f − f*) mod (z^p − 1)
    if e ≤ D then   f** ⟵ f** + cz^e
return f**
```

Definition 13. Let $c^* z^{e^*}$, $e^* \leq D$ be a monomial such that $c^* z^{e^* \bmod p}$ appears in $g \bmod z^p - 1$, and $c^* z^{e^* \bmod pq_i}$ is the unique term of degree congruent to e^* modulo p appearing in $g \bmod (z^{pq_i} - 1)$ for each modulus q_i. If $c^* z^{e^*}$ is not a term of g we call it a deceptive term.

Fortunately, we can detect a collision comprised of only two terms. Namely, if $c_1 z^{e_1} + c_2 z^{e_2}$ collide, there will exist a q_i such that $q_i \nmid (e_1 - e_2)$. That is, $g \bmod (z^{pq_i} - 1)$ will have two terms whose degree is congruent to $e_1 \bmod p$. Once we observe that, we know the term $(c_1 + c_2) z^{e_1 \bmod p}$ appearing in $g \bmod (z^p - 1)$ was not a distinct term, and we can ignore exponents of the congruence class $e_1 \bmod p$ in subsequent images $g \bmod (z^{pq_j} - 1)$.

Thus, supposing $g \bmod (z^p - 1)$ has at most 2γ colliding terms and at least $t - 2\gamma$ non-colliding terms, f^{**} will have the $t - 2\gamma$ non-colliding terms of g, plus potentially an additional $\frac{2}{3}\gamma$ deceptive terms produced by the colliding terms of g. In any case, $g - f^{**}$ has sparsity at most $\frac{8}{3}\gamma$. Choosing $\gamma = \frac{3}{16} T_g$ guarantees that $g - f^{**}$ has sparsity at most $T_g/2$. This would make $\lambda = \lceil \frac{160}{9}(T_g - 1) \ln(D) \rceil$.

Procedure 2 gives a pseudocode description of how we construct f^{**}.

If we find a prospective term in our new approximation f^{**} has degree greater than D, then we know that term must have been a deceptive term and discard it. There are other obvious things we can do to recognize deceptive terms which we exclude here. For instance, we should check that all terms from images modulo $z^{pq} - 1$ whose degrees agree modulo p share the same coefficient.

2.3 Recursively Interpolating $f - f^*$

Procedure Interpolate$(\mathcal{S}_f, T, D, \mu)$

Input:
- $\mathcal{S}_f$, a straight-line program that computes a polynomial f
- T and D, bounds on the sparsity and degree of f, respectively
- μ, an upper bound on the probability of failure

Output: With probability at least $1 - \mu$, we return f

$x \longleftarrow \max(2 \ln(D), 17)$
$\mathcal{Q} \longleftarrow \{p^{\lfloor \log_p x \rfloor} : p \text{ is prime}, p \le x\}$
return $4(\mathcal{S}_f, 0, T, D, \mathcal{Q}, \mu/(\log T + 1))$

Once we have constructed f^{**}, we refine our approximation f^* by adding f^{**} to it, giving us a new difference $g = f - f^*$ containing at most half the terms of the previous polynomial g. We recursively interpolate our new polynomial g. With an updated sparsity bound $\lfloor T_g/2 \rfloor$, we update the values of γ and λ and perform the steps of Sections 2.1 and 2.2. We recurse in this fashion $\log T$ times. Thus, the total number of probes becomes

$$\mathcal{O}\left(\log T(\frac{\log D}{\log \log D} + \log(1/\varepsilon)) \right),$$

of degree at most $\mathcal{O}(T \log^2 D)$.

Procedure InterpolateRecurse$(\mathcal{S}_f, f^*, T_g, D, \mathcal{Q}, \varepsilon)$

Input:
  − $\mathcal{S}_f$, a straight-line program that computes a polynomial f
  − f^*, a current approximation to f
  − T_g and D, bounds on the sparsity and degree of $g = f - f^*$, respectively
  − $\mathcal{Q}$, a set of coprime moduli whose product is at least D
  − ε, a bound on the probability of failure at one recursive step
Output: With probability at least $1 - \mu$, the algorithm outputs f

  if $T_g = 0$ **then return** f^*
  $p \longleftarrow 1(\mathcal{S}_f, f^*, T_g, D, \varepsilon)$
  $f^{**} \longleftarrow 2(\mathcal{S}_f, f^*, D, p, \mathcal{Q})$
  return $4(\mathcal{S}_f, f^* + f^{**}, \lfloor T_g/2 \rfloor, D, \mathcal{Q}, \varepsilon)$

Note now that in order for this method to work we need that, at every recursive call, we in fact get a good prime, otherwise our sparsity bound on the subsequent difference of polynomials could be incorrect. At every stage we succeed with probability $1 - \varepsilon$, thus the probability of failure is $1 - (1 - \varepsilon)^{\lceil \log T \rceil}$. This is less than $\lceil \log T \rceil \varepsilon$. If we want to succeed with probability μ, then we can choose $\varepsilon = \frac{\mu}{\log T + 1} \in \mathcal{O}(\frac{\mu}{\log T})$.

3 pre-computes our set of moduli $\mathcal{Q}$, then makes the first recursive call to 4, which subsequently calls itself.

2.4 A Cost Analysis

We analyse the cost of our algorithm, thereby proving Theorems 1 and 5.

Pre-computation. Using the wheel sieve (Pritchard, 1982), we can compute the set of primes up to $x \in \mathcal{O}(\log D)$ in $\tilde{\mathcal{O}}(\log D)$ bit operations. From this set of primes we obtain $\mathcal{Q}$ by computing $p^{\lfloor \log_p x \rfloor}$ for $p \leq \sqrt{x}$ by way of squaring-and-multiplying. For each such prime, this costs $\tilde{\mathcal{O}}(\log x)$ bit operations, so the total cost of computing $\mathcal{Q}$ is $\tilde{\mathcal{O}}(\log D)$.

Finding ok Primes. In one recursive call, we will look at some $\log 1/\varepsilon = \mathcal{O}(\log 1/\mu \log \log T)$ primes in the range $[\lambda, 2\lambda]$ in order to find an ok prime. Any practical implementation would select such primes by using probabilistic primality testing on random integer values in the range $[\lambda, 2\lambda]$; however, the probabilistic analysis of such an approach, in the context of our interpolation algorithm, becomes somewhat ungainly. We merely note here that we could instead pre-compute primes up to our initial value of $\lambda \in \mathcal{O}(T \log D)$ in $\tilde{\mathcal{O}}(T \log D)$ bit operations by way of the wheel sieve.

Each prime p is of order $T \log D$, and so, per our discussion in Section 1, each probe costs $\tilde{\mathcal{O}}(LT \log D)$ ring operations and similarly many bit operations. Considering the $\mathcal{O}(\log T)$ recursive calls, this totals $\tilde{\mathcal{O}}(LT \log D \log 1/\mu)$ ring and bit operations.

Constructing the New Approximation f^{**}. Constructing f^{**} requires $\mathcal{O}^\sim(\log D)$ probes of degree $\mathcal{O}^\sim(T\log^2 D)$. This costs $\mathcal{O}^\sim(LT\log^3 D)$ ring and bit operations. Performing these probes at each $\mathcal{O}(\log T)$ recursive call introduces an additional factor of $\log T$, which does not affect the "soft-Oh" complexity. This step dominates the cost of the algorithm.

Building a term cz^e of f^{**} amounts to solving a set of congruences. By Theorem 5.8 of Gathen and Gerhard (2003), this requires some $\mathcal{O}(\log^2 D)$ word operations. Thus the total cost of Chinese remaindering to construct f^{**} becomes $\mathcal{O}(T\log^2 D)$. Again, the additional $\log T$ factor due to the recursive calls does not affect the stated complexity.

3 Conclusions

We have presented a recursive algorithm for interpolating a polynomial f given by a straight-line program, using probes of smaller degree than in previously known methods. We achieve this by looking for "ok" primes which separate most of the terms of f, as opposed to "good" primes which separate all of the terms of f. As is seen in Table 1, our algorithm is an improvement over previous algorithms for moderate values of T.

This work suggests a number of problems for future work. We believe our algorithms have the potential for good numerical stability, and could improve on Giesbrecht and Roche's (2011) work on numerical interpolation of sparse complex polynomials, hopefully capitalizing on the lower degree probes. Our Monte Carlo algorithms are now more efficient than the best known algorithms for polynomial identity testing, and hence these cannot be used to make them error free. We would ideally like to expedite polynomial identity testing of straight-line programs, the best known methods currently due to Bläser et al. (2009). Finally, we believe there is still room for improvement in sparse interpolation algorithms. The vector of exponents of f comprises some $T\log D$ bits. Assuming no collisions, a degree-ℓ probe gives us some $t\log\ell$ bits of information about these exponents. One might hope, aside from some seemingly rare degenerate cases, that $\log D$ probes of degree $T\log D$ should be sufficient to interpolate f.

Acknowledgements. We would like to thank Reinhold Burger and Colton Pauderis for their feedback on a draft of this paper.

References

Ben-Or, M., Tiwari, P.: A deterministic algorithm for sparse multivariate polynomial interpolation. In: Proceedings of the Twentieth Annual ACM Symposium on Theory of Computing, pp. 301–309. ACM (1988)

Bläser, M., Hardt, M., Lipton, R.J., Vishnoi, N.K.: Deterministically testing sparse polynomial identities of unbounded degree. Information Processing Letters 109(3), 187–192 (2009)

Bruno, N., Heintz, J., Matera, G., Wachenchauzer, R.: Functional programming concepts and straight-line programs in computer algebra. Mathematics and Computers in Simulation 60(6), 423–473 (2002), doi:10.1016/S0378-4754(02)00035-6

Bürgisser, P., Clausen, M., Shokrollahi, M.A.: Algebraic Complexity Theory. Grundlehren der mathematischen Wissenschaften, vol. 315. Springer (1997)

Cantor, D.G., Kaltofen, E.: On fast multiplication of polynomials over arbitrary algebras. Acta Informatica 28, 693–701 (1991)

de Prony, R.: Essai expérimental et analytique sur les lois de la dilabilité et sur celles de la force expansive de la vapeur de l'eau et de la vapeur de l'alkool, à différentes températures. J. de l'École Polytechnique 1, 24–76 (1795)

Garg, S., Schost, É.: Interpolation of polynomials given by straight-line programs. Theor. Comput. Sci. 410(27-29), 2659–2662 (2009), http://dx.doi.org/10.1016/j.tcs.2009.03.030, doi:10.1016/j.tcs.2009.03.030, ISSN 0304-3975

von zur Gathen, J., Gerhard, J.: Modern Computer Algebra, 2nd edn. Cambridge University Press, New York (2003) ISBN 0521826462

Giesbrecht, M., Roche, D.S.: Diversification improves interpolation. In: ISSAC 2011, pp. 123–130 (2011), http://doi.acm.org/10.1145/1993886.1993909, doi:10.1145/1993886.1993909

Giesbrecht, M., Labahn, G., Lee, W.-S.: Symbolic– numeric sparse interpolation of multivariate polynomials. Journal of Symbolic Computation 44(8), 943–959 (2009)

Kaltofen, E.: Factorization of polynomials given by straight-line programs. In: Randomness and Computation, pp. 375–412. JAI Press (1989)

Kaltofen, E., Lakshman, Y.N., Wiley, J.M.: Modular rational sparse multivariate polynomial interpolation. In: Proceedings of the International Symposium on Symbolic and Algebraic Computation, ISSAC 1990, pp. 135–139. ACM, New York (1990), doi:10.1145/96877.96912

Pritchard, P.: Explaining the wheel sieve. Acta Informatica 17(4), 477–485 (1982)

Barkley Rosser, J., Schoenfeld, L.: Approximate formulas for some functions of prime numbers. Illinois J. Math. 6, 64–94 (2082) ISSN 0019-2082

Sturtivant, C., Zhang, Z.-L.: Efficiently inverting bijections given by straight line programs. In: Proceedings of the 31st Annual Symposium on Foundations of Computer Science, pp. 327–334. IEEE (October 1990), doi:10.1109/FSCS.1990.89551

On Possibility of Additional Solutions of the Degenerate System Near Double Degeneration at the Special Value of the Parameter

Alexander D. Bruno[1] and Victor F. Edneral[2]

[1] Keldysh Institute for Applied Mathematics of RAS
Miusskaya Sq. 4, Moscow, 125047, Russia
abruno@keldysh.ru
[2] Skobeltsyn Institute of Nuclear Physics,
Lomonosov Moscow State University
Leninskie Gory 1, Moscow, 119991, Russia
edneral@theory.sinp.msu.ru

Abstract. We consider an autonomous system of ordinary differential equations, which is resolved with respect to derivatives. To study local integrability of the system near a degenerate stationary point, we use an approach based on Power Geometry and on the computation of the resonant normal form. For the particular non-Hamilton 5-parameter case of concrete planar system, we found previously the almost complete set of necessary conditions on parameters of the system for which the system is locally integrable near a degenerate stationary point. These sets of parameters, satisfying the conditions, consist of 4 two-parameter subsets in this 5-parameter space except 1 special hyper plane $b^2 = 2/3$. We wrote down 4 first integrals of motion as functions of the system parameters. Here we have proved that the limitation $b^2 \neq 2/3$ can be excluded from the previously obtained solutions. Now we have not found the additional first integrals.

Keywords: Ordinary differential equations, Integrability, Resonant normal form, Power geometry, Computer algebra.

1 Introduction

We consider the autonomous system of ordinary differential equations

$$dx_i/dt \stackrel{\text{def}}{=} \dot{x}_i = \varphi_i(X), \quad i = 1, \ldots, n \ , \tag{1}$$

where $X = (x_1, \ldots, x_n) \in \mathbb{C}^n$ and $\varphi_i(X)$ are polynomials.

At the stationary point $X = X^0$, system (1) is *locally integrable* if it has sufficient number m of independent first integrals of the form

$$a_j(X)/b_j(X), \quad j = 1, \ldots, m \ , \tag{2}$$

V.P. Gerdt et al. (Eds.): CASC 2013, LNCS 8136, pp. 75–87, 2013.
© Springer International Publishing Switzerland 2013

where functions $a_j(X)$ and $b_j(X)$ are analytic in a neighborhood of the point $X = X^0$. Otherwise we call the system (1) *locally nonintegrable* at this point. Number m is equal to 1 for a planar ($n = 2$) autonomous system.

In [7], a method was proposed for the analysis of integrability of system (1) based on power transformations [5] and computation of normal forms near stationary solutions of transformed systems.

In the neighborhood of the stationary point $X = 0$, system (1) can be written in the form

$$\dot{X} = A X + \tilde{\Phi}(X),   (3)$$

where $\tilde{\Phi}(X)$ has no terms linear in X.

Let $\lambda_1, \lambda_2, \ldots, \lambda_n$ be the eigenvalues of the matrix A. If at least one of them $\lambda_i \neq 0$, then the stationary point $X = 0$ is called an *elementary* stationary point. In this case, system (1) has a normal form which is equivalent to a system of lower order [4]. If all eigenvalues vanish, then the stationary point $X = 0$ is called a *nonelementary* stationary point. In this case, there is no normal form for the system (1). But by using power transformations, the nonelementary stationary point $X = 0$ can be blown up to a set of elementary stationary points. For each of these elementary stationary points, we can compute the normal form and write conditions of local integrability.

In this paper, we demonstrate how this approach can be applied to study the local and global integrability in the planar case of the system (1) near the stationary point $X^0 = 0$ of high degeneracy

$$\begin{aligned} \dot{x} &= \alpha\, y^3 + \beta\, x^3\, y + (a_0\, x^5 + a_1\, x^2 y^2) + (a_2\, x^4\, y + a_3\, x\, y^3)\ , \\ \dot{y} &= \gamma\, x^2\, y^2 + \delta\, x^5 + (b_0\, x^4 y + b_1\, x\, y^3) + (b_2\, x^6 + b_3\, x^3\, y^2 + b_4\, y^4)\ . \end{aligned}   (4)$$

Systems with a nilpotent matrix of the linear part were thoroughly studied by Lyapunov et al. In system (4), there is no linear part, and the first approximation is nonhomogeneous. This is the simplest case of a planar system without linear part and with Newton's open polygon [4] consisting of a single edge. In general case, such problems have not been studied. However, the system with such support was considered in [1], where authors put $-\alpha = \delta = 1$ and $3\beta + 2\gamma = 0$. Further authors of [1] studied Hamiltonian case of this system with the additional assumption that the Hamiltonian function is expandable into the product of only square-free factors.

We study the problem: what are the conditions on parameters under which the system (4) is locally or globally integrable at the another particular case $-\alpha = \delta = 1$, $b \overset{\text{def}}{=} -\beta = 1/\gamma$ and $a_2 = a_3 = b_2 = b_3 = b_4 = 0$. In this situation, the system is not Hamiltonian.

Further we discuss the first quasi-homogeneous approximation of system(4), recall the main ideas of the method of resonant normal forms, prove the theorem on necessary conditions of local integrability, find some of sufficient conditions of global integrability, and formulate the conclusions. At searching necessary conditions of local integrability it is absolutely impossible to bypass of computer algebra methods.

2   About First Quasi-Homogeneous Approximation

If $n = 2$ then rationality of the ratio λ_1/λ_2 and the condition A (see the next Section) are necessary and sufficient conditions for local integrability of a system near an elementary stationary point. For local integrability of original system (1) near a degenerate (nonelementary) stationary point, it is necessary and sufficient to have local integrability near each of elementary stationary points, which are produced by the blowing up process described above.

So we study the system

$$
\begin{aligned}
\dot{x}_1 &= x_1 \sum \phi_Q X^Q \ , \\
\dot{x}_2 &= x_2 \sum \psi_Q X^Q \ ,
\end{aligned}
\tag{5}
$$

where $Q = (q_1, q_2)$, $X = (x_1, x_2)$, $X^Q = x_1^{q_1} x_2^{q_2}$, $q_1, q_2 \in \mathbb{Z}$; ϕ_Q and ψ_Q are constant coefficients, which are polynomials in parameters of the system.

System (5) has a quasi-homogeneous initial approximation if there exists an integer vector $R = (r_1, r_2) > 0$ and a number s such that the scalar product

$$
\langle Q, R \rangle \stackrel{\text{def}}{=} q_1 r_1 + q_2 r_2 \geq s = \text{const}
$$

for nonzero $|\phi_Q| + |\psi_Q| \neq 0$, and between vectors Q with $\langle Q, R \rangle = s$ there are vectors of the form $(q_1, -1)$ and $(-1, q_2)$. In this case, system (5) takes the form

$$
\begin{aligned}
\dot{x}_1 &= x_1[\phi_s(X) + \phi_{s+1}(X) + \phi_{s+2}(X) + \ldots] \ , \\
\dot{x}_2 &= x_2[\psi_s(X) + \psi_{s+1}(X) + \psi_{s+2}(X) + \ldots] \ ,
\end{aligned}
$$

where $\phi_k(X)$ is the sum of terms $\phi_Q X^Q$ for which $\langle Q, R \rangle = k$. And the same holds for the $\psi_k(X)$. Then the initial approximation (or truncation) of system (5) is the quasi-homogeneous system

$$
\begin{aligned}
\dot{x}_1 &= x_1 \phi_s(X) \ , \\
\dot{x}_2 &= x_2 \psi_s(X) \ .
\end{aligned}
\tag{6}
$$

We will study the case $R = (2, 3)$ and $s = 7$, when the quasi-homogeneous system (6) has the form

$$
\dot{x} = \alpha y^3 + \beta x^3 y, \quad \dot{y} = \gamma x^2 y^2 + \delta x^5 \ ,
\tag{7}
$$

where $\alpha \neq 0$ and $\delta \neq 0$. Using linear transformation $x = \sigma \tilde{x}$, $y = \tau \tilde{y}$ we can fix two nonzero parameters in (7)

$$
\dot{x} = -y^3 - b x^3 y, \quad \dot{y} = c x^2 y^2 + x^5 \ .
\tag{8}
$$

We study the problem: what are the conditions on parameters under which system (5) is locally or globally integrable. For this, the system (7) should be at least locally integrable. But each autonomous planar quasi-homogeneous system like (7) has an integral, but it can have not the form (2) with analytic a_1 and b_1. So we need to have the local integrability of (7) in the sense (2).

Theorem 1. *In the case $D \overset{\text{def}}{=} (3\,b + 2\,c)^2 - 24 \neq 0$, system (8) is locally integrable if and only if the number $(3\,b - 2\,c)/\sqrt{D}$ is rational.*

Proof. After the power transformation

$$x = u\,v^2, \quad y = u\,v^3 \tag{9}$$

and time rescaling

$$dt = u^2 v^7 d\tau \ ,$$

we obtain system (8) in the form

$$\dot{u} = -u\,(3 + (3\,b + 2\,c)u + 2\,u^2), \quad \dot{v} = v\,(1 + (b + c)\,u + u^2) \ .$$

So

$$\frac{d\log v}{d\,u} = -\frac{1 + (b + c)\,u + u^2}{u\,[3 + (3\,b + 2\,c)\,u + 2\,u^2]} \ . \tag{10}$$

The number D is the discriminant of the polynomial $3 + (3\,b + 2\,c)\,u + 2\,u^2$. In our case $D \neq 0$, so the polynomial has two different roots $u_1 \neq u_2$, and the right-hand side of (10) has the form

$$\frac{\xi}{u} + \frac{\eta}{u - u_1} + \frac{\zeta}{u - u_2} \ ,$$

where ξ, η, and ζ are constants. Direct computation shows that

$$\xi = -\frac{1}{3}, \quad \eta + \zeta = -\frac{1}{6}, \quad \zeta = -\frac{1 + (3\,b - 2\,c)/\sqrt{D}}{12} \ . \tag{11}$$

The first integral of (10) is

$$u^{\xi}\,(u - u_1)^{\eta}\,(u - u_2)^{\zeta}\,v^{-1} \ .$$

According to (11), its integral power can have the form (2) if and only if the number

$$\frac{3\,b - 2\,c}{\sqrt{D}} \tag{12}$$

is rational. The same is true for the integral in variables x, y, because according to (9)

$$u = \frac{x^3}{y^2}, \quad v = \frac{y}{x} \ .$$

Proof is finished.

We will study the integrability problem for entire system (5) with the first quasi-homogeneous approximation (8). So we choose (5) in the form

$$\begin{aligned} dx/dt &= -y^3 - b\,x^3y + a_0\,x^5 + a_1\,x^2y^2 \ , \\ dy/dt &= c\,x^2y^2 + x^5 + b_0\,x^4y + b_1\,x\,y^3 \ , \end{aligned} \tag{13}$$

with arbitrary parameters a_i, b_i and $b \neq 0$.

In our system (13), there is no linear part, and the first approximation is not homogeneous but quasi-homogeneous. This is the simplest case of a planar system without linear part and with Newton's open polygon consisting of a single edge. In our case, the system corresponds to the quasi-homogeneous first approximation with $R = (2,3)$, $s = 7$. In general case, such problems have not been studied. Firstly the system with such a support was considered in [1], where authors put $3\,b - 2\,c = 0$, then the order (12) is equal to zero and the integral of system (8) is

$$I = 2\,x^6 + 4\,c\,x^3 y^2 + 3\,y^4. \tag{14}$$

In this case, system (8) is the Hamiltonian system with Hamiltonian function $H = I/12$. Further authors of [1] studied Hamiltonian case of the full system under the additional assumption that the Hamiltonian function is expandable into the product of only square-free factors.

In this paper, we will study another simple particular case, when the order (12) is equal to ± 1, in the case $c = 1/b$ and $2\,x^3 + 3\,b\,y^2$ is the first integral of (8). With respect of Theorem 2.1, the first quasi-homogeneous approximation has an analytic integral but it is not a Hamiltonian system. Also we restrict ourselves to the case $b^2 \neq 2/3$ which should be studied separately (see below).

3   About Normal Form and the Condition A

Let the linear transformation

$$X = BY \tag{15}$$

bring the matrix A to the Jordan form $J = B^{-1}AB$ and system (3) to the form

$$\dot{Y} = JY + \tilde{\Phi}(Y) \ . \tag{16}$$

Let the formal change of coordinates

$$Y = Z + \Xi(Z), \tag{17}$$

where $\Xi = (\xi_1, \ldots, \xi_n)$ and $\xi_j(Z)$ are formal power series, transform the system (16) into the system

$$\dot{Z} = JZ + \Psi(Z) \ . \tag{18}$$

We write it in the form

$$\dot{z}_j = z_j g_i(Z) = z_j \sum g_{jQ} Z^Q \text{ over } Q \in \mathbb{N}_j, \ j = 1, \ldots, n \ , \tag{19}$$

where $Q = (q_1, \ldots, q_n)$, $Z^Q = z_1^{q_1} \ldots z_n^{q_n}$,

$$\mathbb{N}_j = \{Q : Q \in \mathbb{Z}^n, \ Q + E_j \geq 0\}, \ j = 1, \ldots, n \ ,$$

E_j means the unit vector. Denote

$$\mathbb{N} = \mathbb{N}_1 \cup \ldots \cup \mathbb{N}_n \ . \tag{20}$$

The diagonal $\Lambda = (\lambda_1, \ldots, \lambda_n)$ of J consists of eigenvalues of the matrix A.

System (18), (19) is called the *resonant normal form* if:

a) J is the Jordan matrix,

b) under the sum in (19), there are only the *resonant terms*, for which the scalar product

$$\langle Q, \Lambda \rangle \overset{\text{def}}{=} q_1 \lambda_1 + \ldots + q_n \lambda_n = 0 \ . \tag{21}$$

Theorem 2. *[3] There exists a formal change (17) reducing system (16) to its normal form (18), (19).*

In [3], there are conditions on the normal form (19), which guarantee the convergence of the normalizing transformation (17).

Condition A. *In the normal form (19)*

$$g_j = \lambda_j \alpha(Z) + \bar{\lambda}_j \beta(Z), \quad j = 1, \ldots, n \ ,$$

where $\alpha(Z)$ and $\beta(Z)$ are some power series.

Let

$$\omega_k = \min |\langle Q, \Lambda \rangle| \ \text{ over } \ Q \in \mathbb{N}, \ \langle Q, \Lambda \rangle \neq 0, \ \sum_{j=1}^{n} q_j < 2^k, \ k = 1, 2, \ldots \ .$$

Condition ω (on small divisors). *The series*

$$\sum_{k=1}^{\infty} 2^{-k} \log \omega_k > -\infty \ ,$$

i.e., it converges.

It is fulfilled for almost all vectors Λ.

Theorem 3. *[3] If vector Λ satisfies Condition ω and the normal form (19) satisfies Condition A then the normalizing transformation (17) converges.*

The algorithm for calculation of the normal form and of the normalizing transformation together with the corresponding computer program are briefly described in [6].

4   Necessary Conditions of Local Integrability

We consider the system

$$\begin{aligned} dx/dt &= -y^3 - b\,x^3 y + a_0\,x^5 + a_1\,x^2 y^2 \ , \\ dy/dt &= (1/b)\,x^2 y^2 + x^5 + b_0\,x^4 y + b_1\,x\,y^3 \ , \end{aligned} \tag{22}$$

with arbitrary complex parameters a_i, b_i and $b \neq 0$.

Systems with a nilpotent matrix of the linear part were thoroughly studied by Lyapunov et al. In our system (22), there is no linear part, and the first approximation is not homogeneous but quasi-homogeneous. This is the simplest case of a planar system without linear part and with Newton's open polygon consisting of a single edge. In our case, the system corresponds to the quasi-homogeneous first approximation with $R = (2,3)$, $s = 7$. In general case, such problems have not been studied, and the authors do not know of any applications of the system (22).

After the power transformation (9) and the time rescaling $(dt = u^2 v^7 d\tau)$, we obtain system (22) in the form

$$du/d\tau = -3\,u - [3\,b + (2/b)]u^2 - 2\,u^3 + (3\,a_1 - 2\,b_1)u^2 v + (3\,a_0 - 2\,b_0)u^3 v \ ,$$
$$dv/d\tau = v + [b + (1/b)]u\,v + u^2 v + (b_1 - a_1)u\,v^2 + (b_0 - a_0)u^2 v^2 \ .$$
$$(23)$$

Under the power transformation (9), the point $x = y = 0$ blows up into two straight lines $u = 0$ and $v = 0$. Along the line $u = 0$ the system (23) has a single stationary point $u = v = 0$. Along the second line $v = 0$ this system has three elementary stationary points

$$u = 0, \quad u = -\frac{1}{b}, \quad u = -\frac{3b}{2} \ . \tag{24}$$

The necessary and sufficient condition of local integrability of system (22) near the point $x = y = 0$ is local integrability near all stationary points of the system (23).

Lemma 1. *Near the point $u = v = 0$, the system (23) is locally integrable.*

Proof. In accordance with Chapter 2 of the book [4], the support of system (23) consists of five points $Q = (q_1, q_2)$

$$(0,0), \quad (1,0), \quad (2,0), \quad (1,1), \quad (2,1). \tag{25}$$

At the point $u = v = 0$ eigenvalues of system (23) are $\Lambda = (\lambda_1, \lambda_2) = (-3, 1)$. Only for the first point from (25) $Q = 0$, the scalar product $\langle Q, \Lambda \rangle$ is zero, for the remaining four points (25) it is negative, so these four points lie on the same side of the straight line $\langle Q, \Lambda \rangle = 0$. In accordance with the remark at the end of Subsection 2.1 of Chapter 2 of the book [4], in such case the normal form consists only of the terms of a right-hand side of system (23) such that their support Q lies on the straight line $\langle Q, \Lambda \rangle = 0$. But only linear terms of the system (23) satisfy this condition. Therefore, at point $u = v = 0$, the normal form of the system is linear

$$dz_1/d\tau = -3\,z_1, \quad dz_2/d\tau = z_2 \ .$$

It is obvious that this normal form satisfies the condition A. So the normalizing transformation converges, and at point $u = v = 0$, system (23) has the analytic first integral

$$z_1 z_2^3 = \text{const} \ .$$

Proof is finished.

The proof of local integrability at point $u = \infty, v = 0$ is similar.

Thus, if we must find conditions of local integrability at two other stationary points (24), then we will have the conditions of local integrability of system (22) near the point $X = 0$.

Let us consider the stationary point $u = -1/b, v = 0$. Below we restrict ourselves to the case $b^2 \neq 2/3$ when the linear part of system (23), after the shift $u = \tilde{u} - 1/b$, has non-vanishing eigenvalues. At $b^2 = 2/3$, the matrix of the linear part of the shifted system in new variables $\tilde{u}$ and v has a Jordan cell with both zero eigenvalues. This case can be studied by means of one more power transformation and will be studied later.

To simplify eigenvalues, we change the time at this point once more with the factor $d\tau = (2 - 3b^2)/b^2 \, d\tau_1$. After that we obtain the vector of eigenvalues of system (23) at this point as $(\lambda_1, \lambda_2) = (-1, 0)$. So the normal form of the system will become

$$
\begin{aligned}
dz_1/d\tau_1 &= -z_1 + z_1 \, g_1(z_2) \ , \\
dz_2/d\tau_1 &= \qquad z_2 \, g_2(z_2) \ ,
\end{aligned}
\tag{26}
$$

where $g_{1,2}(x)$ are formal power series in x. Coefficients of these series are rational functions of the parameters a_0, a_1, b_0, b_1 and b of the system. It can be proved that denominator of each of these rational functions is proportional to some integer degree $k(n)$ of the polynomial $(2 - 3b^2)$. Their numerators are polynomials in parameters of the system

$$
g_{1,2}(x) = \sum_{n=1}^{\infty} \frac{p_{1,2;n}(b, a_0, a_1, b_0, b_1)}{(2 - 3b^2)^{k(n)}} x^n \ .
$$

Condition A of integrability for system (26) is $g_2(x) \equiv 0$. It is equivalent to the infinite polynomial system of equations

$$
p_{2,n}(b, a_0, a_1, b_0, b_1) = 0, \quad n = 1, 2, \ldots \ .
\tag{27}
$$

According to the Hilbert's theorem on bases in polynomial ideals [10], this system has a finite basis.

We computed the first three polynomials $p_{2,1}, p_{2,2}, p_{2,3}$ by our program [6]. There are 2 solutions of a corresponding subset of equations (27) at $b \neq 0$

$$
a_0 = 0, \quad a_1 = -b_0 \, b, \quad b_1 = 0, \quad b^2 \neq 2/3
\tag{28}
$$

and

$$
a_0 = a_1 \, b, \quad b_0 = b_1 \, b, \quad b^2 \neq 2/3 \ .
\tag{29}
$$

Addition of the fourth equation $p_{2,4} = 0$ to the subset of equations does not change these solutions.

A calculation of polynomials $p_{2,n}(b, a_0, a_1, b_0, b_1)$ in generic case is technically a very difficult problem. But we can verify some of these equations from the set (27) on solutions (28) and (29) for several fixed values of the parameter b. We can conclude from these results about a necessary condition of local integrability and create hypothesis about existence of the sufficient condition.

We verified solutions of subset of equations

$$p_{2,n}(b,\ a_0 = a_1 b,\ a_1,\ b_0 = b_1 b,\ b_1) = 0, \quad n = 1, 2, \ldots, 28 \ .$$

at $b = 1$ and $b = 2$. All equations above are satisfied. Thus, we can conclude that (28) and (29) are necessary and with good probability sufficient conditions on parameters of the system for the satisfaction of the conditions A, i.e., conditions of the local integrability of equation (23) near the stationary point $u = -1/b, v = 0$.

Let us consider the stationary point $u = -3b/2, v = 0$. We rescale time at this point with the factor $d\tau = (2 - 3b^2)\, d\tau_2$. After that we get the vector of eigenvalues of system (23) at this point as $(-1/4, 3/2)$. So the normal form has a resonance of the seventh order

$$
\begin{aligned}
dz_1/d\tau_2 &= -(1/4)\, z_1 + z_1\, r_1(z_1^6\, z_2) \ , \\
dz_2/d\tau_2 &= (3/2)\, z_2 + z_2\, r_2(z_1^6\, z_2) \ ,
\end{aligned}
\tag{30}
$$

where $r_1(x)$ and $r_2(x)$ are also formal power series, and in (30) they depend on single "resonant" variable $z_1^6 z_2$. Coefficients of these series are again rational functions of parameters a_0, a_1, b_0, b_1 and b. The denominator of each of these functions is proportional to some integer degree $l(n)$ of the polynomial $2 - 3b^2$. Their numerators are polynomials in parameters of the system

$$r_{1,2}(x) = \sum_{n=1}^{\infty} \frac{q_{1,2;n}(b, a_0, a_1, b_0, b_1)}{(2 - 3b^2)^{l(n)}} x^n \ .$$

The condition A for the system (30) is $6\, r_1(x) + r_2(x) = 0$. It is equivalent to the infinite system of polynomial equations

$$6\, q_{1,n}(b, a_0, a_1, b_0, b_1) + q_{2,n}(b, a_0, a_1, b_0, b_1) = 0, \quad n = 7, 14, \ldots. \tag{31}$$

We computed polynomials $q_{1,7}, q_{2,7}$ and solved the lowest equation from the set (31) for the parameters of the solution (29). We have found 5 different two-parameter (b and a_1) solutions. With (29) they are

$$
\begin{aligned}
1) \quad & b_1 = -2\, a_1, & a_0 &= a_1 b, & b_0 &= b_1 b, & b^2 &\neq 2/3 \ , \\
2) \quad & b_1 = (3/2)\, a_1, & a_0 &= a_1 b, & b_0 &= b_1 b, & b^2 &\neq 2/3 \ , \\
3) \quad & b_1 = (8/3)\, a_1, & a_0 &= a_1 b, & b_0 &= b_1 b, & b^2 &\neq 2/3
\end{aligned}
\tag{32}
$$

and

$$
\begin{aligned}
4) \quad & b_1 = \tfrac{197 - 7\sqrt{745}}{24} a_1, & a_0 &= a_1 b, & b_0 &= b_1 b, & b^2 &\neq 2/3 \ , \\
5) \quad & b_1 = \tfrac{197 + 7\sqrt{745}}{24} a_1, & a_0 &= a_1 b, & b_0 &= b_1 b & b^2 &\neq 2/3 \ .
\end{aligned}
\tag{33}
$$

We have verified (31) up to $n = 49$ for solutions (32) for $b = 1$ and $b = 2$ and arbitrary a_1. They are correct. We have verified the solution (28) in the same way. It is also correct.

Solutions (33) starting from the order $n = 14$ are correct only for the additional condition $a_1 = 0$. But for this condition, solutions (33) are a special case of solutions (32).

So, we have proved.

Theorem 4. *Equalities (28), and (32) form a full set of necessary conditions of a local integrability of system (23) at all its stationary points and a local integrability of system (22) at stationary point $x = y = 0$.*

5   About Sufficient Conditions of Integrability

The conditions resulted in theorem 3.2 as necessary for local integrability of system (22) at the zero stationary point, can be considered from the point of view of "The experimental mathematics" as sufficient conditions of local integrability, especially considering high enough orders of the checks above. It is, however, necessary to prove the sufficiency of these conditions by independent methods. It is necessary to do it for each of four conditions (28),(32) in each of stationary points $u = -3b/2, v = 0$ and $u = -1/b, v = 0$, and $b^2 \neq 2/3$.

With essential assistance by Prof. V. Romanovski [8] we have found first integrals for all cases (28),(32) mainly by method of the Darboux factor of system (23), see [9].

We have found four families of solutions which exhausted all integrable cases except possibly the case $b^2 = 2/3$.

$$
\begin{aligned}
&\text{At}\quad a_0 = 0, \quad a_1 = -b_0\,b, \quad b_1 = 0: \\
&I_{1uv} = u^2(3\,b + 2\,u)v^6 \ , \\
&I_{1xy} = 2\,x^3 + 3\,b\,y^2 \ .
\end{aligned}
\tag{34}
$$

$$
\begin{aligned}
&\text{At}\quad b_1 = -2a_1,\ a_0 = a_1 b,\ b_0 = b_1 b: \\
&I_{2uv} = u^2\,v^6\,(3\,b + u\,(2 - 6\,a_1\,b\,v)) \ , \\
&I_{2xy} = 2\,x^3 - 6\,a_1\,b\,x^2\,y + 3\,b\,y^2 \ .
\end{aligned}
\tag{35}
$$

$$
\text{At}\quad b_1 = 3a_1/2,\ a_0 = a_1 b,\ b_0 = b_1 b:
$$

$$
I_{3uv} = \frac{4 - 4a_1\,u\,v + 3^{5/6} a_1\,{}_2F_1(2/3,1/6;5/3;-2u/(3b))\,u\,v\,(3+2u/b)^{1/6}}{u^{1/3}v\,(3b+2u)^{1/6}} \ ,
\tag{36}
$$

$$
I_{3xy} = \frac{a_1 x^2\left(-4 + 3^{5/6}\,{}_2F_1\left(2/3,1/6;5/3;-2\,x^3/(3\,b\,y^2)\right)(3+2x^3/(b\,y^2))^{1/6}\right) + 4y}{y^{4/3}(3\,b + 2\,x^3/y^2)^{1/6}} \ ,
$$

$$
\text{At}\quad b_1 = 8a_1/3,\ a_0 = a_1 b,\ b_0 = b_1 b:
$$

$$
\begin{aligned}
I_{4u,v} = {} & \frac{u\,(3 + 2\,a_1^2 bu) + 6\,a_1\,b\,v}{3\,u\,[u^3(6 + a_1^2 b\,u) + 6\,a_1^2 b\,u^2 v + 9\,b\,v^2]^{1/6}} - \\
& \frac{8\,a_1\sqrt{-b}}{3^{5/3}}B_{6 + a_1\,\sqrt{-6\,b}\,u + 3\,v\,\sqrt{-6\,b/u^3}}(5/6, 5/6) \ ,
\end{aligned}
\tag{37}
$$

where B is the incomplete beta function.

In the paper [8], there are printed corresponding solutions for the integrals above. The integrals and solutions have no singularities near the points $b^2 = 2/3$, but the approach in which these solutions were obtained has the limitation $b^2 \neq 2/3$, so theoretically there is possible additional solutions at this point. So we need to study the domains of $b^2 = 2/3$ separately.

6 Case $b^2 = 2/3$

At this value of b, the both stationary points $(u = -3b/2, v = 0)$ and $(u = -1/b, v = 0)$ are collapsing, and we have instead (23) after the shift $u \to w - 1/b = w - \sqrt{3/2}$ the degenerate system again

$$
\begin{aligned}
dw/d\tau &= v(-\tfrac{9}{2}\sqrt{\tfrac{3}{2}}\,a_0 + \tfrac{9}{2}\,a_1 + 3\sqrt{\tfrac{3}{2}}\,b_0 - 3\,b_1) \\
&+ wv(\tfrac{27}{2}\,a_0 - 3\sqrt{6}\,a_1 - 9\,b_0 + 2\sqrt{6}\,b_1) \\
&+ \sqrt{6}\,w^2 + w^2 v(-9\sqrt{\tfrac{3}{2}}\,a_0 + 3\,a_1 + 3\sqrt{6}\,b_0 - 2\,b_1) \\
&- 2\,w^3 + w^3 v(3\,a_0 - 2\,b_0) \ ,
\end{aligned}
\tag{38}
$$
$$
\begin{aligned}
dv/d\tau &= -\tfrac{\sqrt{6}}{6}wv + v^2(-\tfrac{3}{2}\,a_0 + \sqrt{\tfrac{3}{2}}\,a_1 + \tfrac{3}{2}\,b_0 - \sqrt{\tfrac{3}{2}}\,b_1) \\
&+ w^2 v + wv^2((\sqrt{6}\,a_0 - a_1 - \sqrt{6}\,b_0 + b_1) \\
&+ w^2 v^2(-a_0 + b_0) \ .
\end{aligned}
$$

On the other hand, we should apply a power transformation once more. We choose here the transformation

$$
v \to r^2 w, \quad v' \to 2r'rw + r^2 w' \ ,
\tag{39}
$$

with the time scaling by dividing the equations by $u/\sqrt{6}$. We will have from (38)

$$
\begin{aligned}
dw/d\tilde{\tau} &= \tfrac{1}{2}\sqrt{\tfrac{3}{2}}w(4\sqrt{6} + 3(-3\sqrt{6}a_0 + 6a_1 + 2\sqrt{6}b_0 - 4b_1)r + \\
& 2w(-4 + (27a_0 - 6\sqrt{6}a_1 - 18b_0 + 4\sqrt{6}b_1 + \\
& w(-9\sqrt{6}a_0 + 6a_1 + 6\sqrt{6}b_0 - 4b_1 + (6a_0 - 4b_0)w))r)) \ , \\
dr/d\tilde{\tau} &= \tfrac{-1}{2}r(26 - 6(9a_0 - 3\sqrt{6}a_1 - 6b_0 + 2\sqrt{6}b_1)r + \\
& w(-10\sqrt{6} + (57\sqrt{6}a_0 - 78a_1 - 39\sqrt{6}b_0 + 54b_1 + \\
& 2w(-60a_0 + 7\sqrt{6}a_1 + 42b_0 - 5\sqrt{6}b_1 + \sqrt{6}(7a_0 + 5b_0)w))r)) \ .
\end{aligned}
\tag{40}
$$

On the invariant line $u = 0$, this equation has 2 finite singular points: $w = 0$ and $w = 13/(3(9a_0 - 3\sqrt{6}a_1 - 6b_0 + 2\sqrt{6}b_1))$. Near the first point we have a resonance of 19 level $(13 : 6)$ and near the second point of 27 level $(1 : 26)$. The infinite point is integrable.

The calculation of the normal form of equation (40) up to 19th order produced a single equation for the necessary condition of integrability A as polynomial in variables a_0, a_1, b_0, and b_1. This calculation was produced by the REDUCE [11] based system. It takes about 6.5 hours on the 3 GGc Pentium IV processor. The normalizing transformation concludes 226145 terms, and the normal form has 1174 terms.

The obtained condition A is consistent with condition (28) and each from (32), i.e., all four solutions. We have obtained the condition A as the long polynomial $p(a_0, a_1, b_0, b_1)$. This polynomial was sixth-order homogenous polynomial. If you put $a_0 = a_1 b$ and $b_0 = b_1 b$ as in (32) and solve the equation $p = 0$ you will receive exactly solution (32), (33) but without limitation $b^2 \neq 2/3$. Solution (33) is not realized thereafter, of course. Also the $p(a_0, a_1, b_0, b_1)$ is equal to zero under conditions (28). So our result is compatible with previous results, but the result which we have really obtained is that solutions (32) do exist at $b^2 = 2/3$, and there are no more solutions of such a type in any case.

So, we have proved.

Theorem 5. *All solutions (28),(32) are right without limitation* $b^2 \neq 2/3$.

For searching additional first integrals we need to calculate the condition A at the point with the resonance (1 : 26). There are also other transformations with corresponding resonances. Its studying is a very heavy technical problem, and we work on it at present.

7   Conclusions

For the particular 5-parametrical case at $b^2 \neq 2/3$ planar system (13), non-Hamiltonian at $c = 1/b$, we have found all necessary sets of conditions on parameters at which system (22) is locally integrable near degenerate point $X = 0$. They form 4 sets of conditions on parameters of the system. So these sets of conditions are also sufficient for local and global integrability of system (22). The case of point $b^2 = 2/3$ at $c = 1/b$ and different from $c = 1/b$ but with a real rational value of (12) cases should be studied separately. But we have proved here that the limitation $b^2 \neq 2/3$ can be excluded from the solutions obtained previously [8]. At the moment, we have not found the additional first integrals.

Acknowledgements. This work was supported by the Grant 11-01-00023-a– of the Russian Foundation for Basic Research and the Grant 195.2008.2 of the President of the Russian Federation on support of scientific schools.

References

1. Algaba, A., Gamero, E., Garcia, C.: The integrability problem for a class of planar systems. Nonlinearity 22, 395–420 (2009)
2. Bateman, H., Erdélyi, A.: Higher Transcendental Functions, vol. 1. McGraw-Hill Book Company, Inc., New York (1953)
3. Bruno, A.D.: Analytical form of differential equations (I,II). Trudy Moskov. Mat. Obsc. 25, 119–262 (1971); 26, 199–239 (1972); (Russian) = Trans. Moscow Math. Soc. 25, 131–288 (1971); 26, 199–239 (1972) (English)
4. Bruno, A.D.: Local Methods in Nonlinear Differential Equations, Nauka, Moscow (1979); (Russian) = Springer, Berlin (1989) (English)
5. Bruno, A.D.: Power Geometry in Algebraic and Differential Equations, Fizmatlit, Moscow (1998); (Russian) = Elsevier Science, Amsterdam (2000) (English)
6. Edneral, V.F.: An algorithm for construction of normal forms. In: Ganzha, V.G., Mayr, E.W., Vorozhtsov, E.V. (eds.) CASC 2007. LNCS, vol. 4770, pp. 134–142. Springer, Heidelberg (2007)
7. Bruno, A.D., Edneral, V.F.: Algorithmic analysis of local integrability. Dokl. Akademii Nauk. 424(3), 299–303 (2009); (Russian) = Doklady Mathem. 79(1), 48–52 (2009) (English)

8. Edneral, V., Romanovski, V.G.: Normal forms of two $p: -q$ resonant polynomial vector fields. In: Gerdt, V.P., Koepf, W., Mayr, E.W., Vorozhtsov, E.V. (eds.) CASC 2011. LNCS, vol. 6885, pp. 126–134. Springer, Heidelberg (2011)
9. Romanovski, V.G., Shafer, D.S.: The Center and Cyclicity Problems: A Computational Algebra Approach. Birkhäuser, Boston (2009)
10. Siegel, C.L.: Vorlesungen über Himmelsmechanik. Springer, Berlin (1956)
11. Hearn, A.C.: REDUCE. User's Manual. Version 3.8., Santa Monica, CA, USA (February 2004), reduce@rand.org

Efficient Methods to Compute Hopf Bifurcations in Chemical Reaction Networks Using Reaction Coordinates

Hassan Errami[1], Markus Eiswirth[2], Dima Grigoriev[3], Werner M. Seiler[1], Thomas Sturm[4], and Andreas Weber[5]

[1] Institut für Mathematik, Universität Kassel, Kassel, Germany
errami@uni-kassel.de, seiler@mathematik.uni-kassel.de
[2] Fritz-Haber-Institut der Max-Planck-Gesellschaft, Berlin, Germany
and Ertl Center for Electrochemisty and Catalysis, GIST, South Korea
eiswirth@fhi-berlin.mpg.de
[3] CNRS, Mathématiques, Université de Lille, Villeneuve d'Ascq, France
Dmitry.Grigoryev@math.univ-lille1.fr
[4] Max-Planck-Institut für Informatik, Saarbrücken, Germany
sturm@mpi-inf.mpg.de
[5] Institut für Informatik II, Universität Bonn, Bonn, Germany
weber@cs.uni-bonn.de

Abstract. We build on our previous work to compute Hopf bifurcation fixed points for chemical reaction systems on the basis of reaction coordinates. For determining the existence of Hopf bifurcations the main algorithmic problem is to determine whether a single multivariate polynomial has a zero for positive coordinates. For this purpose we provide heuristics on the basis of the Newton polytope that ensure the existence of positive and negative values of the polynomial for positive coordinates. We apply our method to the example of the Methylene Blue Oscillator (MBO).

1 Introduction

In this paper we build on our previous work [1] to compute Hopf bifurcation fixed points for chemical reaction systems on the basis of reaction coordinates. In that previous work algorithmic ideas introduced by El Kahoui and Weber [2], which already had been used for mass action kinetics of small dimension [3], have been combined with methods of stoichiometric network analysis (SNA) introduced by Clarke in 1980 [4], which had been used in several "hand computations" in a semi-algorithmic way for parametric systems. The most elaborate of these computations haven been described in [5]. The algorithmic method presented in [1] uses and combines the ideas of these methods and extends them to a new approach for computing Hopf bifurcation in complex systems using reaction coordinates also allowing systems with linear constraints.

However, the criteria used for determining Hopf bifurcation fixed points with empty unstable manifold involving an equality condition on the principal minor

V.P. Gerdt et al. (Eds.): CASC 2013, LNCS 8136, pp. 88–99, 2013.
© Springer International Publishing Switzerland 2013

$\Delta_{n-1} = 0$ in conjunction with inequality conditions on $\Delta_{n-2} > 0 \wedge \cdots \wedge \Delta_1 > 0$ and positivity conditions on the variables and parameters still turned out to be hard problems for general real quantifier elimination procedures even for moderate dimensions.

In this paper we use the rather basic observation that when using convex coordinates, the condition for the existence of Hopf bifurcation fixed points is given by the single polynomial equation $\Delta_{n-1} = 0$ together with positivity conditions on the convex coordinates. This allows to delay or even drop a test for having empty unstable manifold on already determined witness points for Hopf bifurcations. Hence the main algorithmic problem is to determine whether a single multivariate polynomial has a zero for positive coordinates.

For this purpose we provide in Sect. 2 heuristics on the basis of the Newton polytope, which ensure the existence of positive and negative values of the polynomial for positive coordinates. In Sect. 3 we apply our method to the example of the Methylene Blue Oscillator (MBO).

As we are continuing here the work in [1] we allow ourselves to refer the reader to that publication for basic definitions as well as for an introduction of our software toolchains.

2   Condition for a Hopf Bifurcation

Consider a parameterized autonomous ordinary differential equation of the form $\dot{x} = f(u, x)$ with a scalar parameter u. Due to a classical result by Hopf, this system exhibits at the point (u_0, x_0) a Hopf bifurcation, i.e. an equilibrium transforms into a limit cycle, if $f(u_0, x_0) = 0$ and the Jacobian $D_x f(u_0, x_0)$ has a simple pair of pure imaginary eigenvalues and no other eigenvalues with zero real parts [6, Thm. 3.4.2].[1] The proof of this result is based on the center manifold theorem. From a physical point of view, the most interesting case is that the unstable manifold of the equilibrium (u_0, x_0) is empty. However, for the mere existence of a Hopf bifurcation, this assumption is not necessary.

In [2] it is shown that for a parameterized vector field $f(u, x)$ and the autonomous ordinary differential system associated with it, there is a semi-algebraic description of the set of parameter values for which a Hopf bifurcation (with empty unstable manifold) occurs. Specifically, this semi-algebraic description can be expressed by the following first-order formula:

$$\exists x (f_1(u, x) = 0 \wedge f_2(u, x) = 0 \wedge \cdots \wedge f_n(u, x) = 0 \wedge$$
$$a_n > 0 \wedge \Delta_{n-1}(u, x) = 0 \wedge \Delta_{n-2}(u, x) > 0 \wedge \cdots \wedge \Delta_1(u, x) > 0).$$

In this formula a_n is $(-1)^n$ times the Jacobian determinant of the matrix $Df(u, x)$, and the $\Delta_i(u, x)$'s are the ith Hurwitz determinants of the characteristic polynomial of the same matrix $Df(u, x)$.

---

[1] We ignore here the non-degeneracy condition that this pair of eigenvalues crosses the imaginary axis transversally, as it is always satisfied in realistic models.

The proof uses a formula of Orlando [7], which is discussed also in several monographs, e.g., [8, 9]. However, a closer inspection of the two parts of the proof of [2, Theorem 3.5] shows even the following: for a fixed points (given in possibly parameterized form) the condition that there is a pair of purely complex eigenvalues is given by the condition $\Delta_{n-1}(u, x) = 0$ and the condition that all other eigenvalues have negative real part is given by $\Delta_{n-2}(u, x) > 0 \wedge \cdots \wedge \Delta_1(u, x) > 0$. This statement (without referring to parameters explicitly) is also contained in [10, Theorem 2], in which a different proof technique is used.

Hence if we drop the condition on Hopf bifurcation points that they have empty unstable manifold a semi-algebraic description of the set of parameter values for which a Hopf bifurcation occurs for the system is given by the following formula:

$$\exists x (f_1(u, x) = 0 \wedge \cdots \wedge f_n(u, x) = 0 \wedge a_n > 0 \wedge \Delta_{n-1}(u, x) = 0).$$

Notice that when the quantifier elimination procedure yields sample points for existentially quantified formulae, then the condition $\Delta_{n-2}(u, x) > 0 \wedge \cdots \wedge \Delta_1(u, x) > 0$) can be tested for the sample points later on, i.e. one can then test whether this Hopf bifurcation fixed points has empty unstable manifold. Such sample points are yielded, e.g., by virtual substitution-based methods in Redlog [12–15].

As an example consider the famous *Lorenz System* named after Edward Lorenz at MIT, who first investigated this system as a simple model arising in connection with fluid convection [16, 6, 17]. It is given by the following system of ODEs:

$$\dot{x}(t) = \alpha \, (y(t) - x(t))$$
$$\dot{y}(t) = r \, x(t) - y(t) - x(t) \, z(t)$$
$$\dot{z}(t) = x(t) \, y(t) - \beta \, z(t).$$

After imposing positivity conditions on the parameters the following answer is obtained using a combination of Redlog and SLFQ as described in [18] for the test of a Hopf bifurcation fixed points:

$$(-\alpha^2 - \alpha\beta + \alpha r - 3\alpha - \beta r - r = 0 \vee -\alpha\beta + \alpha r - \alpha - \beta^2 - \beta = 0) \wedge$$
$$- \alpha^2 - \alpha\beta + \alpha r - 3\alpha - \beta r - r \leq 0 \wedge$$
$$\beta > 0 \wedge \alpha > 0 \wedge -\alpha\beta + \alpha r - \alpha - \beta^2 - \beta \geq 0.$$

When testing, in contrast, for Hopf bifurcation fixed points with empty unstable manifold we obtain the following formula, which is not equivalent to the one above:

$$\alpha^2 + \alpha\beta - \alpha r + 3\alpha + \beta r + r = 0 \wedge \alpha r - \alpha - \beta^2 - \beta \geq 0 \wedge$$
$$2\alpha - 1 \geq 0 \wedge \beta > 0.$$

Hence for the case of the Lorenz system not all Hopf bifurcation fixed points have an unstable empty manifold.

This approach using concentration coordinates has been applied in [3].

2.1 Using Reaction Coordinates

In [1] a new approach for computing Hopf bifurcations in complex systems has been given, which uses reaction coordinates in contrast to concentration coordinates and also allows systems with linear constraints. The Jacobian matrix of a subsystem formed by d-faces is given by the following equation, where $\mathcal{S}$, $\mathcal{K}$ and $\mathcal{E}$ denote the stoichiometric matrix, kinetic matrix, and the set of extreme currents, respectively.

$$\mathrm{Jac}(x) = \mathcal{S}\mathrm{diag}(\sum_i^d j_i \mathcal{E}_i)\mathcal{K}^t \mathrm{diag}(1/x_1, ..., 1/x_n).$$

For checking for the existence of Hopf bifurcation fixed points without requiring empty unstable manifolds we have to decide the satisfiability of the following formula:

$$\Delta_{n-1}(j, x) = 0 \wedge j_1 \geq 0 \wedge \cdots j_d \geq 0 \wedge x_1 > 0 \wedge \cdots x_n > 0.$$

Hence the algorithmic task is to determine whether the single multivariate polynomial equation $\Delta_{n-1}(j, x) = 0$ has a solution subject to the given sign conditions on the variables. Note that whenever a satisfiability test provides sample points in the satisfiable case, then those points can be tested for having an empty unstable manifold by substituting them into the condition

$$\Delta_{n-2}(j, x) > 0 \wedge \cdots \wedge \Delta_1(j, x) > 0.$$

In the next section we are going to discuss sufficient conditions and their efficient algorithmic realizations.

2.2 Sufficient Conditions for a Positive Solution of a Single Multivariate Polynomial Equation

The method discussed in this section is summarized in an algorithmic way in Alg. 1, which uses Alg. 2 as a sub-algorithm.

Given $f \in \mathbb{Z}[x_1, \ldots, x_m]$, our goal is to heuristically certify the existence of at least one zero $(z_1, \ldots, z_m) \in {]0, \infty[}^m$ for which all coordinates are strictly positive. To start with, we evaluate $f(1, \ldots, 1) = f_1 \in \mathbb{R}$. If $f_1 = 0$, then we are done. If $f_1 < 0$, then it suffices by the intermediate value theorem to find $p \in {]0, \infty[}^m$ such that $f(p) > 0$. Similarly, if $f_1 > 0$ it suffices to find $p \in {]0, \infty[}^m$ such that $(-f)(p) > 0$. This algorithmically reduces our original problem to finding for given $g \in \mathbb{Z}[x_1, \ldots, x_m]$ at least one $p \in {]0, \infty[}^m$ such that $g(p) = f_2 > 0$.

We are going to accompany the description of our method with the example $g_0 = -2x_1^6 + x_1^3 x_2 - 3x_1^3 + 2x_1 x_2^2 \in \mathbb{Z}[x_1, x_2]$. Fig. 1 shows an implicit plot of this polynomial. In addition to its variety, g_0 has three sign invariant regions, one bounded one and two unbounded ones. One of the unbounded regions contains our initial test point $(1, 1)$, for which we find that $g_0(1, 1) = -2 < 0$. Thus our goal is to find one point $p \in {]0, \infty[}^2$ such that $g_0(p) > 0$.

Algorithm 1. pzerop

Input: $f \in \mathbb{Z}[x_1, \ldots, x_m]$

Output: One of the following:
(A) 1, which means that $f(1, \ldots, 1) = 0$.
(B) (π, ν), where $\nu = (p, f(p))$ and $\pi = (q, f(q))$ for $p, q \in]0, \infty[^m$, which means that $f(p) < 0 < f(q)$. Then there is a zero on $]0, \infty[^m$ by the intermediate value theorem.
(C) $+$, which means that f has been identified as positive definite on $]0, \infty[^m$. Then there is no zero on $]0, \infty[^m$.
(D) $-$, which means that f has been identified as negative definite on $]0, \infty[^m$. Then there is no zero on $]0, \infty[^m$.
(E) $\bot$, which means that this incomplete procedure failed.

```
 1  begin
 2  │   f₁ := f(1, …, 1)
 3  │   if f₁ = 0 then
 4  │   │   return 1
 5  │   else if f₁ < 0 then
 6  │   │   π := pzerop₁(f)
 7  │   │   ν := ((1, …, 1), f₁)
 8  │   │   if π ∈ {⊥, −} then
 9  │   │   │   return π
10  │   │   else
11  │   │   │   return (ν, π)
12  │   else
13  │   │   π := ((1, …, 1), f₁)
14  │   │   ν′ := pzerop₁(−f)
15  │   │   if ν′ = ⊥ then
16  │   │   │   return ⊥
17  │   │   else if ν′ = − then
18  │   │   │   return +
19  │   │   else
20  │   │   │   (p, f(p)) := ν′
21  │   │   │   ν := (p, −f(p))
22  │   │   │   return (ν, π)
```

Algorithm 2. pzerop$_1$

Input: $g \in \mathbb{Z}[x_1, \ldots, x_m]$

Output: One of the following:
(A) $\pi = (q, g(q))$, where $q \in]0, \infty[^m$ with $0 < g(q)$.
(B) $-$, which means that g has been identified as negative definite on $]0, \infty[^m$.
    Then there is no zero on $]0, \infty[^m$.
(C) $\bot$, which means that this incomplete procedure failed.

```
 1 begin
 2 |   F⁺ := { d ∈ frame(g) | sgn(d) = 1 }
 3 |   if F⁺ = ∅ then
 4 |   └   return −
 5 |   foreach (d₁,...,dₘ) ∈ F⁺ do
 6 |   |   L := {d₁n₁ + ··· + dₘnₘ − c = 0}
 7 |   |   foreach (e₁,...,eₘ) ∈ frame(g) \ F⁺ do
 8 |   |   └   L := L ∪ {e₁n₁ + ··· + eₘnₘ − c ≤ −1}
 9 |   |   if L is feasible with solution (n₁,...,nₘ,c) ∈ ℚ^{m+1} then
10 |   |   |   n := the principal denominator of n₁, ..., nₘ
11 |   |   |   (N₁,...,Nₘ) := (nn₁,...,nnₘ) ∈ ℤᵐ
12 |   |   |   ḡ := g[x₁ ← ω^{N₁},...,xₘ ← ω^{Nₘ}] ∈ ℤ(ω)
13 |   |   |   assert lc(ḡ) > 0 when using non-exact arithmetic in the LP solver
14 |   |   |   k := min{ k ∈ ℕ | ḡ(2ᵏ) > 0 }
15 |   |   └   return ((2^{kN1},...,2^{kNm}), ḡ(2ᵏ))
16 |   return ⊥
```

In the spirit of tropical geometry—and we refer to [19] as a standard reference with respect to its applications for polynomial system solving—we take an abstract view of

$$g = \sum_{d \in D} a_d x^d := \sum_{(d_1, \ldots, d_m) \in D} a_{d_1, \ldots, d_m} x_1^{d_1} \cdots x_m^{d_m}$$

as the set $\mathrm{frame}(g) = D \subseteq \mathbb{N}^m$ of all exponent vectors of the contained monomials. For each $d \in \mathrm{frame}(g)$ we are able to determine $\mathrm{sgn}(d) := \mathrm{sgn}(a_d) \in \{-1, 1\}$. The set of vertices of the convex hull of the frame is called the *Newton polytope* $\mathrm{newton}(g) \subseteq \mathrm{frame}(g)$. In fact, the existence of at least one point $d^* \in \mathrm{newton}(g)$ with $\mathrm{sgn}(d^*) = 1$ is sufficient for the existence of $p \in]0, \infty[^m$ with $g(p) > 0$.

In our example we have $\mathrm{frame}(g_0) = \{(6,0), (3,1), (3,0), (1,2)\}$ and $\mathrm{newton}(g_0) = \{(6,0), (3,0), (1,2)\} \subsetneq \mathrm{frame}(g_0)$. We are particularly interested in $d^* = (d_1^*, d_2^*) = (1, 2)$, which is the only point there with a positive sign as it corresponds to the monomial $2x_1 x_2^2$.

In order to understand this sufficient condition, we are now going to compute from d^* and g a suitable point p. We construct a hyperplane $H : n^T x = c$ containing d^* such that all other points of $\mathrm{newton}(g)$ are not contained in H

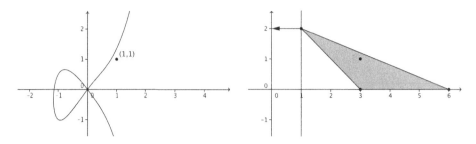

Fig. 1. We consider $g_0 = -2x_1^6 + x_1^3x_2 - 3x_1^3 + 2x_1x_2^2$. The left hand shows the variety $g_0 = 0$. The right hand side shows the frame, the Newton polytope, and a separating hyperplane for the positive monomial $2x_1x_2^2$ with its normal vector.

and lie on the same side of H. We choose the normal vector $n \in \mathbb{Q}^m$ such that it points into the half-space not containing the Newton polytope. The vector $c \in \mathbb{R}^m$ is such that $\frac{c}{|n|}$ is the offset of H from the origin in the direction of n. We may assume w.l.o.g. that $n \in \mathbb{Z}^m$.

In our example H is the line $x = 1$ given by $n = (-1, 0)$ and $c = -1$. Fig. 1 pictures the situation.

Considering the standard scalar product $\langle \cdot | \cdot \rangle$, it turns out that generally $\langle n | d^* \rangle = \max\{ \langle n | d \rangle \mid d \in \text{newton}(g) \}$, and that this maximum is strict. For the monomials of the original polynomial $g = \sum_{d \in D} a_d x^d$ and a new variable ω this observation translates via the following identity:

$$\bar{g} = g[x \leftarrow \omega^n] = \sum_{d \in D} a_d \omega^{\langle n | d \rangle} \in \mathbb{Z}(\omega).$$

Hence plugging into $\bar{g}$ a number $\beta \in \mathbb{R}$ corresponds to plugging into g the point $\beta^n \in \mathbb{R}^m$ and from our identity we see that in $\bar{g}$ the exponent $\langle n | d^* \rangle$ corresponding to our chosen point $d^* \in \text{newton}(g)$ dominates all other exponents so that for large β the sign of $\bar{g}(\beta) = g(\beta^n)$ equals the positive sign of the coefficient a_{d^*} of the corresponding monomial. To find a suitable β we successively compute $\bar{g}(2^k)$ for increasing $k \in \mathbb{N}$.

In our example we obtain $\bar{g} = 2\omega^{-1} - 2\omega^{-3} - 2\omega^{-6}$, we obtain $\bar{g}(1) = -2 \leq 0$, but already $\bar{g}(2) = \frac{23}{32} > 0$. In terms of the original g this corresponds to plugging in the point $p = 2^{(-1,0)} = (\frac{1}{2}, 1) \in]0, \infty[^2$.

It remains to be clarified how to construct the hyperplane H. Consider $\text{frame}(g) = \{ (d_{i1}, \ldots, d_{im}) \in \mathbb{N}^m \mid i \in \{1, \ldots, k\} \}$. If $\text{sgn}(d) = -1$ for all $d \in \text{frame}(g)$, then we know that g is negative definite on $]0, \infty[^m$. Otherwise assume without loss of generality that $\text{sgn}(d_{11}, \ldots, d_{1m}) = 1$. We write down the following linear program:

$$\begin{pmatrix} d_{11} & \ldots & d_{1m} & -1 \end{pmatrix} \cdot \begin{pmatrix} n_1 \\ \vdots \\ n_m \\ c \end{pmatrix} = 0, \quad I = \begin{pmatrix} d_{21} & \ldots & d_{2m} & -1 \\ \vdots & \ddots & \vdots & \vdots \\ d_{k1} & \ldots & d_{km} & -1 \end{pmatrix} \cdot \begin{pmatrix} n_1 \\ \vdots \\ n_m \\ c \end{pmatrix} \le -1.$$

Notice that in our system of inequalities we can use the LP-friendly conditions $I \leq -1$ in favor of the more natural conditions $I < 0$. Since the distance of the points $(d_{21}, \ldots, d_{2m}), \ldots, (d_{k1}, \ldots, d_{km})$ to the desired hyperplane H is scaled by $|(n_1, \ldots, n_m)|$, there is a sufficient degree of freedom in the choice of c in combination with $(n_1, \ldots, n_m)$ to achieve values smaller or equal to -1 in the feasible case. Our program is feasible if and only if $(d_{11}, \ldots, d_{1m}) \in \mathrm{newton}(g)$. In the negative case, we know that $(d_{11}, \ldots, d_{1m}) \in \mathrm{frame}(g) \setminus \mathrm{newton}(g)$, and we iterate with another $d \in \mathrm{frame}(g)$ with $\mathrm{sgn}(d) = 1$. If we finally fail on all such d, then our incomplete algorithm has failed. In the positive case, the solution provides a normal vector $n = (n_1, \ldots, n_m)$ and the offset c for a suitable hyperplane H. Our linear program can be solved using any standard LP solver. For our purposes here we have used Gurobi[2]; it turns out that the dual simplex of Glpsol[3] performs quite similarly on the input considered here.

For our example $g_0 = -2x_1^6 + x_1^3 x_2 - 3x_1^3 + 2x_1 x_2^2$, we generate the linear program

$$n_1 + 2n_2 - c = 0$$
$$6n_1 - c \leq -1$$
$$3n_1 + n_2 - c \leq -1$$
$$3n_1 - c \leq -1,$$

for which Gurobi computes the solution $n = (n_1, n_2) = (-0.5, 0)$, $c = -0.5$. Notice that the solutions obtained from the LP solvers are typically floats, which we lift to integer vectors by suitable rounding and gcd computations.

Note that we do not explicitly construct the convex hull $\mathrm{newton}(g)$ of $\mathrm{frame}(g)$ although there are advanced algorithms and implementations like QuickHull[4] available for this. Instead we favor a linear programming approach for several reasons. Firstly, we do not really need the quite comprehensive information, comprising, e.g. adjacency, obtained from such algorithms. For our purposes, it is rather sufficient to find one vertex with a positive sign the convex hull. Secondly, for the application discussed here it turns out that there typically exist only few (around 10%) such candidate points at all. Finally, it is known that for high dimensions the subset of $\mathrm{frame}(g)$ establishing vertices of the convex hull gets comparatively large. Practical experiments using QuickHull on our data support these theoretical considerations.

2.3 Summarizing the Algorithm for Checking the Existence of Hopf Bifurcations

Computing Hopf bifurcation fixed points for high-dimensional systems and systems with conservation laws had turned out to be difficult in practice. To overcome this difficulty for systems arising from chemical reaction networks we introduced in our previous paper [1] an algorithm based on using reaction coordinates

---

[2] www.gurobi.com

[3] www.gnu.org/software/glpk

[4] www.qhull.org

instead of concentration coordinates and applying real quantifier elimination for testing satisfiability. This enabled us to decide the occurrence of Hopf bifurcation in various chemical systems even with conservation laws. For some chemical networks with complex dynamics, however, it remained difficult to finally process the obtained quantified formulae with the currently available quantifier elimination packages; one hard example is the Methylene Blue Oscillator (MBO) discussed in the next section. To conclude this section, we are going to summarize our new efficient algorithmic approach for checking for Hopf bifurcation in complex chemical systems. Again using reaction coordinates, our approach here improves the previous one by simplifying the formulas expressing Hopf-existence conditions as shown in Subsection 2.1 and solving them by the method described in Subsection 2.2. The pre-processing step and the steps 2-6 presented in [1] remain the same. After computing the characteristic polynomial of each Jacobian matrix, we compute the $(n-1)^{\text{th}}$ Hurwitz determinant of the characteristic polynomial, and we apply Alg. 1 to check for positive solutions of the respective polynomial equations $\Delta_{n-1} = 0$. Alg. 3 outlines our new approach in an algorithmic fashion.

3   Algorithmic Determination of Hopf bifurcations in the Methylene Blue Oscillator System

As a complex example we consider the autocatalytic system *Methylen Blue Oscillator (MBO)*, which is defined by the following reaction equations:

$$MB^+ + HS^- \longrightarrow MB + HS$$
$$H_2O + MB + HS^- \longrightarrow MBH + HS + OH^-$$
$$HS + OH^- + MB^+ \longrightarrow MB + S + H_2O$$
$$H_2O + 2MB \longrightarrow MB^+ + MBH + OH^-$$
$$HS^- + O_2 \longrightarrow HS + O_2^-$$
$$HS + O_2 + OH^- \longrightarrow O_2^- + S + H_2O$$
$$H_2O + HS^- + O_2^- \longrightarrow HO_2^- + HS + OH^-$$
$$O_2^- + HS \longrightarrow HO_2^- + S$$
$$H_2O_2 + 2HS^- \longrightarrow 2HS + 2OH^-$$
$$MB + O_2 \longrightarrow MB^+ + O_2^-$$
$$HS^- + MB + H_2O_2 \longrightarrow MB^+ + HS + 2OH^-$$
$$OH^- + 2HS \longrightarrow HS^- + S + H_2O$$
$$MB + HS \longrightarrow MBH + S$$
$$H_2O + MBH + O_2^- \longrightarrow MB + H_2O_2 + OH^-$$
$$\longrightarrow O_2$$

Algorithm 3. Computing Hopf Bifurcations in Chemical Reaction Networks Using Reaction Coordinates

Input: A chemical reaction network $\mathcal{N}$ with $\dim(\mathcal{N}) = n$.

Output: (L_t, L_f, L_u) as follows: L_t is a list of subsystems containing a Hopf bifurcation, L_f is a list of subsystems in which its occurrence is excluded, and L_u is a list of subsystems for which the incomplete sub-procedure pzerop fails.

```
 1  begin
 2  │   Lt = ∅
 3  │   Lf = ∅
 4  │   Lu = ∅
 5  │   generate the stoichiometric matrix S and the kinetic matrix K of N
 6  │   compute the minimal set E of the vectors generating the flux cone
 7  │   for d = 1 ... n do
 8  │   │   compute all d-faces (subsystems) {Ni}i of the flux cone
 9  │   for each subsystem Ni do
10  │   │   compute from K, S the transformed Jacobian Jaci of Ni in terms of
    │   │   convex coordinates ji
11  │   │   if Jaci is singular then
12  │   │   │   compute the reduced manifold of Jaci calling the result also Jaci
13  │   │   compute the characteristic polynomial χi of Jaci
14  │   │   compute the (n − 1)th Hurwitz determinant Δn−1 of χi
15  │   │   compute Fi := pzerop(Δn−1(j, x)) using Algorithm 1
16  │   │   if Fi = 1 or Fi is of the form (π, ν) then
17  │   │   │   Lt := Lt ∪ {Ni}
18  │   │   else if Fi = + or Fi = − then
19  │   │   │   Lf := Lf ∪ {Ni}
20  │   │   else if Fi = ⊥ then
21  │   │   │   Lu := Lu ∪ {Ni}
22  │   return (Lt, Lf, Lu)
```

The MBO reaction system contains eleven species (not counting water) and fifteen reactions O_2, O_2^-, HS, MB^+, MB, MBH, HS^-, OH^-, S, H_2O_2 and HO_2^-. It may be reduced to a six dimensional system by considering only the essential species O_2, O_2^-, HS, MB^+, MB and MBH. The pre-processing step of our algorithm yields the following two matrices describing the reaction laws: stoichiometric matrix $\mathcal{S}$ and kinetic matrix $\mathcal{K}$.

$$\mathcal{S} = \begin{pmatrix} 1 & -1 & 1 & -2 & 0 & 0 & 0 & 0 & 0 & -1 & -1 & 0 & -1 & 1 & 0 \\ -1 & 0 & -1 & 1 & 0 & 0 & 0 & 0 & 0 & 1 & 1 & 0 & 0 & 0 & 0 \\ 1 & 1 & -1 & 0 & 1 & -1 & 1 & -1 & 2 & 0 & 1 & -2 & -1 & 0 & 0 \\ 0 & 1 & 0 & 1 & 0 & 0 & 0 & 0 & 0 & 0 & 0 & 0 & 1 & -1 & 0 \\ 0 & 0 & 0 & 0 & -1 & -1 & -1 & 0 & 0 & -1 & 0 & 0 & 0 & 0 & 1 \\ 0 & 0 & 0 & 0 & 1 & 1 & 0 & -1 & 0 & 1 & 0 & 0 & 0 & -1 & 0 \end{pmatrix}$$

$$\mathcal{K} = \begin{pmatrix} 0 & 1 & 0 & 2 & 0 & 0 & 0 & 0 & 0 & 1 & 1 & 0 & 1 & 0 & 0 \\ 1 & 0 & 1 & 0 & 0 & 0 & 0 & 0 & 0 & 0 & 0 & 0 & 0 & 0 & 0 \\ 0 & 0 & 1 & 0 & 0 & 1 & 0 & 1 & 0 & 0 & 0 & 2 & 1 & 0 & 0 \\ 0 & 0 & 0 & 0 & 0 & 0 & 0 & 0 & 0 & 0 & 0 & 0 & 0 & 1 & 0 \\ 0 & 0 & 0 & 0 & 1 & 1 & 1 & 0 & 0 & 1 & 0 & 0 & 0 & 0 & 0 \\ 0 & 0 & 0 & 0 & 0 & 0 & 0 & 1 & 0 & 0 & 0 & 0 & 0 & 1 & 0 \end{pmatrix}.$$

The flux cone of this Model is generated by 31 extreme currents. We tried to compute Hopf bifurcation in all subsystems involving 2-faces and 3-faces using our original approach described in [1], but the generated quantified formulae could not be solved by quantifier elimination, even with main memory up to 500 GB and computation times up to one week.

Using our new approach described here, in only 3% of the cases no definite answer could be obtained; in 67% of the cases it could be excluded that the resulting polynomial has a zero, whereas in 30% of the cases it could be verified that the resulting polynomial has a zero. Hence for at least 30% of the 2-faces there are Hopf bifurcations on these faces. Recall that the positive answer for at least one of the cases guarantees the existence of a Hopf bifurcation for the original system in spite of the fact that there are cases without definite answer.

The algorithmic test sketched in Section 2.2 can be parallelized easily for the different faces. Using 60 hyper-cores on a 2.4 GHz Intel Xeon E5-4640 running Debian Linux 64 bit the computation for all instances resulting from 2-faces could be completed in less than 90 seconds of wall clock time, which implies that also the worst-case computation time for the single instances has been at most 90 seconds.

Acknowledgements. This research was supported in part by *Deutsche Forschungsgemeinschaft* within SPP 1489 and by the German Transregional Collaborative Research Center SFB/TR 14 AVACS. Thomas Sturm would like to thank B. Barber for his support with QuickHull and convex hull computation and W. Hagemann and M. Košta for helpful discussions on linear programming aspects.

References

1. Errami, H., Seiler, W.M., Eiswirth, M., Weber, A.: Computing hopf bifurcations in chemical reaction networks using reaction coordinates. In: Gerdt, V.P., Koepf, W., Mayr, E.W., Vorozhtsov, E.V. (eds.) CASC 2012. LNCS, vol. 7442, pp. 84–97. Springer, Heidelberg (2012)
2. El Kahoui, M., Weber, A.: Deciding Hopf bifurcations by quantifier elimination in a software-component architecture. Journal of Symbolic Computation 30(2), 161–179 (2000)
3. Sturm, T., Weber, A., Abdel-Rahman, E.O., El Kahoui, M.: Investigating algebraic and logical algorithms to solve Hopf bifurcation problems in algebraic biology. Mathematics in Computer Science 2(3), 493–515 (2009)
4. Clarke, B.L.: Stability of Complex Reaction Networks. Advances in Chemical Physics, vol. XLIII. Wiley Online Library (1980)
5. Gatermann, K., Eiswirth, M., Sensse, A.: Toric ideals and graph theory to analyze Hopf bifurcations in mass action systems. Journal of Symbolic Computation 40(6), 1361–1382 (2005)
6. Guckenheimer, J., Holmes, P.: Nonlinear Oscillations, Dynamical Systems, and Bifurcations of Vector Fields. Applied Mathematical Sciences, vol. 42. Springer (1990)
7. Orlando, L.: Sul problema di hurwitz relativo alle parti reali delle radici di un'equazione algebrica. Mathematische Annalen 71(2), 233–245 (1911)
8. Gantmacher, F.R.: Application of the Theory of Matrices. Interscience Publishers, New York (1959)
9. Porter, B.: Stability Criteria for Linear Dynamical Systems. Academic Press, New York (1967)
10. Yu, P.: Closed-form conditions of bifurcation points for general differential equations. International Journal of Bifurcation and Chaos 15(4), 1467–1483 (2005)
11. Liu, W.M.: Criterion of Hopf bifurcations without using eigenvalues. Journal of Mathematical Analysis and Applications 182(1), 250–256 (1994)
12. Dolzmann, A., Sturm, T.: Redlog: Computer algebra meets computer logic. ACM SIGSAM Bulletin 31(2), 2–9 (1997)
13. Weispfenning, V.: Quantifier elimination for real algebra—the quadratic case and beyond. Applicable Algebra in Engineering Communication and Computing 8(2), 85–101 (1997)
14. Dolzmann, A., Sturm, T., Weispfenning, V.: Real quantifier elimination in practice. In: Matzat, B.H., Greuel, G.M., Hiss, G. (eds.) Algorithmic Algebra and Number Theory, pp. 221–247. Springer, Heidelberg (1998)
15. Dolzmann, A., Seidl, A., Sturm, T.: Efficient projection orders for CAD. In: Gutierrez, J. (ed.) Proceedings of the 2004 International Symposium on Symbolic and Algebraic Computation (ISSAC 2004), pp. 111–118. ACM Press, New York (2004)
16. Lorenz, E.N.: Deterministic nonperiodic flow. Journal of the Atmospheric Sciences 20(2), 130–141 (1963)
17. Rand, R.H., Armbruster, D.: Perturbation Methods, Bifurcation Theory and Computer Algebra. Applied Mathematical Sciences, vol. 65. Springer (1987)
18. Brown, C.W., El Kahoui, M., Novotni, D., Weber, A.: Algorithmic methods for investigating equilibria in epidemic modeling. Journal of Symbolic Computation 41(11), 1157–1173 (2006)
19. Sturmfels, B.: Solving Systems of Polynomial Equations. AMS, Providence (2002)

Highly Scalable Multiplication for Distributed Sparse Multivariate Polynomials on Many-Core Systems

Mickaël Gastineau and Jacques Laskar

IMCCE-CNRS UMR8028, Observatoire de Paris, UPMC
Astronomie et Systèmes Dynamiques
77 Avenue Denfert-Rochereau
75014 Paris, France
{gastineau,laskar}@imcce.fr

Abstract. We present a highly scalable algorithm for multiplying sparse multivariate polynomials represented in a distributed format. This algorithm targets not only the shared memory multicore computers, but also computers clusters or specialized hardware attached to a host computer, such as graphics processing units or many-core coprocessors. The scalability on the large number of cores is ensured by the lacks of synchronizations, locks and false-sharing during the main parallel step.

1 Introduction

Since the emergence of computers with multiple processors, and nowadays with several cores per processor, computer algebra systems have been trying to take advantage of such computational powers to reduce execution timings. As sparse multivariate polynomials are intensively present in many symbolic computation problems, the algorithms of the basic operations on these objects, such as multiplication, have been designed to use the available processors in workstations. These algorithms depend on the polynomial representation in main memory. The multivariate polynomials are usually stored in a distributed or recursive format. In the distributed format, a polynomial is a list of terms, each term being a tuple of a coefficient and an exponent. In the recursive form, a polynomial is considered as an univariate polynomial whose coefficients are polynomials in the remaining variables.

When the inputs are sparse multivariate polynomials, only the naive schoolbook product, that is the pairwise term products, is usually more optimal in practice than the asymptotically fast multiplication algorithms. Several parallel algorithms have been proposed for modern parallel hardware. An algorithm for the recursive representation has been designed using a work-stealing technique [1]. It scales at least up to 128 cores for large polynomials. Several algorithms [2], [3] have been designed for the distributed format for a parallel processing. The algorithm due to Monagan [2] uses binary heaps to merge and sort produced terms but its scalability depends on the kind of the operands. Indeed, dense

V.P. Gerdt et al. (Eds.): CASC 2013, LNCS 8136, pp. 100–115, 2013.
© Springer International Publishing Switzerland 2013

operands shows a super-linear scalability [1], [2]. If the number of cores becomes large, no improvement is observed because each thread should process at least a fixed number of terms of the input polynomials. This behavior is due to the fact that Monagan's algorithm benefits from the shared cache inside the processor. For sparse operands, a sub-linear speedup on a limited number of cores and a regression above these number of cores is observed but their algorithm only focuses on single processor computers with multiple cores sharing a large cache. The working threads have a private heap to sort their owned results and share a global heap for computing the output polynomial. The access by a thread to the global heap requires a lock statement to avoid a race condition. This lock statement avoids to obtain a good scalability on a large number of cores. The algorithm designed by Biscani [3] and implemented in the Piranha algebraic manipulator [4] has no limitation according to the number of cores. The work is split in closed intervals based on the hashed value of the operands' terms and pushed in a list of available tasks. Since the result of the different tasks may overlap, the access to the two lists of available and busy tasks is controlled by a mutual-exclusion lock to avoid a race condition. This single mutual-exclusion lock becomes a bottleneck when the number of cores becomes very large. Biscani's algorithm assumes that the cost of the access to the global memory for the result is the same for all the cores and does not depend on the memory location while two threads executed on two different processors may write successively the result to the same location.

Several new hardware processing units have appeared in the last decade, such as the multi-core processors in desktop computers or laptops, GPU with hundred or thousands of elementary processing units or specialized accelerators. In these different hybrid architectures, the memory access times depend on the memory location because each processor accesses faster to its own attached global memory. Cluster of nodes embedding all these different processors are available and may be used to perform the multiplication of sparse polynomials. We present a new algorithm for sparse distributed multivariate polynomials targeting these different architectures in Section 2. Our contribution resides in providing a natural lock-free algorithm using any available merge sort algorithm. This algorithm on sparse polynomials is the first one which targets the many-core hardware. Using a small specialization of one step inside this algorithm due the constraints of the different hardware, we adapt it to a cluster of computers in Section 3 and to specialized many-core hardware in Section 4. Benchmarks for these computers are presented in Section 6.

2   Algorithm on Shared Memory Computers

The designed algorithm should minimize the number of synchronizations or locks between threads in order to obtain a good scalability on many cores. Indeed, many synchronizations or locks are required only if different threads compute the terms of the result which have the same exponent. To avoid any lock or synchronization during the computations of resulting terms, a simple strategy is that each thread computes independent terms. Computation of independent terms is very easy if a recursive data structure for the polynomials is used, as shown in Wang [5] and Trip [6] but if a distributed form is used then this task is much more tricky. The proposed algorithm 1 requires two major steps. A preliminary step is required to split the work between threads to avoid any communication between the threads during the computational task.

Let c be the number of available cores and the same number of computational threads. Let us consider the polynomials in m variables $x_1, \ldots, x_m$,

$$A(\mathbf{x}) = \sum_{i=1}^{n_a} a_i \mathbf{x}^{\alpha_i} \text{ and } B(\mathbf{x}) = \sum_{j=1}^{n_b} b_j \mathbf{x}^{\beta_j}$$

where $\mathbf{x}$ corresponds to the variables $x_1, \ldots, x_m$, the a_i and b_j are numerical coefficients, and the m-dimensional integer vectors α_i and β_j are the exponents. These polynomials are stored in a sparse distributed format and their terms are sorted with a monomial order $\prec$.

The product P of A and B is the sum of the terms $P_{i,j} = a_i b_j \mathbf{x}^{\gamma_{i,j}}$ where $\gamma_{i,j} = \alpha_i + \beta_j$ for $i = 1 \ldots n_a$ and $j = 1 \ldots n_b$. We can construct the $n_a \times n_b$ matrix of the sum of exponents, called *pp-matrix* following Horowitz' denomination [7], to understand how the work is split between the threads. In fact, this matrix (Fig. 1) is never stored in memory during the execution of the algorithm due to its size.

$$\begin{bmatrix} \beta_1 & \cdots & \beta_j & \cdots & \beta_{n_b} \end{bmatrix}$$

$$\begin{bmatrix} \alpha_1 \\ \cdots \\ \alpha_i \\ \cdots \\ \alpha_{n_a} \end{bmatrix} \begin{bmatrix} \gamma_{1,1} & \cdots & \gamma_{1,j} & \cdots & \gamma_{1,n_b} \\ \cdots & \cdots & \cdots & \cdots & \cdots \\ \gamma_{i,1} & \cdots & \gamma_{i,j} & \cdots & \gamma_{i,n_b} \\ \cdots & \cdots & \cdots & \cdots & \cdots \\ \gamma_{n_a,1} & \cdots & \gamma_{n_a,j} & \cdots & \gamma_{n_a,n_b} \end{bmatrix}$$

Fig. 1. pp-matrix : matrix of the sum of exponents of $A \times B$

As each thread must compute independent terms, each thread must process all the pairwise term products $P_{i,j}$ of the *pp-matrix* which have the same value for $\gamma_{i,j}$. Since the possible values of all the $\gamma_{i,j}$ are in the interval $[\gamma_{1,1}, \gamma_{n_a,n_b}]$, this interval may be split into subintervals which are processed by the different threads. If this interval is split in equal subintervals, the load-balancing between the threads will be very poor. Several values $\gamma_{i,j}$ of the *pp-matrix* are selected and used to split into subintervals in order to obtain a better load-balancing. Since a simple and fast method to select these values does not guarantee that duplicated values are not selected, duplicated selected values are removed and the remaining values $\gamma_{i,j}$ are the bounds of the subintervals. Left-closed, right-open subintervals are required to guarantee that all the $\gamma_{i,j}$, which have the value of one of the bounds of the interval, remain in a single subinterval. The exponent γ_{end} is introduced in order to have a same interval type (left-closed, right-open) for the last subinterval which must contain the value γ_{n_a,n_b} and γ_{end} is any exponent greater than γ_{n_a,n_b} according to the monomial order.

The algorithm begins with the construction of the set $S^\star$ which consists of the selection of $n_{s^\star}$ values (or exponents) inside Γ. The selection method must always select the exponents $\gamma_{1,1}$ and γ_{end} in order to be sure that all pairwise term products will be processed in the next step of the algorithm. $n_{s^\star}$ needs to be provided as an input of the algorithm and its value must remain very small in front of the size of Γ since the first step of the algorithm needs to be fast. The way to select the $n_{s^\star}$ exponents will be discussed in Section 5. This set $S^\star$ is then sorted according to the monomial order and its duplicate values are removed in order to obtain the new subset S. The number of elements of the set S is noted n_s. This first step is very fast and could be computed by a single thread. The monomial order has only an effect on the number of elements n_s as the number of duplicate values of the set $S^\star$ may change according to the monomial order. If an almost regular grid is used, then the generation of the set $S^\star$ requires to read $2\sqrt{n_s^\star}$ exponents from the main memory and write $n_s^\star$ exponents to it. The sorting step takes $\mathcal{O}(n_s^\star \log n_s^\star)$ time on average using the sequential quicksort algorithm. Of course, this step may be parallelized using a parallel sorting algorithm. If we define the set $\Gamma = \{\gamma_{i,j} \mid 1 \leq i \leq n_a \text{ and } 1 \leq j \leq n_b\} \bigcup \{\gamma_{end}\}$, then the first step produces the following set

$$S = \{S_k \mid 1 \leq k \leq n_s \text{ and } S_k \in \Gamma\} \text{ with } S_1 = \gamma_{1,1}, S_{n_s} = \gamma_{end} \text{ and } S_k < S_{k+1}$$

After this preliminary step, every $\gamma_{i,j}$ could be located inside a single interval $[S_k, S_{k+1}[$ and all the $\gamma_{i,j}$ with the same value are located in the same interval. The threads may process the $n_s - 1$ intervals of exponents at the same time since they compute independent terms of the result. Indeed, if a thread processes the interval $[S_k, S_{k+1}[$, it computes the summation of the selected terms $P_{i,j}$ such that $S_k \leq \gamma_{i,j} < S_{k+1}$. So the second step consists in computing the resulting terms using a parallel loop over all the $n_s - 1$ intervals. As each interval may have different execution times, due to a variable amount of $P_{i,j}$ involved, the work should be balanced between the cores using a number of intervals greater than the number of cores. The load-balancing may be done using a dynamic scheduling, such as work-stealing [8], which does not require any bottleneck synchronization. Even if the number of intervals varies according to the chosen monomial order, the monomial order does not impact the complexity of the algorithm as the number of intervals has only an effect on the quality of the load-balancing.

During this second step, each thread needs to check if the entry $\gamma_{i,j}$ of the *pp-matrix* is included inside its own current interval in order to process it or not. If the thread checks each entry, each thread will perform $n_a n_b$ comparisons which are very inefficient. As the *pp-matrix* has an ordered structure, as $\gamma_{i,j} < \gamma_{i+1,j}$ and $\gamma_{i,j} < \gamma_{i,j+1}$, this property may be exploited to find efficiently the necessary entries of the intervals. For each line of this matrix, only the location of the first and last element, which corresponds to the first and last exponent processed by the thread, should be determined. So each thread needs to find the edge of the area of terms that it should process for the current interval. This edge consists

---

Algorithm 1. $\mathrm{mul}(A, B, n_{s^\star})$. Return $A \times B$ using at most $n_s^\star$ intervals

---

Input: $A = \sum_{i=1}^{n_a} a_i \mathbf{x}^{\alpha_i}$

Input: $B = \sum_{j=1}^{n_b} b_j \mathbf{x}^{\beta_j}$

Input: $n_{s^\star}$: integer number of intervals

Input: monomial order $\prec$

Output: $C = \sum_{k=1}^{n_c} c_k \mathbf{x}^{\gamma_k}$

   // First step

1 $S^\star \leftarrow$ Compute $n_{s^\star}$ exponents $\gamma_{i,j} = \alpha_i + \beta_j$ using an almost regular grid over the *pp-matrix* associated to A and B

2 $S \leftarrow$ sort $S^\star$ using the monomial order $\prec$

3 remove duplicate values from S

   // S has now n_s sorted elements

   // Second step

4 Initialize an array D of n_s empty containers for the result

5 **for** $k \leftarrow 1$ **to** $n_s - 1$ **do in parallel**

6     $(L_{min}, L_{max}) \leftarrow$ FindEdge (A, B, S_k, S_{k+1})

7     $D_k \leftarrow$ MergeSort (A, B, L_{min}, L_{max})

8 **end**

9 $C \leftarrow$ concatenate all containers of D using ascending order

---

of two lines, which corresponds to the first exponents and last exponents on each line, as shown in the figure 2(b). This work is done by the function FindEdge. This algorithm consists in storing the location of the first, respectively last, column j where $S_k \leq \gamma_{i,j}$, respectively $\gamma_{i,j} < S_{k+1}$, in two arrays L_{min} and L_{max} of size n_a. Its time complexity is $\mathcal{O}(n_a + n_b)$ because, when the thread processes the line $i+1$ of the matrix, it does not start at the column 1 but at the found column in the previous line i, as $\gamma_{i,j} < \gamma_{i+1,j}$. Using the ideal distributed cache model [9], the computation of L_{min} and L_{max} incurs $((\mathcal{E} + 2)n_a + \mathcal{E}n_b)/\mathcal{L}$ cache misses in the worst case if each core has a private cache of $\mathcal{Z}$ words (*cache size*) partitioned into cache lines of $\mathcal{L}$ words and if each exponent of the polynomial is stored on $\mathcal{E}$ words. $n_{s^\star}$ needs to be kept small because each thread will have to process $(n_a + n_b)(n_s - 1)/c$ exponents to compute these arrays if the work is well balanced. As each thread has its own arrays L_{min} and L_{max}, the additional memory usage requirement for these arrays is only $2n_a c$ integers during the second step. However, the storage of these arrays is not required if it is possible to combine this function with the function MergeSort but this depends on the algorithm used in that function.

Using its own arrays L_{min} and L_{max}, each thread computes the summation of its own terms $P_{i,j} = a_i b_j \mathbf{x}^{\gamma_{i,j}}$ using any sequential comparison-based sorting algorithm (function MergeSort in the algorithm) and store them in a container D_k associated to the corresponding interval. No concurrent writing or reading access occurs to the same container because threads need to read or write data only about their own current interval. Johnson proposes a sequential algorithm [10] which computes the result using a binary heap. If the multiplication produces

$\mathcal{O}(n_a + n_b)$ terms, only $\mathcal{O}(n_a n_b \log \min(n_a, n_b))$ comparisons of exponents are required. Monagan and Pearce have improved this algorithm with a chained heap [11]. When all threads have finished to process all the intervals, a simple concatenation of the containers is performed to obtain the canonical form of the polynomial as the containers of D are already sorted according to the sorted intervals.

The Monagan's and Biscani's algorithms use a global read-write container which prevents from having a good scalability on a large number of cores. While our algorithm does not suffer from such limitation, Monagan's algorithm always benefits from the cache effect for the private heap of each thread whereas our algorithm benefits from this cache effect only if the used comparison-based sorting method has that property. Indeed, the size of their private heap is fitted for the cache size of the processor. The time complexity of our algorithm is largely dominated by the number of exponents comparisons inside the sorting method and by the coefficient's multiplication. Other steps have a linear complexity with the number of terms of the input polynomials and a linearithmic complexity with the value n_{s^*}.

3   Adaptation to Computer Cluster

As the second step could be computed in independent parallel tasks, our algorithm could be easily adapted to a cluster of computational nodes. Cluster of computer nodes offers a distributed memory architecture where the access time to the memory located on the other nodes is several magnitude order greater than the access to the local memory. A message passing paradigm, such as MPI standard, should be used to perform the communications between the nodes. But a pure MPI application does not take advantage of the multiple cores available inside a node. An hybrid (multi-threading+MPI) approach must be used in order to reduce the cost of the communication and to improve the parallel scheduling of the second step of the previous algorithm. We assume that the operands are located on a single node and the result should be stored on this node. So the operands should be broadcast to the other nodes. If the operands are located on different nodes, each node broadcasts its content of the operands to the other nodes. A simple parallel scheduling could use the master-slave paradigm where a node is dedicated to be the master and other nodes request intervals to this master node, process the intervals and send the result to the master. Good load-balancing in this context requires to have many intervals which involve many communications. Furthermore, the result may generate large messages which require to use the Rendezvous protocol and imply a waiting for the other slaves.

In order to limit the number of communications, a node should process consecutive intervals and send a single result for all these sorted intervals. Inside this node, the same parallel scheduling as in the shared memory context may be chosen to distribute the work between the threads. To reduce to the minimal number of communications, the number of group of consecutive intervals is chosen to be equal to the number of nodes. But an extra step is introduced to

---

Algorithm 2. $\mathrm{mul}(A, B, n_{s^\star})$. Return $A \times B$ using at most $n_s^\star$ intervals on a cluster of N computer nodes.

Input: $A = \sum_{i=1}^{n_a} a_i \mathbf{x}^{\alpha_i}$
Input: $B = \sum_{j=1}^{n_b} b_j \mathbf{x}^{\beta_j}$
Input: $n_{s^\star}$: integer number of intervals
Input: monomial order $\prec$
Output: $C = \sum_{k=1}^{n_c} c_k \mathbf{x}^{\gamma_k}$

---

node 0	node $1 \ldots N - 1$

Perform step 1 of Alg. 1

Broadcast A, B, n_s and S $\qquad\qquad \Longrightarrow$ Receive A, B, n_s and S

```
                    // do in parallel on all nodes
```
$(L_{min}, L_{max}) \leftarrow$ FindEdge (A, B, S_k, S_{k+1}) for $k = 1 \ldots n_s$
$O_k \leftarrow$ number of operations for $[S_k, S_{k+1}[$ from (L_{min}, L_{max})

Gather O_k from all nodes $\qquad\qquad \Longleftarrow$ Send O_k
Split S in N consecutive intervals $[S_{l_1}, S_{l_2}[$
using the cumulative summation of O_k
Send the N tuples l_1, l_2 $\qquad\qquad \Longrightarrow$ Receive l_1, l_2

```
                    // similar to the step 2 of Alg. 1
```
Initialize an array D $\qquad\qquad\qquad\qquad$ Initialize an array $D_{l_1 \ldots l_2}$
```
                    // do in parallel on all nodes
```
for $k \leftarrow l_1$ **to** l_2 **do in parallel**
  $\quad | \quad (L_{min}, L_{max}) \leftarrow$ FindEdge (A, B, S_k, S_{k+1})
  $\quad | \quad D_k \leftarrow$ MergeSort (A, B, L_{min}, L_{max})
end

Gather D_k from all nodes $\qquad\qquad \Longleftarrow$ Send $D_{l_1 \ldots l_2}$
$C \leftarrow$ concatenate all containers of D

---

perform a good load-balancing between all the nodes. This extra step requires to compute the number of multiplications or operations required to process each interval. The cumulative summation of the number of operations is performed in order to create the group of consecutive intervals with almost the same number of operations. The master node will also compute a part of the result. Other nodes send their results back to the master node. The algorithm 2 shows the processing steps required to perform the multiplication on the cluster of nodes.

4   Adaptation to Specialized Many-Core Hardware

GPU and other dedicated cards are able to perform general-purpose computations but they have dedicated memory. So the same adaptation as for the cluster of computers is done for the data transfer between the host memory and the GPU memory. The scheduling is easier than on the cluster since it can be done by the processor of the host computer. The values O_k are not computed to perform the scheduling. If several many-core hardware are connected to the host computer, only the bounds l_1 and l_2 are sent to the different cards. As the memory transfer may be expensive, host processor may compute other data, e.g. put the result in a canonical form, in order to overlap the memory communication.

Available specialized many-core hardware are able to schedule a large number of threads at the same time but they only have a small cache shared among a group of threads. An interval is processed by a group of threads in order to benefit from this shared memory and to avoid divergence of the execution path in the threads inside the group. As the arrays L_{min} and L_{max} cannot be stored on the device memory due to the large number of groups, the computation of these arrays must be merged with the function MergeSort. The available sequential comparison-based sorting algorithms designed for the CPU are not well adapted to these hardware constraints. The parallel sorting algorithm among the group of threads depends too much on the targeted hardware to be designed generically.

5   Choice of the Set $S^\star$

The set $S^\star$ should be chosen in order to balance the work as better as possible between the intervals, even if a perfect work balancing is impossible without computing all the elements of the *pp-matrix*. The choice of the elements of $S^\star$ could be done using an almost regular grid over the *pp-matrix*. This method is very fast and very simple to implement. In order to obtain the fixed $n_{s^\star}$ elements, our grid is defined using the following rules.

$$S_k^\star = \begin{cases} \alpha_i + \beta_j & \text{for } i = 1 \text{ to } n_a \text{ step } \lfloor \frac{n_a}{l} \rfloor, \\ & \text{and for } j = j_{0,i} \text{ to } n_b \text{ step } \lfloor \frac{n_b}{l} \rfloor \\ \alpha_{n_a} + \beta_j & \text{for } j = 1 \text{ to } n_b \text{ step } \lfloor \frac{n_b}{l} \rfloor \\ \alpha_i + \beta_{n_b} & \text{for } i = 1 \text{ to } n_a \text{ step } \lfloor \frac{n_a}{l} \rfloor \end{cases}$$

$$\text{with} \quad \begin{cases} n_{s^\star} = (l+1)^2 \\ j_{0,i} = 1 + \left((i/\frac{n_a}{l}) \mod 2 \right) \lfloor \frac{n_b}{2l} \rfloor \end{cases}$$

The value $\lfloor \frac{n_a}{l} \rfloor$ and $\lfloor \frac{n_b}{l} \rfloor$ are the distances between two selected points on the same line or column in the *pp-mtarix*. The value $j_{0,i}$ is used to avoid the effect of the large strip. Due to the integer division, some elements in the last column and in the last line are selected to avoid a too large strip in the last part of the matrix. Figure 2(a) shows an example of a computed grid and the figure 2(b) shows the edge of the intervals computed by the different threads. Other sort of grids may be used instead of our selected grid but they have insignificant impact on the performance of the algorithm. For example, the *pp-matrix* may be divided in $n_{s^\star}$ submatrices and a random exponent may be chosen inside each submatrix. Instead of selecting equidistant points on the line i of our grid, non-equidistant points may be selected on the line i to generate another sort of grid. We have tested these grids and the differences of the execution time of the

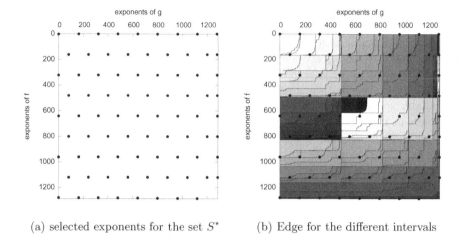

(a) selected exponents for the set $S^\star$ (b) Edge for the different intervals

Fig. 2. Grid and intervals computed for the multiplication of $f = (1 + x + y + 2z^2 + 3t^3 + 5u^5)^8$ by $g = (1 + u + t + 2z^2 + 3y^3 + 5x^5)^8$ for $l = 8$ or $n_{s^\star} = 81$

product are less than 1.5% on a multicore multiprocessor computer using these grids.

To achieve maximal performance, the value $n_{s^\star}$ or l should be chosen dynamically according to the number of available cores and/or to the number of terms of the polynomials. As $n_{s^\star}$ should remain small in order to reduce the time spent in the first step and in the function FindEdge of the algorithm, the parameter should be fitted only to the number of available cores. The parameter l is preferred instead of $n_{s^\star}$ for the tuning because a simple linear variation on this parameter is possible. Its value must be tuned only once, for example at the installation of the software. Of course, for small polynomials, the tuned value l may be too large and must be reduced in order to have enough work for each thread.

6   Benchmarks

Three examples are selected to test the implementation of our algorithm. The two first examples are due to Fateman in [12] and Monagan and Pearce in [2].

- Example 1 : $f_1 \times g_1$ with $f_1 = (1 + x + y + z + t)^{40}$ and $g_1 = f_1 + 1$. This example is very dense. f_1 and g_1 have 135751 terms and the result contains 1929501 terms.
- Example 2 : $f_2 \times g_2$ with $f = (1 + x + y + 2z^2 + 3t^3 + 5u^5)^{25}$ and $g_2 = (1 + u + t + 2z^2 + 3y^3 + 5x^5)^{25}$. As shown in [2] and [1], a linear speedup is quite difficult to obtain on this very sparse example. f_2 and g_2 have 142506 terms and the result contains 312855140 terms.

– Example 3 : $f_3 \times g_3$ with $f_3 = (1 + u^2 + v + w^2 + x - y^2)^{28}$ and $g_3 = (1 + u + v^2 + w + x^2 + y^3)^{28} + 1$. f_3 and g_3 have 237336 terms and the result contains 144049555 terms.

The scalability of our algorithm depends on the number of intervals n_{s^*}, the size of operands (n_a, n_b), and the number of cores (c) available on the computer. We have implemented two kinds of MergeSort algorithms for the parallel step to show its independence with respect to this algorithm. As in Monagan and Pearce, a chained heap algorithm is implemented to perform the summation of the terms but it does not include any lock as the binary heap is accessed only by one thread. This algorithm is noted *heap* in the tables and figures. The second sorter algorithm, noted *tree*, uses a tree data structure in which each internal node has exactly 16 children. At each level of this tree, four bits of the exponents are used to index the next children. If the exponents are encoded on 2^d bits, our tree will have 2^{d-2} levels. The tree associated to each interval is converted to a distributed representation at the end of the algorithm in order to obtain a canonical distributed form of the polynomial. This container uses more memory but its complexity to insert all the elements is only in $\mathcal{O}(2^{d-2}n_an_b)$. This practical complexity is better if the exponents are packed since many inserted terms have common bits inside their exponents. The exponents of the polynomials are packed on a 64-bit unsigned integer in the implemented algorithm.

To fit the value l, and so n_{s^*}, on the available hardware, we generate randomly two sets of 280 sparse polynomials in several variables with different numbers of terms. The number of variables of these polynomials is from 4 to 8 and the number of terms varies from 10000 to 60000 terms. The products of the two sets of polynomials are performed with different values of l. An histogram is built with the values of l, whose time of execution does not differ more than 10% from the best time for each product.

6.1   Shared Memory Multiprocessors

As processors with multiple cores are now widely available in any computer, the three examples are executed on a computer with 8 Intel Xeon processors X7560 running at 2.27Ghz under the Linux operating system. Each processor has 8 physical cores sharing 24 Mbytes of cache. This computer has a total of 256Gbytes of RAM shared by its 64 cores. The parallel dynamic scheduling of the second step of the algorithm is performed by the OpenMP API [13] of the compiler. As the memory management could be a bottleneck in a multi-threading multiplication of sparse polynomials, the memory management is performed by the Intel threading building blocks library [14], noted *TBB*, which provides a scalable allocator instead of the operating system C library.

The first step has been to tune the parameter n_{s^*} or l on this hardware. Figure 3 shows the histogram of the number of best execution time according to the parameter l using the *heap* algorithm. For small value of l, not enough parallelism is provided to get good execution time. We fix the parameter l to 64 in order to perform the benchmarks on this computer.

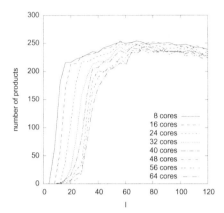

Fig. 3. Number of products of the two sets of 280 randomly generated sparse multivariate polynomials using different values l, whose execution time does not differ more than 10% from the best time

Our algorithm, noted *DMPMC*, is compared to the computer algebra systems Maple 16 [15], Piranha [4] and Trip [6]. In all these software excepted Piranha, the coefficients of the polynomials are represented with integers using a mixed representation. For the integers smaller than $2^{63} - 1$ on 64-bit computers, hardware integers are used instead of integers' type of the GMP library [16]. The multiplication and additions of the terms use a three word-sized integers accumulator (a total of 192 bits) for the small integers. The same optimization is used in Maple [11] and Trip. The timings for Maple 16 are the timings reported by the multiplication step of the SDMP which excludes the DAG reconstruction of the polynomial. Piranha uses only the GMP integers and allocates memory with the same scalable memory allocator *TBB*. Two times are reported for Trip. The *dense* time is for the optimized dense recursive polynomial data structure (POLPV) and the *sparse* time is for the optimized sparse recursive polynomial data structure (POLYV).

Table 1 shows the execution times of the three examples on the 64 cores computer. Even if our chained heap is less tuned for the dense polynomials on single core, our algorithm for distributed representation scales with the same behavior as the recursive algorithms of Trip. We define the speedup as T_1/T_c and the efficiency as $T_1/(c \times T_c)$ where T_c is the execution time on c cores. Figure 4 shows the speedup for the different implementations and confirms that the SDMP algorithm of Maple 16 [15] focuses only on a single multi-core processor with large shared memory cache. Indeed, the efficiency of the SDMP algorithm drops to less than 0.6 above 16 cores for the three examples while the efficiency of our algorithm remains above 0.7 on 64 cores. The limited scalability of Piranha is confirmed due to the access to its two global lists shared between all the threads. The kind of MergeSort algorithm has only a significant impact on the execution time but not on the scalability of our algorithm. The small differences observed on the speedup between the *tree* and *heap* algorithm come

Table 1. Execution time in seconds of the examples on the shared memory computer with integer coefficients. DMPMC uses the tuned parameter $l = 64$ or $n_{s^*} = 4225$.

Software		example 1			example 2			example 3		
		cores			cores			cores		
		1	16	64	1	16	64	1	16	64
DMPMC	heap	1843.2	116.2	32.3	1317	83.8	23.9	2081	126.1	35.0
	tree	878.5	55.4	16.1	1394	90.4	30.2	1632	102.6	29.7
Maple 16		1226.7	358.5	262.4	1364	625.0	900.5	3070	317.4	609.9
Piranha		677.5	57.0	45.9	1576	138.4	174.6	2466	826.8	816.4
Trip 1.2	dense	649.9	40.3	10.4	1227	75.6	19.8	2738	164.9	42.9
	sparse	705.5	43.7	11.5	1071	65.6	19.9	2874	177.7	45.8

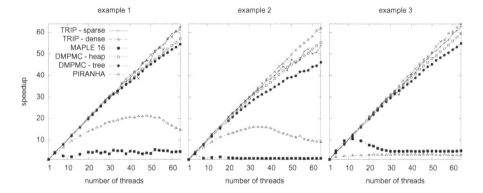

Fig. 4. Speedup on the shared memory computer with integer coefficients' polynomial

from the different number of memory allocations since the *tree* version requires more memory allocation. Although we have used a sequential implementation of the first step, its duration remains insignificant. The computation of the edge (function FindEdge) by each thread takes only a few percent of the total time.

6.2 Distributed Memory Computers

In this second set of experiments, algorithm 2 is implemented using the hybrid approach OpenMP+MPI on the MesoPSL cluster with 64 nodes interconnected with a QDR InfiniBand network for a total of 1024 cores. Each node embeds two Intel E5-2670 processors sharing a total of 64 Gbytes of RAM between the 16 cores of the node. Quadruple precision floating point numbers have been used for the polynomials coefficients instead of variable size integers coefficients in order to simplify the exchange of the coefficients between the nodes. Figure 6.2 shows the speedup of the algorithm on this cluster with the tuned parameter $l = 8\sqrt{c}$ where c is the total number of cores. The speedup is defined as $T_{1,\mathrm{OpenMP}}/T_{n,\mathrm{MPI}}$ where $T_{1,\mathrm{OpenMP}}$ is the execution time on one core of a single node using the OpenMP implementation of the algorithm 1 and $T_{n,\mathrm{MPI}}$ is the execution time on n nodes using the hybrid OpenMP+MPI implementation of algorithm 2. The limitation of the 2 GB maximum message size in the MPI implementation of the cluster requires to implement a custom gather operation using the send/receive messages in order to collect the result on the root node which have significant impact on the timings. This bottleneck is especially visible on the large result of the second example which contains more than 300 million terms since half of the time is spent to transfer the result from the slave nodes to the master node. However, the algorithm always continues to scale according to the number of available cores. In all cases, the algorithm scales well up to at least two hundred cores. Similar behaviors are obtained if double precision coefficients are used instead of quadruple precision coefficients.

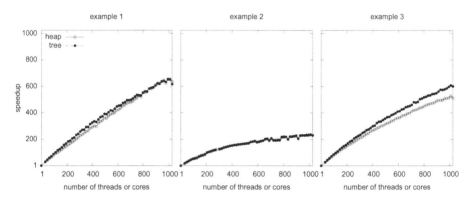

Fig. 5. Speedup of the DMPMC algorithm on a cluster of 64 nodes (a total of 1024 cores) interconnected with a QDR InfiniBand network. The coefficients of polynomials are quadruple precision floating point numbers.

6.3   Specialized Many-Core Hardware

To test the algorithm on a many-core hardware, the benchmarks are performed on a Nvidia Tesla S2050 computing System based on the Fermi architecture interconnected through two links to a host computer using a PCI Express 2 16x controller. The host computer is the same computer as for the shared memory benchmark. The Nvidia Tesla S2050 consists of four Fermi graphics processing units. So two GPU cards share the same links to the host controller. Each GPU embeds 14 streaming multiprocessors and 3 GBytes DDR5 memory. Each streaming processor has 32 cores running at 1.15 Ghz and is able to schedule two groups of 32 threads simultaneously. A total of 448 cores is available per GPU. The kernel function running on the GPU is implemented using the CUDA programming model [17].

Table 2. Execution time in seconds and speedup of the examples on the Nvidia Tesla S2050 with double precision floating-point coefficients. Each example is executed with a different number of threads in a group with a tuned value of the parameter l for the corresponding group.

group of t threads	l	example 1				example 2				example 3			
			gpu				gpu				gpu		
		1	2	3	4	1	2	3	4	1	2	3	4
32	90	42	24	19	16	98	61	43	36	181	103	83	67
			1.7x	2.2x	2.5x		1.6x	2.2x	2.7x		1.8x	2.2x	2.7x
64	90	32	18	14	12	71	42	30	25	129	71	55	44
			1.7x	2.3x	2.7x		1.7x	2.3x	2.8x		1.8x	2.3x	2.9x
128	60	35	20	15	13	73	42	31	25	141	79	63	49
			1.7x	2.3x	2.7x		1.7x	2.4x	2.8x		1.8x	2.2x	2.9x
256	40	40	22	17	14	79	43	32	26	149	91	68	57
			1.8x	2.3x	2.7x		1.8x	2.5x	3x		1.6x	2.2x	2.6x
512	25	48	29	23	20	93	52	38	31	174	106	77	67
			1.6x	2x	2.4x		1.8x	2.5x	2.9x		1.6x	2.3x	2.6x

Table 3. Execution time in seconds of the examples on the shared memory computer with double precision floating-point coefficients. DMPMC uses the tuned parameter $l = 64$ or $n_{s^*} = 4225$.

Software		example 1			example 2			example 3		
		cores			cores			cores		
		8	16	32	8	16	32	8	16	32
DMPMC	tree	44.7	22.6	11.4	69.9	35.6	18.1	167.5	84.5	42.6
Piranha		8.6	6.5	20.0	62.8	43.2	59.5	134.4	128.9	166.9
Trip 1.2	sparse	11.49	5.8	3.0	30.9	15.4	7.9	86.6	43.5	22.3

Only the *tree* algorithm for the MergeSort function is implemented since the *heap* version is not well adapted for the GPU architecture. Indeed, an interval is not processed by a single thread but it is processed by a group of t threads, called a *block* in the CUDA terminology. So for each interval, a group of threads constructs a temporary tree in order to merge and sort the terms. To avoid the divergence of the execution path of the threads inside a group, this one handles the terms line after line. A cache blocking technique is used to process the input data in order to minimize the access to the global memory. We try to minimize the global memory access to the input polynomials but not for the output polynomial. Indeed, at the same time, only t exponents and coefficients of the polynomial A and t elements of the array L_{min} are stored in the shared memory of the GPU. So $\lceil n_a/t \rceil$ coalesced global memory accesses are required to process the polynomial A. The construction of tree requires an explicit synchronization between threads for the allocation of the tree nodes. In the worst case, each group of threads performs $\mathcal{O}(l(n_a + n_b)/t)$ global memory accesses to insert the elements in the tree. Due to the shared memory limits, more investigations are required in order to reduce this number of global transactions and tune the implementation of the merge sort algorithm.

In our implementation, the reconstruction of the canonical distributed representation from the tree is performed on the host processor with one host thread per GPU and overlaps the computations by the GPU. A simple static scheduling is performed : $\frac{n_s}{5g}$ intervals are processed at a same time by a GPU if g GPUs are used for the computations. So each GPU executes 5 times the kernel function. More sophisticated scheduling may be done by overlapping memory transfer between the host and GPU memory. As no version of the GMP library is available for the GPU side, double-precision floating point numbers have been used for the coefficients on CPU and GPU. Table 2 shows the execution time and speedup obtained on the GPU with different number of threads (t) inside the group. As the kernel function needs to be transferred on the card by the CUDA driver, the timings are the average of eight executions without two first useless executions. The group of 64 threads has the best execution time for the three examples. The scalability on several GPUs is not linear for several reasons. The major reason is that the four cards share the two links to the host. A better dynamic scheduling should be done using an optimized heterogeneous scheduler, such as StarPU [18], and the computations on the card are not overlapped by the memory transfer. For comparison, the execution time of Trip, Piranha and DMPMC on the same host computer are reported in Table 3 with the same kind of numerical coefficients.

7   Conclusions

The presented algorithm for the multiplication of sparse multivariate polynomials stored in a distributed format does not have any bottleneck related to the numbers of cores due to the lack of synchronization or locks during the main parallel step. But it requires a preliminary one time step to tune the size of the grid to the targeted hardware. The range of targeted processor units is wide for our algorithm. The only drawback comes from the time to transfer data between nodes on the distributed memory systems due to the limited performance of the interconnection network. It can use any available fastest sequential merge and sort algorithm to generate the terms of the result and can benefit from any efficient dynamic scheduling. A more appropriate algorithm for this merge and sort step may be designed for the GPU hardware in order to take into account all features of these specialized hardware.

Acknowledgements. The authors thank the computing centre MesoPSL of the PSL Research University for providing the necessary computational resources for this work.

References

1. Gastineau, M.: Parallel operations of sparse polynomials on multicores: I. multiplication and poisson bracket. In: Moreno Maza, M., Roch, J.L. (eds.) PASCO 2010: Proceedings of the 4th International Workshop on Parallel and Symbolic Computation, pp. 44–52. ACM, New York (2010)
2. Monagan, M., Pearce, R.: Parallel sparse polynomial multiplication using heaps. In: Johnson, J., Park, H., Kaltofen, E. (eds.) ISSAC 2009: Proceedings of the 2009 International Symposium on Symbolic and Algebraic Computation, pp. 263–270. ACM, New York (2009)
3. Biscani, F.: Parallel sparse polynomial multiplication on modern hardware architectures. In: van der Hoeven, J., van Hoeij, M. (eds.) Proceedings of the 37th International Symposium on Symbolic and Algebraic Computation. ISSAC 2012, pp. 83–90. ACM, New York (2012)
4. Biscani, F.: Design and implementation of a modern algebraic manipulator for Celestial Mechanics. PhD thesis, Centro Interdipartimentale Studi e Attivita Spaziali,Universita degli Studi di Padova, Padova (May 2008)
5. Wang, P.S.: Parallel polynomial operations on smps: an overview. J. Symb. Comput. 21(4-6), 397–410 (1996)
6. Gastineau, M., Laskar, J.: Trip: a computer algebra system dedicated to celestial mechanics and perturbation series. ACM Commun. Comput. Algebra 44(3/4), 194–197 (2011)
7. Horowitz, E.: A sorting algorithm for polynomial multiplication. J. ACM 22(4), 450–462 (1975)
8. Blumofe, R.D., Leiserson, C.E.: Scheduling multithreaded computations by work stealing. J. ACM 46(5), 720–748 (1999)
9. Frigo, M., Strumpen, V.: The cache complexity of multithreaded cache oblivious algorithms. In: Gibbons, P.B., Vishkin, U. (eds.) Proceedings of the Eighteenth Annual ACM Symposium on Parallelism in Algorithms and Architectures, SPAA 2006, pp. 271–280. ACM, New York (2006)
10. Johnson, S.C.: Sparse polynomial arithmetic. SIGSAM Bull. 8(3), 63–71 (1974)
11. Monagan, M., Pearce, R.: Parallel sparse polynomial division using heaps. In: Moreno Maza, M., Roch, J.L. (eds.) Proceedings of the 4th International Workshop on Parallel and Symbolic Computation, PASCO 2010, pp. 105–111. ACM, New York (2010)
12. Fateman, R.: Comparing the speed of programs for sparse polynomial multiplication. SIGSAM Bull. 37(1), 4–15 (2003)
13. OpenMP Architecture Review Board: OpenMP application program interface version 3.0 (May 2008)
14. Reinders, J.: Intel threading building blocks, 1st edn. Reilly & Associates, Inc., Sebastopol (2007)
15. Monagan, M., Pearce, R.: Sparse polynomial multiplication and division in maple 14. ACM Commun. Comput. Algebra 44(3/4), 205–209 (2011)
16. Granlund, T.: GNU multiple precision arithmetic library 4.2.4 (September 2008), http://swox.com/gmp/
17. Sanders, J., Kandrot, E.: CUDA by Example: An Introduction to General-Purpose GPU Programming, 1st edn. Addison-Wesley Professional (2010)
18. Augonnet, C., Thibault, S., Namyst, R., Wacrenier, P.A.: Starpu: a unified platform for task scheduling on heterogeneous multicore architectures. Concurrency and Computation: Practice and Experience 23(2), 187–198 (2011)

Simulation of Quantum Error Correction
with *Mathematica*

Vladimir P. Gerdt[1] and Alexander N. Prokopenya[2,3]

[1] Joint Institute for Nuclear Research
141980 Dubna, Russia
gerdt@jinr.ru
[2] Warsaw University of Life Sciences – SGGW
Nowoursynowska 166, 02-787, Warsaw, Poland
alexander_prokopenya@sggw.pl
[3] Collegium Mazovia Innovative University
Sokolowska 161, 08-110 Siedlce, Poland

Abstract. In this paper we consider the problem of quantum error correction and its simulation with the computer algebra system *Mathematica*. Basic ideas of constructing the quantum error correcting codes are discussed, and some examples of error correction by means of quantum circuits constructed with application of the *Mathematica* package *QuantumCircuit* are presented.

1 Introduction

Quantum computing attracts much attention in the past three decades because it promises a major performance speedup in certain types of computational problems (see [1–4]). At the same time, a realistic quantum computer is still not available, and the majority of studies in this field are theoretical ones. This stimulates the development of classical simulators of quantum computation which help to better understand existing quantum algorithms and can be used for searching and testing new efficient algorithms.

In papers [5, 6], we presented our *Mathematica* package *QuantumCircuit* for the simulation of quantum computation based on the circuit model [1]. Remind that quantum computation may be represented as a unitary transformation of a 2^n-dimensional vector describing a state of n-qubit memory register. Therefore, simulation of quantum computation on a classical computer is nothing else than construction of the unitary (circuit) matrix. Such matrix is implemented by a sequence of quantum gates (also unitary transformations in the state space) acting on one or several qubits of the memory register according to a quantum algorithm, which is represented as a diagram called quantum circuit. It is a quantum analogue of classical circuit implementing Boolean functions in terms of classical logical gates.

Our package *QuantumCircuit* provides a user-friendly interface to specify a general n-qubit quantum circuit, to draw it, and to construct the corresponding unitary matrix for quantum computation defined by the circuit. Using this

V.P. Gerdt et al. (Eds.): CASC 2013, LNCS 8136, pp. 116–129, 2013.
© Springer International Publishing Switzerland 2013

matrix, one can determine the final state of the quantum memory register for its given initial state. Thereby, a user of the package can analyse the quantum algorithm determined by the circuit. By varying a circuit one can study and design quantum algorithms.

In our recent paper [7], we demonstrated an application of the package *Quantum Circuit* to the simulation of quantum circuits implementing two best known quantum algorithms, namely, the Grover search algorithm [3] and the Shor algorithm for order finding [2]. Then we presented an algorithmic extension of the *Quantum Circuit* package designed for modeling quantum error correction by means of codes that correct errors [8]. There we considered the case when the errors may arise due to noise in transmission of quantum states through transmission channels, while encoding and decoding of the quantum states is performed without errors. Using the *Quantum Circuit* package, we constructed quantum circuits identifying and correcting simple errors such as bit-flip error and phase error and illustrated their work.

In this paper, we continue simulation of quantum error correction with the help of the *Quantum Circuit* package. We discuss basic ideas of quantum error correction proposed firstly by P. Shor [9] and A. Steane [10] and illustrate its implementation by way of concrete examples. The main aim of the paper is to demonstrate step by step, on a classical computer and by using facilities of *Mathematica*, how quantum errors arising during transmission of quantum states can be diagnosed and corrected without destroying these states.

It should be noted that the possibility of quantum error correction is not obvious, although similar problem is solved successfully in case of classical computation, and there are some reasons for that. First, arbitrary state of a qubit cannot be copied due to no-cloning theorem (see [1]), and so direct application of an idea of redundant information used in classical computation is impossible. Second, measurement of a qubit state with the aim of diagnosing an error as it is done in classical computation is also not possible because measurement destroys the qubit state and original information cannot be recovered afterwards. Finally, a classical bit has only two possible states, and an error arises only if the bit is flipped without any control, while a qubit state changes continuously due to interaction of the qubit with environment, and its final state is not predictable. It is very important to understand how all these difficulties may be overcome to be able to design new effective algorithms for quantum error correction, and the simulation seems to be a good way to do this.

The paper is organized as follows. In Section 2, we briefly describe basic ideas of constructing the codes correcting quantum errors and illustrate them by the example of a bit-flip error that may occur only in a single qubit of the three-qubit codeword encoding a single informational qubit. Such an error is simulated by means of applying the Pauli-X gate to one of the three qubits chosen randomly. Inspecting the codeword with a special quantum circuit detects the corrupted qubit, and it is corrected afterwards. In Section 3, we discuss a five-qubit error correcting code that can be used to detect and correct a general error occurring in a single qubit of the five-qubit codeword. Finally, we conclude in Section 4.

2   Basic Concepts of Quantum Error Correction

General state $|\psi\rangle$ of a qubit that is an elementary unit of information used in quantum computation and quantum communication is represented in the form

$$|\psi\rangle = \alpha|0\rangle + \beta|1\rangle \ , \tag{1}$$

where $|0\rangle$, $|1\rangle$ are two basis states of the qubit, and α, β are two arbitrary complex numbers satisfying the condition $|\alpha|^2 + |\beta|^2 = 1$. The state (1) is assumed to be transformed only during computation when the qubit is acted on by some quantum gates, and it is maintained without change between operations or during transmission. However, complete isolation of a qubit from its environment cannot be achieved in reality, and their interaction necessarily occurs. As a result, quantum states of the qubit and environment become entangled, and the qubit state (1) is corrupted what means a loss of quantum information.

Assuming that initially the environment state $|e\rangle$ is not entangled with the qubit state (1), one can represent general interaction between the qubit and its environment as the following transformation

$$|e\rangle|\psi\rangle \rightarrow (|e_I\rangle I + |e_X\rangle X + |e_Y\rangle Y + |e_Z\rangle Z)\,|\psi\rangle \ , \tag{2}$$

where $|e_I\rangle$, $|e_X\rangle$, $|e_Y\rangle$, $|e_Y\rangle$ are possible final states of the environment, I is the identity operator, and X, Y, Z are the logical operators Pauli-X, Pauli-Y, Pauli-Z, respectively (see, for example, [11, 12]). Therefore, under an influence of environment, initial qubit state (1) is transformed in general to the state (2) that is a superposition of the uncorrupted state $I|\psi\rangle = |\psi\rangle$ and three states with a bit-flip error $X|\psi\rangle$, a phase error $Z|\psi\rangle$, and a combined bit-flip and phase error $Y|\psi\rangle$. And to restore initial state (1) we need to construct a quantum circuit that can identify an error using any of the four states $|\psi\rangle$, $X|\psi\rangle$, $Y|\psi\rangle$, $Z|\psi\rangle$ as an input and correct it. As unitary transformation executed by any quantum circuit is linear, the quantum circuit we are going to construct will be able to correct an error using any superposition of the four states $|\psi\rangle$, $X|\psi\rangle$, $Y|\psi\rangle$, $Z|\psi\rangle$ as an input.

To clarify basic ideas of constructing such quantum circuit let us assume that the qubit state (1) is acted on by the operator Pauli-X for example, simulating a bit-flip error, and takes a form

$$X|\psi\rangle = \alpha|1\rangle + \beta|0\rangle \ , \tag{3}$$

where we have taken into account an action of the Pauli-X operator on the basis states: $X|0\rangle = |1\rangle$, $X|1\rangle = |0\rangle$. One can readily see from (1) and (3) that measuring these states gives either $|0\rangle$ or $|1\rangle$ in both cases, and although probabilities of getting these results are different for the states (1) and (3), in general we cannot distinguish them in this way. Besides, measurement destroys a qubit state, and its initial state cannot be restored afterwards.

However, according to the postulates of quantum physics (see, for example, [1]), measurement of some observable U does not destroy a quantum state if this state is an eigenvector of the corresponding Hermitian operator U. Therefore, if the states (1) and (3) would be two eigenvectors of some Hermitian operator U corresponding to its different eigenvalues then one could distinguish these states without their destruction by means of measuring U. But the problem is that any vector $|\psi\rangle$ belonging to the two-dimensional Hilbert space spanned by the basis vectors $|0\rangle$, $|1\rangle$ should be an eigenvector of U with the same eigenvalue. As the states (1) and (3) belong to the same Hilbert space both of them should correspond to the same eigenvalue of U and cannot be distinguished.

The only way to separate an uncorrupted state and states with error is to increase dimension of the Hilbert space and to set these states into different subspaces. It means we have to increase number of qubits used for encoding the state (1), let this number be equal to n. Of course, due to interaction with environment each of the n qubits may be corrupted but it is quite reasonable to assume that errors occur in different qubits independently of one another. Therefore, if probability of error occurrence is small enough then the cases of simultaneous errors in two or more qubits may be excluded from the consideration, and one can assume that an error occurs only in one of the n qubits. As a state of each qubit is described by a vector in two-dimensional Hilbert space and operator Pauli-X, simulating a bit-flip error, can be applied to each of them we need $(n+1)$ different two-dimensional subspaces to separate the uncorrupted state and n states with bit-flip error in different qubits. Obviously, total dimension of the n-qubit Hilbert space being equal to 2^n must be greater or equal than $2(n+1)$. Solving the inequality $2^n \geq 2(n+1)$ gives $n \geq 3$ and, hence, minimum number of qubits we need to encode the state (1) is equal to 3 provided that a bit-flip error can occur only in one of the three qubits.

Let us now consider two commuting Hermitian operators U_1 and U_2 satisfying the condition $U_k^2 = I$, $(k = 1, 2)$. Obviously, such operators can have the eigenvalues only ± 1, since operator U_k acting twice on its eigenvector must act as the identity operator I. Assume that uncorrupted 3-qubit state $|\Psi\rangle$ encoding the state (1) is the eigenvector of both operators U_1 and U_2 corresponding to the same eigenvalue $+1$. Assume also that operator U_1 commutes with operator X_0 and anti-commutes with operators X_1 and X_2, while operator U_2 commutes with X_2 and anti-commutes with operators X_0 and X_1, where index $k = 0, 1, 2$ in the notation X_k indicates the number of the qubit to which the operator Pauli-X simulating a bit-flip error is applied. Recall that qubits in a quantum circuit are usually numbered from bottom to top beginning with zero. Then any corrupted state $X_k|\Psi\rangle$ will be an eigenvector of the operators U_1 and U_2, as well. Indeed, the following equalities take place

$$U_1 X_0|\Psi\rangle = X_0 U_1|\Psi\rangle = X_0|\Psi\rangle , \quad U_2 X_0|\Psi\rangle = -X_0 U_2|\Psi\rangle = -X_0|\Psi\rangle ,$$

$$U_1 X_1|\Psi\rangle = -X_1 U_1|\Psi\rangle = -X_1|\Psi\rangle , \quad U_2 X_1|\Psi\rangle = -X_1 U_2|\Psi\rangle = -X_1|\Psi\rangle , \quad (4)$$

$$U_1 X_2|\Psi\rangle = -X_2 U_1|\Psi\rangle = -X_2|\Psi\rangle , \quad U_2 X_2|\Psi\rangle = X_2 U_2|\Psi\rangle = X_2|\Psi\rangle .$$

One can readily see from (4) that corrupted states $X_0|\Psi\rangle$, $X_1|\Psi\rangle$, $X_2|\Psi\rangle$ are eigenvectors of the operators U_1 and U_2 corresponding to the eigenvalues

$(+1, -1)$, $(-1, -1)$, $(-1, +1)$, while uncorrupted state $|\Psi\rangle$ is eigenvector of these operators corresponding to the eigenvalue $(+1, +1)$, according to our assumption. Therefore, measuring U_1 and U_2 diagnoses errors in the considered case of bit-flip errors in one qubit and indicates the corrupted qubit.

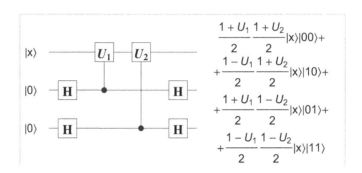

Fig. 1. Quantum circuit with two ancillary qubits set initially in the state $|0\rangle$ for measurement of U_1 and U_2

Note that either of the two operators U_1 and U_2 can be measured without destroying an input state $|x\rangle$ by means of measuring ancillary qubits entangled with this state, the corresponding quantum circuit, consisting of four Hadamard gates and controlled U_1 and U_2 operators, is shown in Fig. 1. One can readily see that if $|x\rangle$ is eigenvector of U_1 and U_2 with eigenvalues $(+1, +1)$ then the final qubit state will be $\frac{1}{4}(1 + U_1)(1 + U_2)|x\rangle = |x\rangle$ and measuring the ancillary qubits indicates $|00\rangle$. But if $|x\rangle$ is eigenvector of U_1 and U_2 with eigenvalues $(+1, -1)$, for example, then measuring the ancillary qubits indicates $|01\rangle$, and so on. It should be emphasized that ancillary qubits are used here specially to indicate an error and their measurement does not destroy the input state $|x\rangle$.

Remind that the Pauli operators X, Y, Z anti-commutes if they act on the same qubit and commute otherwise. Therefore, the operators U_1 and U_2 having the above-described properties can be chosen in the form $U_1 = Z_2 Z_1 I_0$, $U_2 = I_2 Z_1 Z_0$. As quantum circuits implementing these operators are encoded in the *QuantumCircuit* package by two symbolic vectors $(Z, Z, 1)^T$ and $(1, Z, Z)^T$ (see [6, 7]), we can easily write two matrix equations determining their eigenvectors corresponding to eigenvalues $(+1, +1)$. The corresponding commands written in the language of the *Mathematica* system are

$$vec\Psi = Table[a_j, \{j, 8\}] \ ;$$

$$matrixU[\{\{Z\}, \{Z\}, \{1\}\}] \, . \, vec\Psi == vec\Psi \ ;$$

$$matrixU[\{\{1\}, \{Z\}, \{Z\}\}] \, . \, vec\Psi == vec\Psi \ ; \tag{5}$$

where the built-in function *matrixU[mat]* computes unitary matrix determined by symbolic matrix *mat*, and function *Table* generates vector $vec\Psi$ having eight components a_j, $(j = 1, 2, ..., 8)$. Using the built-in *Mathematica* function *Solve*, one can easily find a solution of the system (5) that is

$$vec\Psi = (a_1, 0, 0, 0, 0, 0, 0, a_8)^T .$$ (6)

Therefore, three-qubit state $|\Psi\rangle$ being eigenvector of both operators U_1, U_2 with eigenvalues $(+1, +1)$, may be represented as superposition of two basis vectors

$$|\Psi\rangle = a_1|0\rangle_3 + a_8|7\rangle_3 \equiv a_1|000\rangle + a_8|111\rangle ,$$

where complex numbers a_1 and a_8 must satisfy the normalization condition $|a_1|^2 + |a_8|^2 = 1$. Assuming $a_1 = \alpha$, $a_8 = \beta$, one can readily check that three-qubit state

$$|\Psi\rangle = \alpha|000\rangle + \beta|111\rangle .$$ (7)

is generated for arbitrary state (1) by the quantum circuit shown in Fig. 2.

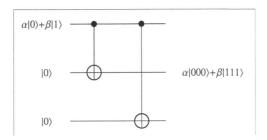

Fig. 2. Quantum circuit for encoding the state (1) by three-qubit state (7)

Thus, to diagnose a bit-flip error arising in arbitrary qubit state (1) we can encode this state by three-qubit state (7), using quantum circuit shown in Fig. 2. This state is taken as input vector $|\Psi\rangle$ in the quantum circuit shown in Fig. 3 with two ancillary qubits set in state $|0\rangle$ and used for identification of error. Pauli-X operator in the left-hand side of the circuit is applied to the upper qubit (it is applied to any of the tree qubits in the state (7) chosen randomly) and simulates bit-flip error. Arising error is corrected then by means of application of the Pauli-X operator to corrupted qubit.

3 General Case of Five-Qubit Error Correcting Code

One can readily check that operator $U_1 = Z_2 Z_1 I_0$ considered above commutes with Pauli operator Y_0 and anti-commutes with Y_1 and Y_2, while operator $U_2 = I_2 Z_1 Z_0$ commutes with Y_2 and anti-commutes with operators Y_0 and Y_1. It means that quantum circuit shown in Fig. 3 can be also used to diagnose the combined bit-flip and phase error simulated by means of application of the Pauli-Y operator to one of the three qubits of the state (7) (see Fig. 4, where the Pauli-Y operator

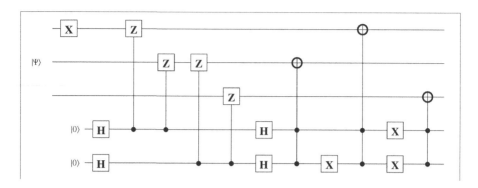

Fig. 3. Quantum circuit correcting bit-flip error in one qubit of three-qubit codeword encoding state (1)

is applied to the upper qubit in the left-hand side of the circuit). To correct such error we need to apply another Pauli-Y operator to the corrupted qubit (see three controlled Pauli-Y operators in the right-hand side) but the circuit indicating corrupted qubit remains the same as in case of bit-flip error. Therefore, the circuits shown in Figs. 3 and 4 indicate arising bit-flip or combined bit-flip and phase error simulated by means of application of the Pauli-X and Y operators, respectively, but they cannot distinguish these two errors.

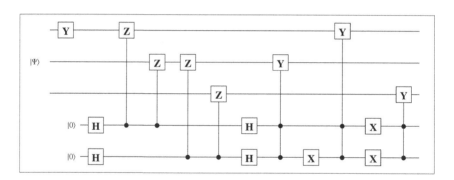

Fig. 4. Quantum circuit correcting combined bit-flip and phase error in one qubit of three-qubit codeword encoding state (1)

Remind that in general case of n-qubit codeword, each qubit can be acted on by any of the three Pauli operators or the codeword remains uncorrupted (see (2)). Therefore, to distinguish all possible errors we need $(3n+1)$ two-dimensional subspaces in the Hilbert space of dimension 2^n. Obviously, the condition $2^n \geq 2(3n+1)$ should be fulfilled and, hence, we need at least five-qubit codeword for encoding the state (1). Besides, we need four ancillary qubits and four mutually commuting Hermitian operators U_j, $(j = 1, 2, 3, 4)$, obeying the condition

$U_j^2 = I$ to indicate the corrupted qubit. As in the case of three-qubit codeword, such operators can be constructed from the Pauli operators X, Y, and Z and are defined as follows (see, for example, [11]):

$$U_1 = Z_4 X_3 X_2 Z_1 I_0 , \quad U_2 = X_4 X_3 Z_2 I_1 Z_0 ,$$

$$U_3 = X_4 Z_3 I_2 Z_1 X_0 , \quad U_4 = Z_4 I_3 Z_2 X_1 X_0 . \tag{8}$$

One can readily see that these operators U_j are mutually commuting because anti-commuting Pauli operators X_j and Z_j with the same index encounter exactly two times in any pair U_j and U_k. As $U_j^2 = I$ each operator has two eigenvalues ± 1, and there are exactly $2^4 = 16$ possible combinations of their eigenvalues $(\pm 1, \pm 1, \pm 1, \pm 1)$.

Let us assume that five-qubit state $|\Psi\rangle \equiv |x_4 x_3 x_2 x_1 x_0\rangle$ is the eigenvector of all four operators U_j corresponding to the set of eigenvalues $(+1, +1, +1, +1)$. Pauli operator X_j, Y_j, or Z_j, acting on one of the five qubits $|x_4 x_3 x_2 x_1 x_0\rangle$, either commutes or anti-commutes with operators U_k and, hence, any corrupted state $X_j|\Psi\rangle$, $Y_j|\Psi\rangle$, or $Z_j|\Psi\rangle$, $(j = 0, 1, 2, 3, 4)$, will be the eigenvector of operators U_k, $(k = 1, 2, 3, 4)$ as well but with different set of eigenvalues. For example, Pauli operator Z_4 acting on the qubit $|x_4\rangle$ and simulating phase error in this qubit commutes with operators U_1, U_4 and anti-commutes with U_2, U_3. Then the following equalities take place

$$U_1 Z_4 |\Psi\rangle = Z_4 U_1 |\Psi\rangle = Z_4 |\Psi\rangle ,$$

$$U_2 Z_4 |\Psi\rangle = -Z_4 U_2 |\Psi\rangle = -Z_4 |\Psi\rangle ,$$

$$U_3 Z_4 |\Psi\rangle = -Z_4 U_3 |\Psi\rangle = -Z_4 |\Psi\rangle ,$$

$$U_4 Z_4 |\Psi\rangle = Z_4 U_4 |\Psi\rangle = Z_4 |\Psi\rangle .$$

Therefore, the state $Z_4|\Psi\rangle$ is eigenvector of operators U_k with the set of eigenvalues $(+1, -1, -1, +1)$. Similarly one can show that all 15 corrupted five-qubit states are eigenvectors of operators U_k with different sets of eigenvalues. And measuring operators U_k can diagnose error in one of the five qubits and indicate corrupted qubit and kind of error. To implement such measurement we need four ancillary qubits set initially in the state $|0\rangle$. The corresponding quantum circuit is constructed similarly to the case of three-qubit codeword considered in Section 2 and is shown in Fig. 5.

After measuring the operators U_k and identifying the error we have to apply the corresponding Pauli operator to the corrupted qubit to restore it. A circuit that corrects any of the possible 15 errors is quite large, and due to this we separate it into two parts shown in Fig. 6 and 7.

Finally, we have to construct a five-qubit codeword $|\Psi\rangle$ that is eigenvector of all four operators U_k with the same eigenvalue $+1$. Using the *QuantumCircuit* package (see [6, 7]), we can write four matrix equations determining eigenvectors of operators U_k corresponding to eigenvalues $(+1, +1, +1, +1)$. The corresponding commands written in the language of the *Mathematica* system are

$$vec\Psi = Table[a_j, \{j, 32\}] ;$$

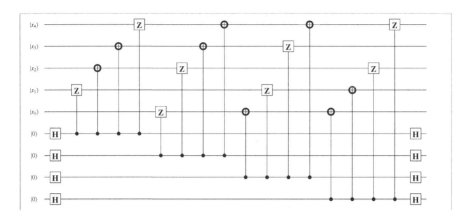

Fig. 5. Quantum circuit diagnosing error in one qubit of five-qubit codeword encoding state (1)

$$matrixU[\{\{Z\}, \{X\}, \{X\}, \{Z\}, \{1\}\}] \cdot vec\Psi == vec\Psi \ ;$$

$$matrixU[\{\{X\}, \{X\}, \{Z\}, \{1\}, \{Z\}\}] \cdot vec\Psi == vec\Psi \ ;$$

$$matrixU[\{\{X\}, \{Z\}, \{1\}, \{Z\}, \{X\}\}] \cdot vec\Psi == vec\Psi \ ;$$

$$matrixU[\{\{Z\}, \{1\}, \{Z\}, \{X\}, \{X\}\}] \cdot vec\Psi == vec\Psi \ ; \tag{9}$$

where we have used the built-in function $matrixU[mat]$ and four symbolic vectors mat corresponding to the operators (8). Solving the system (9), we obtain the following vector with 32 components:

$$\Psi = (a_1, a_2, a_2, a_1, a_2, -a_1, a_1, -a_2, a_2, -a_1, -a_1, a_2, a_1, a_2, -a_2, -a_1, a_2, a_1,$$

$$-a_1, -a_2, -a_1, a_2, a_2, -a_1, a_1, -a_2, a_2, -a_1, -a_2, -a_1, -a_1, -a_2) \ , \tag{10}$$

where two numbers a_1 and a_2 are constrained by the normalization condition $|\Psi|^2 = 1$. Denoting $a_1 = \alpha/4$, $a_2 = -\beta/4$, we can rewrite vector (10) in the form

$$|\Psi\rangle = \frac{\alpha}{4}(|00000\rangle + |00011\rangle - |00101\rangle + |00110\rangle - |01001\rangle - |01010\rangle +$$

$$+|01100\rangle - |01111\rangle + |10001\rangle - |10010\rangle - |10100\rangle - |10111\rangle +$$

$$+|11000\rangle - |11011\rangle - |11101\rangle - |11110\rangle) -$$

$$-\frac{\beta}{4}(|00001\rangle + |00010\rangle + |00100\rangle - |00111\rangle + |01000\rangle + |01011\rangle +$$

$$+|01101\rangle - |01110\rangle + |10000\rangle - |10011\rangle + |10101\rangle + |10110\rangle -$$

$$-|11001\rangle + |11010\rangle - |11100\rangle - |11111\rangle) \ . \tag{11}$$

One can readily check that vector (11) is normalized ($\langle\Psi|\Psi\rangle = 1$) if the numbers α and β satisfy the condition $|\alpha|^2 + |\beta|^2 = 1$ that is exactly the same as in the

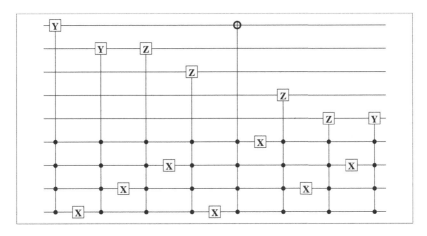

Fig. 6. First part of quantum circuit correcting errors in one qubit of five-qubit code-word encoding state (1)

case of vector (1) we want to encode. Obviously, vector (11) is a superposition of all 32 basis states in the five-qubit Hilbert space, and some quantum circuit is needed to encode the state (1) into the five-qubit codeword (11). An example of such a circuit is depicted in Fig. 8 (see [11]).

To check whether the circuit shown in Fig. 8 really encodes the state (1) into five-qubit codeword (11) let us remind that all information about this circuit is encoded in the *Quantum Circuit* package in symbolic matrix $mat1$ given by

$$mat1 = \begin{pmatrix} Z & H & Z & C & H & C & 1 & 1 & C & 1 & H & C & 1 & 1 & H \\ 1 & 1 & 1 & X & H & 1 & C & 1 & 1 & 1 & 1 & C & 1 & H \\ 1 & 1 & 1 & 1 & 1 & X & X & H & 1 & C & 1 & 1 & 1 & C & 1 \\ 1 & 1 & 1 & 1 & 1 & 1 & 1 & 1 & X & X & H & 1 & 1 & 1 & 1 \\ 1 & 1 & 1 & 1 & 1 & 1 & 1 & 1 & 1 & 1 & X & X & X & 1 \end{pmatrix} . \qquad (12)$$

The input state $|\psi 0000\rangle = \alpha|00000\rangle + \beta|10000\rangle$ is a vector in the 2^5-dimensional state space of five qubits, which has only two nonzero components (the first and seventeenth). Using built-in function *SparseArray*, we can represent such vector in codes of the *Mathematica* system as

$$vecInput[n\_] := SparseArray[\{1 \to \alpha, 2^{n-1} + 1 \to \beta\}, 2^n] ;$$

Then we can easily compute a final state *vecOutput1* for the circuit encoded by matrix (12), it is given by

$$vecOutput1 = matrixU[mat1].vecInput[5] ; \qquad (13)$$

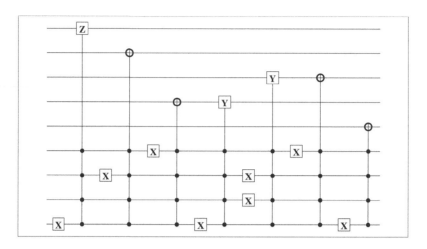

Fig. 7. Second part of quantum circuit correcting errors in one qubit of five-qubit codeword

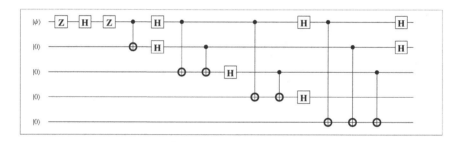

Fig. 8. Quantum circuit encoding the state (1) into five-qubit codeword

Representing the state (13) in the form of superposition of the basis states in the five-qubit Hilbert space gives a vector

$$|\Psi\rangle = \frac{\alpha}{4} \left(|0\rangle_5 + |3\rangle_5 - |5\rangle_5 + |6\rangle_5 - |9\rangle_5 - |10\rangle_5 + |12\rangle_5 - |15\rangle_5 + |17\rangle_5 - \right.$$

$$- |18\rangle_5 - |20\rangle_5 - |23\rangle_5 + |24\rangle_5 - |27\rangle_5 - |29\rangle_5 - |30\rangle_5 \left. \right) +$$

$$+ \frac{\beta}{4} \left(-|1\rangle_5 - |2\rangle_5 - |4\rangle_5 + |7\rangle_5 - |8\rangle_5 - |11\rangle_5 - |13\rangle_5 + |14\rangle_5 - |16\rangle_5 + \right.$$

$$\left. + |19\rangle_5 - |21\rangle_5 - |22\rangle_5 + |25\rangle_5 - |26\rangle_5 + |28\rangle_5 + |31\rangle_5 \right) \; ,$$

that coincides exactly with vector (11).

To simulate error correction we need to apply any of the three Pauli operators X, Y, Z to any of the five qubits in (11) and to use the vector obtained as input for quantum circuit diagnosing error that is shown in Fig. 5. As this circuit contains four ancillary qubits for measuring operators U_j we need to represent the state (11) in the form of vector in the 2^9-dimensional Hilbert space. This can easily be done by means of adding four rows of the units at the bottom of symbolic matrix (12) and repeating the calculation of $vecOutput1$ (see (13)) with $vecInput[9]$.

The circuit simulating error may be encoded by symbolic vector with 9 components. One of the first five symbols may be X, Y or Z while the rest eight symbols are units. For example, circuit simulating phase error in the qubit $|x_1\rangle$ is encoded by the matrix that can be written in codes of the *Mathematica* system as

$$matErrorZ1 = \{\{1\}, \{1\}, \{1\}, \{Z\}, \{1\}, \{1\}, \{1\}, \{1\}, \{1\}\};$$

Symbolic matrix encoding information about the circuit shown in Fig. 5 may be represented in the form

$$mat2 = \begin{pmatrix} 1 & 1 & 1 & 1 & Z & 1 & 1 & 1 & X & 1 & 1 & 1 & X & 1 & 1 & 1 & Z & 1 \\ 1 & 1 & 1 & X & 1 & 1 & 1 & X & 1 & 1 & 1 & Z & 1 & 1 & 1 & 1 & 1 & 1 \\ 1 & 1 & X & 1 & 1 & Z & 1 & 1 & 1 & 1 & 1 & 1 & 1 & 1 & Z & 1 & 1 \\ 1 & Z & 1 & 1 & 1 & 1 & 1 & 1 & 1 & Z & 1 & 1 & 1 & X & 1 & 1 & 1 \\ 1 & 1 & 1 & 1 & 1 & Z & 1 & 1 & 1 & X & 1 & 1 & 1 & X & 1 & 1 & 1 & 1 \\ H & C & C & C & C & 1 & 1 & 1 & 1 & 1 & 1 & 1 & 1 & 1 & 1 & 1 & 1 & H \\ H & 1 & 1 & 1 & 1 & C & C & C & C & 1 & 1 & 1 & 1 & 1 & 1 & 1 & 1 & H \\ H & 1 & 1 & 1 & 1 & 1 & 1 & 1 & 1 & C & C & C & C & 1 & 1 & 1 & 1 & H \\ H & 1 & 1 & 1 & 1 & 1 & 1 & 1 & 1 & 1 & 1 & 1 & 1 & C & C & C & C & H \end{pmatrix}.$$

Similarly one can define symbolic matrix determining the circuit correcting errors that is shown in Figs. 6 and 7. This matrix is quite cumbersome, and we do not show it here, let us denote this matrix as $mat3$. Then we can calculate a final 9-qubit state according to the following command written in codes of the *Mathematica* system:

$$vecOutput = matrixU[mat3].matrixU[mat2].$$

$$matrixU[matErrorZ1].vecOutput1 ; \qquad (14)$$

Vector (14) is easily transformed to the form

$$vecOutput \rightarrow |\Psi\rangle|1010\rangle , \qquad (15)$$

where vector $|\Psi\rangle$ is given by (11). One can readily see that action of the Pauli-Z operator is eliminated, and initial state (11) is restored, while ancillary qubits are in the state $|1010\rangle$. Changing matrix $matError$ and repeating calculation (14), one can simulate all possible errors in one of the five qubits $|x_4x_3x_2x_1x_0\rangle$. The calculation shows that any of the 15 possible errors in original state (11) encoding the state (1) is restored.

It should be emphasized that quantum circuit described above can diagnose and correct an arbitrary error arising in one qubit. Such error may be simulated by means of application of an arbitrary superposition of four operators I, X, Y, Z

to one qubit of the state (11) encoding the state (1). Actually, let us define the following three matrices

$$matError0 = \{\{1\}, \{1\}, \{1\}, \{1\}, \{1\}, \{1\}, \{1\}, \{1\}, \{1\}\};$$

$$matErrorX1 = \{\{1\}, \{1\}, \{1\}, \{X\}, \{1\}, \{1\}, \{1\}, \{1\}, \{1\}\};$$

$$matErrorY1 = \{\{1\}, \{1\}, \{1\}, \{Y\}, \{1\}, \{1\}, \{1\}, \{1\}, \{1\}\};$$

The first one corresponds to the case of error absence, while the second and third matrices simulate the bit-flip and combined bit-flip and phase error in the qubit $|x_1\rangle$. Then the state (11) with the qubit $|x_1\rangle$ corrupted arbitrarily is given by

$$vecCorrupted = (a * matrixU[matError0] + b * matrixU[matErrorX1] +$$

$$+c * matrixU[matErrorY1] + d * matrixU[matErrorZ1]) \, . \, vecOutput1 \; ; \; (16)$$

Here a, b, c, d are arbitrary complex numbers constrained only by the condition that vector (16) is normalized by 1. After diagnosing and correcting errors we obtain the vector

$$vecOutput = matrixU[mat3] \, . \, matrixU[mat2] \, . \, vecCorrupted \; ;$$

which can be represented in the form

$$vecOutput \rightarrow |\Psi\rangle (a|1011\rangle + b|0001\rangle + c|0000\rangle + d|1010\rangle) \, .$$

Thus, the vector $|\Psi\rangle$ encoding vector (1) turns out to be restored, while ancillary qubits are in some superposition of the basis states. Note that final state of the ancillary qubits is not essential because they should be set into state $|0000\rangle$ afterwards to prepare the circuit for diagnosing the next qubit.

4   Conclusion

In this paper, we discussed basic ideas of quantum error correction of noise-induced errors arising upon transmission of quantum states through communication channel. These ideas were illustrated first by means of simulation of bit-flip error arising in one qubit of the three-qubit codeword encoding an arbitrary qubit state (1).

It was shown that correction of general type error arising in one qubit requires at least five-qubit codeword encoding the informational qubit. Nevertheless, our simulation confirms that quantum error correction is possible if the noise level in the transmission channel is sufficiently small, and probability of error arising in one qubit is sufficiently small as well. At the same time, quantum circuit (see Fig. 8) used for encoding an arbitrary qubit state (1) into five-qubit codeword is quite cumbersome, and this makes difficult a practical usage of the five-qubit error-correcting codes. The seven-qubit error-correcting code (see [10]) turns out to be much more efficient and simpler for implementation, and we are planning to consider it in our next paper.

For constructing and modeling quantum circuits that correct errors and for performing all necessary calculations we used here our *Mathematica* package *QuantumCircuit*.

Acknowledgement. This work was supported in part by the Russian Foundation for Basic Research, project no. 13-01-00668.

References

1. Nielsen, M., Chuang, I.: Quantum Computation and Quantum Information, 10th ed., Cambridge University Press (2010)
2. Shor, P.W.: Polynomial-time algorithms for prime factorization and discrete logarithms on a quantum computer. SIAM J. Comp. 26(5), 1484–1509 (1997)
3. Grover, L.K.: Quantum mechanics helps in searching for a needle in a haystack. Phys. Rev. Lett. 79, 325–328 (1997)
4. Phoenix, S.J.D., Townsend, P.D.: Quantum cryptography: how to beat the code breakers using quantum mechanics. Contemp. Phys. 36, 165–195 (1995)
5. Gerdt, V.P., Kragler, R., Prokopenya, A.N.: A Mathematica package for simulation of quantum computation. In: Gerdt, V.P., Mayr, E.W., Vorozhtsov, E.V. (eds.) CASC 2009. LNCS, vol. 5743, pp. 106–117. Springer, Heidelberg (2009)
6. Gerdt, V.P., Prokopenya, A.N.: Some algorithms for calculating unitary matrices for quantum circuits. Programming and Computer Software 36(2), 111–116 (2010)
7. Gerdt, V.P., Prokopenya, A.N.: The circuit model of quantum computation and its simulation with mathematica. In: Adam, G., Buša, J., Hnatič, M. (eds.) MMCP 2011. LNCS, vol. 7125, pp. 43–55. Springer, Heidelberg (2012)
8. Gerdt, V.P., Prokopenya, A.N.: Simulation of quantum error correction by means of QuantumCircuit package. Programming and Computer Software 39(3), 143–149 (2013)
9. Shor, P.W.: Scheme for reducing decoherence in quantum memory. Phys. Rev. A 52, R2493–R2496 (1995)
10. Steane, A.M.: Quantum error correction. In: Lo, H.-K., Popescu, S., Spiller, T. (eds.) Introduction to Quantum Computation and Information, pp. 181–212. World Scientific, Singapore (1998)
11. Mermin, N.D.: Quantum Computer Science. An Introduction. Cambridge University Press (2007)
12. Preskill, J.: Lecture Notes for Physics 229: Quantum Information and Computation. California Institute of Technology (1998)

From the Product Example to PDE Adjoints, Algorithmic Differentiation and Its Application (Invited Talk)

Andreas Griewank

Mathematics Department, Humboldt University, Berlin

One Origin of the Reverse Mode

At the last conference on *algorithmic* or *automatic differentiation* (AD) in July 2012, Bert Speelpenning gave a very entertaining account of his pioneering work in the field. After finishing his thesis titled *Compiling fast partial derivatives of functions given by algorithms* at Urbana Champain in 1981 he had dropped from sight of the scientific community and spent a few decades in the software industry working amongst others for Oracle and Microsoft.

I came across his thesis in 1987 when I took up a position at the mathematics and computer science division of Argonne National University. Jorge Moré had picked up his thesis and the associated source transformation tool JAKE, which he had translated from C to to Fortran, then apparently considered the language for real numerical types. The official supervisor of Bert S. had been the well known numerical analyst Bill Gear, and this lead to the plausible story that Bill G. had asked Bert S. to find a way of accurately and efficiently calculating the Jacobians of the right hand sides of stiff ODEs for their numerical integration by Gear's BDF methods. I propagated that legend as late as June 2012, when I submitted the article *Who invented the reverse mode of differentiation* for the book Optimization Stories edited by Martin Grötschel for the 21. IMSP at Berlin in August 2012 [Gri12].

Occasionally, I also have spread the notion that Bert had used compiler optimization techniques to go from the straight forward application of the chain rule, known as the forward mode in the AD community, to the reverse mode, which is much more efficient if there are fewer dependent then independent variables. Experiments in the nineties with expression simplification algorithms in Reduce showed, that for the calculation of a determinant $y = \det(X)$ as a function of $X \in R^{n \times n}$, the transition from the forward to the reverse mode could be achieved only up to an n of about 8. That limit may be considerably higher now. Nevertheless the janitorial approach, as Bruce Char called it, of first creating an extensive mess of formulas through forward differentiation and then cleaning it up by identifying internal redundancies, is certainly still not a recommended technique.

V.P. Gerdt et al. (Eds.): CASC 2013, LNCS 8136, pp. 130–135, 2013.
© Springer International Publishing Switzerland 2013

Arthur Sedgwick's Hunch

In contrast to the plausible myths I perpetuated, Bert told us in July 2012, that Bill hardly knew him and only took over the supervision formally, when his actual supervisor Arthur Sedgwick returned to his native Canada to build up the statistics department at Dalhousie University. Bert related furthermore that the young professor Arthur S. had been inspired by the gradient calculation for the extremely simple product example

$$f(\mathbf{x}) = \prod_{i=1}^{n} x_i \quad \Rightarrow \quad g_j(x) \equiv \frac{\partial f(x)}{\partial x_j} = \frac{f(x)}{x_j} \; .$$

The interesting observation here is that once f has been evaluated at the cost of $n - 1$ multiplications, the whole gradient $g(x) = \nabla f(\mathbf{x})$ can be obtained using n divisions. In other words the operations count for the gradient is only a small multiple of that for evaluating the function by itself. The objection that divisions are more expensive than multiplications and may lead to a NaN if some component x_j is zero may be easily overcome by noting that

$$g_j(x) = y_j z_j \quad \text{with} \quad y_j \equiv \prod_{i=1}^{j-1} x_i \quad \text{and} \quad z_j \equiv \prod_{i=j}^{n-1} x_{i+1}$$

Since the partial products y_j can be accumulated forward and the z_j backward one sees that

$$\mathtt{OPS}\{\nabla f(\mathbf{x})\} \leq 3 \, \mathtt{OPS}\{f(\mathbf{x})\} \tag{1}$$

with the gap between between the gradient cost and its upper bound being very small in this case. Here OPS represents the classical count of multiplications, but it may be substituted by other temporal measures of computational complexity that might also take into account memory access costs. Accordingly, the constant 3 may have to be changed to another small and platform but not problem depedent constant.

   Arthus S. had the daring thought that gradients might always be obtainable at a similarly cheap cost and asked Bert S. to look into that. And indeed Speelpenning demonstrated that this holds generally true by analysing and implementing what we now call the reverse mode of algorithmic differentiation. It requires absolutely no human intervention or algebraic simplifications, though efficiency gains can usually be made by manual or automatic post processing. For the product function above, which is known as the Speelpenning example in the AD literature, but probably should be called the Sedgwick example now, the input to the AD tool would be simply the loop for evaluating the partial products y_j for $j = 1 \ldots n$. The reverse loop for evaluating the z_j and computig the g_j is then generated by source transformation, or operator overloading, or a combination thereof.

Other Origins and Destinations

Many computer scientists know the reverse mode as the *Baur-Strassen* method, actually published two years later (BS83), for computing gradients of rational functions that are evaluated by a sequence of arithmetic operations. Here and throughout references that are not listed in the present bibliography are noted in round brackets and can be found in the book [GW08]. For the particular case of matrix algorithms Miller et al proposed the corresponding roundoff analysis [MW80]. Much more general, Kim, Nesterov et al (KN+84) considered the composition of elementary functions from an arbitrary library with bounded gradient complexity. From a memory and numerical stability point of view the most difficult aspect of the reverse mode is the reversal of a program. This problem was discussed in the context of Turing Machines by Benett (Ben73), who foreshadowed the use of checkpointing as a tradeoff between numerical computational effort and memory requirement. From the computer algebra community Erich Kaltofen [KS91] has been one of the first who understood and applied the reverse mode. For more historical and anecdotal background on the reverse mode see [Gri12].

Early Applications and Tool Developments

One of the earliest uses of the reverse mode has been in data assimilation in weather forecasting and oceanography. This is really just a history match by a weighted least squares calculation on a time-dependent evolution, where the parameters to be approximated include the present state of the atmosphere. The recurrent substantial effort of writing an adjoint code for geophysical models eventually spawned activities to generate adjoint compilers such as Tapenade (HP04) and TAF (GK98).

In the 80's there was also a considerable effort to analyse the safety of nuclear reactors using forward simulations and adjoints. The corresponding tool GRESS/ADGEN [Gre89] used at first only the forward mode but later also the reverse mode. Unfortunately, that promising tool development effort was abandoned in the late nineties. Starting in the mid nineties the tool ADIFOR was developed and immediately applied to aerodynamical codes at Nasa Langley [GNH96]. ADIFOR (BC+96) was and is a very stable in the forward mode but its reverse pendant ADJIFOR never got beyond an early beta stage, mainly due to the fact that its main designer Alan Carle has to stop working for health reasons.

The first implementations of the reverse mode based on the alternative software technology of operator overloading was done in PASCAL-SC, an extension of PASCAL for the purposes of interval computation. Rall had already used Pascal-SC for the forward mode of AD (Rall81).The verified computing community has later included the revers mode in their analysis and some but not all of the software [Kea09].

Forward Mode Based Applications

Of course, there is more to AD than the reverse mode. In fact the idea of propagating first and higher derivatives forward through a code list of elementary instructions dates back to the very beginning of programmable computers (Wen64). This theory was comprehensively described in the monograph (Rall84). The forward mode of AD yields derivatives with much better accuracy than the classical approximation by differences. Moreover, using the fact that all intrinsic functions are solutions of linear ODE's one can propagate univariate Taylor series of length d with a computational complexity of order d^2.

This low cost has been exploited in various tools for the numerical integration of ODEs (CC94), especially also for the computation of periodic orbits. It has been reported [Phi03] that if one wishes to achieve rather high accuracy such Taylor expansion based methods are more efficient than Runge Kutta and other classical integrators. The optimal control of initial value problems in ODEs leads via the Pontryagin principle to so-called differential algebraic equations. They can also be efficiently analyzed and integrated by Taylor series methods as demonstrated in a series of papers by Nedialkov and Pryce [NP07]. There are many other promising applications, for example in numerical bifurcation, where higher derivatives need to be calculated selectively.

AD and Adjoints

The *cheap gradient principle* (1) continues to be the most intriguing result of AD. It also provides the most serious challenges for an efficient implementation, especially with respect to the memory management. It may be viewed as a discrete analog of so called adjoint equations or evolutions. The most serious competitor of the reverse mode of AD is certainly the hand coding of derivatives. In the context of ODEs and PDEs, that may mean first writing down an adjoint equation in an appropriate function space and then discretizing it, hopefully consistently with the discretization of the *original* or *primal* equation.

For an ODE on the unit time interval $0 \leq t \leq 1$, we may have the primal dual pair of evolutions

$$\dot{\mathbf{u}}(t) \equiv \partial \mathbf{u}(t)/\partial t = \mathbf{F}(\mathbf{u}(t)) \qquad \text{with} \quad \mathbf{u}(0) = \mathbf{x},$$
$$\dot{\bar{\mathbf{u}}}(t) \equiv \partial \bar{\mathbf{u}}(t)/\partial t = \mathbf{F}'(\mathbf{u}(t))^{\top} \bar{\mathbf{u}}(t) \qquad \text{with} \quad \bar{\mathbf{u}}(1) = \nabla f(\mathbf{u}(1)).$$

Here the state $\mathbf{u}$ belongs to some Euclidean or Banach space and the adjoint vector $\bar{\mathbf{u}}$ to its topological dual. Correspondingly, the right-hand side $\mathbf{F}(\mathbf{u})$ and its dual $\mathbf{F}'(\mathbf{u})^{\top} \bar{\mathbf{u}}$ may be strictly algebraic or involve differential operators.

There is an ongoing debate whether it is better to first form the adjoint equations and then discretize them, or to first discretize the primal equations and then differentiate them in the reverse mode. Both approaches have fervent supporters and balancing their advantages and disadvantages is not always easy. For practical examples of a range of options see the volume [ea12] reporting results of the DFG priority program 1253.

I was recently asked to judge a large scale and long term industrial project, where adjoints of a complex CFD code had ben written and maintained by hand for several years. However, it became harder and harder to keep the primal and adjoint code versions consistent so that eventually a time lag of more than a year arose between the introduction of improvements and refinements of the primal model and the corresponding changes on the adjoint code. Since it was forseeable that the adjoint code could not be kept up to date at all the management decided to set things up such that the primal code was was developed in a way that allows the almost instant generation of the primal or dual model using AD tools.

Future Challenges

Of course there is more to AD than the forward and reverse mode. It is well known that a combination of the two, specifically so-called cross-country elimination on the computational graph can lead to lower operations counts for the calculation of Jacobian matrices. However, finding the optimal elimination order is certainly NP hard (Nau06). While this result lends some air of sophistication and possibly even respectability to the whole AD enterprise, it is not clear whether it is terribly relevant from a practical point of view. Storing and manipulating the whole computational graph is usually not advisable as it destroys locality of the memory accesses. Of course all this may make sense on the subroutine or block level, where some peephole optimization can be performed at a reasonable cost.

Another issue which has been known for a long time but never really tackled by the AD community is that of nonsmoothness. Many codes, especially space discretizations of stationary or unstationary PDEs contain **abs**, min and max as elementary operations so that the resulting composite function is merely Lipschitz but no longer differentiable. While by Rademacher's theorem the function will still be almost everywhere differentiable this means that local linearizations or second order Taylor models will have a very limited range of validity. In my view piecewise linearizations, which can be generated in an AD fashion open the door to constructive nonsmooth numerics, for optimization, equation solving and the integration of dynamical system.

References

[ea12]   Gauger, N., et al.: In: Leugering, G., Engell, S., Griewank, A., Hinze, M., Rannacher, R., Schulz, V., Ulbrich, M., Ulbrich, S. (eds.) Constrained optimization and optimal control for partial differential equations. International Series of Numerical Mathematics, pp. 99–122. Springer, Heidelberg (2012)

[GNH96] Green, L.L., Newman, P.A., Haigler, K.J.: Sensitivity derivatives for advanced CFD algorithm and viscous modeling parameters via automatic differentiation. J. Comp. Physics 125, 313–325 (1996)

[Gre89] Gress, O.O.: Rsic peripheral shielding routine collection. Technical report, Oak Ridge National Laboratory, Oak Ridge National Labority, Oak Ridge, Tennessee (February 1989)

[Gri12] Griewank, A.: Who invented the reverse mode of differentiation. In: Grötschel, M. (ed.) Optimization Stories, pp. 389–400. Deutsche Mathematikervereinigung, Bielefeld (2012)

[GW08] Griewank, A., Walther, A.: Principles and Techniques of Algorithmic Differentiation, 2nd edn. SIAM (2008)

[Kea09] Kearfott, R.B.: GlobSol user guide. Optimization Methods and Software 24(4-5), 687–708 (2009)

[KS91] Kaltofen, E., Singer, M.F.: Size efficient parallel algebraic circuits for partial derivatives. In: Shirkov, D.V., Rostovtsev, V.A., Gerdt, V.P. (eds.) IV International Conference on Computer Algebra in Physical Research, pp. 133–145. World Scientific Publ. (1991)

[MW80] Miller, W., Wrathall, C.: Software for Roundoff Analysis of Matrix Algorithms. Academic Press (1980)

[NP07] Nedialkov, N.S., Pryce, J.D.: Solving differential-algebraic equations by Taylor series (I): Computing the system jacobian. BIT 47(1), 121–135 (2007)

[Phi03] Phipps, E.T.: Taylor Series Integration of Differential-Algebraic Equations: Automatic Differentiation as a Tool for Simulating Rigid Body Mechanical Systems. PhD thesis, Cornell University (2003)

Polynomial Complexity of Solving Systems of Few Algebraic Equations with Small Degrees

Dima Grigoriev

CNRS, Mathématiques, Université de Lille, Villeneuve d'Ascq, 59655, France
Dmitry.Grigoryev@math.univ-lille1.fr
http://logic.pdmi.ras.ru/~grigorev

Abstract. An algorithm is designed which tests solvability of a system of k polynomial equations in n variables with degrees d within complexity polynomial in $n^{d^{3k}}$. If the system is solvable then the algorithm yields one of its solutions. Thus, for fixed d, k the complexity of the algorithm is polynomial.

Keywords: polynomial complexity, solving systems of few equations, small degrees.

Introduction

Consider a system of polynomial equations

$$f_1 = \cdots = f_k = 0, \tag{1}$$

where $f_1, \ldots, f_k \in \mathbb{Z}[X_1, \ldots, X_n]$, $\deg f_i \leq d$, $1 \leq i \leq k$. The algorithm from [1], [2] (see also [3]) solves (1) within complexity polynomial in M, k, d^{n^2}, where M denotes the bound on bit-sizes of (integer) coefficients of polynomials $f_1, \ldots, f_k$. Moreover, this algorithm finds the irreducible components of the variety in $\mathbb{C}^n$ determined by (1). We mention also that in [4] an algorithm is designed which tests solvability of (1) reducing it to a system of equations over $\mathbb{R}$, within a better complexity polynomial in M, $(k \cdot d)^n$. We note that the algorithm from [4] tests solvability of (1) and outputs a solution, provided that (1) is solvable, rather than finds the irreducible components as the algorithms from [1], [2].

In the present paper we design an algorithm which tests solvability of (1) within complexity polynomial in $M \cdot \binom{n+d^{3k}}{n} \leq M \cdot n^{d^{3k}}$, which provides polynomial (in the size $M \cdot k \cdot \binom{n+d}{n}$ of the input system (1)) complexity when d, k being fixed. If (1) is solvable then the algorithm yields one of its solutions. Note that the algorithm from [4] has a polynomial complexity when, say $d > n^2$ and k being polynomial in n; when d is close to n the complexity is subexponential, while for small d the complexity is exponential.

We mention that in [5] an algorithm was designed testing solvability of (1) over $\mathbb{R}$ (and finding a real solution, provided that it does exist) within the complexity polynomial in M, n^{2k} for quadratic equations ($d = 2$), and moreover, one can replace equations by inequalities.

V.P. Gerdt et al. (Eds.): CASC 2013, LNCS 8136, pp. 136–139, 2013.
© Springer International Publishing Switzerland 2013

It would be interesting to clarify, for which relations between n, k, and d the complexity of solvability of (1) is polynomial. In particular, when $d = 2$ and k is close to n the problem of solvability is NP-hard.

1   Testing Points for Sparse Polynomials

Recall (see [6]) a construction of testing points for sparse polynomials in n variables. Let p_i denote the ith prime and $s_j = (p_1^j, \ldots, p_n^j) \in \mathbb{Z}^n, j \geq 0$ be a point. A polynomial $f \in \mathbb{C}[X_1, \ldots, X_n]$ is called t-sparse if it contains at most t monomials.

Lemma 1. *For a t-sparse polynomial f there exists $0 \leq j < t$ such that $f(s_j) \neq 0$.*

The proof follows from the observation that writing $f = \sum_{1 \leq l \leq t} a_l \cdot X^{I_l}$ where coefficients $a_l \in \mathbb{C}$ and X^{I_l} are monomials, the equations $f(s_j) = 0, 0 \leq j < t$ lead to a $t \times t$ linear system with Vandermonde matrix and its solution $(a_1, \ldots, a_t)$. Since Vandermonde matrix is nonsingular, the obtained contradiction proves the lemma.

The substitution of points s_j was first introduced in the proof of theorem 1 [6].

Corollary 1. *Let $\deg f \leq D$. There exists $0 \leq j < \binom{n+D}{n}$ such that $f(s_j) \neq 0$.*

2   Reduction of Solvability to Systems in Few Variables

The goal of this section is to reduce testing solvability of (1) to testing solvability of several systems in k variables.

Let $V \subset \mathbb{C}^n$ be an irreducible (over $\mathbb{Q}$) component of the variety determined by (1). Observe that the algorithm described in the next Section does not need to produce V. Then $\dim V =: m \geq n - k$ and $\deg V \leq d^{n-m} \leq d^k$ due to Bezout inequality [7].

Let variables $X_{i_1}, \ldots, X_{i_m}$ constitute a transcendental basis over $\mathbb{C}$ of the field $\mathbb{C}(V)$ of rational functions on V, clearly such $i_1, \ldots, i_m$ do exist. Then the degree of fields extension $e := [\mathbb{C}(V) : \mathbb{C}(X_{i_1}, \ldots, X_{i_m})] \leq \deg V$ equals the typical (and at the same time, the maximal) number of points in the intersections $V \cap \{X_{i_1} = c_1, \ldots, X_{i_m} = c_m\}$ for different $c_1, \ldots, c_m \in \mathbb{C}$, provided that this intersection being finite. Observe that for almost all vectors $(c_1, \ldots, c_m) \in \mathbb{C}^n$ the intersection is finite and consists of e points.

There exists a primitive element $Y = \sum_{i \neq i_1, \ldots, i_m} b_i \cdot X_i$ of the extension $\mathbb{C}(V)$ of the field $\mathbb{C}(X_{i_1}, \ldots, X_{i_m})$ for appropriate integers b_i [8] (moreover, one can take integers $0 \leq b_i \leq e$ for all i, see e. g. [1], [2], but we do not need here these bounds). Moreover, there exist $n - m$ linearly over $\mathbb{C}$ independent primitive elements $Y_1, \ldots, Y_{n-m}$ of this form. One can view $Y_1, \ldots, Y_{n-m}, X_{i_1}, \ldots, X_{i_m}$ as new coordinates.

Consider a linear projection $\pi_l : \mathbb{C}^n \rightarrow \mathbb{C}^{m+1}$ onto the coordinates $Y_l, X_{i_1}, \ldots, X_{i_m}, 1 \leq l \leq n-m$. Then the closure $\overline{\pi_l(V)} \subset \mathbb{C}^{m+1}$ is an irreducible hypersurface, so $\dim \overline{\pi_l(V)} = m$. Denote by $g_l \in \mathbb{Q}[Y_l, X_{i_1}, \ldots, X_{i_m}]$ the minimal polynomial providing the equation of $\overline{\pi_l(V)}$. Then $\deg g_l = \deg \pi_l(V) \leq \deg V$ [7] and $\deg_{Y_l} g_l = e$, taking into account that Y_l is a primitive element.

Rewriting $g_l = \sum_{q \leq e} Y_l^q \cdot h_q$, $h_q \in \mathbb{Q}[X_{i_1}, \ldots, X_{i_m}]$ as a polynomial in a distinguished variable Y_l, we denote $H_l := h_e \cdot \mathrm{Disc}_{Y_l}(g_l) \in \mathbb{Q}[X_{i_1}, \ldots, X_{i_m}]$, where Disc_{Y_l} denotes the discriminant with respect to the variable Y_l (the discriminant does not vanish identically since Y_l is a primitive element). We have $\deg H_l \leq d^k + d^{2k}$. Consider the product $H := \prod_{1 \leq l \leq n-m} H_l$, then $D := \deg H \leq (n - m) \cdot (d^k + d^{2k}) \leq d^{3k}$.

Due to Corollary 1 there exists $0 \leq j < \binom{D+m}{D} \leq m^{d^{3k}}$ such that $H(s_j) = H(p_1^j, \ldots, p_m^j) \neq 0$. Observe that the projective intersection $\overline{V} \cap \{X_{i_1} = p_1^j \cdot X_0, \cdots, X_{i_m} = p_m^j \cdot X_0\}$ in the projective space $\mathbb{PC}^n \supset \mathbb{C}^n$ with the coordinates $[X_0 : X_1 : \cdots : X_n]$ consists of e points, where $\overline{V}$ denotes the projective closure of V. On the other hand, coordinate Y_l of the points of the affine intersection $V \cap \{X_{i_1} = p_1^j, \ldots, X_{i_m} = p_m^j\}$ attains e different values, taking into account that $H_l(s_j) \neq 0$, $1 \leq l \leq n - m$. Therefore, all e points from the projective intersection lie in the affine chart $\mathbb{C}^n$. Consequently, the intersection $V \cap \{X_{i_1} = p_1^j, \ldots, X_{i_m} = p_m^j\}$ is not empty.

3 Test of Solvability and Its Complexity

Thus, to test solvability of (1) the algorithm chooses all possible subsets $\{i_1, \ldots, i_m\} \subset \{1, \ldots, n\}$ with $m \geq n - k$ treating $X_{i_1}, \ldots, X_{i_m}$ as a candidate for a transcendental basis of some irreducible component V of the variety determined by (1). The number of these choices is bounded by $\binom{n}{m} < n^k$. After that for each $0 \leq j < \binom{D+m}{D}$ where $D \leq d^{3k}$, the algorithm substitutes $X_{i_1} = p_1^j, \ldots, X_{i_m} = p_m^j$ into polynomials $f_1, \ldots, f_k$ and solves the resulting system of polynomial equations in $n - m \leq k$ variables applying the algorithm from [1], [2]. The complexity of each of these applications does not exceed a polynomial in $M \cdot \binom{D+m}{D} \cdot d^{(n-m)^2}$, i. e., a polynomial in $M \cdot n^{d^{3k}}$. Moreover, the algorithm from [1], [2] yields an algebraic numbers solution of a system, provided that it does exist, in the symbolic way as follows. The algorithm produces an irreducible over $\mathbb{Q}$ polynomial $\phi(Z) \in \mathbb{Q}[Z]$ with degree $\deg(\phi) \leq d^{n-m}$ and polynomials $\phi_i(Z) \in \mathbb{Q}[Z]$, $1 \leq i \leq n$, $i \neq i_1, \ldots, i_m$ such that for a root $\theta \in \overline{\mathbb{Q}}$ of $\phi(\theta) = 0$ the point $(x_1, \ldots, x_n) \in \overline{\mathbb{Q}}^n$ with $x_{i_1} = p_1^j, \ldots, x_{i_m} = p_m^j$ and $x_i = \phi_i(\theta)$, $1 \leq i \leq n$, $i \neq i_1, \ldots, i_m$ is a solution of (1).

Summarizing, we obtain the following theorem.

Theorem 1. *One can test solvability over* $\mathbb{C}$ *of a system (1) of* k *polynomials* $f_1, \ldots, f_k \in \mathbb{Z}[X_1, \ldots, X_n]$ *with degrees* d *within complexity polynomial in* $M \cdot \binom{n+d^{3k}}{n} \leq M \cdot n^{d^{3k}}$, *where* M *bounds the bit-sizes of (integer) coefficients of* $f_1, \ldots, f_k$. *If (1) is solvable then the algorithm yields one of its solutions.*

Corollary 2. *For fixed* d, k *the complexity of the algorithm is polynomial.*

The construction and the Theorem extend literally to polynomials with coefficients from a field F of characteristic zero (for complexity bounds one needs that the elements of F are given in an efficient way). For F of a positive characteristic one can obtain similar results replacing the zero test from Section 1 by the zero test from [9].

Acknowledgements. The author is grateful to the Max-Planck Institut für Mathematik, Bonn for its hospitality during writing this paper and to Labex CEMPI (ANR-11-LABX-0007-01).

References

1. Chistov, A.: An algorithm of polynomial complexity for factoring polynomials, and determination of the components of a variety in a subexponential time. J. Soviet Math. 34, 1838–1882 (1986)
2. Grigoriev, D.: Polynomial factoring over a finite field and solving systems of algebraic equations. J. Soviet Math. 34, 1762–1803 (1986)
3. Chistov, A., Grigoriev, D.: Complexity of quantifier elimination in the theory of algebraically closed fields. In: Chytil, M.P., Koubek, V. (eds.) Mathematical Foundations of Computer Science 1984. LNCS, vol. 176, pp. 17–31. Springer, Heidelberg (1984)
4. Renegar, J.: On the computational complexity and geometry of the first-order theory of the reals. I. Introduction. Preliminaries. The geometry of semi-algebraic sets. The decision problem for the existential theory of the reals. J. Symbolic Comput. 13, 255–299 (1992)
5. Grigoriev, D., Pasechnik, D.: Polynomial-time computing over quadratic maps I. Sampling in real algebraic sets. Computational Complexity 14, 20–52 (2005)
6. Grigoriev, D., Karpinski, M.: The matching problem for bipartite graphs with polynomially bounded permanents is in NC. In: Proc. 28 Symp. Found. Comput. Sci., pp. 166–172. IEEE, New York (1987)
7. Shafarevich, I.: Foundations of algebraic geometry. MacMillan Journals (1969)
8. Lang, S.: Algebra. Springer (2002)
9. Grigoriev, D., Karpinski, M., Singer, M.: Fast parallel algorithms for sparse multivariate polynomial interpolation over finite fields. SIAM J. Comput. 19, 1059–1063 (1990)

Computing Divisors and Common Multiples of Quasi-linear Ordinary Differential Equations

Dima Grigoriev[1] and Fritz Schwarz[2]

[1] CNRS, Mathématiques, Université de Lille, Villeneuve d'Ascq, 59655, France
Dmitry.Grigoryev@math.univ-lille1.fr
http://logic.pdmi.ras.ru/~grigorev
[2] Fraunhofer Gesellschaft, Institut SCAI 53754 Sankt Augustin, Germany
fritz.schwarz@scai.fraunhofer.de
http://www.scai.fraunhofer.de/schwarz.html

Abstract. If solutions of a non-linear differential equation may be obtained by specialization of solutions of another equation we say that the former equation is a generalized divisor of the latter one. We design an algorithm which finds first-order quasi-linear generalized divisors of a second-order quasi-linear ordinary differential equation. If solutions of an equation contain solutions of a pair of equations we say that the equation is a common multiple of the pair. We prove that a quasi-linear common multiple of a pair of quasi-linear equations always exists and design an algorithm which yields a quasi-linear common multiple.

Keywords: quasi-linear ordinary differential equations, divisor of equations, multiple of equations.

Introduction

The problem of factoring linear ordinary differential operators $L = T \circ Q$ was studied in [1]. Algorithms for this problem were designed in [2], [3] (in [2] a complexity bound better than for the algorithm from [1] was established). In [4] an algorithm is exhibited for factoring a partial linear differential operator in two variables with a separable symbol. In [5] an algorithm is constructed for finding all first-order factors of a partial linear differential operator in two variables. A generalization of factoring for D-modules (in other words, for systems of linear partial differential operators) was considered in [6]. A particular case of factoring for D-modules is the Laplace problem [7], [8] (one can find a short exposition of the Laplace problem in [9]).

The meaning of factoring for search of solutions is that any solution of operator Q is a solution of operator L, thus factoring allows one to diminish the order of operators or its differential dimension.

Much less is known for factoring non-linear (even ordinary) differential equations. In Section 1, we design an algorithm for finding (first-order) generalized divisors of a second-order quasi-linear differential equation.

We note that our definition of generalized divisors is in the framework of differential ideals [12] rather than the definition of factorization from [10], [11]

V.P. Gerdt et al. (Eds.): CASC 2013, LNCS 8136, pp. 140–147, 2013.
© Springer International Publishing Switzerland 2013

being in terms of a composition of nonlinear ordinary differential polynomials. A decomposition algorithm is designed in [11].

One can also introduce a different (from [10], [11]) concept of composition (which yields generalized divisors) as follows. For a differential field K consider an operator $A = \sum_{0 \leq i \leq n} a_i \cdot \frac{d^i}{dx^i}$ acting on the algebra $K\{y\}$ of differential polynomials in y [12] where the coefficients $a_i \in K\{y\}$, the result of the action we denote by $A * z \in K\{y\}$ for $z \in K\{y\}$. Clearly, z is a generalized divisor of $A * z$. In Section 1, we show, conversely, that if a quasi-linear $z \in K\{y\}$ is a generalized divisor of $v \in K\{y\}$ then $v = A * z$ for an appropriate operator A.

Easy examples demonstrate that the two considered compositions differ from each other. In the sense of [10], [11] we have $1 = 1 \circ z$ for an arbitrary $z \in K\{y\}$, while 1 cannot be represented as $A * y$. On the other hand, $y \cdot y' = y * y'$, while one cannot represent $y \cdot y'$ as $g \circ y'$ for any $g \in K\{y\}$.

In Section 2, we define a common multiple of a pair of equations as an equation satisfied by the solutions of both equations. We prove the existence of a quasi-linear common multiple for any pair of quasi-linear differential equations, design an algorithm for computing a quasi-linear common multiple and bound its complexity in terms of Grzegorczyk's hierarchy of primitive-recursive functions.

It would be interesting to extend the algorithm from Section 1 to equations of arbitrary order and from quasi-linear to arbitrary non-linear ordinary equations, and then possibly to partial differential equations.

1   A Bound on the Degree and an Algorithm for Generalized Divisors

We study second-order non-linear ordinary differential equations of the form

$$y'' = f(y', y, x) \tag{1}$$

for a polynomial $f \in \overline{\mathbb{Q}}[z, y, x]$. We assume the coefficients of polynomials to be algebraic since we are interested in algorithms, although for the purpose of bounds (see below) one can consider coefficients from an arbitrary field with characteristic zero.

Definition 1. *We say that a first-order equation*

$$y' = p(y, x) \tag{2}$$

is a generalized divisor of (1), where $p \in \overline{\mathbb{Q}}[y, x]$, if any y satisfying (2) is a solution of (1).

It suffices to verify the condition in the definition just for *generic* y [12], i. e., y satisfying only the differential polynomials from the differential ideal generated by $y' - p(y, x)$. In particular, y is algebraically independent of x over $\mathbb{Q}$.

Lemma 1. *If (2) is a generalized divisor of (1) then the differential polynomial $y'' - f(y', y, x)$ has the form $A * (y' - p(y, x))$ for an operator $A = \frac{d}{dx} + a_0$ with a suitable differential polynomial $a_0 \in \overline{\mathbb{Q}}[y', y, x]$.*
The converse statement is evident.

Proof. Dividing with remainder (with respect to y') the differential polynomial

$$u := y'' - f(y', y, x) - (y' - p(y, x))' \in \overline{\mathbb{Q}}[y', y, x]$$

by $y' - p(y, x)$, we get a differential polynomial $a_0 \in \overline{\mathbb{Q}}[y', y, x]$ such that $u = a_0 \cdot (y' - p(y, x)) + v$ for suitable $v \in \overline{\mathbb{Q}}[y, x]$. Any solution y of (2) is a solution of v, hence $v \equiv 0$ since y is algebraically independent of x.

Remark 1. i) The proof of Lemma 1 provides an algorithm to test whether (2) is a generalized divisor of (1).

ii) Lemma 1 holds for an arbitrary quasi-linear differential polynomial of the form $y^{(n)} - p_n(y^{(n-1)}, \ldots, y, x)$ in place of (2) and for an arbitrary differential polynomial (not necessary quasi-linear) in place of (1).

From (2) we have

$$y'' = \frac{\partial p}{\partial y} \cdot p + \frac{\partial p}{\partial x}.$$

Substituting this into (1) and rewriting

$$f(y', y, x) = \sum_{0 \le i \le l} f_i \cdot (y')^i \tag{3}$$

where $f_i \in \overline{\mathbb{Q}}[y, x]$, we get

$$\frac{\partial p}{\partial y} \cdot p + \frac{\partial p}{\partial x} = \sum_{0 \le i \le l} f_i \cdot p^i. \tag{4}$$

Observe that (4) is equivalent to that (2) is a generalized divisor of (1).
Then (4) implies that

$$p \mid (f_0 - \frac{\partial p}{\partial x}). \tag{5}$$

Hence either $\deg_x p \le \deg_x f_0$ or $f_0 = \frac{\partial p}{\partial x}$. Indeed, expand $p = \sum_{0 \le j \le k} a_j \cdot x^j$ for certain polynomials $a_j \in \overline{\mathbb{Q}}[y]$, $a_k \ne 0$. If $k = \deg_x p > \deg_x f_0$ then $\deg(f_0 - \frac{\partial p}{\partial x}) < k$, therefore, $f_0 = \frac{\partial p}{\partial x}$ due to (5).

In a similar way, we claim that either $\deg_y p \le \deg_y f_0$ or $f_0 = \frac{\partial p}{\partial x}$. Indeed, expand $p = \sum_{0 \le i \le m} b_i \cdot y^i$ for certain polynomials $b_i \in \overline{\mathbb{Q}}[x]$, $b_m \ne 0$. If $m = \deg_y p > \deg_y f_0$ then the coefficient of $f_0 - \frac{\partial p}{\partial x}$ at monomial y^m equals $-\frac{\partial b_m}{\partial x}$. If $\frac{\partial b_m}{\partial x} \ne 0$, and thereby y^m is the leading monomial of $f_0 - \frac{\partial p}{\partial x}$ with respect to the expansion in y, we get a contradiction with (5). Therefore, $\frac{\partial b_m}{\partial x} = 0$ and $f_0 = \frac{\partial p}{\partial x}$ due to (5), which proves the claim.

So, it remains to consider the case $f_0 = \frac{\partial p}{\partial x}$. Then (4) entails that

$$\frac{\partial p}{\partial y} = \sum_{0 \le i \le l-1} f_{i+1} \cdot p^i,$$

hence $p|(f_1 - \frac{\partial p}{\partial y})$. Arguing as above, we deduce that either $\deg_x p \le \deg_x f_1$, $\deg_y p \le \deg_y f_1$ or $f_1 = \frac{\partial p}{\partial y}$.

We note that in the latter case $f_1 = \frac{\partial p}{\partial y}$, $f_0 = \frac{\partial p}{\partial x}$; hence $\sum_{2 \le i \le l} f_i \cdot p^{i-2} = 0$ because of (4), and p is determined uniquely up to an additive constant. Moreover, in this case $\deg_x p \le 1 + \deg_x f_0$, $\deg_y p \le 1 + \deg_y f_1$.

Summarizing, we conclude with the following theorem.

Theorem 1. *i) If (2) is a generalized divisor of (1) then either $\deg_x p \le \deg_x f$, $\deg_y p \le \deg_y f$ or $f_1 = \frac{\partial p}{\partial y}$, $f_0 = \frac{\partial p}{\partial x}$, see (3); this determines p up to an additive constant; in the latter case $\deg_x p \le 1 + \deg_x f_0$, $\deg_y p \le 1 + \deg_y f_1$.*

ii) An algorithm which either constructs a generalized divisor (2) of (1) or finds out that it does not exist, looks for polynomial p with indeterminate coefficients from $\overline{Q}$ satisfying the degree bounds

$$\deg_x p \le \max\{\deg_x f, 1 + \deg_x f_0\}, \quad \deg_y p \le \max\{\deg_y f, 1 + \deg_y f_1\}$$

from item i), solving (4) as a system of polynomial equations in the indeterminate coefficients of p.

Example 1. Consider the equation

$$E \equiv y'' + (x + 3y)y' + y^3 + xy^2 = 0. \tag{6}$$

According to the above theorem $\deg_x E = 1$ and $\deg_y E = 3$, i.e., $\deg_x p \le 1$ and $\deg_y p \le 3$. The second alternative $\frac{\partial p}{\partial x} = f_0 = -xy^2 - y^3$, $\frac{\partial p}{\partial y} = f_0 = -x - 3y$ does not apply because this system for p is inconsistent. Proceeding as described above, two divisors are obtained and the representations

$$E \equiv (y' + y^2)' + (y + x)(y' + y^2) \quad \text{and} \quad E = (y' + y^2 + xy - 1)' + y(y' + y^2 + xy - 1)$$

follow. They yield the two one-parameter solutions

$$y = \frac{1}{x + C} \quad \text{and} \quad y = \frac{1}{x} + \frac{1}{x^2} \frac{\exp\left(-\frac{1}{2}x^2\right)}{\int \exp\left(-\frac{1}{2}x^2\right)\frac{dx}{x^2} + C}$$

respectively.

An extension of the definition of a generalized divisor is the definition of a first integral.

Definition 2. *We say that $y' - p(y, x)$ is a first integral of (1) if for any constant c any solution y of equation $y' - p(y, x) = c$ is also a solution of (1).*

Denote by $A(c)$ a formula obtained from (4) by means of replacing p with $p + c$. Arguing as above, we get that $y - p(y, x)$ is a first integral of (1) iff $A(c)$ holds for any constant c. We obtain the same bound on $\deg p$ as in Theorem i). The algorithm for finding first integrals applies a quantifier elimination procedure to the following formula of the first-order theory of algebraically closed fields:

$$\exists P \forall c A(c)$$

where P denotes the vector of indeterminate coefficients of polynomial p. Thus, the algorithm finds the constructible set of all vectors P for which $\forall c A(c)$ holds. These vectors P provide all first integrals of (1).

2   Computing Common Multiples of Quasi-linear Differential Equations

Definition 3. *We say that a differential equation $f = 0$ is a common multiple of equations $f_1 = 0$ and $f_2 = 0$ if solutions of $f = 0$ contain solutions of both $f_1 = 0$ and $f_2 = 0$.*

The goal of this Section is to design an algorithm which for a given pair of quasi-linear ordinary differential equations yields a quasi-linear common multiple. To simplify the notation we assume that the equations are of first order: $y' = p(y, x)$ and $y' = q(y, x)$ where polynomials $p, q \in \mathbb{Q}[y, x]$, although one can extend the algorithm to equations of arbitrary orders almost literally.

Treating y as a generic solution [12] of either of two given equations, one can assume that y is algebraically independent of x over $\mathbb{Q}$.

First, the algorithm looks for a common multiple being a quasi-linear second-order equation $y'' = s(y', y, x)$ for a suitable polynomial $s(z, y, x) \in \mathbb{Q}[z, y, x]$. Hence

$$s(p, y, x) = p' = \frac{\partial p}{\partial y} \cdot p + \frac{\partial p}{\partial x}, \; s(q, y, x) = q' = \frac{\partial q}{\partial y} \cdot q + \frac{\partial q}{\partial x}.$$

Therefore

$$s(z, y, x) = r \cdot (z - p) + \frac{\partial p}{\partial y} \cdot p + \frac{\partial p}{\partial x} = t \cdot (z - q) + \frac{\partial q}{\partial y} \cdot q + \frac{\partial q}{\partial x}$$

for appropriate polynomials $r, t \in \mathbb{Q}[z, y, x]$, whence

$$(t - r) \cdot (z - q) + r \cdot (p - q) = (\frac{\partial p}{\partial y} \cdot p + \frac{\partial p}{\partial x}) - (\frac{\partial q}{\partial y} \cdot q + \frac{\partial q}{\partial x}). \tag{7}$$

There exist $r, t \in \mathbb{Q}[z, y, x]$ which fulfil (7) iff

$$(p - q) \mid (\frac{\partial p}{\partial y} \cdot p + \frac{\partial p}{\partial x}) - (\frac{\partial q}{\partial y} \cdot q + \frac{\partial q}{\partial x}).$$

If the latter relation holds, i. e., $r \cdot (p - q) = (p - q)'$ for a suitable $r \in \mathbb{Q}[y, x]$, one can put $t := r$ to get (7) and take $s(z, y, x) := r \cdot (z - p) + p' = r \cdot (z - q) + q'$ to obtain a quasi-linear common multiple $y'' = s(y', y, x)$. Conversely, when (7) holds, we substitute in it $z = q$. Thus, there exists a quasi-linear common multiple of the second order of a pair of equations $y' = p(y, x)$, $y' = q(y, x)$ iff $(p - q)' \in \langle p - q \rangle$, where $\langle p - q \rangle$ denotes the ideal generated by $p - q$.

More generally, following the same argument one can prove that

Lemma 2. *There exists a quasi-linear common multiple of the order $n + 1$ of a pair of first-order equations $y' = p(y, x)$, $y' = q(y, x)$ iff n-th derivative*

$$(p - q)^{(n)} \in \langle p - q, (p - q)^{(1)}, \ldots, (p - q)^{(n-1)} \rangle.$$

More explicitly, if the latter relation holds, i. e., $(p - q)^{(n)} = \sum_{0 \leq i < n} r_i \cdot (p - q)^{(i)}$ for some polynomials $r_i \in \mathbb{Q}[y, x]$, $0 \leq i < n$ then for polynomial

$$s_n(z_n, \ldots, z_1, y, x) := \sum_{0 \leq i < n} r_i \cdot (z_{i+1} - p^{(i)}) + p^{(n)} = \sum_{0 \leq i < n} r_i \cdot (z_{i+1} - q^{(i)}) + q^{(n)}$$

equation $y^{(n+1)} = s_n(y^{(n)}, \ldots, y', y, x)$ is a required quasi-linear common multiple.

For the proof we observe that $y^{(n+1)} = s_n(y^{(n)}, \ldots, y', y, x)$ for a polynomial $s_n \in \mathbb{Q}[z_n, \ldots, z_1, y, x]$ is a common multiple iff

$$s_n(p^{(n-1)}, \ldots, p, y, x) = p^{(n)}, \; s_n(q^{(n-1)}, \ldots, q, y, x) = q^{(n)}.$$

Therefore,

$$s_n(z_n, \ldots, z_1, y, x) - p^{(n)} \in \langle z_n - p^{(n-1)}, \ldots, z_1 - p \rangle,$$

$$s_n(z_n, \ldots, z_1, y, x) - q^{(n)} \in \langle z_n - q^{(n-1)}, \ldots, z_1 - q \rangle.$$

Subtracting two latter equalities we complete the proof of the lemma.

One can directly extend the lemma to a quasi-linear common multiple of a pair of quasi-linear equations of an arbitrary order.

Employing Hilbert's Idealbasissatz we obtain

Corollary 1. *Any pair of ordinary quasi-linear differential equations has a quasi-linear common multiple.*

Moreover, from the explicit bound on the Idealbasissatz [13] we obtain

Corollary 2. *Any pair of ordinary quasi-linear differential equations*

$$y^{(k)} = p_k(y^{(k-1)}, \ldots, y, x), \quad y^{(k)} = q_k(y^{(k-1)}, \ldots, y, x)$$

of order k with polynomials of degrees $\deg(p_k)$, $\deg(q_k) \leq d$ has a quasi-linear common multiple of order $g(d)$, where g is a primitive-recursive function from the class $\mathcal{E}^{k+2}$ of Grzegorczyk's hierarchy [14], [15].

This provides also a complexity bound of similar order of magnitude of the algorithm which looks for a quasi-linear common multiple by trying consecutively increasing orders of a candidate and solving the membership problem to an ideal (see Lemma 2), say, with the help of Gröbner basis.

In particular, in case of first-order equations $(k = 1)$ the function $g(d)$ grows exponentially.

Example 2. Let $E_1 \equiv y' + y^2 = 0$ and $E_2 \equiv y' + y = 0$. By Lemma 2 a multiple of order 2 does not exist; however, there is the following multiple of order 3 involving a parameter C:

$$E_3 \equiv y''' + (C-4)yy'' + (C+1)y'' + (2C-2)y'^2 + (2C+2)yy' + Cy' + Cy^2.$$

For $C = 0$ it simplifies to

$$E_0 \equiv y''' + 4yy'' + y'' - 2y'^2 + 2yy' = 0.$$

Applying again Theorem 1 the factors $y' + y^2$, $y' + y$ and y' are obtained.

The next example is interesting because it allows to determine the general solution of all equations involved.

Example 3. Let $E_1 \equiv y' + y^2 = 0$ and $E_2 \equiv y' = 0$ with solutions $y = \frac{1}{x+C}$ and $y = C$, respectively. The multiple of E_1 and E_2 yields $y'' + 2yy' = 0$ with the first integral $y' + y^2 = C$. Its general solution is $y = C_1 \tan(C_2 - C_1 x)$. It is not obvious how the latter solution is related to the two solutions involving a single parameter.

Remark 2. The general solution of the second-order equation in the preceding example may also be written as $y = C_1 \tanh(C_2 + C_1 x)$; the two representations are transformed into each other by multiplying the constants with the complex unit i and representing them in terms of exponentials. From the latter representation the constant solution may be obtained by taking the limit $C_2 \to \infty$. The first integral $y' + y^2 = C$ generalizes the divisor E_1; its existence simplifies the solution procedure because it provides already one of the constants involved in the general solution of the second-order equation.

Acknowledgements. The first author is grateful to the Max-Planck Institut für Mathematik, Bonn for its hospitality during writing this paper and to Labex CEMPI (ANR-11-LABX-0007-01). Both authors are thankful to an anonymous referee whose remarks have encouraged to improve the exposition.

References

1. Schlesinger, L.: Handbuch der Theorie der linearen Differentialgleichungen II. Teubner, Leipzig (1897)
2. Grigoriev, D.: Complexity of factoring and GCD calculating of ordinary linear differential operators. J. Symb. Comput. 10, 7–37 (1990)
3. Schwarz, F.: A factorization algorithm for linear ordinary differential equations. In: Proc. Intern. Symp. Symbol. Algebr. Comput., pp. 17–25. ACM Press, New York (1989)
4. Grigoriev, D., Schwarz, F.: Factoring and solving linear partial differential equations. Computing 73, 179–197 (2004)
5. Grigoriev, D.: Analogue of Newton-Puiseux series for non-holonomic D-modules and factoring. Moscow Math. J. 9, 775–800 (2009)
6. Grigoriev, D., Schwarz, F.: Loewy and primary decompositions of D-modules. Adv. Appl. Math. 38, 526–541 (2007)
7. Goursat, E.: Leçons sur l'intégration des équations aux dérivées partielles du 2nd ordre, Hermann, Paris, vol. II (1898)
8. Tsarev, S.: Generalized Laplace transformations and integration of hyperbolic systems of linear partial differential equations. In: Proc. Intern. ACM Symp. Symbol. Algebr. Comput., pp. 325–331. ACM Press, New York (2005)
9. Grigoriev, D., Schwarz, F.: Non-holonomic ideal in the plane and absolute factoring. In: Proc. Intern. ACM Symp. Symbol. Algebr. Comput., pp. 93–97. ACM Press, New York (2010)
10. Tsarev, S.: On factorization of nonlinear ordinary differential equations. In: Proc. Intern. ACM Symp. Symbol. Algebr. Comput., pp. 159–164. ACM Press, New York (1999)
11. Gao, X.-S., Zhang, M.: Decomposition of ordinary differential polynomials. Appl. Alg. Eng. Commun. Comput. 19, 1–25 (2008)
12. Kolchin, E.: Differential Algebra and Algebraic Groups. Academic Press, New York (1973)
13. Seidenberg, A.: On the length of a Hilbert ascending chain. Proc. Amer. Math. Soc. 29, 443–450 (1971)
14. Grzegorczyk, A.: Some classes of recursive functions. Rozprawy Matematiczne 4, 1–44 (1953)
15. Wagner, K., Wechsung, G.: Computational Complexity. Mathematics and its Applications 21 (1986)

Complexity in Tropical Algebra
(Invited Talk)

Dima Grigoriev

CNRS, Mathématiques, Université de Lille, Villeneuve d'Ascq, 59655, France
Dmitry.Grigoryev@math.univ-lille1.fr
http://logic.pdmi.ras.ru/~grigorev

Abstract. We give a survey on complexity results in tropical algebra.

Keywords: tropical semi-ring, solving tropical linear systems, complexity, dual Nullstellensatz.

1   Introduction and Basic Concepts

The basic concept of tropical algebra is the *tropical semi-ring* (see e. g. [1], [2]) T endowed with operations $\oplus$, $\otimes$. Thus, a tropical semi-ring is a semi-group w.r.t. each of the two operations.

If T is an ordered semi-group then T is a tropical semi-ring with inherited operations $\oplus := \min$, $\otimes := +$. If T is an ordered group then T is a *tropical semi-skew-field* w.r.t. the tropical division $\oslash := -$. If, in addition, T is abelian, then it is a *tropical semi-field*.

Examples • $\mathbb{Z}^+ := \{0 \le a \in \mathbb{Z}\}$, $\mathbb{Z}^+_\infty := \mathbb{Z}^+ \cup \{\infty\}$ are commutative tropical semi-rings. The element ∞ plays a role of 0, in its turn 0 plays a role of 1;

• $\mathbb{Z}$, $\mathbb{Z}_\infty$ are semi-fields;

• $n \times n$ matrices over $\mathbb{Z}_\infty$ form a non-commutative tropical semi-ring with the tropical multiplication: $(a_{ij}) \otimes (b_{kl}) := (\oplus_{1 \le j \le n} a_{ij} \otimes b_{jl})$.

Now we recall the concept of a tropical polynomial. First we define a *tropical monomial* $x^{\otimes i} := x \otimes \cdots \otimes x$, $Q = a \otimes x_1^{\otimes i_1} \otimes \cdots \otimes x_n^{\otimes i_n}$, its *tropical degree* $trdeg = i_1 + \cdots + i_n$. Then $Q = a + i_1 \cdot x_1 + \cdots + i_n \cdot x_n$ is a linear function in the classical notations. *A tropical polynomial* $f = \bigoplus_j (a_j \otimes x_1^{i_{j1}} \otimes \cdots \otimes x_n^{i_{jn}}) = \min_j\{Q_j\}$ is a convex piecewise linear function in the classical language.

A vector $x = (x_1, \ldots, x_n)$ is called a **tropical zero** of f if minimum $\min_j\{Q_j\}$ is attained for at least two different values of j.

If T is an ordered semi-group then a tropical linear function over T^n can be written as $\min_{1 \le i \le n}\{a_i + x_i\}$.

Tropical linear system

$$\min_{1 \le j \le n}\{a_{i,j} + x_j\}, \; 1 \le i \le m \tag{1}$$

V.P. Gerdt et al. (Eds.): CASC 2013, LNCS 8136, pp. 148–154, 2013.
© Springer International Publishing Switzerland 2013

(or $(m \times n)$-matrix $A = (a_{i,j})$) has a *tropical solution* $x = (x_1 \ldots, x_n)$ if for every row $1 \leq i \leq m$ there are two columns $1 \leq k < l \leq n$ such that

$$a_{i,k} + x_k = a_{i,l} + x_l = \min_{1 \leq j \leq n} \{a_{i,j} + x_j\}.$$

Coefficients $a_{i,j} \in \mathbb{Z}_\infty := \mathbb{Z} \cup \{\infty\}$. Not all $x_j = \infty$. For $a_{i,j} \in \mathbb{Z}$ we assume $0 \leq a_{i,j} \leq M$.

When $m < n$ system (1) always has a solution (due to a tropical analogue of the Cramer's rule).

An $n \times n$ matrix $A = (a_{i,j})$ is **tropically non-singular** if $\min_{\pi \in S_n} \{a_{1,\pi(1)} + \cdots + a_{n,\pi(n)}\}$ is attained for a *unique* permutation π. System (1) with $n \times n$ matrix has a tropical linear solution iff matrix A is tropically singular.

2 Complexity of Solving Tropical Linear Systems

Theorem 1. *One can solve an $m \times n$ tropical linear system (1) within complexity polynomial in n, m, M.*

Moreover, the algorithm either finds a solution over $\mathbb{Z}_\infty$ or produces an $n \times n$ tropically nonsingular submatrix of A ([3], [4]).

Note that the size of the input (1) of the algorithms is polynomial in n, m, $\log M$, thereby, the complexity of the algorithms in Theorem 1 is exponential (in $\log M$).

Corollary 1. *The problem of solvability of tropical linear systems is in the complexity class $NP \cap coNP$.*

Remark 1. i) The algorithm from [4] has also a complexity bound polynomial in $2^{nm}, \log M$ (improved in [5] to $\binom{m+n}{n}, \log M$) as well as an obvious algorithm which invokes linear programming.

ii) In [5] a family of matrices A with $M \asymp 2^n \asymp 2^m$ was constructed for which the algorithm from [4] runs with exponential complexity $\Omega(M)$.

iii) In [6] an example of $A = a_{ij}$ with $m = 2$, $n = 3$ where $a_{11} = a_{12} = a_{21} = 0$, $a_{22} = 1$, $a_{13} = a_{23} = M$ such that the algorithm from [3] runs with exponential complexity $\Omega(M)$. Observe that a similar example with a constant size matrix for the algorithm from [4] would be impossible due to i).

Question. Are tropical linear systems solvable within polynomial (in n, m, $\log M$) complexity (i. e., in the complexity class P)?

3 Tropical, Kapranov and Barvinok Ranks

Unlike the classical linear algebra in the tropical linear algebra at least three different concepts of rank of a matrix are considered [7]. The first concept is similar to its classical counterpart. The **tropical rank** $trk(A)$ of matrix A is the maximal size of its tropically nonsingular square submatrices.

Now we introduce the second concept. A **lifting** of A is a matrix $F = (f_{i,j})$ over the field of Newton-Puiseux series $K = R((t^{1/\infty}))$ for a field R such that the order $ord_t(f_{i,j}) = a_{i,j}$ where $f_{i,j} = b_1 \cdot t^{q_1} + b_2 \cdot t^{q_2} + \cdots$ with rational exponents $a_{i,j} = q_1 < q_2 < \cdots$ having common denominator, or $f_{i,j} = 0$ when $a_{i,j} = \infty$.

Kapranov rank $Krk_R(A)$ equals the minimum of ranks (over K) of liftings of A.

Finally, the third concept of **Barvinok rank** $Brk(A)$ is defined as the minimal q such that $A = (u_1 \otimes v_1) \oplus \cdots \oplus (u_q \otimes v_q)$ for suitable vectors $u_1, \ldots, v_q$ over T.

It is known that $trk(A) \le Krk_R(A) \le Brk(A)$, and both inequalities can be strict [7].

The following is known on the complexity of computing ranks.

• For $n \times n$ matrix B testing $trk(B) = n$ ($\Leftrightarrow B$ is tropically nonsingular) has polynomial complexity [8];

• Testing $trk(A) = r$ is NP-hard, testing $trk(A) \ge r$ is NP-complete [9];

• Solvability of polynomial equations over R is reducible to $Krk_R(A) = 3$ [10]. Thus, $Krk_R(A)$ can be incomputable, say when $R = GF[p](t)$;

• Testing $Brk(A) \le r$ is NP-complete [10].

Similar to the classical linear algebra we get from the algorithm designed in the proof of Theorem 1 a criterion on solvability of a tropical linear system.

Corollary 2. *The following statements are equivalent*
1) a tropical linear system with $m \times n$ matrix A has a solution;
2) $trk(A) < n$;
3) $Krk_R(A) < n$.

Remark 2. • The corollary holds for matrices over $\mathbb{R}_\infty$ [4].
• For matrices A over $\mathbb{R}$ the corollary was proved in [7].
• Equivalence of 1) and 2) was established in [11].

The set of solutions of (1) called a *tropical linear prevariety* is a contractible union of convex polyhedra in $\mathbb{R}^n$ [2]. One can treat a tropical linear prevariety as a subset of the *tropical projective space* in which two vectors from $\mathbb{R}^n$ are identified iff their difference is a mulptiple of vector $(1, \ldots, 1)$. Unlike the classical linear algebra it is difficult to compute the dimension of a tropical linear prevariety.

Proposition 1. *One can test uniqueness (in the tropical projective space) of a solution of a tropical linear system (i. e., whether the dimension of a tropical linear prevariety equals 0) within complexity polynomial in n, m, M [4].*

Theorem 2. *Computing the dimension of a tropical linear prevariety is NP-complete [6].*

System (1) is homogeneous, one can solve also tropical nonhomogeneous linear systems.

Proposition 2. *One can test solvability of a tropical nonhomogeneous linear system*

$\min_{1 \le j \le n} \{a_{i,j} + x_j, a_i\}$, $1 \le i \le m$
within complexity $(n \cdot m \cdot M)^{O(1)}$ *[4].*

We say that two tropical linear systems are equivalent if their prevarieties coincide.

Theorem 3. *One can reduce within* $(n \cdot m \cdot \log M)^{O(1)}$ *complexity testing equivalence of a pair of tropical linear systems to solving tropical linear systems ([12], [6]). The inverse reduction is evident.*

Similar to tropical linear systems are min-plus linear systems [1]

$$\min_{1 \le j \le n_1} \{a_{i,j} + x_j\} = \min_{1 \le j \le n_2} \{b_{i,j} + y_j\}, 1 \le i \le m. \tag{2}$$

Theorem 4. *One can test solvability of a min-plus linear system within complexity polynomial in* M, n_1, n_2, m. *If the system is solvable the algorithm yields one of its solutions ([13], [3]).*

We say that two min-plus linear systems are equivalent if they have the same sets of solutions.

Theorem 5. *Complexities of the following 4 problems coincide up to a polynomial: solvability and equivalence of both min-plus and of tropical linear systems ([12], [6]).*

We also mention two other problems close to tropical linear systems. **Min-atom problem** is a system of inequalities of the form $\min\{x, y\} + c \le z$, $c \in \mathbb{Z}$ (it could be viewed as a min-plus linear programming).

The second problem is a **mean payoff game**. Let a bipartite graph (V, W, E) with integer weights a_{ij} on edges $e_{ij} \in E$ be given. Two players in turn move a token between nodes $V \cup W$ of the graph. The first player moves from a (current) node $i \in V$ to a node $j \in W$ (respectively, the second player moves from W to V). Weight a_{ij} is assigned to this move. The mean sum of assigned weights after k moves is computed: $(\sum a_{ij})/k$.

If $\liminf_{k \to \infty} (\sum a_{ij})/k > 0$ then the first player wins. The problem of mean payoff games consists in whether the first player has a winning strategy?

Theorem 6. *The following 4 problems are equivalent up to a polynomial complexity: mean payoff games, min-atom, min-plus linear systems and tropical linear systems ([14], [3]).*

4   Tropical Polynomial Systems and Nullstellensatz

Similar to (1) one can consider tropical polynomial systems $f_1, \ldots, f_m$ where $f_1, \ldots, f_m$ being tropical polynomials and similar to (2) min-plus polynomial systems $f_1 = g_1, \ldots, f_m = g_m$ where $f_1, \ldots, f_m, g_1, \ldots, g_m$ being tropical polynomials.

Unlike the classical algebra factoring tropical univariate polynomials is NP-hard [9]. On the contrary, for solving polynomial systems the following result on tropical polynomials is similar to its classical counterpart.

Theorem 7. *Solvability of tropical polynomial systems is NP-complete ([15]).*

Theorem 8. *Solvability of min-plus polynomial systems $f_i = 0, 1 \leq i \leq m$ where f_i are min-plus polynomials, is NP-complete ([16]).*

How to reduce tropical polynomial systems to tropical linear ones?

In the classical algebra, Nullstellensatzfor serves this aim.

In the tropical world, the direct version of Nullstellensatz is false even for linear univariate polynomials: $X \oplus 0$, $X \oplus 1$ do not have a common tropical solution, while their (tropical) ideal does not contain 0 or any other monomial (tropical monomials are the only polynomials without tropical zeroes).

That is why we study a *dual* Nullstellensatz [17], and first we start with a classical dual Nullstellensatz. For polynomials $g_1, \ldots, g_s \in \mathbb{C}[X_1, \ldots, X_k]$ consider an infinite Cayley matrix C with the columns indexed by monomials X^I and the rows by shifts $X^J \cdot g_i$.

Nullstellensatz states that system $g_1 = \cdots = g_s = 0$ has no solution iff a linear combination of the rows of a suitable *finite* submatrix C_0 of C (generated by a set of rows of C) equals vector $(1, 0, \ldots, 0)$ where the first coordinate corresponds to monomial 1.

Effective Nullstellensatz provides a bound polynomial in $(\max_i\{\deg(g_i)\})^n$ on the size of C_0 ([18], [19]).

Dual Nullstellensatz (being equivalent to the classical Nullstellensatz) states that $g_1 = \cdots = g_s = 0$ has a solution iff for any finite submatrix C_0 of C linear system $C_0 \cdot (y_0, \ldots, y_N) = 0$ has a solution with $y_0 \neq 0$.

Infinite dual Nullstellensatz (also being equivalent to the classical Nullstellensatz) states that $g_1 = \cdots = g_s = 0$ has a solution iff infinite linear system $C \cdot (y_0, \ldots) = 0$ has a solution with $y_0 \neq 0$.

Remark 3. Nullstellensatz deals with ideal $\langle g_1, \ldots, g_s \rangle$, while dual Nullstellensatz forgets the ideal, therefore, gives a hope to hold in the tropical setting.

Now we proceed to the *tropical* dual Nullstellensatz. Assume w.l.o.g. that for tropical polynomials $h = \bigoplus_J (a_J \otimes X^{\otimes J})$ in k variables which we consider, function $J \to a_J$ is concave on $\mathbb{R}^k$. This assumption does not change tropical prevarieties.

For tropical polynomials $h_1, \ldots, h_s$ consider (infinite in all 4 directions) Cayley matrix H with the rows indexed by $X^{\otimes I} \otimes h_i$ for $I \in \mathbb{Z}^k$, $1 \leq i \leq s$ and with the columns indexed by $X^{\otimes I}$, $I \in \mathbb{Z}^k$. Now we formulate a **tropical dual effective Nullstellensatz** ([20], for univariate tropical polynomials established in [17]).

Theorem 9. *Tropical polynomials $h_1, \ldots, h_s$ have a zero iff tropical linear system $H_0 \otimes (z_0, \ldots, z_N)$ has a solution with $z_0 \neq \infty$ where H_0 is (finite) submatrix of H generated by its rows $X^{\otimes I} \otimes h_i$ for*
$$0 \leq |I| \leq (n+2) \cdot (trdeg(h_1) + \cdots + trdeg(h_s)), \ 1 \leq i \leq s.$$

The conjecture is that one can replace the latter bound by
$$O(trdeg(h_1) + \cdots + trdeg(h_s)).$$

For two tropical univariate $(n = 1)$ polynomials $(s = 2)$ the bound $trdeg(h_1) + trdeg(h_2)$ holds using the classical resultant and Kapranov's theorem [21].

Now we give a convex-geometrical rephrasing of the tropical (infinite) dual Nullstellensatz. For a tropical polynomial $h = \bigoplus_J (a_J \otimes X^{\otimes J})$ consider its extended Newton polyhedron G being the convex hull of the graph $\{(J, a) : a \leq -a_J\} \subset \mathbb{R}^{k+1}$. As vertices of G consider all the points of the form (I, c), $I \in \mathbb{Z}^k$ on the boundary of G. Let G_i correspond to h_i, $1 \leq i \leq s$. Denote by $G^{(I)} := G + (I, 0)$ a horizontal shift of G.

We treat the solution $Y := \{(J, y_J)\} \subset \mathbb{R}^{k+1}$ of a tropical linear system $H \otimes Y$ also as a graph on $\mathbb{R}^k$.

The tropical (infinite) dual Nullstellensatz is equivalent to the following.

For any I, i take the maximal $b := b_{I,i}$ such that a vertical shift $G_i^{(I)} + (0, b) \leq Y$ (pointwise as graphs). Assume that $G_i^{(I)} + (0, b)$ has at least two common points with Y. Then there is a hyperplane in $\mathbb{R}^{k+1}$ (not containing the vertical line) which supports (after a parallel shift) each G_i, $1 \leq i \leq s$ at least at two points.

We note that in [22] the tropical (customary) Nullstellensatz was established for a "ghost" semi-ring introduced there. In [23] the radical of a tropical ideal was explicitly described.

We also mention that in [24] an example of a linear polynomial ideal is exhibited with an exponential lower bound of the size of its tropical bases.

Acknowledgements. The author is grateful to the Max-Planck Institut für Mathematik, Bonn for its hospitality during writing this paper and to Labex CEMPI (ANR-11-LABX-0007-01).

References

1. Butkovic, P.: Max-linear systems: theory and algorithms. Springer, Berlin (2010)
2. Itenberg, I., Mikhalkin, G., Shustin, E.: Tropical algebraic geometry. In: Oberwolfach Seminars, Birkhäuser, Basel (2009)
3. Akian, M., Gaubert, S., Guterman, A.: Tropical polyhedra are equivalent to mean payoff games. Int. J. Algebra Comput. 22, 1793–1835 (2012)
4. Grigoriev, D.: Complexity of solving tropical linear systems. Comput. Complexity 22, 71–88 (2013)
5. Davydow, A.: Upper and lower bounds for Grigoriev's algorithm for solving integral tropical linear systems. Zap. Nauchn. Sem. POMI St.Petersbourg 402, 69–82 (2012)
6. Grigoriev, D., Podolskii, V.: Complexity of tropical and min-plus prevarieties. Preprint MPIM, 2013-23 (2012)
7. Develin, M., Santos, F., Sturmfels, B.: On the rank of a tropical matrix. Math. Sci. Res. Inst. Publ. 52, 213–242 (2005)
8. Butkovic, P., Hevery, F.: A condition for the strong regularity of matrices in the minimax algebra. Discr. Appl. Math. 11, 209–222 (1985)
9. Kim, K., Roush, F.: Factorization of polynomials in one variable over the tropical semiring. arXiv:math/050116/v2

10. Kim, K., Roush, F.: Kapranov rank vs. tropical rank. Proc. Amer. Math. Soc. 134, 2487–2494 (2006)
11. Izhakian, Z., Rowen, L.: The tropical rank of a tropical matrix. Communic. Algebra 37, 3912–3927 (2009)
12. Allamigeon, X., Gaubert, S., Katz, R.: Tropical polar cones, hypergraph transversals, and mean payoff games. Lin. Alg. and Its Appl. 435, 1549–1574 (2011)
13. Butkovic, P., Hegeduš, G.: An elimination method for finding all solutions of the system of linear equations over an extremal algebra. Ekonom.-Mat. Obzor 20, 203–215 (1984)
14. Bezem, M., Nieuwenhuis, R., Rodriguez-Carbonell, E.: Hard problem in max-algebra, control theory, hypergraphs and other areas. Inf. Procss. Lett. 110, 133–138 (2010)
15. Theobald, T.: On the frontiers of polynomial computations in tropical geometry. J. Symbolic Comput. 41, 1360–1375 (2006)
16. Grigoriev, D., Shpilrain, V.: Tropical cryptography. Preprint MPIM, Bonn (2011)
17. Grigoriev, D.: On a tropical dual Nullstellensatz. Adv. Appl. Math. 48, 457–464 (2012)
18. Giusti, M., Heintz, J., Sabia, J.: On the efficiency of effective Nullstellensaetze. Comput. Complexity 3, 56–95 (1993)
19. Kollár, J.: Sharp effective Nullstellensatz. J. Amer. Math. Soc. 1, 963–975 (1988)
20. Grigoriev, D., Podolskii, V.: Tropical dual effective Nullstellensatz (in preparation)
21. Tabera, L.: Tropical resultants for curves and stable intersection. Rev. Mat. Iberoam. 24, 941–961 (2008)
22. Izhakian, Z.: Tropical algebraic sets, ideals and an algebraic Nullstellensatz. Internat. J. Algebra Comput. 18, 1067–1098 (2008)
23. Shustin, E., Izhakian, Z.: A tropical Nullstellensatz. Proc. Amer. Math. Soc. 135, 3815–3821 (2007)
24. Bogart, T., Jensen, A., Speyer, D., Sturmfels, B., Thomas, R.: Computing tropical varieties. J. Symb. Comput. 42, 54–73 (2007)

Symbolic-Numerical Algorithm for Generating Cluster Eigenfunctions: Identical Particles with Pair Oscillator Interactions

Alexander Gusev[1], Sergue Vinitsky[1], Ochbadrakh Chuluunbaatar[1],
Vitaly Rostovtsev[1], Luong Le Hai[1,2], Vladimir Derbov[3],
Andrzej Góźdź[4], and Evgenii Klimov[5]

[1] Joint Institute for Nuclear Research, Dubna, Moscow Region, Russia
gooseff@jinr.ru
[2] Belgorod State University, Belgorod, Russia
[3] Saratov State University, Saratov, Russia
[4] Department of Mathematical Physics, Institute of Physics,
University of Maria Curie–Skłodowska, Lublin, Poland
[5] Tver State University, Tver, Russia

Abstract. The quantum model of a cluster, consisting of A identical particles, coupled by the internal pair interactions and affected by the external field of a target, is considered. A symbolic-numerical algorithm for generating $A-1$-dimensional oscillator eigenfunctions, symmetric or antisymmetric with respect to permutations of A identical particles in the new symmetrized coordinates, is formulated and implemented using the MAPLE computer algebra system. Examples of generating the symmetrized coordinate representation for $A-1$ dimensional oscillator functions in one-dimensional Euclidean space are analyzed. The approach is aimed at solving the problem of tunnelling the clusters, consisting of several identical particles, through repulsive potential barriers of a target.

1 Introduction

Quantum harmonic oscillator wave functions have a lot of applications in modern physics, particularly, as a basis for constructing the wave functions of a quantum system, consisting of A identical particles, totally symmetric or antisymmetric with respect to permutations of coordinates of the particles [1]. Various special methods, algorithms, and programs (see, e.g., [1–9]) were used to construct the desired solutions in the form of linear combinations of the eigenfunctions of an $A-1$-dimensional harmonic oscillator that are totally symmetric (or antisymmetric) with respect to the coordinate permutations. However, the implementation of this procedure in closed analytical form is still an open problem [10].

A promising approach to the construction of oscillator basis functions for four identical particles was proposed in [2–4]. It was demonstrated that a clear algorithm for generating symmetric (S) and antisymmetric (A) states can be obtained using the symmetrized coordinates instead of the conventional Jacobi

V.P. Gerdt et al. (Eds.): CASC 2013, LNCS 8136, pp. 155–168, 2013.
© Springer International Publishing Switzerland 2013

coordinates. However, until now this approach was not generalized for a quantum system comprising an arbitrary number A of identical particles.

We intend to develop this approach in order to describe the tunnelling of clusters, consisting of several coupled identical particles, through repulsive potential barriers of a target. Previously this problem was solved only for a pair of coupled particles [11, 12]. The developed approach will be also applicable to the microscopic study of tetrahedral- and octahedral-symmetric nuclei [13] that can be considered in the basis of seven-dimensional harmonic oscillator eigenfunctions [14]. *The aim of this paper is to present a convenient formulation of the problem stated above and the calculation methods, algorithms, and programs for solving it.*

In this paper, we consider the quantum model of a cluster, consisting of A identical particles with the internal pair interactions, under the influence of the external field of a target. We assume that the spin part of the wave function is known, so that only the spatial part of the wave function is to be considered, which can be either symmetric or antisymmetric with respect to a permutation of A identical particles [15–17]. The initial problem is reduced to the problem for a composite system whose internal degrees of freedom describe an $(A - 1) \times d$-dimensional oscillator, and the external degrees of freedom describe the center-of-mass motion of A particles in the d-dimensional Euclidean space. For simplicity, we restrict our consideration to the so-called s-wave approximation [11] corresponding to one-dimensional Euclidean space ($d = 1$). It is shown that the reduction is provided by using appropriately chosen symmetrized coordinates rather than the conventional Jacoby coordinates.

The main goal of introducing the symmetrized coordinates is to provide the invariance of the Hamiltonian with respect to permutations of A identical particles. This allows construction not only of basis functions, symmetric or antisymmetric under permutations of $A - 1$ relative coordinates, but also of basis functions, symmetric (S) or antisymmetric (A) under permutations of A Cartesian coordinates of the initial particles. We refer the expansion of the solution in the basis of such type as the Symmetrized Coordinate Representation (SCR).

The paper is organized as follows. In Section 2, we present the statement of the problem in the conventional Jacobi and the symmetrized coordinates. In Section 3, we introduce the SCR of the solution of the considered problem and describe the appropriate algorithm implemented using the MAPLE computer algebra system. In Section 4, we analyze some examples of generating the symmetrized coordinate representation for $A - 1$-dimensional oscillator functions in one-dimensional Euclidean space. In Conclusion, we summarize the results and discuss briefly the prospects of application of the developed approach.

2   Problem Statement

Consider the system of A identical quantum particles with the mass m and the set of Cartesian coordinates $x_i \in \mathbf{R}^d$ in the d-dimensional Euclidean space, considered as the vector $\tilde{\mathbf{x}} = (\tilde{x}_1, ..., \tilde{x}_A) \in \mathbf{R}^{A \times d}$ in the $A \times d$-dimensional configuration space. The particles are coupled by the pair potential $\tilde{V}^{pair}(\tilde{x}_{ij})$ depending on the relative positions, $\tilde{x}_{ij} = \tilde{x}_i - \tilde{x}_j$, similar to that of a harmonic oscillator $\tilde{V}^{hosc}(\tilde{x}_{ij}) = \frac{m\omega^2}{2}(\tilde{x}_{ij})^2$ with the frequency ω. The whole system is subject to the influence of the potentials $\tilde{V}(\tilde{x}_i)$ describing the external field of a target. The system is described by the Schrödinger equation

$$\left[-\frac{\hbar^2}{2m} \sum_{i=1}^{A} \frac{\partial^2}{\partial \tilde{x}_i^2} + \sum_{i,j=1;i<j}^{A} \tilde{V}^{pair}(\tilde{x}_{ij}) + \sum_{i=1}^{A} \tilde{V}(\tilde{x}_i) - \tilde{E} \right] \tilde{\Psi}(\tilde{\mathbf{x}}) = 0,$$

where $\tilde{E}$ is the total energy of the system of A particles and $\tilde{P}^2 = 2m\tilde{E}/\hbar^2$, $\tilde{P}$ is the total momentum of the system, and $\hbar$ is Planck constant. Using the oscillator units $x_{osc} = \sqrt{\hbar/(m\omega\sqrt{A})}$, $p_{osc} = \sqrt{(m\omega\sqrt{A})/\hbar} = x_{osc}^{-1}$, and $E_{osc} = \hbar\omega\sqrt{A}/2$ to introduce the dimensionless coordinates $x_i = \tilde{x}_i/x_{osc}$, $x_{ij} = \tilde{x}_{ij}/x_{osc} = x_i - x_j$, $E = \tilde{E}/E_{osc} = P^2$, $P = \tilde{P}/p_{osc} = \tilde{P}x_{osc}$, $V^{pair}(x_{ij}) = \tilde{V}^{pair}(x_{ij}x_{osc})/E_{osc}$, $V^{hosc}(x_{ij}) = \tilde{V}^{hosc}(x_{ij}x_{osc})/E_{osc} = \frac{1}{A}(x_{ij})^2$ and $V(x_i) = \tilde{V}(x_ix_{osc})/E_{osc}$, one can rewrite the above equation in the form

$$\left[-\sum_{i=1}^{A} \frac{\partial^2}{\partial x_i^2} + \sum_{i,j=1;i<j}^{A} \frac{1}{A}(x_{ij})^2 + \sum_{i,j=1;i<j}^{A} U^{pair}(x_{ij}) + \sum_{i=1}^{A} V(x_i) - E \right] \Psi(\mathbf{x}) = 0, (1)$$

where $U^{pair}(x_{ij}) = V^{pair}(x_{ij}) - V^{hosc}(x_{ij})$, i.e., if $V^{pair}(x_{ij}) = V^{hosc}(x_{ij})$, then $U^{pair}(x_{ij}) = 0$.

Our goal is to find the solutions $\Psi(x_1, ..., x_A)$ of Eq. (1), totally symmetric (or antisymmetric) with respect to the permutations of A particles that belong to the permutation group S_n [16]. The permutation of particles is nothing but a permutation of the Cartesian coordinates $x_i \leftrightarrow x_j$, $i, j = 1, ..., A$. First we introduce the Jacobi coordinates, $y = Jx$, following one of the possible definitions:

$$y_0 = \frac{1}{\sqrt{A}} \left(\sum_{t=1}^{A} x_t \right), \quad y_s = \frac{1}{\sqrt{s(s+1)}} \left(\sum_{t=1}^{s} x_t - sx_{s+1} \right), \quad s = 1, .., A-1. (2)$$

In the matrix form Eqs. (2) read as

$$\begin{pmatrix} y_0 \\ y_1 \\ y_2 \\ y_3 \\ \vdots \\ y_{A-1} \end{pmatrix} = J \begin{pmatrix} x_1 \\ x_2 \\ x_3 \\ \vdots \\ x_{A-1} \\ x_A \end{pmatrix}, \quad J = \begin{pmatrix} 1/\sqrt{A} & 1/\sqrt{A} & 1/\sqrt{A} & 1/\sqrt{A} & \cdots & 1/\sqrt{A} \\ 1/\sqrt{2} & -1/\sqrt{2} & 0 & 0 & \cdots & 0 \\ 1/\sqrt{6} & 1/\sqrt{6} & -2/\sqrt{6} & 0 & \cdots & 0 \\ 1/\sqrt{12} & 1/\sqrt{12} & 1/\sqrt{12} & -3/\sqrt{12} & \cdots & 0 \\ \vdots & \vdots & \vdots & \vdots & \ddots & \vdots \\ \frac{1}{\sqrt{A^2-A}} & \frac{1}{\sqrt{A^2-A}} & \frac{1}{\sqrt{A^2-A}} & \frac{1}{\sqrt{A^2-A}} & \cdots & -\frac{A-1}{\sqrt{A^2-A}} \end{pmatrix},$$

The inverse coordinate transformation $x = J^{-1}y$ is implemented using the transposed matrix $J^{-1} = J^T$, i.e., J is an orthogonal matrix with pairs of complex conjugate eigenvalues, the absolute values of which are equal to one. The Jacobi coordinates have the property $\sum_{i=0}^{A-1}(y_i \cdot y_i) = \sum_{i=1}^{A}(x_i \cdot x_i) = r^2$. Therefore,

$$\sum_{i,j=1}^{A}(x_{ij})^2 = 2A\sum_{i=0}^{A-1}(y_i)^2 - 2(\sum_{i=1}^{A}x_i)^2 = 2A\sum_{i=1}^{A-1}(y_i)^2,$$

so that Eq. (1) takes the form

$$\left[-\frac{\partial^2}{\partial y_0^2} + \sum_{i=1}^{A-1}\left(-\frac{\partial^2}{\partial y_i^2} + (y_i)^2\right) + U(y_0, ..., y_{A-1}) - E\right]\Psi(y_0, ..., y_{A-1}) = 0,$$

$$U(y_0, ..., y_{A-1}) = \sum_{i,j=1;i<j}^{A} U^{pair}(x_{ij}(y_1, ..., y_{A-1})) + \sum_{i=1}^{A} V(x_i(y_0, ..., y_{A-1})),$$

which, as follows from Eq. (2), is *not invariant* with respect to permutations $y_i \leftrightarrow y_j$ at $i, j = 1, ..., A - 1$.

Symmetrized Coordinates

The transformation from the Cartesian coordinates to one of the possible choices of the symmetrized ones ξ_i has the form, $\xi = Cx$ and $x = C\xi$:

$$\xi_0 = \frac{1}{\sqrt{A}}\left(\sum_{t=1}^{A} x_t\right), \ \ \xi_s = \frac{1}{\sqrt{A}}\left(x_1 + \sum_{t=2}^{A} a_0 x_t + \sqrt{A}x_{s+1}\right), \ \ s = 1, ..., A-1,$$

$$x_1 = \frac{1}{\sqrt{A}}\left(\sum_{t=0}^{A-1}\xi_t\right), \ \ x_s = \frac{1}{\sqrt{A}}\left(\xi_0 + \sum_{t=1}^{A-1} a_0 \xi_t + \sqrt{A}\xi_{s-1}\right), \ \ s = 2, ..., A,$$

or, in the matrix form,

$$\begin{pmatrix} \xi_0 \\ \xi_1 \\ \xi_2 \\ \vdots \\ \xi_{A-2} \\ \xi_{A-1} \end{pmatrix} = C \begin{pmatrix} x_1 \\ x_2 \\ x_3 \\ \vdots \\ x_{A-1} \\ x_A \end{pmatrix}, \ \ C = \frac{1}{\sqrt{A}}\begin{pmatrix} 1 & 1 & 1 & 1 & \cdots & 1 & 1 \\ 1 & a_1 & a_0 & a_0 & \cdots & a_0 & a_0 \\ 1 & a_0 & a_1 & a_0 & \cdots & a_0 & a_0 \\ 1 & a_0 & a_0 & a_1 & \cdots & a_0 & a_0 \\ \vdots & \vdots & \vdots & \vdots & \ddots & \vdots & \vdots \\ 1 & a_0 & a_0 & a_0 & \cdots & a_1 & a_0 \\ 1 & a_0 & a_0 & a_0 & \cdots & a_0 & a_1 \end{pmatrix}, \quad (3)$$

where $a_0 = 1/(1 - \sqrt{A}) < 0$, $a_1 = a_0 + \sqrt{A}$. The inverse coordinate transformation is performed using the same matrix $C^{-1} = C$, $C^2 = I$, i. e., $C = C^T$ is a symmetric orthogonal matrix with the eigenvalues $\lambda_1 = -1$, $\lambda_2 = 1, ..., \lambda_A = 1$ and $\det C = -1$. For $A = 2$, the symmetrized variables (3) are within normalization factors similar to the symmetrized Jacobi coordinates (2) considered in [9], while at $A = 4$ they correspond to another choice of symmetrized coordinates $(\ddot{x}_4, \ddot{x}_1, \ddot{x}_2, \ddot{x}_3)^T = C(x_4, x_1, x_2, x_3)^T$ considered in [2–4] and mentioned earlier in [5, 18]. We could not find a general definition of symmetrized coordinates for A-identical particles like (3) in the available literature, so we believe that in the present paper it is introduced for the first time. With the relations $a_1 - a_0 = \sqrt{A}$, $a_0 - 1 = a_0\sqrt{A}$ taken into into account, the relative coordinates $x_{ij} \equiv x_i - x_j$ of a pair of particles i and j are expressed in terms of the internal $A - 1$ symmetrized coordinates only:

$$x_{ij} \equiv x_i - x_j = \xi_{i-1} - \xi_{j-1} \equiv \xi_{i-1,j-1},$$

$$x_{i1} \equiv x_i - x_1 = \xi_{i-1} + a_0 \sum_{i'=1}^{A-1} \xi_{i'}, \quad i, j = 2, ..., A. \tag{4}$$

So, if only the absolute values of x_{ij} are to be considered, then there are $(A - 1)(A - 2)/2$ old relative coordinates transformed into new relative ones and $A - 1$ old relative coordinates expressed in terms of $A - 1$ internal symmetrized coordinates. These important relations essentially simplify the procedures of symmetrization (or antisymmetrization) of the oscillator basis functions and the calculations of the corresponding pair-interaction integrals $V^{pair}(x_{ij})$. The symmetrized coordinates are related to the Jacobi ones as $y = B\xi$, $B = JC$:

$$
\begin{pmatrix} y_0 \\ y_1 \\ y_2 \\ \vdots \\ y_{A-2} \\ y_{A-1} \end{pmatrix} = B \begin{pmatrix} \xi_0 \\ \xi_1 \\ \xi_2 \\ \vdots \\ \xi_{A-2} \\ \xi_{A-1} \end{pmatrix}, \quad B = \begin{pmatrix} 1 & 0 & 0 & 0 & 0 & \cdots & 0 & 0 \\ 0 & b_1^0 & b_1^- & b_1^- & b_1^- & \cdots & b_1^- & b_1^- \\ 0 & b_2^+ & b_2^0 & b_2^- & b_2^- & \cdots & b_2^- & b_2^- \\ 0 & b_3^+ & b_3^+ & b_3^0 & b_3^- & \cdots & b_3^- & b_3^- \\ 0 & b_4^+ & b_4^+ & b_4^+ & b_4^0 & \cdots & b_4^- & b_4^- \\ \vdots & \vdots & \vdots & \vdots & \vdots & \ddots & \vdots & \vdots \\ 0 & b_{A-1}^+ & b_{A-1}^+ & b_{A-1}^+ & b_{A-1}^+ & \cdots & b_{A-1}^+ & b_{A-1}^0 \end{pmatrix}, \tag{5}
$$

where $b_s^+ = 1/((\sqrt{A} - 1)\sqrt{s(s+1)})$, $b_s^- = \sqrt{A}/((\sqrt{A} - 1)\sqrt{s(s+1)})$, and $b_s^0 = (1 + s - s\sqrt{A})/((\sqrt{A} - 1)\sqrt{s(s+1)})$. One can see that for the center of mass the symmetrized and Jacobi coordinates are equal, $y_0 = \xi_0$, while the relative coordinates are related via the $(A - 1) \times (A - 1)$ matrix M with the elements $M_{ij} = B_{i+1,j+1}$ and $\det M = (-1)^{A \times d}$, i.e., the matrix, obtained by cancelling the first row and the first column. The inverse transformation $\xi = B^{-1}y$ is given by the matrix $B^{-1} = (JC)^{-1} = CJ^T = B^T$, i.e., B is also an orthogonal matrix.

In the symmetrized coordinates Eq. (1) takes the form

$$\left[-\frac{\partial^2}{\partial \xi_0^2} + \sum_{i=1}^{A-1} \left(-\frac{\partial^2}{\partial \xi_i^2} + (\xi_i)^2 \right) + U(\xi_0, ..., \xi_{A-1}) - E \right] \Psi(\xi_0, ..., \xi_{A-1}) = 0, \tag{6}$$

$$U(\xi_0, ..., \xi_{A-1}) = \sum_{i,j=1;i<j}^{A} U^{pair}(x_{ij}(\xi_1, ..., \xi_{A-1})) + \sum_{i=1}^{A} V(x_i(\xi_0, ..., \xi_{A-1})),$$

which is *invariant* under permutations $\xi_i \leftrightarrow \xi_j$ at $i, j = 1, ..., A - 1$, as follows from Eq. (3), i.e., the *invariance* of Eq. (1) under permutations $x_i \leftrightarrow x_j$ at $i, j = 1, ..., A$ survives.

Table 1. The first few eigenvalues E_j^S and the oscillator S-eigenfunctions (13) at $E_j^S - E_1^S \leq 10$, $E_1^S = A - 1$. We use the notations $|[i_1, i_2, ..., i_{A-1}]\rangle \equiv \Phi_{[i_1,i_2,...,i_{A-1}]}^s(\xi_1, ..., \xi_{A-1})$ from Eqs. (8) and (10), i.e., $[i_1, i_2, ..., i_{A-1}]$ assumes the summation over permutations of $[i_1, i_2, ..., i_{A-1}]$ in the layer $2\sum_{k=1}^{A-1} i_k + A - 1 = E_i^{s(a)}$.

A=2		A=3		A=4		$E_j^S - E_1^S$
j	$\Phi_j^S(\xi_1)$	j	$\Phi_j^S(\xi_1, \xi_2)$	j	$\Phi_j^S(\xi_1, \xi_2, \xi_3)$	
1	$|[0]\rangle$	1	$|[0,0]\rangle$	1	$|[0,0,0]\rangle$	0
2	$|[2]\rangle$	2	$|[0,2]\rangle$	2	$|[0,0,2]\rangle$	4
		3	$\frac{1}{2}|[0,3]\rangle - \frac{\sqrt{3}}{2}|[1,2]\rangle$	3	$|[1,1,1]\rangle$	6
3	$|[4]\rangle$	4	$\frac{\sqrt{3}}{2}|[0,4]\rangle + \frac{1}{2}|[2,2]\rangle$	4	$|[0,0,4]\rangle$	8
				5	$|[0,2,2]\rangle$	8
		5	$\frac{\sqrt{5}}{4}|[0,5]\rangle - \frac{3}{4}|[1,4]\rangle - \frac{\sqrt{2}}{4}|[2,3]\rangle$	6	$|[1,1,3]\rangle$	10

Table 2. The first few eigenvalues E_j^A and the oscillator A-eigenfunctions (13) at $E_j^A - E_1^A \leq 10$, $E_1^A = A^2 - 1$. We use the notations $|[i_1, i_2, ..., i_{A-1}]\rangle \equiv \Phi_{[i_1,i_2,...,i_{A-1}]}^a(\xi_1, ..., \xi_{A-1})$ from Eq. (11), i.e., $[i_1, i_2, ..., i_{A-1}]$ assumes the summation over the multiset permutations of $[i_1, i_2, ..., i_{A-1}]$ in the layer $2\sum_{k=1}^{A-1} i_k + A - 1 = E_i^{s(a)}$.

$A = 2,\ E_1^A = 3$		$A = 3,\ E_1^A = 8$		$A = 4,\ E_1^A = 15$		$E_j^A - E_1^A$
j	$\Phi_j^A(\xi_1)$	j	$\Phi_j^A(\xi_1, \xi_2)$	j	$\Phi_j^A(\xi_1, \xi_2, \xi_3)$	
1	$|[1]\rangle$	1	$\frac{1}{2}|[0,3]\rangle + \frac{\sqrt{3}}{2}|[1,2]\rangle$	1	$|[0,2,4]\rangle$	0
2	$|[3]\rangle$	2	$\frac{\sqrt{5}}{4}|[0,5]\rangle + \frac{3}{4}|[1,4]\rangle - \frac{\sqrt{2}}{4}|[2,3]\rangle$	2	$|[0,2,6]\rangle$	4
		3	$\frac{1}{4}|[0,6]\rangle - \frac{\sqrt{15}}{4}|[2,4]\rangle$	3	$|[1,3,5]\rangle$	6
3	$|[5]\rangle$	4	$\frac{\sqrt{21}}{8}|[0,7]\rangle + \frac{3\sqrt{3}}{8}|[1,6]\rangle - \frac{1}{8}|[2,5]\rangle + \frac{\sqrt{5}}{8}|[3,4]\rangle$	4	$|[0,4,6]\rangle$	8
				5	$|[0,2,8]\rangle$	8
		5	$\frac{\sqrt{2}}{4}|[0,8]\rangle - \frac{\sqrt{14}}{4}|[2,6]\rangle$	6	$|[1,3,7]\rangle$	10

3 The SCR Algorithm: Symmetrized Coordinate Representation

For simplicity, consider the solutions of Eq. (6) in the internal symmetrized coordinates $\{\xi_1, ..., \xi_{A-1}\} \in \mathbf{R}^{A-1}$, $x_i \in \mathbf{R}^1$, in the case of 1D Euclidean space $(d = 1)$. The relevant equation describes an $(A - 1)$-dimensional oscillator with the eigenfunctions $\Phi_j(\xi_1, ..., \xi_{A-1})$ and the energy eigenvalues E_j:

$$\left[\sum_{i=1}^{A-1}\left(-\frac{\partial^2}{\partial \xi_i^2} + (\xi_i)^2\right) - E_j\right]\Phi_j(\xi_1, ..., \xi_{A-1}) = 0, \quad E_j = 2\sum_{k=1}^{A-1} i_k + A - 1, \quad (7)$$

where the numbers i_k, $k = 1, ..., A - 1$ are integer, $i_k = 0, 1, 2, 3, ...$. The eigenfunctions $\Phi_j(\xi_1, ..., \xi_{A-1})$ can be expressed in terms of the conventional eigenfunctions of individual 1D oscillators as

$$\Phi_j(\xi_1, ..., \xi_{A-1}) = \sum_{2 \sum_{k=1}^{A-1} i_k + A - 1 = E_j} \beta_{j[i_1, i_2, ..., i_{A-1}]} \bar{\Phi}_{[i_1, i_2, ..., i_{A-1}]}(\xi_1, ..., \xi_{A-1}), \quad (8)$$

$$\bar{\Phi}_{[i_1, i_2, ..., i_{A-1}]}(\xi_1, ..., \xi_{A-1}) = \prod_{k=1}^{A-1} \bar{\Phi}_{i_k}(\xi_k), \quad \bar{\Phi}_{i_k}(\xi_k) = \frac{\exp(-\xi_k^2/2) H_{i_k}(\xi_k)}{\sqrt[4]{\pi} \sqrt{2^{i_k}} \sqrt{i_k!}},$$

where $H_{i_k}(\xi_k)$ are Hermite polynomials [19]. Generally the energy level $E_f = 2f + A - 1$, $f = \sum_{k=1}^{A-1} i_k$, of an $(A - 1)$-dimensional oscillator is known [20] to possess the degeneracy multiplicity $p = (A + f - 2)!/f!/(A - 2)!$ with respect to the conventional oscillator eigenfunctions $\bar{\Phi}_{[i_1, i_2, ..., i_{A-1}]}(\xi_1, ..., \xi_{A-1})$. This degeneracy allows further symmetrization by choosing the appropriate coefficients $\beta_{[i_1, i_2, ..., i_{A-1}]}^{(j)}$. Degeneracy multiplicity p of all states with the given energy E_j defined by formula

$$p = \sum_{2 \sum_{k=1}^{A-1} i_k + A - 1 = E_j} N_\beta, \quad N_\beta = (A - 1)! / \prod_{k=1}^{N_v} v_k!, \quad (9)$$

where N_β is the number of multiset permutations (m.p.) of $[i_1, i_2, ..., i_{A-1}]$, and $N_v \leq A - 1$ is the number of different values i_k in the multiset $[i_1, i_2, ..., i_{A-1}]$, and v_k is the number of repetitions of the given value i_k.

Step 1. Symmetrization with respect to permutation of A–1 particles
For the states $\Phi_j^s(\xi_1, ..., \xi_{A-1}) \equiv \Phi_{[i_1, i_2, ..., i_{A-1}]}^s(\xi_1, ..., \xi_{A-1})$, symmetric with respect to permutation of $A - 1$ particles $i = [i_1, i_2, ..., i_{A-1}]$, the coefficients $\beta_{i[i_1', i_2', ..., i_{A-1}']}$ in Eq. (8) are

$$\beta_{i[i_1', i_2', ..., i_{A-1}']} = \begin{cases} \frac{1}{\sqrt{N_\beta}}, & \text{if } [i_1', i_2', ..., i_{A-1}'] \text{ is a m. p. of } [i_1, i_2, ..., i_{A-1}], \\ 0, & \text{otherwise.} \end{cases} \quad (10)$$

The states $\Phi_j^a(\xi_1, ..., \xi_{A-1}) \equiv \Phi_{[i_1, i_2, ..., i_{A-1}]}^a(\xi_1, ..., \xi_{A-1})$, antisymmetric with respect to permutation of $A - 1$ particles are constructed in a conventional way

$$\Phi_j^a(\xi_1, ..., \xi_{A-1}) = \frac{1}{\sqrt{(A-1)!}} \begin{vmatrix} \bar{\Phi}_{i_1}(\xi_1) & \bar{\Phi}_{i_2}(\xi_1) & \cdots & \bar{\Phi}_{i_{A-1}}(\xi_1) \\ \bar{\Phi}_{i_1}(\xi_2) & \bar{\Phi}_{i_2}(\xi_2) & \cdots & \bar{\Phi}_{i_{A-1}}(\xi_2) \\ \vdots & \vdots & \ddots & \vdots \\ \bar{\Phi}_{i_1}(\xi_{A-1}) & \bar{\Phi}_{i_2}(\xi_{A-1}) & \cdots & \bar{\Phi}_{i_{A-1}}(\xi_{A-1}) \end{vmatrix}, \quad (11)$$

i.e., the coefficients $\beta_{[i_1', i_2', ..., i_{A-1}']}^{(i)}$ in (8) are expressed as

$$\beta_{[i_1', i_2', ..., i_{A-1}']}^{(i)} = \varepsilon_{i_1', i_2', ..., i_{A-1}'} / \sqrt{(A-1)!},$$

where $\varepsilon_{i_1', i_2', ..., i_{A-1}'}$ is a totally antisymmetric tensor. This tensor is defined as follows: $\varepsilon_{i_1', i_2', ..., i_{A-1}'} = +1(-1)$, if $i_1', i_2', ..., i_{A-1}'$ is an even (odd) permutation of

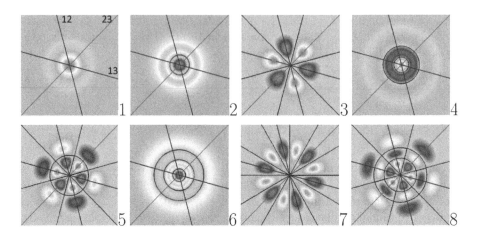

Fig. 1. Profiles of the first eight oscillator S-eigenfunctions $\Phi^S_{[i_1,i_2]}(\xi_1,\xi_2)$, at $A =$ 3 in the coordinate frame (ξ_1,ξ_2). The lines correspond to pair collision $x_2 = x_3$, $x_1 = x_2$ and $x_1 = x_3$ of the projection $(x_1,x_2,x_3) \rightarrow (\xi_1,\xi_2)$, marked only in the left upper panel with '23', '12', and '13', respectively. The additional lines are nodes of the eigenfunctions $\Phi^S_{[i_1,i_2]}(\xi_1,\xi_2)$.

the numbers $i_1 < i_2 < ... < i_{A-1}$, and $\varepsilon_{i'_1,i'_2,...,i'_{A-1}} = 0$ otherwise, i.e., when some two numbers in the set $i'_1, i'_2, ..., i'_{A-1}$ are equal. Therefore, for antisymmetric states the numbers i_k in Eq. (7) take the integer values $i_k = k - 1, k, k + 1, ...,$ $k = 1, ..., A - 1$.

Here and below s and a are used for the functions, symmetric (antisymmetric) under permutations of $A - 1$ relative coordinates, constructed at the first step of the procedure. On the contrary, S and A are used for the functions, symmetric (asymmetric) under permutations of A initial Cartesian coordinates. This is actually the symmetry with respect to permutation of identical particles themselves; in this sense, S and A states may be attributed to boson- and fermion-like particles. However, we prefer to use the S (A) notation as more rigorous.

Step 2. Symmetrization with respect to permutation of A particles

For $A = 2$, the symmetrized coordinate ξ_1 corresponds to the difference $x_2 - x_1$ of Cartesian coordinates, so that a function even (odd) with respect to ξ_1 appears to be symmetric (antisymmetric) with respect to the permutation of two particles $x_2 \leftrightarrow x_1$. Hence, even (odd) eigenfunctions with corresponding eigenvalues $E^s_j = 2(2n) + 1$ ($E^a_j = 2(2n + 1) + 1$) describe S (A) solutions.

For $A \geq 3$, the functions, symmetric (antisymmetric) with respect to permutations of Cartesian coordinates $x_{i+1} \leftrightarrow x_{j+1}$, $i, j = 0, ..., A - 1$:

$$\Phi^{S(A)}(..., x_{i+1}, ..., x_{j+1}, ...) \equiv \Phi^{S(A)}(\xi_1(x_1, ..., x_A), ..., \xi_{A-1}(x_1, ..., x_A))$$

$$= \pm\Phi^{S(A)}(..., x_{j+1}, ..., x_{i+1}, ...)$$

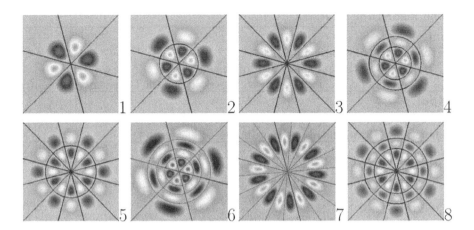

Fig. 2. The same as in Fig. 1, but for the first eight oscillator A-eigenfunctions $\Phi^A_{[i_1,i_2]}(\xi_1,\xi_2)$, at $A = 3$

become symmetric (antisymmetric) with respect to permutations of symmetrized coordinates $\xi_i \leftrightarrow \xi_j$, $i, j = 1, ..., A - 1$:

$$\Phi^{S(A)}(...,\xi_i,...,\xi_j,...) = \pm\Phi^{S(A)}(...,\xi_j,...,\xi_i,...),$$

as follows from Eq. (4). However, the converse statement is not valid,

$$\Phi^{s(a)}(...,\xi_i,...,\xi_j,...) = \pm\Phi^{s(a)}(...,\xi_j,...,\xi_i,...)$$

$$\not\Rightarrow \Phi^{s(a)}(x_1,...,x_{i+1},...) = \pm\Phi^{s(a)}(x_{i+1},...,x_1,...),$$

because we deal with a projection map

$$(\xi_1,...,\xi_{A-1})^T = \hat{C}(x_1,...,x_A)^T, \tag{12}$$

which is implemented by the $(A - 1) \times (A)$ matrix $\hat{C}$ with the matrix elements $\hat{C}_{ij} = C_{i+1,j}$, obtained from (3) by cancelling the first row. Hence, the functions, symmetric (antisymmetric) with respect to permutations of symmetrized coordinates, are divided into two types, namely, the S (A) solutions, symmetric (antisymmetric) with respect to permutations $x_1 \leftrightarrow x_{j+1}$ at $j = 1, ..., A - 1$:

$$\Phi^{S(A)}(x_1,...,x_{j+1},...) = \pm\Phi^{S(A)}(x_{j+1},...,x_1,...)$$

and the other s (a) solutions, $\Phi^{s(a)}(x_1,...,x_{i+1},...) \neq \pm\Phi^{s(a)}(x_{i+1},...,x_1,...)$, which should be eliminated. These requirements are equivalent to only one permutation $x_1 \leftrightarrow x_2$, as follows from (4), which simplifies their practical implementation. With these requirements taken into account in the Gram–Schmidt process, implemented in the *symbolic algorithm SCR*, we obtained the required characteristics of S and A eigenfunctions,

$$\Phi^{S(A)}_i(\xi_1,...,\xi_{A-1}) = \sum_{2\sum_{k=1}^{A-1} i_k+A-1=E^{s(a)}_i} \alpha^{S(A)}_{i[i_1,i_2,...,i_{A-1}]}\Phi^{s(a)}_{[i_1,i_2,...,i_{A-1}]}(\xi_1,...,\xi_{A-1}). \tag{13}$$

The algorithm SCR:

Input:

A is the number of identical particles;

i_{max} is defined by the maximal value of the energy $E_{i_{max}}$;

$(\xi_1, ..., \xi_{A-1})$ and $(x_1, ..., x_A)$ are the symmetrized and the Cartesian coordinates;

Output:

$\Phi_i^{S(A)}(\xi_1, ..., \xi_{A-1})$ and $\Phi_i^{S(A)}(x_1, ..., x_A)$ are the total symmetric (antisymmetric) functions (13) in the above coordinates connected by (12);

Local:

$E_i^{s(a)} \equiv E_i^{S(A)} = 2\sum_{k=1}^{A-1} i_k + A - 1$ is the $(i+1)^{th}$ eigenenergy;

$i_{min} = 0$ for the symmetric and $i_{min} = (A-1)^2$ for the antisymmetric case;

$\Phi_j \equiv \Phi_{[i_1,i_2,...,i_{A-1}]}^{s(a)}(\xi_1, ..., \xi_{A-1})$ and $\Phi_j \equiv \Phi_{[i_1,i_2,...,i_{A-1}]}^{s(a)}(x_1, ..., x_A)$ are the functions, symmetric (antisymmetric) with respect to $A-1$ Cartesian coordinates;

$p_{s(a)} \equiv p_{i;s(a)}$ and $p_{S(A)} \equiv p_{i;S(A)}$ are the degeneracy factors of the energy levels $E_i^{s(a)}$ and $E_i^{S(A)}$ for s(a) and S(A) functions, respectively;

$p_{i;min}$ $(p_{i;max})$ and $P_{i;min}$ $(P_{i;max})$ are the lowest (highest) numbers of s(a) and S(A) functions, belonging to the energy levels $E_i^{s(a)}$ and $E_i^{S(A)}$, respectively;

$\{\bar{\alpha}_j\}$ and $\{\alpha_{pj}^{S(A)}\}$ are the sets of intermediate and desired coefficients;

1.1 $j := 0$;

for i from i_{min} to i_{max} **do**;

1.2: $p_{i;min} := j + 1$;

1.3: **for each** sorted $i_1, i_2, ..., i_{A-1}$, $2\sum_{k=1}^{A-1} i_k + A - 1 = E_i^{s(a)}$ **do**

$\qquad j := j + 1$;

$\qquad$ construction $\Phi_j(\xi_1, ..., \xi_{A-1}) = \Phi_j^s(\xi_1, ..., \xi_{A-1})$ from (8), (10)

$\qquad$ or $\Phi_j(\xi_1, ..., \xi_{A-1}) = \Phi_j^a(\xi_1, ..., \xi_{A-1})$ from (11)

$\qquad \Phi_j(x_1, ..., x_A) =$ subs$((\xi_1, ..., \xi_{A-1}) \rightarrow (x_1, ..., x_A), \Phi_j(\xi_1, ..., \xi_{A-1}))$;

end for

1.4: $p_{i;max} := j$; $\quad p_{i;s(a)} = p_{i;max} - p_{i;min} + 1$;

end for

2.1.: $P_{min} = 1$;

for i from i_{min} to i_{max} **do**

2.2.: $P_{i;min} = P_{min}$;

2.3.: $\Phi(\xi_1, ..., \xi_{A-1}) = \sum_{j=p_{i;min}}^{p_{i;max}} \bar{\alpha}_j \Phi_j(\xi_1, ..., \xi_{A-1})$;

$\qquad \Phi(x_1, ..., x_A) = \sum_{j=p_{i;min}}^{p_{i;max}} \bar{\alpha}_j \Phi_j(x_1, ..., x_A)$;

2.4.: $\Phi(x_2, x_1, ..., x_A) :=$ change$(x_1 \leftrightarrow x_2, \Phi(x_1, x_2, ..., x_A)))$;

2.5.: $\Phi(x_2, x_1, ..., x_A) \mp \Phi(x_1, x_2, ..., x_A) = 0$,

$\qquad \rightarrow \quad (\bar{\alpha}_{pj}, j = p_{i;min}, ..., p_{i;max}, p = 1, ..., p_{i;S(A)})$;

2.6.: $P_{i;max} = P_{i;min} - 1 + p_{i;S(A)}$;

2.7.: Gram–Schmidt procedure for $\Phi(\xi_1, ..., \xi_{A-1})$ $\qquad \rightarrow$

$\qquad \Phi_p^{S(A)}(x_1, x_2, ..., x_A) = \sum_{j=p_{i;min}}^{p_{i;max}} \alpha_{pj}^{S(A)} \Phi_j(x_1, x_2, ..., x_A)$;

$\qquad \Phi_p^{S(A)}(\xi_1, ..., \xi_{A-1}) = \sum_{j=p_{i;min}}^{p_{i;max}} \alpha_{pj}^{S(A)} \Phi_j(\xi_1, ..., \xi_{A-1})$,

$\qquad$ at $p = P_{i;min}, ..., P_{i;max}$;

end for

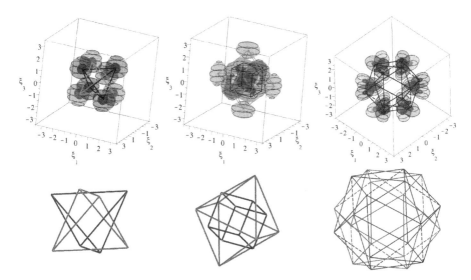

Fig. 3. Upper panel: Profiles of the oscillator S-eigenfunctions $\Phi^S_{[1,1,1]}(\xi_1, \xi_2, \xi_3)$, $\Phi^S_{[0,0,4]}(\xi_1, \xi_2, \xi_3)$ and A-eigenfunction $\Phi^A_{[0,2,4]}(\xi_1, \xi_2, \xi_3)$, at $A = 4$ (left, middle, and right panels, respectively). Some maxima and minima positions of these functions are connected by black and gray lines and duplicated in lower panels: two tetrahedrons forming a *stella octangula* for $\Phi^S_{[1,1,1]}(\xi_1, \xi_2, \xi_3)$, a cube and an octahedron for $\Phi^S_{[0,0,4]}(\xi_1, \xi_2, \xi_3)$, and a polyhedron with 20 triangle faces (only 8 of them being equilateral triangles) for $\Phi^A_{[0,2,4]}(\xi_1, \xi_2, \xi_3)$.

4   Examples of the SCR Generation

The SCR algorithm was implemented in MAPLE 14 on Intel Core i5 CPU 660 3.33GHz, 4GB 64 bit, to generate first 11 symmetric (antisymmetric) functions up to $\Delta E_j = 12$ at $A = 6$ with CPU time 10 seconds (600 seconds), that together with a number of functions in dependence of number of particles given in Table 3 demonstrates efficiency and complexity of the algorithm.

The examples of generated total symmetric and antisymmetric $(A - 1)$-dimensional oscillator functions are presented in Tables 1 and 2. Note that for $A = 4$, the first four states from Table 1 are similar to those of the translation-invariant model without excitation of the center-of-mass variable [3].

As an example, in Figs. 1 and 2 we show isolines of the first eight S and A oscillator eigenfunctions $\Phi^S_{[i_1,i_2]}(\xi_1, \xi_2)$ and $\Phi^A_{[i_1,i_2]}(\xi_1, \xi_2)$ for $A = 3$, calculated at the second step of the algorithm. One can see that the S (A) oscillator eigenfunctions are symmetric (antisymmetric) with respect to reflections from three straight lines. The first line (labelled '23') corresponds to the permutation (x_2, x_3) and is rotated by $\pi/4$ counterclockwise with respect to the axis ξ_1. The second and the third lines (labelled '12' and '13') correspond to the permutations (x_1, x_2) and (x_1, x_3) and are rotated by $\pi/3$ clockwise and counterclockwise with respect to the first line. These lines divide the plane into six sectors, while the symmetric (antisymmetric) oscillator eigenfunctions, calculated at the first

Table 3. The degeneracy multiplicities p from (9), $p_s = p_a$ and $p_S = p_A$ of s-, a-, S-, and A-eigenfunctions of the oscillator energy levels $\Delta E_j = E_j^\bullet - E_1^\bullet$, $\bullet = \emptyset, s, a, S, A$

A=3			A=4			A=5			A=6			ΔE_j
p	p_s,p_a	p_S,p_A	p	p_s,p_a	p_S,p_A	p	p_s,p_a	p_S,p_A	p	p_s,p_a	p_S,p_A	
1	1	1	1	1	1	1	1	1	1	1	1	0
2	1	0	3	1	0	4	1	0	5	1	0	2
3	2	1	6	2	1	10	2	1	15	2	1	4
4	2	1	10	3	1	20	3	1	35	3	1	6
5	3	1	15	4	2	35	5	2	70	5	2	8
6	3	1	21	5	1	56	6	2	126	7	2	10
7	4	2	28	7	3	84	9	3	210	10	4	12

step of the algorithm, which are symmetric (or antisymmetric) with respect to reflections from the first line, generate the division of the plane into two parts. This illustrates the isomorphism between the symmetry group of an equilateral triangle D_3 in $\mathbf{R}^2$ and the 3-body permutation group S_3 (A = 3).

Figure 3 shows examples of profiles of S and A oscillator eigenfunctions for $A = 4$. Note that four maxima (black) and four minima (grey) of the S eigenfunction $\Phi_{[1,1,1]}^S(\xi_1, \xi_2, \xi_3)$ are positioned at the vertices of two tetrahedrons forming a *stella octangula*, with the edges shown by black and grey lines, respectively. Eight maxima and six outer minima for S eigenfunction $\Phi_{[0,0,4]}^S(\xi_1, \xi_2, \xi_3)$ are positioned at the vertices of a cube and an octahedron, the edges of which are shown by black and grey lines, respectively. The positions of twelve maxima of the A oscillator eigenfunction, $\Phi_{[0,2,4]}^A(\xi_1, \xi_2, \xi_3)$ coincide with the vertices of a polyhedron with 20 triangle faces (only 8 of them being equilateral triangles) and 30 edges, 6 of them having the length 2.25 and the other having the length 2.66. The above shapes of eigenfunctions illustrate the isomorphism between the tetrahedron group T_d in $\mathbf{R}^3$ and the 4-particle permutation group S_4 (A = 4), discussed in [2] in the case of $d = 3$.

The degeneracy multiplicity (9), i.e., number p of all states with the given energy E_j of low part of spectra, the numbers p_s (p_a) of the states, symmetric (antisymmetric) under permutations of $A - 1$ relative coordinates together with the total numbers p_S (p_A) of the states, symmetric (antisymmetric) under permutations of A initial Cartesian coordinates are summarized in Table 3. Note that the S and A states with $E' = E_1^{S,A} + 2$ do not exist. The numbers p_s (p_a) are essentially smaller than the total number p of all states, which simplifies the procedure of constructing S (A) states with possible excitation of the center-of-mass degree of freedom and allows the use of a compact basis with the reduced degeneracy p_S (p_A) of the S (A) states in our final calculations. For clarity, in the case $A = 3$, $d = 1$, the S(A)-type functions generated by the *SCR* algorithm, in polar coordinates $\xi_1 = \rho \cos \varphi$, $\xi_2 = \rho \sin \varphi$ are expressed as:

$$\Phi_{k,m}^{S(A)}(\rho, \varphi) = C_{km}(\rho^2)^{3m/2} \exp(-\rho^2/2) L_k^{3m}(\rho^2) \frac{\cos}{\sin}(3m(\varphi + \pi/12)),$$

where C_{km} is the normalization constant, $L_k^{3m}(\rho^2)$ are the Laguerre polynomials [19], $k = 0, 1, ..., m = 0, 1, ...$ for S states, while $m = 1, 2, ...$ for A states, that are classified by irr of the D_{3m}-symmetry group. The corresponding energy levels $E_{k,m}^{S(A)} = 2(2k+3m+1) = E_{[i_1,i_2]}^{s(a)} = 2(i_1+i_2+1)$ have the degeneracy multiplicity $K+1$, if the energy $E_{k,m}^{S(A)} - E_1^{S(A)} = 12K + K'$, where $K' = 0, 4, 6, 8, 10, 14$. For example, in Figs. 1 and 2 we show the wave functions $\Phi_{3,0}^S(\rho, \varphi)$ and $\Phi_{0,2}^S(\rho, \varphi)$ (or $\Phi_{3,1}^A(\rho, \varphi)$ and $\Phi_{0,3}^A(\rho, \varphi)$) labelled with 6 and 7, corresponding to the energy levels $E_{k,m}^{S(A)} - E_1^{S(A)} = 12$ with the degeneracy $K = 2$, while the functions labelled with $1, 2, 3, 4, 5, 8$ are nondegenerate $(K = 1)$. So, the eigenfunctions of the A-identical particle system in one dimension are degenerate in accordance with [21], and this result disagrees with nondegenerate ansatz solutions [10].

5   Conclusion

We considered a model of A identical particles bound by the oscillator-type potential under the influence of the external field of a target in the new symmetrized coordinates. The constructive SCR algorithm of symmetrizing or antisymmetrizing the $A - 1$-dimensional harmonic oscillator basis functions with respect to permutations of A identical particles was described. One can see that the transformations of $(A - 1)$-dimensional oscillator basis functions from the symmetrized coordinates to the Jacobi coordinates, reducible to permutations of coordinates and $(A - 1)$-dimensional finite rotation (5), are implemented by means of the $(A - 1)$-dimensional oscillator Wigner functions [23]. Typical examples were analyzed, and a correspondence between the representations of the symmetry groups D_3 and T_d for $A = 3$ and $A = 4$ shapes is displayed. It is shown that one can use the presented SCR algorithm, implemented using the MAPLE computer algebra system, to construct the basis functions in the closed analytical form. However, for practical calculations of matrix elements between the basis states, belonging to the lower part of the spectrum, this is not necessary. The application of the developed approach and algorithm for solving the problem of tunnelling clusters through barrier potentials of a target is considered in our forthcoming paper [22]. The proposed approach can be adapted to the analysis of tetrahedral-symmetric nuclei, quantum diffusion of molecules and micro-clusters through surfaces, and the fragmentation in producing neutron-rich light nuclei.

The authors thank Professor V.P. Gerdt for collaboration. The work was supported partially by grants 13-602-02 JINR, 11-01-00523 and 13-01-00668 RFBR and the Bogoliubov-Infeld program.

References

1. Moshinsky, M., Smirnov, Y.F.: The harmonic oscillator in modern physics. Informa Health Care, Amsterdam (1996)
2. Kramer, P., Moshinsky, M.: Group theory of harmonic oscillators (III). States with permutational symmetry. Nucl. Phys. 82, 241–274 (1966)
3. Aguilera-Navarro, V.C., Moshinsky, M., Yeh, W.W.: Harmonic-oscillator states and the α particle I. Form factor for symmetric states in configuration space. Ann. Phys. 51, 312–336 (1969)

4. Aguilera-Navarro, V.C., Moshinsky, M., Kramer, P.: Harmonic-oscillator states and the α particle II. Configuration-space states of arbitrary symmetry. Ann. Phys. 54, 379–393 (1969)
5. Lévy-Leblond, J.-M.: Global and democratic methods for classifying N particle states. J. Math. Phys. 7, 2217–2229 (1966)
6. Neudatchin, V.G., Smirnov, Y.F.: Nucleon clusters in the light nuclei, Nauka, Moscow (1969) (in Russian)
7. Novoselsky, A., Katriel, J.: Non-spurious harmonic oscillator states with arbitrary symmetry. Ann. Phys. 196, 135–149 (1989)
8. Wildermuth, K., Tang, Y.C.: A unified theory of the nucleus. Academic Press, New York (1977)
9. Kamuntavičius, G.P., Kalinauskas, R.K., Barrett, B.R., Mickevičius, S., Germanas, D.: The general harmonic-oscillator brackets: compact expression, symmetries, sums and Fortran code. Nucl. Phys. A 695, 191–201 (2001)
10. Wang, Z., Wang, A., Yang, Y., Xuechao, L.: Exact eigenfunctions of N-body system with quadratic pair potential. arXiv:1108.1607v4 (2012)
11. Pen'kov, F.M.: Metastable states of a coupled pair on a repulsive barrier. Phys. Rev. A 62, 44701–44701 (2000)
12. Gusev, A.A., Vinitsky, S.I., Chuluunbaatar, O., Gerdt, V.P., Rostovtsev, V.A.: Symbolic-numerical algorithms to solve the quantum tunneling problem for a coupled pair of ions. In: Gerdt, V.P., Koepf, W., Mayr, E.W., Vorozhtsov, E.V. (eds.) CASC 2011. LNCS, vol. 6885, pp. 175–191. Springer, Heidelberg (2011)
13. Dobrowolski, A., Góźdź, A., Mazurek, K., Dudek, J.: Tetrahedral symmetry in nuclei: new predictions based on the collective model. International Journal of Modern Physics E 20(2), 500–506
14. Dobrowolski, A., Szulerecka, A., Góźdź, A.: Electromagnetic transitions in hypothetical tetrahedral and octahedral bands. In: Góźdź, A. (ed.) Hidden symmetries in intrinsic frame Proc. 19th Nuclear Physics Workshop in Kazimierz Dolny (September 2012), http://kft.umcs.lublin.pl/wfj/archive/2012/proceedings.php
15. Fock, V.A.: Näherungsmethode zur Lösung des quantenmechanischen Mehrkörperproblems. Zs. Phys. 61, 126–148 (1930)
16. Hamermesh, M.: Group theory and its application to physical problems. Dover, New York (1989)
17. Kanada-En'yo, Y., Hidaka, Y.: α-cluster structure and density waves in oblate nuclei. Phys. Rev. C 84, 014313-1–014313-16 (2011)
18. Jepsent, D.W., Hirschfelder, J.O.: Set of co-ordinate systems which diagonalize the kinetic energy of relative motion. Proc. Natl. Acad. Sci. U.S.A. 45, 249–256 (1959)
19. Abramovits, M., Stegun, I.A.: Handbook of Mathematical Functions, p. 1037. Dover, New York (1972)
20. Baker Jr., G.A.: Degeneracy of the n-dimensional, isotropic, harmonic oscillator. Phys. Rev. 103, 1119–1120 (1956)
21. Lévy-Leblond, J.M.: Generalized uncertainty relations for many-fermion system. Phys. Lett. A 26, 540–541 (1968)
22. Vinitsky, S., Gusev, A., Chuluunbaatar, O., Rostovtsev, V., Le Hai, L., Derbov, V., Krassovitskiy, P.: Symbolic-numerical algorithm for generating cluster eigenfunctions: quantum tunneling of clusters through repulsive barriers. In: Gerdt, V.P., Koepf, W., Mayr, E.W., Vorozhtsov, E.V. (eds.) CASC 2013. LNCS, vol. 8136, pp. 427–442. Springer, Heidelberg (2013)
23. Pogosyan, G.S., Smorodinsky, Y.A., Ter-Antonyan, V.M.: Oscillator Wigner functions. J. Phys. A 14, 769–776 (1981)

Symbolic-Numerical Investigation of Gyrostat Satellite Dynamics

Sergey A. Gutnik[1] and Vasily A. Sarychev[2]

[1] Moscow State Institute of International Relations (University) 76, Prospekt Vernadskogo, Moscow, 119454, Russia
s.gutnik@inno.mgimo.ru
[2] Keldysh Institute of Applied Mathematics (Russian Academy of Sciences)
4, Miusskaya Square, Moscow, 125047, Russia
vas31@rambler.ru

Abstract. Dynamics of gyrostat satellite moving along a circular orbit in the central Newtonian gravitational field is studied. A symbolic-numerical method for determination of all equilibrium orientations of gyrostat satellite in the orbital coordinate system with given gyrostatic torque and given principal central moments of inertia is proposed. A computer algebra method based on the algorithm for the construction of a Groebner basis and the resultant concept for solving the problem is used. All bifurcation values of parameters at which there is a change of numbers of equilibrium orientations are determined. Evolution of domains in the space of parameters of the system which correspond to various numbers of equilibria are carried out numerically in detail. The stability analysis of equilibria is performed on the basis of Lyapunov's theorem. It is shown that the number of equilibria of the gyrostat satellite in general case is no less than 8 and no more than 24, and the number of stable equilibria changes from 4 to 2.

1   Introduction

The important aspect of the development of space technology is the design of systems of orientation of the satellites. The orientation of the satellite can be carried out with the use of active or passive methods. In the design of passive control system of satellite orientation, it is possible to use the properties of the gravitational and magnetic fields, the effect of atmospheric drag forces and solar radiation pressure, gyroscopic properties of rotating bodies et cetera. An important property of passive systems orientation is that these systems can operate for a long time without spending energy. Among the passive methods of orientation, the most widespread are the gravity orientation systems of the satellite. These systems are based on the fact that a satellite with different moments of inertia in the central Newtonian force field in a circular orbit has 24 equilibrium orientations, and four of them are stable [1]. Adding inside the body of the satellite statically and dynamically balanced rotors rotating with a constant angular velocity relative to the body of the satellite leads to new equilibrium

V.P. Gerdt et al. (Eds.): CASC 2013, LNCS 8136, pp. 169–178, 2013.
© Springer International Publishing Switzerland 2013

orientations of satellite, interesting for practical applications. Therefore, it is necessary to study the joint action of gravitational and gyrostatic moments and, in particular, to analyze all possible satellite's equilibria in a circular orbit. Such solutions can be used in practical space technology in the design of passive control systems of the satellites. In this work, a symbolic–numerical investigation of satellite's dynamics under the influence of gravitational and gyrostatic moments is presented. The symbolic-numerical method for determination of equilibrium orientation of a gyrostat satellite had been successfully used for the analysis of equilibrium orientations of a satellite in a circular orbit under the influence of gravitational and aerodynamic forces [2].

Many publications are dedicated to the problem of determination of the equilibrium orientations of the gyrostat satellite. The basic problems of satellite's dynamics with a gyrostatic attitude control system have been presented in [1]. In [3], [4], [5], [6], and [7], all equilibrium orientations were found in some special cases when the vector of gyrostatic moment is located along a satellite's principal central axis of inertia and in the satellite's principal central plane of inertia. The general case of the problem of gyrostat for general values of gyrostatic moment was first studied in [8]. In the present work, the problem of determination of the classes of equilibrium orientations for the general case is considered. The equilibrium orientations are determined by real roots of the system of nonlinear algebraic equations. The investigation of equilibria was possible due to application of Computer Algebra Groebner basis and resultant methods. Evolution of domains with a fixed number of equilibria is investigated numerically in dependence of four dimensionless system parameters. Sufficient conditions for stability of all equilibrium orientations are obtained using the generalized integral of energy. The stability of the equilibrium orientations is analyzed numerically.

2   Equations of Motion

Consider the attitude motion of the gyrostat satellite (further we call it as the gyrostat), which is a rigid body with statically and dynamically balanced rotors inside the satellite body. The rotors angular velocities relative to the satellite body are constant. To write the equations of motion we introduce two right Cartesian coordinate systems with origin in the satellite's center of mass O. $OXYZ$ is the orbital coordinate system whose OZ axis is directed along the radius vector connecting the centers of mass of the Earth and of the gyrostat satellite; the OX axis is directed along the vector of linear velocity of the center of mass O. $Oxyz$ is the gyrostat-fixed coordinate system; Ox, Oy, and Oz are the principal central axes of inertia of the gyrostat satellite. The orientation of the gyrostat-fixed coordinate system $Oxyz$ with respect to the orbital coordinate system is determined by means of the Euler angles ψ (precession), ϑ (nutation), and φ (spin). The direction cosines in transformation matrix between the coordinate systems $OXYZ$ and $Oxyz$ are represented by the following expressions [8]:

$$a_{11} = \cos(x, X) = \cos\psi\cos\varphi - \sin\psi\cos\vartheta\sin\varphi,$$
$$a_{12} = \cos(y, X) = -\cos\psi\sin\varphi - \sin\psi\cos\vartheta\cos\varphi,$$
$$a_{13} = \cos(z, X) = \sin\psi\sin\vartheta,$$
$$a_{21} = \cos(x, Y) = \sin\psi\cos\varphi + \cos\psi\cos\vartheta\sin\varphi,$$
$$a_{22} = \cos(y, Y) = -\sin\psi\sin\varphi + \cos\psi\cos\vartheta\cos\varphi, \qquad (1)$$
$$a_{23} = \cos(z, Y) = -\cos\psi\sin\vartheta,$$
$$a_{31} = \cos(x, Z) = \sin\vartheta\sin\varphi,$$
$$a_{32} = \cos(y, Z) = \sin\vartheta\cos\varphi,$$
$$a_{33} = \cos(z, Z) = \cos\vartheta.$$

Then equations of the satellite's attitude motion can be written in the Euler form [1], [8]:

$$A\dot{p} + (C - B)qr - 3\omega_0^2(C - B)a_{32}a_{33} - \bar{H}_2 r + \bar{H}_3 q = 0,$$
$$B\dot{q} + (A - C)rp - 3\omega_0^2(A - C)a_{31}a_{33} - \bar{H}_3 p + \bar{H}_1 r = 0, \qquad (2)$$
$$C\dot{r} + (B - A)pq - 3\omega_0^2(B - A)a_{31}a_{32} - \bar{H}_1 q - \bar{H}_2 p = 0,$$

$$p = \dot{\psi}a_{31} + \dot{\vartheta}\cos\varphi + \omega_0 a_{21} = \bar{p} + \omega_0 a_{21},$$
$$q = \dot{\psi}a_{32} - \dot{\vartheta}\sin\varphi + \omega_0 a_{22} = \bar{q} + \omega_0 a_{22}, \qquad (3)$$
$$r = \dot{\psi}a_{33} + \dot{\varphi} + \omega_0 a_{23} = \bar{r} + \omega_0 a_{23}.$$

In equations (2), (3) $\bar{H}_1 = \sum_{i=1}^n J_i\alpha_i\dot{\varphi}_i$, $\bar{H}_2 = \sum_{i=1}^n J_i\beta_i\dot{\varphi}_i$, $\bar{H}_3 = \sum_{i=1}^n J_i\gamma_i\dot{\varphi}_i$, J_i is the axial moment of inertia of the ith rotor; α_i, β_i, and γ_i are the constant direction cosines of the symmetry axis of the ith rotor in the coordinate system $Oxyz$; $\dot{\varphi}_i$ is the constant angular velocity of the ith rotor relative to the gyrostat; n is the number of rotors; A, B, and C are the principal central moments of inertia of the gyrostat; p, q, and r are the projections of the inertial angular velocity of the gyrostat satellite onto the Ox, Oy, and Oz axes; ω_0 is the angular velocity of the orbital motion of the gyrostat satellite center of mass. The dot designates differentiation with respect to time t.

Further it will be more convenient to use parameters $H_1 = \bar{H}_1/\omega_0, H_2 = \bar{H}_2/\omega_0, H_3 = \bar{H}_3/\omega_0$.

For the systems of equations (2) and (3), the generalized energy integral exists in the form

$$\frac{1}{2}(A\bar{p}^2 + B\bar{q}^2 + C\bar{r}^2) + \frac{3}{2}\omega_0^2[(A - C)a_{31}^2 + (B - C)a_{32}^2] +$$

$$+\frac{1}{2}\omega_0^2[(B - A)a_{21}^2 + (B - C)a_{23}^2] - \omega_0^2(H_1 a_{21} + H_2 a_{22} + H_3 a_{23}) = \text{const.} \quad (4)$$

3   Equilibrium Orientations

Setting in (2) and (3) $\psi = \psi_0 = \text{const}$, $\vartheta = \vartheta_0 = \text{const}$, $\varphi = \varphi_0 = \text{const}$ we obtain at $A \neq B \neq C$ the equations

$$(C - B)(a_{22}a_{23} - 3a_{32}a_{33}) = H_2a_{23} - H_3a_{22},$$
$$(A - C)(a_{21}a_{23} - 3a_{31}a_{33}) = H_3a_{21} - H_1a_{23}, \qquad (5)$$
$$(B - A)(a_{21}a_{22} - 3a_{31}a_{32}) = H_1a_{22} - H_2a_{21}$$

allowing us to determine the gyrostat satellite equilibria in the orbital coordinate system.

Substituting the expressions for the direction cosines from (1) in terms of Euler angles into Eqs. (5), we obtain three equations with three unknowns ψ, ϑ, and φ. The second procedure for closing Eqs. (5) is to add the following six orthogonality conditions for the direction cosines

$$a_{i1}a_{j1} + a_{i2}a_{j2} + a_{i3}a_{j3} = \delta_{ij} \qquad (6)$$

where δ_{ij} is the Kronecker delta and $(i, j = 1, 2, 3)$. Equations (5) and (6) form a closed system with respect to the direction cosines, which also specifies the equilibrium solutions of the satellite.

We state the following problem for the system of equations (5), (6): determine all nine direction cosines, i.e., to find all the equilibrium orientations of the satellite when system parameters $A, B, C, \bar{h}_1, \bar{h}_2$, and $\bar{h}_3$ are given. The problem has been solved only for some specific cases when the vector of gyrostatic moment is located along the satellite's principal central axis of inertia Oy, when $H_1 = 0$, $H_2 \neq 0$, $H_3 = 0$ ([3], [4], [7]), and when the vector of gyrostatic moment locates in the satellite's principal central plane of inertia Oxz of the frame $Oxyz$ and $H_1 \neq 0$, $H_2 = 0$, $H_3 \neq 0$ ([5], [6]).

In the case $H_1 = H_2 = H_3 = 0$, it has been proved that the system (5), (6) has 24 solutions describing the equilibrium orientations of a satellite-rigid body [1].

Here we consider the general case of determination of the equilibria of the satellite when $H_1 \neq 0$, $H_2 \neq 0$, $H_3 \neq 0$. This problem was first studied in [8]. A Computer Algebra approach to determination of equilibrium orientations of the satellite was used. Projecting Eqs. (5) onto the axis of the orbiting frame $OXYZ$, we get the algebraic system, using the method given in [8]:

$$Aa_{11}a_{31} + Ba_{12}a_{32} + Ca_{13}a_{33} = 0,$$
$$Aa_{11}a_{21} + Ba_{12}a_{22} + Ca_{13}a_{23} + (H_1a_{11} + H_2a_{12} + H_3a_{13}) = 0, \qquad (7)$$
$$4(Aa_{21}a_{31} + Ba_{22}a_{32} + Ca_{23}a_{33}) + (H_1a_{31} + H_2a_{32} + H_3a_{33}) = 0.$$

The main task of symbolic calculation is to reduce the algebraic system (6), (7) to a single algebraic equation with one variable. Solution of the system (6), (7) can be obtained using the algorithm for the construction of Groebner bases [9]. The method of Groebner bases is used to solve systems of nonlinear algebraic equations. It comprises an algorithmic procedure for reducing the problem involving polynomials of several variables to investigation of a polynomial of one variable. Using the computer algebra system Maple [10] Groebner[gbasis] package with *tdeg* option, we calculate the Groebner basis of the system (6), (7) of nine polynomials in nine variables a_{ij} $(i, j = 1, 2, 3)$ under the ordering on the total power of the variables. In the list of variables in the Maple Groebner package, we use nine direction cosines, and in the list of polynomials, we include the polynomials from the left-hand sides f_i $(i = 1, 2, ...9)$ of the algebraic equations (6), (7):

map(factor,Groebner[gbasis]([f1,f2,f3, ... f9],tdeg(a11, a12, a13, ... a33))).

Here we write the polynomials in the Groebner basis that depend only on variables a_{31}, a_{32}, a_{33}:

$$16[(B - C)^2 a_{32}^2 a_{33}^2 + (C - A)^2 a_{31}^2 a_{33}^2 + (A - B)^2 a_{31}^2 a_{32}^2] =$$
$$= (H_1 a_{31} + H_2 a_{32} + H_3 a_{33})^2 (a_{31}^2 + a_{32}^2 + a_{33}^2),$$
$$4(B - C)(C - A)(A - B)a_{31}a_{32}a_{33} + [H_1(B - C)a_{32}a_{33} + \qquad (8)$$
$$+ H_2(C - A)a_{31}a_{33} + H_3(A - B)a_{31}a_{32}](H_1 a_{31} + H_2 a_{32} + H_3 a_{33}) = 0,$$
$$a_{31}^2 + a_{32}^2 + a_{33}^2 = 1.$$

Introducing the new variables $x = a_{31}/a_{33}$, $y = a_{32}/a_{33}$, $h_i = H_i/(B - C)$, $\nu = (B - A)/(B - C)$, we deduce two equations for determination of x and y

$$a_0 y^2 + a_1 y + a_2 = 0,$$
$$b_0 y^4 + b_1 y^3 + b_2 y^2 + b_3 y + b_4 = 0, \qquad (9)$$

where

$$a_0 = h_2(h_1 - \nu x h_3),$$
$$a_1 = h_1 h_3 + [4\nu((1 - \nu) + h_1^2 - (1 - \nu)h_2^2 - \nu h_3^2]x - \nu h_1 h_3 x^2,$$
$$a_2 = -(1 - \nu)h_2(h_1 x + h_3)x,$$
$$b_0 = h_2^2,$$
$$b_1 = 2h_2(h_1 x + h_3),$$
$$b_2 = (h_1^2 + h_2^2 - 16) + 2h_1 h_3 x + (h_1^2 + h_2^2 - 16\nu^2)x^2,$$
$$b_3 = 2h_2(h_1 x + h_3)(1 + x^2),$$
$$b_4 = (h_1 x + h_2)^2(1 + x^2) - 16(1 - \nu)^2 x^2.$$

Using the resultant concept we eliminate the variable y from the equations (9). Expanding the determinant of resultant matrix of Eqs.(9) with the help of Maple symbolic matrix function, we obtain a twelfth order algebraic equation in x

$$p_0 x^{12} + p_1 x^{11} + p_2 x^{10} + p_3 x^9 + p_4 x^8 + p_5 x^7 +$$
$$+ p_6 x^6 + p_7 x^5 + p_8 x^4 + p_9 x^3 + p_{10} x^2 + p_{11} x + p_{12} = 0, \qquad (10)$$

the coefficients of which depend in a rather complicated way on the parameters ν, h_1, h_2, h_3:

$$p_0 = -h_1^4 h_3^4 \nu^6, \quad p_1 = 2h_1^3 h_3^3 \nu^5 [2h_1^2 - h_2^2(\nu - 1) - 2\nu(h_3^2 + 2\nu - 2)], \dots (11)$$
$$p_{11} = -2h_1^3 h_3^3 [(2h_1^2 - h_2^2(\nu - 1) - 2\nu(h_3^2 + 2\nu - 2)], \quad p_{12} = -h_1^4 h_3^4.$$

By the definition of resultant, to every root x of Eq.(10) there corresponds a common root y of the system (9). It can easily be shown that to every real root x of Eq.(10), there correspond 2 solutions for (5), (6). Since the number of real roots of Eq.(10) does not exceed 12, the gyrostat satellite in a circular orbit can have at most 24 equilibria in the orbital coordinate system. Using Eq.(10), (11), we can determine numerically all the relative equilibrium orientations of the gyrostat satellite and analyze their stability. We have analyzed numerically dependence of the number of real solutions of Eq.(10) on the parameters, using Mathematica 8.0 factorization package. It is possible to provide the numerical calculations, without breaking a generality for the case when $B > A > C$. From these inequalities it follows that $0 < \nu < 1$. The parameters h_1, h_2, and h_3 can take on any nonzero values. The coefficients of Eq.(10) depend on 4 dimensionless parameters ν, h_1, h_2, h_3. The system of stationary equations (5) depends on 6 dimensional parameters. For the numerical calculations, reduction of the number of system parameters is very essential. Let us consider the properties of the algebraic equation (10) in detail. It is possible to show that the number of real roots of Eq. (10) does not depend on the sign of the parameters h_1, h_2, and h_3. It is evident that coefficients of Eq.(10) with odd x degree depend only on odd degree of the parameters h_1, h_2, h_3. For the coefficients with even x degree, we can represent them, using factorization to the form of two factors - one factor equals $h_1 h_3$ and the second factor depends only on odd degree of the parameters h_1, h_2, and h_3. Thus, changing sign of h_1, h_2, and h_3 we will change only the sign of the factor $h_1 h_3$ and, therefore, the sign of real root of polynomial (10). Therefore, the number of real roots does not change.

Hence, the numerical analysis of the number of real roots of Eq.(10) is possible to do with positive values of h_1, h_2, h_3 and $0 < \nu < 1$ condition. Thus, the numerical investigation of real roots of Eq.(10) will be simplified. The numerical calculations were made for fixed values of ν and h_3, the number of real roots was determined at the nodes of a uniform grid in the plane (h_1, h_2). The direct calculations for h_2 with step equal to 0.0001 are very complicated. In this case, we have for the size 4x4 (h_1, h_2) region about 10^9 nodes. The calculation task was divided in two parts. In the first place, the number of real roots in the 10^7 nodes (0.001 step value for h_2) was calculated. Secondly, the number of real roots was calculated in the vicinity of the border between two regions with the fixed number of real roots (0.0001 step value for h_2).

Then for the fixed value of h_2, it was defined a more precise border value of h_1 between two regions with the fixed number of real roots with the determined accuracy, using the bisection method realized in Mathematica language as a package. Equation (10) was derived under the conditions $h_1 \neq 0$, $h_2 \neq 0$, $h_3 \neq 0$, so we use h_1, h_2, h_3 in the vicinity of zero with the higher accuracy equal to 0.000001.

Calculations were made for the inertia parameters $\nu = 0.01$, $\nu = 0.1$, $\nu = 0.2$, $\nu = 0.3$, $\nu = 0.4$, $\nu = 0.5$, $\nu = 0.6$, $\nu = 0.7$, $\nu = 0.8$, $\nu = 0.9$, $\nu = 0.99$.

The example of numerical calculations of the borders between the regions with the fixed number of real roots at the plane (h_1, h_2) for $\nu = 0.2$ and $h_3 = 0.25$

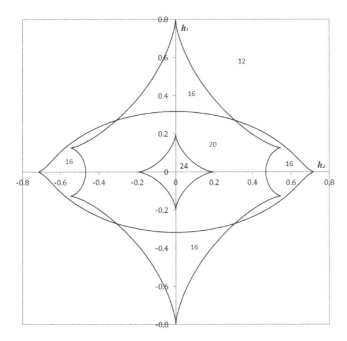

Fig. 1. The regions with the fixed number of equilibria for $\nu = 0.2$, $h_3 = 0.25$

is presented in Fig. 1. At the center of this picture, there is a region with 24 equilibria (12 real roots).

¿From the analysis of all calculations for the above-mentioned inertia parameters ν, it follows that with an increase of the h_3 values, several regions with the fixed number of equilibria become narrowed until they completely disappear. The point in the space of parameters, where a region with the fixed number of equilibria vanishes, was defined as bifurcation point. Calculated bifurcation values of the parameters are presented in Table 1.

Table 1. Bifurcational ν, h_3 values

ν	$h_3(24/20)$	$h_3(20/16)$	$h_3(16/12)$	$h_3(12/8)$
0.01	0.99	0.999	3.959	4.0
0.1	0.90	1.021	3.610	4.0
0.2	0.80	1.048	3.264	4.0
0.3	0.70	1.082	2.950	4.0
0.4	0.60	1.124	2.669	4.0
0.5	0.50	1.182	2.412	4.0
0.6	0.40	1.186	2.167	4.0
0.7	0.30	1.105	1.915	4.0
0.8	0.20	0.909	1.629	4.0
0.9	0.10	0.676	1.245	4.0
0.99	0.01	0.168	0.997	4.0

All bifurcation points from Table 1 were calculated numerically, and it is possible to see that the h_3 bifurcation values for regions with 24 equilibria (12 real roots) vanish in accordance with the equation $h_3 = 1 - \nu$.

For the regions with 20 equilibria (10 real roots), the h_3 bifurcation values increase with the increase of ν up to $\nu = 0.6$, and decrease after that with the further increase of ν.

For the regions with 16 equilibria (8 real roots) the bifurcation values always decrease with the decreasing of ν.

The regions with 12 equilibria become smaller with the increase of h_3 values. These regions are vanishing at the center of coordinate system for $h_3 = 4$. For $h_3 \geq 4$, there are small regions of 12 equilibria near h_2 axis with the size along h_1 and h_2 axes less than 10^{-1}. And with increasing h_3 value, these small regions take the farther position from the center of coordinate system along the h_2 axis.

For the interval $0.01 \leq \nu \leq 0.99$, the evolution of regions with 24, 20, 16, 12, and 8 equilibria was investigated numerically. For example, if $\nu = 0.2$ then analysis of the numerical results shows that five regions with 24, 20, 16, 12, and 8 equilibria exist in the plane (h_1, h_2) for the interval $h_3 < 0.8$. When we pass through the bifurcation value $h_3 = 0.8$ the region with 24 equilibria vanishes, and in the interval $0.8 < h_3 < 1.048$, only four regions with 20, 16, 12, and 8 equilibria exist. When we pass through the bifurcation value $h_3 = 1.048$ the region with 20 equilibria vanishes.

In the interval $1.048 < h_3 < 3.264$, only three regions with 16, 12, and 8 equilibria exist. The value $h_3 = 3.264$ is bifurcational. When we pass through this bifurcation value the region with 16 equilibria vanishes. In the interval $3.264 < h_3 < 4$, only two regions with 12 and 8 equilibria exist near the center of coordinate system.

When the values of parameter h_3 of the gyrostatic torque are more than 4, the satellite has 8 equilibrium orientations, which correspond to four real roots of Eq.(10). There are only small regions of 12 equilibria outside the center of the plane (h_1, h_2) near the h_2 axis.

4   Stability Analysis of Equilibria

To investigate the stability of equilibrium solutions $\psi = \psi_0 = \text{const}$, $\vartheta = \vartheta_0 = \text{const}$, $\varphi = \varphi_0 = \text{const}$ satisfying Eqs. (5), we can use the Jacobi integral of energy (4) as the Lyapunov function in order to obtain sufficient conditions of stability of the equilibrium orientations of the gyrostat satellite. After replacement $\psi \to \psi + \psi_0$, $\vartheta \to \vartheta + \vartheta_0$, $\varphi \to \varphi + \varphi_0$ where ψ, ϑ, φ are small deviations from the satellite's equilibria ψ_0, ϑ_0, φ_0, the energy integral takes the form

$$\frac{1}{2}(A\bar{p}^2 + B\bar{q}^2 + C\bar{r}^2) + \frac{1}{2}(B - C)(A_{11}\psi^2 + A_{22}\vartheta^2 + A_{33}\varphi^2 +$$
$$+ 2A_{12}\psi\vartheta + 2A_{13}\psi\varphi + 2A_{23}\vartheta\varphi) + O_3(\psi, \vartheta, \varphi) = \text{const}, \quad (12)$$

where coefficients A_{ij} depend on the parameters $\nu, h_1, h_2, h_3, \psi_0, \vartheta_0, \varphi_0$ in the form

$$A_{11} = \nu(a_{11}^2 - a_{21}^2) + (a_{13}^2 - a_{23}^2) + h_1 a_{21} + h_2 a_{22} + h_3 a_{23},$$

$$A_{22} = (3 + \cos^2 \psi_0)(1 - \nu \sin^2 \varphi_0) \cos 2\vartheta_0 - \frac{1}{4} \nu \sin 2\psi_0 \cos \vartheta_0 \sin 2\varphi_0 +$$
$$+ (h_1 \sin \varphi_0 + h_2 \cos \varphi_0) \cos \psi_0 \cos \vartheta_0 + h_3 a_{23},$$

$$A_{33} = \nu[(a_{22}^2 - a_{21}^2) - 3(a_{32}^2 - a_{31}^2)] + h_1 a_{21} + h_2 a_{22}, \tag{13}$$

$$A_{12} = -\frac{1}{2} \sin 2\psi_0 \sin 2\vartheta_0 + \nu(a_{11}a_{23} + a_{13}a_{21}) -$$
$$- (h_1 a_{31} + h_2 a_{32} + h_3 a_{33}) \sin \psi_0,$$

$$A_{13} = \nu(a_{11}a_{22} + a_{12}a_{21}) - h_1 a_{12} + h_2 a_{11},$$

$$A_{23} = -\frac{3}{2} \nu \sin 2\vartheta_0 \sin 2\varphi_0 + \nu(a_{21} \cos \varphi_0 + a_{22} \sin \varphi_0)a_{23} -$$
$$- (h_1 \cos \varphi_0 - h_2 \sin \varphi_0)a_{23}.$$

It follows from Lyapunov theorem that the equilibrium solution is stable if the quadratic form (12),(13) is positive definite, i.e., the following inequalities take place:

$$A_{11} > 0,$$
$$A_{11} A_{22} - A_{12}^2 > 0, \tag{14}$$
$$A_{11} A_{22}A_{33} + 2A_{12}A_{23}A_{13} - A_{11}A_{23}^2 - A_{22}A_{13}^2 - A_{33}A_{12}^2 > 0.$$

Substituting the expressions for A_{ij} from (13) for the corresponding equilibrium solution into (14), we obtain the conditions for stability of this solution. Using integral (12), we have analyzed numerically stability conditions (14) for the equilibrium solutions.

¿From the analysis of all calculations for the above-indicated parameters it follows that for $|h_3| < (1 - \nu)$, there are 24 equilibrium solutions, and for 4 equilibrium solutions, stability conditions (14) are valid. There are also 4 stable equilibria for $\nu > 0.5$ and $|h_3| \geq (1 - \nu)$. When the values of parameters $|h_i| > 4$ ($i = 1, 2, 3$) there are 8 equilibrium solutions, and only two of them are stable.

5   Conclusion

In this work, the attitude motion of the gyrostat satellite under the action of gravitational torque in a circular orbit has been investigated. The main attention was given to determination of the satellite equilibrium orientation in the orbital coordinate system and to analysis of their stability. The symbolic-numerical method of determination of all satellite equilibria is suggested in general case when $h_1 \neq 0$, $h_2 \neq 0$, and $h_3 \neq 0$. The symbolic computation system Maple is applied to reduce the satellite stationary motion system of nine algebraic equations with nine variables to a single algebraic equation of the twelfth degree with one variable, using the Groebner package for the construction of a Groebner basis and the resultant approach. It was shown that the equilibrium orientations are determined by real roots of algebraic equation of the twelfth degree. Using this result of symbolic calculations we conclude that the gyrostat satellite can have no more than 24 equilibrium orientations in a circular orbit.

The evolution of domains with fixed number of equilibrium orientations was investigated numerically in the plane of two parameters h_1 and h_2 for different values of parameters ν and h_3. The bifurcation values of h_3 corresponding to the qualitative change of domains with fixed number of equilibria were determined. On the basis of numerical calculation we can conclude that the number of satellite's isolated equilibria is no less than 8. Using the Lyapunov theorem, the sufficient conditions of stability of the equilibrium orientations are investigated numerically at different values of gyrostatic parameters. Analysis of the numerical simulation shows that the number of stable equilibria is no less than 2 and no more than 4. All calculations considered here were implemented with the computer algebra systems Maple and Mathematica.

The results of the study can be used at the stage of preliminary design of the satellite with gravitational control system.

References

1. Sarychev, V.A.: Problems of Orientation of Satellites, Itogi Nauki i Tekhniki. Ser. Space Research, vol. 11. VINITI, Moscow (1978)
2. Gutnik, S.A.: Symbolic-numeric investigation of the aerodynamic forces influence on satellite dynamics. In: Gerdt, V.P., Koepf, W., Mayr, E.W., Vorozhtsov, E.V. (eds.) CASC 2011. LNCS, vol. 6885, pp. 192–199. Springer, Heidelberg (2011)
3. Sarychev, V.A., Mirer, S.A.: Relative equilibria of a gyrostat satellite with internal angular momentum along a principal axis. Acta Astronautica 49, 641–644 (2001)
4. Sarychev, V.A., Mirer, S.A., Degtyarev, A.A.: The dynamics of a satellite gyrostat with a single nonzero component of the vector of gyrostatic moment. Cosmic Research 43, 268–279 (2005)
5. Sarychev, V.A., Mirer, S.A., Degtyarev, A.A.: Dynamics of a gyrostat satellite with the vector of gyrostatic moment in the principal plane of inertia. Cosmic Research 46, 61–74 (2008)
6. Longman, R.W.: Gravity-gradient stabilization of gyrostat satellites with rotor axes in principal planes. Celestial Mechanics 3, 169–188 (1971)
7. Longman, R.W., Hagedorn, P., Beck, A.: Stabilization due to gyroscopic coupling in dual-spin satellites subject to gravitational torques. Celestial Mechanics 25, 353–373 (1981)
8. Sarychev, V.A., Gutnik, S.A.: Relative equilibria of a gyrostat satellite. Cosmic Research 22, 323–326 (1984)
9. Buchberger, B.: Theoretical basis for the reduction of polynomials to canonical forms. SIGSAM Bulletin, 19–29 (1976)
10. Char, B.W., Geddes, K.O., Gonnet, G.H., Monagan, M.B., Watt, S.M.: Maple Reference Manual. Watcom Publications Limitid, Waterloo (1992)

On Stationary Sets of Euler's Equations on $so(3,1)$ and Their Stability

Valentin Irtegov and Tatyana Titorenko

Institute for System Dynamics and Control Theory SB RAS,
134, Lermontov str., Irkutsk, 664033, Russia
irteg@icc.ru

Abstract. With the use of computer algebra methods we investigate two recent found cases of integrability (in the Liouville sense) of Euler's equations on the Lie algebra $so(3,1)$ when the equations possess additional polynomial first integrals of degrees 3 and 6. The problems of obtaining stationary sets of the equations and investigation of their stability are considered. In addition to the sets obtained earlier [1], we have found new zero-dimensional and nonzero-dimensional stationary sets. For a number of the sets we have derived sufficient conditions of their stability and instability.

Keywords: Euler's equations, stationary sets, stability.

1 Introduction

It is known that configuration space in dynamics of a rigid body is, as a rule, some natural Lie group $SO(3), E(3), SO(4)$, etc. In [2], the connection of classical Euler's equations, describing the motion of a rigid body, with Lie algebras is shown. Many problems of mechanics, mathematical physics, etc. [3] reduce to Euler's equations on Lie algebras. These equations are also a good model for the study of singularities of integrable Hamiltonian systems. Therefore, an increased interest takes place in the systems of such a type, in particular, finding new cases of integrability of the systems and qualitative analysis of these cases. So, the works [6]–[8] are devoted to investigation of new integrable cases of Euler's equations on Lie algebras $e(3), so(4)$ [4], [5] when the equations admit additional polynomial first integral of 4th degree. In these works, a topological analysis of the cases has been conducted.

The goal of our paper is investigation of some qualitative properties of two recent found cases of integrability of Euler's equations on the Lie algebra $so(3,1)$ when the equations possess additional polynomial first integrals of degrees 3 and 6 [9], [10]. We find peculiar solutions (stationary sets) of these equations and investigate their stability. For solving these problems, we apply a technique based on computer algebra methods. The latter allows us not only to obtain the desired solutions in an analytical form but also to investigate their stability, e.g., by Lyapunov's methods. In [1], some general analysis of stationary sets of the equations considered has been conducted. Nonzero-dimensional

V.P. Gerdt et al. (Eds.): CASC 2013, LNCS 8136, pp. 179–193, 2013.
© Springer International Publishing Switzerland 2013

stationary sets (stationary invariant manifolds (IMs)) have been found, their stability and bifurcations have been investigated. It was shown that the equations have zero-dimensional stationary sets in capacity of their solutions. In the given paper, we have found these sets and investigated their stability. We have also found a family of complex IMs which possess, in our opinion, an interesting property: under some additional constraints, the motion on the elements of this family is described by real functions of time. For the purpose of solving computational problems, we used the computer algebra systems (CAS) *Mathematica* and *Maple* [11].

2   On Stationary Sets of Euler's Equations with an Additional Cubic First Integral

2.1   Problem Formulation

In [10], the problem of revealing integrable equations in the family of Euler's equations with Hamiltonians of the form

$$H = c_1\,\alpha_3(M_1^2 + M_2^2 + M_3^2) + c_2 M_3(\alpha_1 M_1 + \alpha_3 M_3) + M_1\gamma_2 - M_2\gamma_1 \qquad (1)$$

is considered. Such problems arise, e.g., in classification of integrable Hamiltonian systems [12]. Here M_i, γ_i $(i = 1, 2, 3)$ are the components of the two 3-dimensional vectors, α_j, c_j are some constants.

Hamiltonians (1) belong to a wide class of quadratic Hamiltonians which have numerous applications (two-spin interactions, motion of a rigid body in a constant-curvature space or in an ideal fluid, motion of a body with ellipsoidal cavity filled with fluid around a fixed point, etc.). A number of works are devoted to the study of the Hamiltonians of such a type (see, e.g., [3]–[10]).

In [10], several integrable cases of the equations discussed were found. The Hamiltonians corresponding to them are separated from family (1) by the constraints imposed on parameters c_1 and c_2: $(a)\,c_1$ is arbitrary, $c_2 = 0$; $(b)\,c_1 = 1, c_2 = -2$; $(c)\,c_1 = 1, c_2 = -1$; $(d)\,c_1 = 1, c_2 = -1/2$. In the present paper, cases (b) and (d) are studied.

According to [10], in integrable case (b) there exists an additional cubic integral of the form

$$F = \{2[\alpha_3(M_1\gamma_2 - M_2\gamma_1) + \alpha_1(M_2\gamma_3 - M_3\gamma_2)] - k(M_1^2 + M_2^2 + M_3^2)$$
$$+ \gamma_1^2 + \gamma_2^2 + \gamma_3^2\}M_3 = h_1 = \text{const.} \qquad (2)$$

The latter integral has been found earlier in [9].

Euler's equations corresponding to Hamiltonian (b) write

$$\dot{M}_1 = 2M_2(\alpha_1 M_1 + 2\alpha_3 M_3) - (M_3\gamma_1 - M_1\gamma_3),$$
$$\dot{M}_2 = M_2\gamma_3 - M_3\gamma_2 - 2[\alpha_1(M_1^2 - M_3^2) + 2\alpha_3 M_1 M_3], \quad \dot{M}_3 = -2\alpha_1 M_2 M_3,$$
$$\dot{\gamma}_1 = 2(\alpha_1 M_1 + \alpha_3 M_3)\,\gamma_2 + (2\alpha_3 M_2 - \gamma_1)\,\gamma_3 + kM_1 M_3, \qquad (3)$$
$$\dot{\gamma}_2 = 2[(\alpha_1 M_3 - \alpha_3 M_1)\,\gamma_3 - (\alpha_1 M_1 + \alpha_3 M_3)\,\gamma_1] + kM_2 M_3 - \gamma_2\gamma_3,$$
$$\dot{\gamma}_3 = 2[\alpha_3(M_1\gamma_2 - M_2\gamma_1) - \alpha_1 M_3\gamma_2] + \gamma_1^2 + \gamma_2^2 - k(M_1^2 + M_2^2).$$

The rest of the integrals of equations (3) has the form:

$$V_1 = \sum_{i=1}^{3} M_i \gamma_i = h_2 = \text{const}, \quad V_2 = \sum_{i=1}^{3} (kM_i^2 + \gamma_i^2) = h_3 = \text{const}. \qquad (4)$$

We state the problem of finding stationary sets of equations (3) and investigation of their stability in the Lyapunov sense.

2.2   Finding Stationary Sets

In order to solve the stated problem we shall apply the Routh–Lyapunov method [13] and some its generalizations (see [14]). This method in combination with computer algebra tools allows one not only to find the desired solutions but also to investigate their stability. According to the method, stationary invariant sets of the above differential equations are called solutions of conditional extremum problem for the elements of algebra of the first integrals of these equations. To obtain these sets, some linear or nonlinear combination from the problem's first integrals (a family of the first integrals) is constructed, and necessary conditions for this family to have an extremum with respect to phase variables are written. The problem of finding stationary invariant sets for the system of differential equations with polynomial first integrals is reduced thereby to obtaining solutions of some algebraic system. In our case, it will be a system of nonlinear equations.

Following the technique chosen, we construct the complete linear combination

$$K = \lambda_0 H - \lambda_1 V_1 - \frac{\lambda_2}{2} V_2 - \lambda_3 F \qquad (5)$$

from the first integrals of the problem, and write down the necessary conditions for the integral K to have an extremum with respect to phase variables M_i, γ_i:

$$\partial K / \partial M_1 = \lambda_0 \gamma_2 - \lambda_1 \gamma_1 + (2\alpha_3 \lambda_0 - k\lambda_2) M_1 - 2\left[\alpha_1 \lambda_0 + \lambda_3 (\alpha_3 \gamma_2 - kM_1)\right] M_3 = 0,$$
$$\partial K / \partial M_2 = -\lambda_0 \gamma_1 - \lambda_1 \gamma_2 + (2\alpha_3 \lambda_0 - k\lambda_2) M_2 + 2\lambda_3 (\alpha_3 \gamma_1 + kM_2 - \alpha_1 \gamma_3) M_3 = 0,$$
$$\partial K / \partial M_3 = -\lambda_1 \gamma_3 - 2\alpha_1 \lambda_0 M_1 - (2\alpha_3 \lambda_0 + k\lambda_2) M_3 + \lambda_3 \left[k(M_1^2 + M_2^2 + 3M_3^2)\right.$$
$$\left. +2(\alpha_3 \gamma_1 - \alpha_1 \gamma_3) M_2 + 2(2\alpha_1 M_3 - \alpha_3 M_1)\gamma_2 - (\gamma_1^2 + \gamma_2^2 + \gamma_3^2)\right] = 0, \quad (6)$$
$$\partial K / \partial \gamma_1 = -\lambda_1 M_1 - \lambda_0 M_2 - \lambda_2 \gamma_1 + 2\lambda_3 (\alpha_3 M_2 - \gamma_1) M_3 = 0,$$
$$\partial K / \partial \gamma_2 = \lambda_0 M_1 - \lambda_1 M_2 - \lambda_2 \gamma_2 + 2\lambda_3 \left[(\alpha_1 M_3 - \alpha_3 M_1) - \gamma_2\right] M_3 = 0,$$
$$\partial K / \partial \gamma_3 = -\lambda_1 M_3 - \lambda_2 \gamma_3 - 2\lambda_3 (\alpha_1 M_2 + \gamma_3) M_3 = 0.$$

Here $\lambda_i = \text{const}$ are the family parameters of the integral K.

Equations (6) (the conditions of stationarity for the integral K) represent a system of polynomial equations of 2nd degree with parameters $\lambda_0, \lambda_1, \lambda_2, \lambda_3,$ α_1, α_3. It should be noted that if some part of parameters λ_i in K assumes zero values then we obtain an "incomplete" combination of the integrals. In this case, both stationary equations and solutions of these equations correspond to this "incomplete" combination of the integrals.

We applied computer algebra methods for qualitative analysis of the solution set of equations (6) and for finding solutions of these equations.

It was shown in [1] that equations (6) have a finite number of solutions (6 solutions) over field $Q(\lambda_i, \alpha_j)[M_1, M_2, M_3, \gamma_1, \gamma_2, \gamma_3]$. The latter was revealed by the programs *IsZeroDimensional, NumberOfSolutions* which are included in the *Maple*-program package *PolynomialIdeals*.

The program *IsPrime* (which is also part of this package) allowed us to find out that system (6) can be decomposed into "simpler" subsystems over the above field. To this end, we applied triangular sets method [15]. The method decomposes the algebraic variety of a polynomial system into subvarieties which correspond to one or several solutions of the system. A special form, called "triangular set", is used for representing solutions. In the problem considered, application of this method has not caused any computational difficulties. We used the *Maple*-program *Triangularize*. The result of application of this program to system (6) writes

$$
\begin{aligned}
&\big[\,[\,\gamma_1, \gamma_2, \gamma_3, M_1, M_2, M_3], \\
&[\,M_2,\ \gamma_1(2\lambda_3 M_3 + \lambda_2) + \lambda_1 M_1,\ \gamma_3(2\lambda_3 M_3 + \lambda_2) + \lambda_1 M_3, \\
&\quad \gamma_2(2\lambda_3 M_3 + \lambda_2) + M_1(2\alpha_3\lambda_3 M_3 - \lambda_0) - 2\alpha_1\lambda_3 M_3^2, \\
&\quad 2M_1\left(\alpha_1^2(4\lambda_3^2 M_3^2 - \lambda_2^2) - (\lambda_0 + \alpha_3\lambda_2)^2 - \lambda_1^2\right) \\
&\quad +\alpha_1(4\alpha_3\lambda_3^2 M_3^2 + \lambda_0(2\lambda_3 M_3 + \lambda_2))M_3, \\
&\quad a_0 M_3^5 + a_1 M_3^4 + a_2 M_3^3 + a_3 M_3^2 + a_4 M_3 + a_5\,]\,\big].
\end{aligned}
\tag{7}
$$

Here $a_0, \ldots, a_5$ are some expressions of λ_i, α_j.

The program result is a list of the system solutions represented in "triangular" form. The first element of the list defines a trivial solution of system (6), the 2nd element defines five solutions of this system because the polynomial of variable M_3 has five roots. We have found multiple roots of this polynomial. The conditions of existence of such roots can be obtained as conditions under which the resultant of two polynomials (the polynomial considered and its derivative) vanishes. The found conditions in the form of restrictions imposed on parameters λ_i, α_j and the solutions corresponding to them are given below.

i) $\lambda_1 = 0, \alpha_3 = 0$:
$$
\gamma_1 = 0,\ \gamma_2 = -2\alpha_1\lambda_2\lambda_3^{-1},\ \gamma_3 = 0,\ M_1 = 0,\ M_2 = 0,\ M_3 = -\lambda_2\lambda_3^{-1};\tag{8}
$$

ii) $\lambda_1 = 0, \alpha_3 = \alpha_1$:
$$
\begin{aligned}
&\gamma_1 = 0,\ \gamma_2 = 2(\alpha_1\lambda_2 + \lambda_0)\lambda_3^{-1},\ \gamma_3 = 0,\ M_1 = -2(\alpha_1\lambda_2 + \lambda_0)\,\alpha_1^{-1}\lambda_3^{-1}, \\
&M_2 = 0,\ M_3 = (\alpha_1\lambda_2 + \lambda_0)\,\alpha_1^{-1}\lambda_3^{-1};
\end{aligned}
\tag{9}
$$

iii) $\lambda_2 = 0, \alpha_1 = 0$:
$$
\gamma_1 = 0,\ \gamma_2 = 0,\ \gamma_3 = -\lambda_1\lambda_3^{-1}/2,\ M_1 = 0,\ M_2 = 0,
$$
$$
M_3 = -(2\lambda_0 \pm \sqrt{4\lambda_0^2 + 3\lambda_1^2})\,\alpha_3^{-1}\lambda_3^{-1}/6;
\tag{10}
$$

iv) $\lambda_0 = \dfrac{\lambda_1^2}{2\alpha_3\lambda_2} + \dfrac{1}{2}\alpha_3\lambda_2,\ \alpha_1 = 0$:

$$\gamma_1 = 0, \ \gamma_2 = 0, \ \gamma_3 = -(2\lambda_1 \pm \sqrt{\lambda_1^2 - 3\alpha_3^2\lambda_2^2}\,)\lambda_3^{-1}/2, \ M_1 = 0,$$

$$M_2 = 0, \ M_3 = -(\lambda_1^2 + 3\alpha_3^2\lambda_2^2 \mp \lambda_1\sqrt{\lambda_1^2 - 3\alpha_3^2\lambda_2^2}\,)\,\alpha_3^{-2}\lambda_2^{-1}\lambda_3^{-1}/6. \tag{11}$$

The above expressions are the families of the stationary solutions of differential equations (3) parameterized by λ_i. The latter is proved by direct substitution of (8)–(11) into equations (3) and (6). As a result, these equations turn into identities.

Geometrically, the elements of the families of real solutions (8)-(11) correspond to the families of points in R^6. According to [10], from a mechanical point of view, under corresponding interpretation of the parameters and the phase variables, these elements correspond to helical motions of a rigid body in fluid (generalized Kirchhoff's model), and permanent rotations of a rigid body with cavity filled with fluid (generalized Poincaré's model).

Next, we apply the triangular sets method for finding nonzero-dimensional stationary sets. Similarly to [1], we consider some part of the phase variables and some part of the parameters, e.g., $\gamma_1, \gamma_2, \gamma_3, M_1, M_2, \lambda_1$, in capacity of unknowns. The *Maple*-programs *IsZeroDimensional, HilbertDimention, IsPrime* have revealed that equations (6) have infinite number of solutions with respect to the unknowns, the dimension of these solutions is 1, and the algebraic variety of the equations can be decomposed into subvarieties. Below, the result of application of the program *Triangularize* to system (6), when $\gamma_1, \gamma_2, \gamma_3, M_1, M_2, \lambda_1$ are the unknowns, is given.

$$\begin{aligned}
&[\,[\,\gamma_1(2\lambda_3 M_3 + \lambda_2) + \lambda_1 M_1 - M_2(2\alpha_3\lambda_3 M_3 - \lambda_0), \\
&\quad \gamma_2(2\lambda_3 M_3 + \lambda_2) + +M_1(2\alpha_3\lambda_3 M_3 - \lambda_0) + \lambda_1 M_2 - 2\alpha_1\lambda_3 M_3^2, \\
&\quad \gamma_3(2\lambda_3 M_3 + \lambda_2) + (2\alpha_1\lambda_3 M_2 + \lambda_1)M_3, \\
&\quad (2\alpha_1\lambda_3^2 M_1 + \lambda_0\lambda_3)M_3 + 2\alpha_3\lambda_3^2 M_3^2 + \lambda_0\lambda_2, \\
&\quad \lambda_1^2 + (\lambda_0 + \alpha_3\lambda_2)^2 + \alpha_1^2\lambda_2^2\,], \\
&[\,M_2, \ \gamma_1(2\lambda_3 M_3 + \lambda_2) + \lambda_1 M_1, \ \gamma_3(2\lambda_3 M_3 + \lambda_2) + \lambda_1 M_3, \\
&\quad \gamma_2(2\lambda_3 M_3 + \lambda_2) + M_1(2\alpha_3\lambda_3 M_3 - \lambda_0) - 2\alpha_1\lambda_3 M_3^2, \\
&\quad -M_1((\lambda_0 + \alpha_3\lambda_2)^2 + \lambda_1^2 + \alpha_1^2\lambda_2^2 - 4\alpha_1^2\lambda_3^2 M_3^2) \\
&\quad +2\alpha_1(\lambda_3(2\alpha_3\lambda_3 M_3 + \lambda_0)M_3 + \lambda_0\lambda_2)M_3), \\
&\quad b_0\lambda_1^4 + b_1\lambda_1^2 + b_0\,]\,]. \tag{12}
\end{aligned}$$

Here b_0, b_1, and b_2 are some expressions of $\lambda_0, \lambda_2, \lambda_3, \alpha_j, M_3$.

The first element of list (12) defines two solutions of equations (6) which write

$$M_1 = -\frac{\lambda_3(2\alpha_3\lambda_3 M_3 + \lambda_0)M_3 + \lambda_0\lambda_2}{2\alpha_1\lambda_3^2 M_3}, \quad M_2 = -\frac{(2\lambda_3 M_3 + \lambda_2)\gamma_3 + \lambda_1 M_3}{2\alpha_1\lambda_3 M_3},$$

$$\gamma_1 = \frac{\lambda_0(\gamma_3\lambda_3 + \lambda_1) - 2\alpha_3\lambda_3^2 M_3\gamma_3}{2\alpha_1\lambda_3^2 M_3},$$

$$\gamma_2 = \frac{\lambda_3((\alpha_1^2 + \alpha_3^2)(2\lambda_3 M_3 - \lambda_2)M_3 + \lambda_1\gamma_3) - \lambda_0^2}{2\alpha_1\lambda_3^2 M_3}, \tag{13}$$

$$\lambda_1 = \pm\sqrt{-((\lambda_0 + \alpha_3\lambda_2)^2 + \alpha_1^2\lambda_2^2)}. \tag{14}$$

Expressions (13), taking into account (14), correspond to two families of stationary IMs of the initial differential equations. The latter is proved by substitution of these expressions into (6) and by IMs definition (the derivative of (13) calculated by virtue of equations (3) vanishes on these sets). The calculations are trivial and are not presented here.

Expression λ_1 (14) is the first integral of a vector field on the found IMs that is proved by first integral definition (the derivative of λ_1 calculated by virtue of equations of the vector field is identically equal to zero). In the case considered, the integral is trivial.

The 2nd element of list (12) defines four solutions:

$$M_1 = \frac{2\alpha_1(2\alpha_3\lambda_3^2 M_3^2 + \lambda_0(\lambda_3 M_3 + \lambda_2))M_3}{\lambda_1^2 + (\alpha_3\lambda_2 + \lambda_0)^2 + \alpha_1^2(\lambda_2^2 - 4\lambda_3^2 M_3^2)}, \ M_2 = 0,$$

$$\gamma_1 = -\frac{2\alpha_1\lambda_1(2\alpha_3\lambda_3^2 M_3^2 + \lambda_0(\lambda_3 M_3 + \lambda_2))M_3}{(2\lambda_3 M_3 + \lambda_2)(\lambda_1^2 + (\alpha_3\lambda_2 + \lambda_0)^2 + \alpha_1^2(\lambda_2^2 - 4\lambda_3^2 M_3^2))},$$

$$\gamma_2 = \frac{2\alpha_1\left[\lambda_0^2(2\lambda_3 M_3 + \lambda_2) + \lambda_3 M_3(\lambda_1^2 + (\alpha_1^2 + \alpha_3^2)(\lambda_2^2 - 4\lambda_3^2 M_3^2))\right]M_3}{(2\lambda_3 M_3 + \lambda_2)[\lambda_1^2 + (\alpha_3\lambda_2 + \lambda_0)^2 + \alpha_1^2(\lambda_2^2 - 4\lambda_3^2 M_3^2)]},$$

$$\gamma_3 = -\frac{\lambda_1 M_3}{2\lambda_3 M_3 + \lambda_2}, \tag{15}$$

$$\lambda_1 = \pm\frac{1}{\sqrt{2}}\sqrt{p_0 \pm \frac{[2\alpha_3\lambda_3^2 M_3^2 + \lambda_0(\lambda_2 + \lambda_3 M_3)]\sqrt{z_0}}{\lambda_3 M_3 + \lambda_2}}, \tag{16}$$

$$\text{where } p_0 = -\lambda_0(\lambda_0 - 6\alpha_3\lambda_3 M_3) + \frac{2\alpha_3\lambda_3^2 M_3^2(\lambda_0 + 2\alpha_3\lambda_3 M_3)}{\lambda_3 M_3 + \lambda_2}$$

$$-2(\alpha_1^2 + \alpha_3^2)(\lambda_2^2 - 4\lambda_3^2 M_3^2),$$

$$z_0 = 4\alpha_3(3\lambda_3 M_3 + 2\lambda_2)[2\alpha_3(3\lambda_3 M_3 + 2\lambda_2) + \lambda_0]$$
$$+16\alpha_1^2(\lambda_2 + \lambda_3 M_3)(\lambda_2 + 2\lambda_3 M_3) + \lambda_0^2.$$

Expressions (15), taking into account (16), define four families of one-dimensional stationary IMs of equations (3) which deliver a stationary value to already nonlinear combination of the basic integrals. Expression λ_1 (16) is the first integral of a vector field on the found IMs. These statements are proved as above.

2.3   Motions on the Invariant Manifolds

To analyze solutions on the above IMs, we investigate the differential equations on these IMs. Computer algebra tools play an auxiliary role here.

The equations of the vector field on the elements of families IMs (13) write

$$\dot{M}_3 = \frac{(2\lambda_3 M_3 + \lambda_2)\gamma_3 + \lambda_1 M_3}{\lambda_3},$$

$$\dot{\gamma}_3 = \frac{(\lambda_3 M_3 + \lambda_2)\gamma_3^2}{\lambda_3 M_3} + \frac{\lambda_1}{\lambda_3}\gamma_3 - (\alpha_1^2 + \alpha_3^2)M_3^2 + \frac{\lambda_0^2 \lambda_2}{\lambda_3^3 M_3}. \tag{17}$$

These are derived from equations (3) by eliminating variables $\gamma_1, \gamma_2, M_1, M_2$ from them with the help of (13). Here λ_1 has the form (14).

Next, we consider the case when $\lambda_1 = -\sqrt{-((\lambda_0 + \alpha_3\lambda_2)^2 + \alpha_1^2\lambda_2^2)}$.

Equations (17) admit first integrals which are obtained from initial first integrals by eliminating variables $\gamma_1, \gamma_2, M_1, M_2$ from them with the help of (13). The integrals found by this technique will be, generally speaking, dependent. Take one of them, e.g.,

$$\tilde{V} = \frac{1}{4\alpha_1^2 \lambda_3^4 M_3^2}\Big((\alpha_3\lambda_2 + \lambda_0)\lambda_0^2\lambda_2 + \lambda_2\lambda_3^2(\lambda_0 + \alpha_3\lambda_2)\gamma_3^2$$

$$-2(\lambda_0 + \alpha_3\lambda_2)\lambda_3^2\sqrt{-((\lambda_0 + \alpha_3\lambda_2)^2 + \alpha_1^2\lambda_2^2)}M_3\gamma_3 + 2\lambda_3^3(\lambda_0 + \alpha_3\lambda_2)M_3\gamma_3^2$$

$$+\lambda_3^2(2\alpha_3\lambda_0^2 + (3\alpha_1^2 + \alpha_3^2)\lambda_0\lambda_2 - \alpha_3(\alpha_1^2 + \alpha_3^2)\lambda_2^2)M_3^2$$

$$+2\lambda_3^3(\alpha_1^2 + \alpha_3^2)(\lambda_0 + \alpha_3\lambda_2)M_3^3\Big) = \tilde{c}_1 = const. \tag{18}$$

Eliminate variable γ_3 from equations (17) by expression (18). As a result, we have the differential equations (written in corresponding maps) with respect to M_3:

$$\dot{M}_3 = \pm\frac{\sqrt{z}}{\sqrt{\lambda_0 + \alpha_3\lambda_2}\lambda_3^2},$$

$$z = -4\lambda_3^4(\alpha_1^2 + \alpha_3^2)(\lambda_0 + \alpha_3\lambda_2)M_3^4$$

$$+\lambda_3^3(8\alpha_1^2\tilde{c}_1\lambda_3^2 - 4\lambda_0(2\alpha_1^2\lambda_2 + \alpha_3(\lambda_0 + \alpha_3\lambda_2)))M_3^3$$

$$+\lambda_3^2(4\alpha_1^2\tilde{c}_1\lambda_2\lambda_3^2 - \lambda_0(\lambda_0(\lambda_0 + 5\alpha_3\lambda_2) + 4(\alpha_1^2 + \alpha_3^2)\lambda_2^2))M_3^2$$

$$-\lambda_0^2\lambda_2^2(\lambda_0 + \alpha_3\lambda_2) - 2\lambda_0^2\lambda_2(\lambda_0 + \alpha_3\lambda_2)\lambda_3 M_3 - \lambda_0^2\lambda_2^2(\lambda_0 + \alpha_3\lambda_2).$$

The above equations are integrated in elliptic functions. We have the analogous result when we take λ_1 with positive sign. Hence, the motion on invariant submanifolds of IMs (13), the equations of which are obtained by addition of integral (18) to equations (13), is described by elliptic functions of time.

The families of IMs (13) are complex because the equations of the IMs contain complex coefficients. When $\lambda_0 = \lambda_1 = \lambda_2 = 0$ these families have the real invariant submanifold

$$M_1 = -\frac{\alpha_3 M_3}{\alpha_1}, \ M_2 = -\frac{\gamma_3}{\alpha_1}, \ \gamma_1 = -\frac{\alpha_3\gamma_3}{\alpha_1}, \ \gamma_2 = \frac{(\alpha_1^2 + \alpha_3^2)M_3}{\alpha_1},$$

the motion on which is described by real elementary functions. Note that the submanifold delivers a stationary value to integral F (2).

When $\lambda_0 = 0$, $\lambda_1 = -\sqrt{-(\alpha_1^2 + \alpha_3^2)}\lambda_2$ the family of IMs (13) has the subfamily of complex IMs

$$M_1 = -\frac{\alpha_3 M_3}{\alpha_1}, \; M_2 = \frac{\sqrt{-\alpha_1^2 - \alpha_3^2}\,\lambda_2 M_3 - (2\lambda_3 M_3 + \lambda_2)\gamma_3}{2\alpha_1\lambda_3 M_3}, \; \gamma_1 = -\frac{\alpha_3\gamma_3}{\alpha_1},$$

$$\gamma_2 = -\frac{\sqrt{-\alpha_1^2 - \alpha_3^2}\,\lambda_2\gamma_3 - (\alpha_1^2 + \alpha_3^2)(2\lambda_3 M_3 - \lambda_2)M_3}{2\alpha_1\lambda_3 M_3}, \tag{19}$$

which delivers a stationary value to integral $\tilde{K} = -\sqrt{-(\alpha_1^2 + \alpha_3^2)}\,V_1 - \lambda_2 V_2/2 - \lambda_3 F$.

The motion on the elements of subfamily (19) is defined by the differential equations

$$\dot{M}_3 = \pm\frac{2M_3\sqrt{\alpha_1^2\lambda_2\tilde{c}_1 + 2\alpha_1^2\lambda_3\tilde{c}_1 M_3 - \alpha_3(\alpha_1^2 + \alpha_3^2)\lambda_2 M_3^2}}{\sqrt{\alpha_3\lambda_2}}. \tag{20}$$

Equations (20) are integrated in the elementary functions

$$M_3(t) = \frac{2\alpha_1^2\tilde{c}_1\,\lambda_2}{e^{\mp\alpha_1\sqrt{\tilde{c}_1}(\frac{2t}{\sqrt{\alpha_3}}\pm\sqrt{\lambda_2}\tilde{c}_2)} + \alpha_1^2\tilde{c}_1[e^{\pm\alpha_1\sqrt{\tilde{c}_1}(\frac{2t}{\sqrt{\alpha_3}}\pm\sqrt{\lambda_2}\tilde{c}_2)}(\alpha_1^2\tilde{c}_1\lambda_3^2 - \alpha_3\lambda_2^2 k) - 2\lambda_3]}.$$

Here $k = -(\alpha_1^2 + \alpha_3^2)$, $\tilde{c}_2$ is a constant of integration.

As obvious from the latter expression, it assumes real values when $\alpha_3 > 0, \lambda_2 > 0, \tilde{c}_1 > 0$ or $\alpha_3 < 0, \lambda_2 < 0, \tilde{c}_1 < 0$. Hence, under the above restrictions imposed on parameters $\alpha_3, \lambda_2, \tilde{c}_1$, the motion on the elements of the submanifold of complex IMs (19) is described by the real functions of time.

Finally, let us consider the motion on the elements of families (15).

The vector field on the elements of these families is defined by equation $\dot{M}_3 = 0$. Hence, geometrically, the elements of the families of real solutions (15) correspond to curves in R^6, each point of which is the degenerated stationary solution of the initial differential equations.

3 On Stationary Sets of Euler's Equations with an Additional First Integral of 6th Degree

3.1 Problem Formulation

According to [10], in integrable case (d) there exists an additional first integral of 6th degree of the form

$$F = M_3^2\Big[(M_1^2 + M_3^2)[(\alpha_3 M_1 + \gamma_2)^2 + \alpha_1 M_3(\alpha_1 M_3 - 2(\alpha_3 M_1 + \gamma_2))]$$

$$+ ((\alpha_1 M_2 + \gamma_3)M_3 - (\alpha_3 M_2 - \gamma_1)M_1)^2\Big] = h_1 = const$$

Euler's equations corresponding to Hamiltonian (d) write

$$\dot{M}_1 = 2(M_3\gamma_1 - M_1\gamma_3) - M_2(\alpha_1 M_1 + 2\alpha_3 M_3),$$
$$\dot{M}_2 = 2(M_2\gamma_3 - M_3\gamma_2) + \alpha_1(M_1^2 - M_3^2) + 2\alpha_3 M_1 M_3, \quad \dot{M}_3 = \alpha_1 M_2 M_3,$$
$$\dot{\gamma}_1 = (2\alpha_3 M_3 - \alpha_1 M_1)\gamma_2 - 2(2\alpha_3 M_3 - \gamma_1)\gamma_3 - 2k M_1 M_3, \tag{21}$$
$$\dot{\gamma}_2 = \alpha_1(M_1\gamma_1 - M_3\gamma_3) + 2\alpha_3(2M_1\gamma_3 - M_3\gamma_1) - 2(k M_2 M_3 + \gamma_2\gamma_3),$$
$$\dot{\gamma}_3 = 4\alpha_3(M_2\gamma_1 - M_1\gamma_2) + \alpha_1 M_3\gamma_2 + 2[k(M_1^2 + M_2^2) - (\gamma_1^2 + \gamma_2^2)].$$

The rest of the integrals of these equations has the form (4).

Once again let us state the problem of finding stationary sets, now for equations (21), and investigation of their stability in the Lyapunov sense.

3.2 Finding Stationary Sets

Following the technique chosen, we construct the complete linear combination

$$2K = 4\lambda_0 H - 2\lambda_1 V_1 - \lambda_2 V_2 - \lambda_3 F \quad (\lambda_i = \text{const}) \tag{22}$$

from the problem's first integrals, and write down the necessary conditions for the integral K to have an extremum with respect to phase variables M_i, γ_i:

$$\partial K/\partial M_1 = \lambda_0(2\eta - \alpha_1 M_3) - \lambda_1\gamma_1 - k\lambda_2 M_1 + \lambda_3\Big[\alpha_3(\alpha_1(M_1^2 + M_3^2) - \gamma_2 M_3)M_3$$
$$+ k M_1 M_3^2 + [(2\alpha_1 M_3 - \alpha_3 M_1)\sigma - (\rho^2 + \sigma^2)]M_1 - \rho\chi M_3\Big]M_3^2 = 0,$$
$$\partial K/\partial M_2 = 2\lambda_0 x 5 - \lambda_1\gamma_2 - k\lambda_2 M_2 + \lambda_3(\alpha_3 M_1 - \alpha_1 M_3)(\rho M_1 + \chi M_3)M_3^2 = 0,$$
$$\partial K/\partial M_3 = \lambda_0(2\alpha_3 M_3 - \alpha_1 M_1) - \lambda_1\gamma_3 - k\lambda_2 M_3 - \lambda_3\Big[(\rho^2 + \sigma^2)M_1^2$$
$$+ 3\rho\chi M_1 M_3 + 2\chi^2 M_3^2 + (2\alpha_3^2 M_1^2 + \alpha_1^2(2M_1^2 + 3M_3^2) + 2\eta\gamma_2)M_3^2 \tag{23}$$
$$- \alpha_1(3M_1^2 + 5M_3^2)\sigma M_3\Big]M_3 = 0,$$
$$\partial K/\partial\gamma_1 = -2\lambda_0 M_2 - \lambda_1 M_1 - \lambda_2\gamma_1 - \lambda_3(\rho M_1 + \chi M_3)M_1 M_3^2 = 0,$$
$$\partial K/\partial\gamma_2 = 2\lambda_0 M_1 - \lambda_1 M_2 - \lambda_2\gamma_2 + \lambda_3(M_1^2 + M_3^2)(\alpha_1 M_3 - \sigma)M_3^2 = 0,$$
$$\partial K/\partial\gamma_3 = -\lambda_1 M_3 - \lambda_2\gamma_3 - \lambda_3(\rho M_1 + \chi M_3)M_3^3 = 0.$$

Here the following notations $\eta = 2\alpha_3 M_1 + \gamma_2, \rho = \gamma_1 - \alpha_3 M_2, \sigma = \alpha_3 M_1 + \gamma_2, \chi = \alpha_1 M_2 + \gamma_3$ are used.

The above equations (the conditions of stationarity for the integral K) represent a system of polynomial algebraic equations of 5th degree with parameters λ_i, α_j.

Likewise the previous problem, some qualitative analysis of the solution set of equations (23) by the *Maple*-programs *IsZeroDimensional, NumberOfSolutions, HilbertDimendion* was conducted [1]. It was revealed that system (23) has infinite number of solutions with respect to the phase variables, and the dimension of these solutions is 1.

The above system was decomposed into two subsystems. To this end, Gröbner basis and polynomial factorization methods were used. The 1st subsystem consists of 20 polynomial equations of degrees from 2 to 9. The 2nd subsystem consists of 16 polynomial equations of degrees from 1 to 9. The coefficients of equations belong to field $Q(\lambda_i, \alpha_j)$. The solutions, the dimension of which is 1, form the solution set of the 1st subsystem. The 2nd subsystem has a finite number of solutions which is 33. Each of solutions of the subsystem corresponds to a single point.

The solutions of the 1st subsystem obtained with respect to the phase variables, and the solutions of the 2nd subsystem obtained with respect to some part of the phase variables and some part of the parameters can be found in [1]. These represent IMs of equations (21).

In the given work, we stated the problem of finding solutions for the 2nd subsystem with respect to the phase variables. We tried to obtain solutions corresponding to the complete linear combination of the problem's first integrals (the integral K).

It is known that the lexicographic basis is the most suitable for computing the roots of a polynomial algebraic system. We have not managed to construct a lexicographic basis for the subsystem considered without imposing any additional restrictions on parameters λ_i, α_j neither with the use of Gröbner basis method (the programs *GroebnerBasis* and *gbasis* of CAS *Mathematica* and *Maple* were applied) nor with the use of the *Maple*-program *FGLM*. The computations were executed for over 12 hours on Intel CPU 3.6 GHz, 32 GB RAM running under Windows 7 Professional.

To find the desired solutions we have constructed the Gröbner basis with respect to elimination monomial order. It writes:

$$M_2 = 0, \ f_1(\gamma_1, M_1, M_3, \lambda_i, \alpha_i) = 0, \ f_2(\gamma_2, M_1, M_3, \lambda_i, \alpha_i) = 0,$$
$$f_3(\gamma_3, M_1, M_3, \lambda_i, \alpha_i) = 0, \tag{24}$$

$$g_r(M_1, M_3, \lambda_i, \alpha_i) = 0 \ (r = 1, \ldots, 7). \tag{25}$$

Here f_s are the 5th degree polynomials of variables $M_1, M_3, \gamma_1, \gamma_2, \gamma_3, \ g_r$ are the 6-9 degree polynomials of variables M_1, M_3. The coefficients of the polynomials f_s, g_r represent some expressions of λ_i, α_j.

A number of solutions of equations (25) was obtained by computing resultants for polynomials g_r. Indeed, having computed the resultant of any two polynomials g_l, g_m (25) with respect to, e.g., variable M_1, we obtain some polynomial of variable M_3. Each root M_3 of this polynomial corresponds to common roots M_1 of polynomials g_l, g_m. Since system (25) is compatible, some of the common roots of polynomials g_l, g_m are the roots of all the rest of the polynomials g_r.

Following the technique chosen, compute the resultant for two polynomials of system (25) (e.g., for polynomials having the least degree with respect to variable M_1). It writes:

$$Res = \alpha_1((2\lambda_0 + \alpha_3\lambda_2^2)^2 + \alpha_1^2\lambda_2^2) D M_3 (a_0 M_3^{12} + a_1 M_3^8 + a_2 M_3^4 + a_4)$$
$$\times (b_0 M_3^{20} + b_1 M_3^{16} + b_2 M_3^{12} + b_3 M_3^8 + b_4 M_3^4 + b_5) \tag{26}$$

Here D, a_σ, and b_ρ are some expressions of λ_i, α_j.

As obvious from (26), the resultant expression is factorized and, consequently, computing the roots of the polynomials chosen is reduced to finding the roots of 12th and 20th degree bipolynomials. Under some restrictions imposed on parameters λ_i, α_j, we have found several solutions of these equations. Substituting the obtained values of M_3 into the polynomials chosen, we find common roots of these polynomials and equations (25).

Next, we substitute the found values of M_1, M_3 into equations (24) and find the values of $\gamma_1, \gamma_2, \gamma_3$ corresponding to them. Some of the solutions of equations (24), (25) obtained by the technique described are given below.

i) $\lambda_1 = 0, \lambda_2 = \lambda_3^5, \alpha_3 = 0$:

$$\gamma_1 = 0, \gamma_2 = \pm \frac{1}{2}\alpha_1\lambda_3, \gamma_3 = 0, M_1 = 0, M_2 = 0, M_3 = \pm\lambda_3;$$

ii) $\lambda_1 = 0, \lambda_2 = \lambda_5^3, \alpha_3 = 0$:

$$\gamma_1 = 0, \gamma_2 = \pm\frac{1}{2}\alpha_1\sqrt{-\lambda_3}, \gamma_3 = 0, M_1 = 0, M_2 = 0, M_3 = \pm\sqrt{-\lambda_3};$$

iii) $\lambda_1 = 0, \lambda_2 = -\lambda_3$:

$$\gamma_1 = \frac{\alpha_1\sqrt{[(2\lambda_0 - \alpha_3\lambda_3)^2(4\alpha_3\lambda_0 - (3\alpha_1^2 + 2\alpha_3^2)\lambda_3) - \alpha_1^4\lambda_3^3]\,\lambda_3}}{2(\alpha_3\lambda_3 - 2\lambda_0)^{3/2}[(2\lambda_0 - \alpha_3\lambda_3)^2 + \alpha_1^2\lambda_3^2]^{1/4}},$$

$$\gamma_2 = \frac{\alpha_1((\alpha_1^2 + \alpha_3^2)\lambda_3^2) - 4\lambda_0^2}{2(\alpha_3\lambda_3 - 2\lambda_0)^{3/2}[(2\lambda_0 - \alpha_3\lambda_3)^2 + \alpha_1^2\lambda_3^2]^{1/4}},$$

$$\gamma_3 = \frac{\sqrt{(2\lambda_0 - \alpha_3\lambda_3)^2(4\alpha_3\lambda_0 - (3\alpha_1^2 + 2\alpha_3^2)\lambda_3) - \alpha_1^4\lambda_3^3}}{2\sqrt{(\alpha_3\lambda_3 - 2\lambda_0)\,\lambda_3}\,[(2\lambda_0 - \alpha_3\lambda_3)^2 + \alpha_1^2\lambda_3^2]^{1/4}},$$

$$M_1 = -\frac{\alpha_1\lambda_3}{\sqrt{(\alpha_3\lambda_3 - 2\lambda_0)\,\lambda_3}\,[(2\lambda_0 - \alpha_3\lambda_3)^2 + \alpha_1^2\lambda_3^2]^{1/4}},$$

$$M_2 = 0, \quad M_3 = -\frac{\sqrt{\alpha_3\lambda_3 - 2\lambda_0}}{[(2\lambda_0 - \alpha_3\lambda_3)^2 + \alpha_1^2\lambda_3^2]^{1/4}}. \tag{27}$$

These represent the families of stationary solutions of equations (23) parameterized by λ_i. Likewise above, the latter is proved by direct substitution of expressions (27) into the stationary equations and the initial differential ones (i.e., equations (21) and (23)).

¿From a mechanical point of view and geometrically, the above solutions are interpreted similarly to the stationary solutions of integrable case (b).

4   On Stability of the Stationary Sets

Now, we shall consider the problem of stability for the above stationary solutions. We shall investigate stability of the solutions on the base of Lyapunov's stability theorems: the 2nd method [16], in particular, the Routh-Lyapunov method [13] which is its modification, theorems for linear approximation [16] and theorems for stability with respect to part of variables [17].

First, let us investigate a trivial solution by the Routh–Lyapunov method. This method allows one to obtain sufficient conditions of stability.

The family of Hamiltonians (1) corresponds to a family of Euler's equations on Lie algebras $e(3)$, $so(4)$ and $so(3,1)$. We shall investigate stability of a trivial solution of equations of this family. This solution is peculiar stationary one because it delivers a stationary value to all the above first integrals, and it is the solution for all equations of the family. Hence, investigation of its stability has a special interest.

Application of the Routh–Lyapunov method, in the case considered, reduces the stability problem to analysing the sign definiteness of variation of integral $\tilde{K} = \lambda_0 H - \lambda_1 V_1 - \lambda_2 V_2$ obtained in the neighborhood of the given solution. It writes

$$\Delta \tilde{K} = -\lambda_2(z_1^2 + z_2^2 + z_3^2) - \lambda_1(z_1 z_4 + z_2 z_5 + z_3 z_6) + \lambda_0(z_2 z_4 - z_1 z_5$$
$$+\alpha_1 c_2 z_4 z_6) + (\alpha_3 c_1 \lambda_0 + (\alpha_1^2 + \alpha_3^2)\lambda_2)z_4^2 + (\alpha_3 c_1 \lambda_0 + (\alpha_1^2 + \alpha_3^2)\lambda_2)z_5^2$$
$$+(\alpha_3(c_1 + c_2)\lambda_0 + (\alpha_1^2 + \alpha_3^2)\lambda_2)z_6^2.$$

Here z_i are deviations of the perturbed solution from the unperturbed one.

According to Sylvester's criterion, the quadratic form $\Delta \tilde{K}$ is sign definite when the following conditions

$$\Delta_1 = -\lambda_2 > 0, \quad \Delta_2 = -(\lambda_0^2 + \lambda_1^2 + 4\alpha_3 c_1 \lambda_0 \lambda_2 + 4(\alpha_1^2 + \alpha_3^2)\lambda_2^2) > 0,$$
$$\Delta_3 = \Delta_2(4\alpha_3(c_1 + c_2)\lambda_0^3\lambda_2 + 4\alpha_3(2c_1 + c_2)\lambda_0\lambda_2(\lambda_1^2 + 4(\alpha_1^2 + \alpha_3^2)\lambda_2^2)$$
$$+(\lambda_1^2 + 4(\alpha_1^2 + \alpha_3^2)\lambda_2^2)^2 + \lambda_0^2(\lambda_1^2 + 4(\alpha_3^2(4c_1(c_1 + c_2) + 1) - \alpha_1^2(c_2^2 - 1))\lambda_2^2)).$$

hold.

The above inequalities are compatible under the following restrictions:

$$\Big(\alpha_3 > 0 \wedge \alpha_1 \neq 0 \wedge \lambda_2 < 0 \wedge \big((\lambda_0 < 0 \wedge c_1 > A_1 \wedge B_2 < c_2 < B_1)$$
$$\vee(\lambda_0 > 0 \wedge c_1 < A_1 \wedge B_2 < c_2 < B_1)\big)\Big) \vee \Big(\alpha_3 < 0 \wedge \alpha_1 \neq 0 \wedge \lambda_2 < 0$$
$$\wedge\big((\lambda_0 < 0 \wedge c_1 < A_1 \wedge B_2 < c_2 < B_1) \vee (\lambda_0 > 0 \wedge c_1 > A_1 \wedge B_2 < c_2 < B_1)\big)\Big)$$
$$\vee\Big(\alpha_3 > 0 \wedge \alpha_1 = 0 \wedge \lambda_2 < 0 \wedge \big((\lambda_0 < 0 \wedge c_1 > A_2 \wedge c_2 > B_3)$$
$$\vee(\lambda_0 > 0 \wedge c_1 < A_2 \wedge c_2 < B_3)\big)\Big) \vee \Big(\alpha_3 < 0 \wedge \alpha_1 = 0 \wedge \lambda_2 < 0$$
$$\wedge\big((\lambda_0 < 0 \wedge c_1 < A_2 \wedge c_2 < B_3) \vee (\lambda_0 > 0 \wedge c_1 > A_2 \wedge c_2 > B_3)\big)\Big). \quad (28)$$

These have been obtained with the *Mathematica*-program *Reduce*. Here $A_1 = -(\lambda_0^2 + \lambda_1^2 + 4(\alpha_1^2 + \alpha_3^2)\lambda_2^2)/(4\alpha_3\lambda_0\lambda_2)$, $A_2 = -(\lambda_0^2 + \lambda_1^2 + 4\alpha_3^2\lambda_2^2)/(4\alpha_3\lambda_0\lambda_2)$, $B_{1,2} = 1/(2\alpha_1^2\lambda_0\lambda_2)\,[\alpha_3(\lambda_0^2 + \lambda_1^2 + 4\alpha_3 c_1 \lambda_0 \lambda_2 + 4(\alpha_1^2 + \alpha_3^2)\lambda_2^2) \pm \sqrt{D}]$, $D = (\lambda_0^2 + \lambda_1^2 + 4\alpha_3 c_1 \lambda_0 \lambda_2 + 4(\alpha_1^2 + \alpha_3^2)\lambda_2^2)(\alpha_1^2\lambda_1^2 + \alpha_3^2(\lambda_0^2 + \lambda_1^2) + 4\alpha_3(\alpha_1^2 + \alpha_3^2)c_1 \lambda_0 \lambda_2 + 4(\alpha_1^2 + \alpha_3^2)^2\lambda_2^2)$, $B_3 = -(\lambda_1^2 + 4\alpha_3 c_1 \lambda_0 \lambda_2 + 4\alpha_3^2\lambda_2^2)/(4\alpha_3\lambda_0\lambda_2)$.

Conditions (28) separate from the family of Euler's equations with Hamiltonians (1) some subfamily of systems of the equations, a trivial solution of which is stable.

To verify whether or not integrable cases $(a) - (d)$ (section 2.1) enter into this subfamily it is sufficient, e.g., to test the compatibility of the conditions of integrability with conditions (28) by applying the program *Reduce*. As a result, we have that one integrable case (a) enters into the above subfamily only.

Let us illustrate the above results by a graphical constructing the stability regions for the trivial solution considered. We shall construct a stability region for the trivial solution in integrable case (a). According to (28), we can take, e.g., $\lambda_0 = -1, \lambda_1 = 1, \lambda_2 = -1, \alpha_3 = 1$. Such a choice is relevant because λ_i are the family parameters of integral K (22), and there are no constraints on α_1, α_3 in the integrable case considered. Under the above numerical values for λ_i, α_3, and $c_2 = 0$, inequalities (28) have the following form: $(\alpha_1 \neq 0 \vee c_1 < -3/2) \wedge (\alpha_1 = 0 \vee c_1 < -(\alpha_1^2 + 3/2))$.

The stability region defined by the latter inequalities and plotted with *Mathematica*-program *RegionPlot* is shown in Fig.1 (the grey region). Hence, when λ_i, α_3, and c_2 have the above numerical values and c_1, α_1 assume the values from the constructed region, the trivial solution is stable.

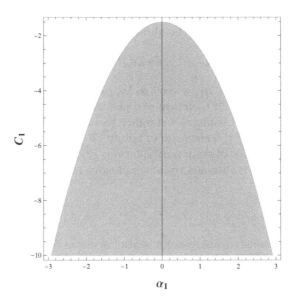

Fig. 1.

We have managed to obtain conditions of stability for the trivial solution of system (21) (integrable case (d)) with respect to some part of the phase variables [17]. To this end, we analyzed the variation of integral $\bar{K} = -\lambda_2(2\alpha_3 H + V_2)$ (where $\alpha_1 = 0$) written in the neighbourhood of this solution. It writes:

$$2\Delta\tilde{K} = -\lambda_2 \left[z_3^2 + (z_2 + \alpha_3 z_4)^2 + (z_1 - \alpha_3 z_5)^2 \right].$$

Introduce variables $y_1 = z_2 + \alpha_3 z_4$, $y_2 = z_1 - \alpha_3 z_5$ and write down $\Delta \tilde{K}$ in terms of y_1, y_2, z_3:

$$2\Delta \tilde{K} = -\lambda_2 (y_1^2 + y_2^2 + z_3^2). \tag{29}$$

As obvious from (29), the quadratic form is sign definite with respect to y_1, y_2, z_3. Hence, the conditions $\alpha_1 = 0$, $\lambda_2 \neq 0$ are sufficient for the stability of the trivial solution with respect to variables y_1, y_2, z_3.

Next, we investigate one of the families of stationary solutions (27), e.g.,

$$\gamma_1 = 0, \ \gamma_2 = \frac{1}{2}\alpha_1\lambda_3, \ \gamma_3 = 0, \ M_1 = 0, \ M_2 = 0, \ M_3 = \lambda_3. \tag{30}$$

We show that the elements of family (30) give instability when $\alpha_1 \neq 0$. To this end, we consider this solution in capacity of the unperturbed one and write down the equations of the first approximation:

$$\dot{z}_1 = \frac{3}{2}\alpha_1^2\lambda_3 z_4, \ \dot{z}_2 = 2\alpha_1^2\lambda_3 z_5, \ \dot{z}_3 = \frac{1}{2}\alpha_1\lambda_3(-2z_2 + \alpha_1 z_6),$$

$$\dot{z}_4 = 2\lambda_3 z_1, \ \dot{z}_5 = \lambda_3(2z_2 - \alpha_1 z_6), \ \dot{z}_6 = \alpha_1\lambda_3 z_5.$$

The characteristic equation of the above linear system writes

$$\mu^2(\mu^2 - 3\alpha_1^2\lambda_3^2)^2 = 0$$

and has four nonzero roots $\mu_1 = -\sqrt{3}\,\alpha_1\lambda_3$, $\mu_2 = -\sqrt{3}\,\alpha_1\lambda_3$, $\mu_3 = \sqrt{3}\,\alpha_1\lambda_3$, $\mu_4 = \sqrt{3}\,\alpha_1\lambda_3$, among which there are real positive roots when $\alpha_1 \neq 0$, $\lambda_3 \neq 0$. The latter, according to Lyapunov's stability theorem for linear approximation [16], means instability of the elements of family (30).

The rest of the families of stationary solutions (27) have been investigated by this technique. We have obtained results analogous to the above.

5   Conclusion

Practically, the completed analysis for stationary sets of Euler's equations on the Lie algebra $so(3,1)$ when the equations possess additional polynomial first integrals of degrees 3 and 6 has been performed. We considered the case when the sets correspond to the complete linear combination of the problem's first integrals. The obtained results can be a base for further qualitative analysis of the considered integrable cases. The approach applied in this work to investigation of integrable systems may be of interest for the study of new integrable cases of equations of such a type when problem's algebraic first integrals have degree higher than 2. This approach may also be of interest for the problems of parametric analysis in which properties of solutions of nonlinear differential equations in relation to continuous change of parameters of these solutions are investigated.

The work was supported by the Presidium of the Russian Academy of Sciences, basic research program No. 17.1.

References

1. Irtegov, V., Titorenko, T.: On invariant manifolds of dynamical systems in Lie algebras. In: Gerdt, V.P., Mayr, E.W., Vorozhtsov, E.V. (eds.) CASC 2009. LNCS, vol. 5743, pp. 142–154. Springer, Heidelberg (2009)
2. Bogoyavlenskii, O.I.: Integrable Euler equations on Lie algebras arising in problems of mathematical physycs. Math. USSR-Izv. 25(2), 207–257 (1985)
3. Borisov, A.V., Mamayev, I.S.: Dynamics of a Rigid Body. R&C Dynamics, Izhevsk (2001)
4. Sokolov, V.V.: A new integrable case for the Kirchhoff equation. Theoret. and Math. Phys. 129(1), 1335–1340 (2001)
5. Borisov, A.V., Mamayev, I.S., Sokolov, V.V.: A new integrable case on so(4). Dokl. Phys. 46(12), 888–889 (2001)
6. Morozov, V.P.: The topology of Liouville foliations for the cases of Steklov's and Sokolov's integrability. Sb. Math. 195(3), 369–412 (2004)
7. Ryabov, P.E.: Bifurcations of first integrals in the Sokolov case. Theoret. and Math. Phys. 134, 181–197 (2003)
8. Kozlov, I.K., Ratiu, T.S.: Bifurcation diagram for the Kovalevskaya case on the Lie algebra so(4). Dokl. Math. 86(3), 827 (2012)
9. Tsiganov, A.V., Goremykin, O.V.: Integrable systems on so(4) related with XXX spin chains with boundaries. J. Phys. A: Math. Gen. 37, 4843–4849 (2004)
10. Sokolov, V.V., Wolf, T.: Integrable quadratic classical Hamiltonians on so(4) and so(3,1). J. Phys. A: Mat. Gen. 39, 1915–1926 (2006)
11. Grabmeier, J., Kaltofen, E., Weispfenning, V.: Handbook in Computer Algebra. Foundations, Applications, Systems. Springer, Berlin (2003)
12. Efimovskaya, O.V., Wolf, T.: Classification of integrable quadratic Hamiltonians on e(3). Reg. and Chaot. Dyn. 8(2), 155–162 (2003)
13. Lyapunov, A.M.: On Permanent Helical Motions of a Rigid Body in Fluid. Collected Works, vol. 1. USSR Acad. Sci., Moscow–Leningrad (1954)
14. Irtegov, V.D., Titorenko, T.N.: The invariant manifolds of systems with first integrals. J. Appl. Math. Mech. 73(4), 379–384 (2009)
15. Aubry, P.A., Lazard, D., Maza, M.M.: On the theories of triangular sets. J. Symb. Comp. 28, 105–124 (1999)
16. Lyapunov, A.M.: Stability of Motion. Academic Press, New York (1966)
17. Rumyantsev, V.V., Oziraner, A.S.: Motion Stability and Stabilization with Respect to Part of Variables. Nauka, Moscow (1987)

An Effective Implementation of a Special Quantifier Elimination for a Sign Definite Condition by Logical Formula Simplification

Hidenao Iwane[1,2], Hiroyuki Higuchi[1], and Hirokazu Anai[1,2]

[1] IT Systems Laboratories, Fujitsu Laboratories Ltd.
Kamikodanaka 4-1-1, Nakahara-ku, Kawasaki 211-8588, Japan
{iwane,h-higuchi,anai}@jp.fujitsu.com
[2] Graduate School of Mathematics, Kyushu University
744 Motooka, Nishi-ku, Fukuoka 819-0395, Japan

Abstract. This paper presents an efficient quantifier elimination algorithm tailored for a sign definite condition (SDC). The SDC for a polynomial $f \in \mathbb{R}[x]$ with parametric coefficients is written as $\forall x(x \geq 0 \rightarrow f(x) > 0)$. To improve the algorithm, simplification of an output formula is needed. We show a necessary condition for the SDC and an approach to simplify formulae by using a logic minimization method. Experimental results show that our approach significantly simplify formulae.

1 Introduction

In science and engineering, parametric treatment of constraint solving and optimization problems has garnered considerable concern. Quantifier elimination (QE) over the reals provides one of the effective means to accomplish parametric optimization. Various algorithms and deep analysis on the complexity about QE have been studied during the last several decades [6]. Moreover, practically efficient software systems of QE have been developed and also applied to many nontrivial application problems (see [15,11,16,12]).

A general-purpose symbolic method aiming for QE is cylindrical algebraic decomposition (CAD). QE based on CAD has a high computational complexity. To circumvent the inherent computational complexity of a QE algorithm based on CAD, several researchers have focused on developing QE algorithms specialized to particular types of input formulae in order to make good use of their speciality. This direction is quite promising in practice since a number of important problems in engineering have been successfully reduced to such particular input formulae and resolved efficiently by using the specialized QE algorithms.

For a special QE problem $\forall x(f(x) > 0)$ where f is a polynomial with parametric coefficients, a combinatorial algorithm based on the Sturm-Habicht sequence is proposed in [9]. The real root classification method [17] of a univariate polynomial with parameter coefficients can also apply the special QE problem. These methods are much more effective than a native direct use of CAD.

V.P. Gerdt et al. (Eds.): CASC 2013, LNCS 8136, pp. 194–208, 2013.
© Springer International Publishing Switzerland 2013

In this paper we focus on one particular input formula

$$\forall x \ (x \geq 0 \rightarrow f(x) > 0) \tag{1}$$

where $f(x)$ is a univariate polynomial with real parameters, which we call a *sign definite condition (SDC)*, because quite a lot of practical engineering problems such as control system design problems can be recast as the SDCs [2]. We note that we mainly consider the case where the coefficients of f contain some parameters. An effective QE algorithm for the SDCs was proposed in [11] and it is based on a combinatorial approach using a real root counting technique. As a real root counting method the Sturm-Habicht sequence [10] is employed.

In order to improve the efficiency of the proposed method in [11], *simplification* of output logical formulae is a critical issue. Our focus is on developing an effective algorithm which produces as simple formulae as possible.

For the purpose we propose two improvement approaches: First we use a necessary condition for the SDC to simplify the output formula algebraically. The necessary condition enables us to eliminate extraneous sign combinations derived from real root counting using the Sturm-Habicht sequence. Second, for further simplification we utilize Boolean function manipulation. We obtain simple formulae by using the idea of *don't cares* for handling a sign condition such that no real numbers satisfy. A *don't care* is an input where a function is not specified. By these improvements we can significantly simplify the output formula of the specialized QE algorithm for the SDC. We also present the experimental results for demonstrating the effectiveness of our proposed method.

The organization of this paper is as follows. Section 2 explains the Sturm-Habicht sequence and also an outline of the specialized QE algorithm for the SDC using the Sturm-Habicht sequence. In Section 3 we show a necessary condition for the SDC. Simplification of Boolean expressions based on Boolean function manipulation is shown in Section 4. The proposed improvements for speeding up the specialized QE algorithm for the SDC are discussed based on experimental results in Section 5. The concluding remarks are made in Section 6.

2 Quantifier Elimination for Sign Definite Condition

In this section we describe the definition of a sign definite condition and a specialized QE algorithm based on the Sturm-Habicht sequence. Let us denote the fields of real numbers by $\mathbb{R}$.

2.1 Sign Definite Condition and Real Root Counting by Sturm-Habicht Sequence

We first introduce a sign definite condition.

Definition 1. *Let $f(x)$ be a polynomial in x over $\mathbb{R}$. We call the following condition a* sign definite condition *(SDC) for $f(x)$:*

$$\forall x \ (x \geq 0 \rightarrow f(x) > 0).$$

In addition, we call the n-th SDC problem if the leading coefficient of f is not equal to zero and its degree is equal to n.

Many important design specifications such as H_∞ norm constraints, stability margins etc, which are frequently used as induces of the robustness, reduce to SDCs [2]. One of the typical examples is the (frequency restricted) H_∞ norm constraint: An H_∞ norm constraint of a strictly proper transfer function $P(s) = n(s)/d(s)$ expressed as

$$\|P(s)\|_\infty := \sup_\omega |P(j\omega)| < \gamma$$

is equivalent to $\forall \omega \ (\gamma^2 d(j\omega)d(-j\omega) > n(j\omega)n(-j\omega))$. Since we can find a function $f(\omega^2)$ which satisfies $f(\omega^2) = \gamma^2 d(j\omega)d(-j\omega) - n(j\omega)n(-j\omega) > 0$, letting $x = \omega^2$ lead to a SDC. Similarly, a finite frequency H_∞ norm defined by

$$\|P(s)\|_{[\omega_1,\omega_2]} := \sup_{\omega_1 \le \omega \le \omega_2} |P(j\omega)| < \gamma$$

can be recast as the condition $f(x) \ne 0$ in $[-\omega_2^2, -\omega_1^2]$, which is reduced to a SDC for $f(z)$ by a bilinear transformation $z = -(x+\omega_2^2)/(x+\omega_1^2)$. A specialized QE algorithm for SDCs has been proposed in [11]. The algorithm utilizes the following proposition and realizes QE by real root counting using the Sturm-Habicht sequence [10].

Proposition 1. *The SDC for a polynomial in $\mathbb{R}[x]$ with the positive leading coefficient is equivalent to a condition that the polynomial has no real root in $x \ge 0$.*

We next describe the Sturm-Habicht sequence.

Definition 2. *Let $f(x)$ be a polynomial in $\mathbb{R}[x]$ with the degree n. The Sturm-Habicht sequence associated to f is defined as the sequence of polynomial $\mathsf{SH}(f) = \{\mathsf{SH}_j(f)\}_{j=0,\dots,n}$ such that $\mathsf{SH}_n(f)=f$, $\mathsf{SH}_{n-1}(f)=\frac{df}{dx}$, $\mathsf{SH}_j(f)=\delta_{n-j}\,\mathrm{Sres}_j(f,\frac{df}{dx})$ for $j \in \{0,\dots,n-2\}$, where $\delta_k = (-1)^{\frac{k(k-1)}{2}}$ and $\mathrm{Sres}_k(f,g)$ is the k-th subresultant which is defined as the determinant of the k-th Sylvester-Habicht matrix of f and g.*

Definition 3. *Let $A = \{a_m, \dots, a_0\}$ be a finite sequence of real numbers. We define the sign of a real number is 1, 0, or -1 if the number is positive, zero, or negative, respectively. In addition, we define the number of sign variations $V(A)$ in the following rules:*

  – *we count 1 sign variation for the groups: $\{-1,+1\}$, $\{+1,-1\}$, $\{-1,0,+1\}$, $\{+1,0,-1\}$, $\{-1,0,0,+1\}$, $\{+1,0,0,-1\}$,*
  – *we count 2 sign variations for the groups: $\{+1,0,0,+1\}$, $\{-1,0,0,-1\}$.*

  Let $S(x) = \{S_n(x), S_{n-1}(x), \dots, S_0(x)\}$ be a finite sequence of polynomials in $\mathbb{R}[x]$ and let α be a real number. We construct a sequence $\{h_s, \dots, h_0\}$ of polynomials in $\mathbb{R}[x]$ obtained from $S(x)$ by deleting the polynomial identically zero. The number of sign variations $V_\alpha(S)$ is defined by $V(\{h_s(\alpha), \dots, h_0(\alpha)\})$.

We note that the sign sequences $\{+1, 0, +1\}$ and $\{-1, 0, -1\}$ cannot appear by [10].

Theorem 1. *(real root counting by the Sturm-Habicht sequence [10]) Let f be a polynomial in $\mathbb{R}[x]$ and α, β in $\mathbb{R} \cup \{-\infty, +\infty\}$ with $\alpha < \beta$ and $f(\alpha)f(\beta) \neq 0$. Then $V_\alpha(\mathsf{SH}(f)) - V_\beta(\mathsf{SH}(f))$ is equal to the number of real roots of $f(x)$ in the interval $[\alpha, \beta]$.*

By using Theorem 1 we can obtain the number of real roots of a polynomial in the interval $[0, +\infty)$.

Here is an additional notation. In the rest of this paper we denote the sign of $\mathsf{SH}_k(f)$ at $x = \infty$ and $x = 0$ by s_k and c_k, respectively.

Remark 1. Let $\mathsf{SH}_k(f) = a_{k,k}x^k + a_{k,k-1}x^{k-1} + \cdots + a_{k,0}$. Then $s_k = 0$ is equivalent to $a_{k,i} = a_{k,k-1} = \cdots = a_{k,0} = 0$ and $s_k > 0$ is equivalent to $(a_{k,k} > 0) \vee (a_{k,k} = 0 \wedge a_{k,k-1} > 0) \vee \cdots \vee (a_{k,k} = a_{k,k-1} = \cdots = a_{k,1} = 0 \wedge a_{k,0} > 0)$. That is to say $s_k = 0$ if and only if $\mathsf{SH}_k(f)$ is identically zero. We note that c_k is equivalent to the sign of $a_{k,0}$.

Example 1. Let $f(x) = 25x^5 + 25x^4 + 10x^3 + 2x^2 + 25x + 1$. The Sturm-Habicht sequence associated to f is as follows:

$$
\begin{aligned}
\mathsf{SH}_5(f) &= f(x) &&= 25x^5 + 25x^4 + 10x^3 + 2x^2 + 25x + 1, \\
\mathsf{SH}_4(f) &= \tfrac{df}{dx}(x) &&= 125x^4 + 100x^3 + 30x^2 + 4x + 25, \\
\mathsf{SH}_3(f) &= \delta_{5-3}\mathrm{Sres}_3(f, \tfrac{df}{dx}) &&= -(310000x), \\
\mathsf{SH}_2(f) &= \delta_{5-2}\mathrm{Sres}_2(f, \tfrac{df}{dx}) &&= -(0), \\
\mathsf{SH}_1(f) &= \delta_{5-1}\mathrm{Sres}_1(f, \tfrac{df}{dx}) &&= +(1906624000000x), \\
\mathsf{SH}_0(f) &= \delta_{5-0}\mathrm{Sres}_0(f, \tfrac{df}{dx}) &&= +(945685504000000).
\end{aligned}
$$

Thus

$$
\begin{aligned}
\{s_5, s_4, s_3, s_2, s_1, s_0\} &= \{+1, +1, -1, 0, +1, +1\}, \\
\{c_5, c_4, c_3, c_2, c_1, c_0\} &= \{+1, +1, 0, 0, 0, +1\}.
\end{aligned}
$$

Therefore $f(x)$ has no real root in $x \geq 0$ because of $V_0(\mathsf{SH}(f)) = V_\infty(\mathsf{SH}(f)) = 2$.

Remark 2. We note that $s_n = s_{n-1}$ and $s_0 = c_0$ for a polynomial with degree n.

Definition 4. $\mathsf{SH}_k(f)$ *is* regular *when the degree of $\mathsf{SH}_k(f)$ is equal to k.*

Theorem 2. *(Sturm-Habicht Structure Theorem [10]) Let f be a polynomial in $\mathbb{R}[x]$ with degree n. Then for every $k \in \{1, \ldots, n-1\}$ such that $\mathsf{SH}_{k+1}(f)$ is regular and $\deg(\mathsf{SH}_k(f)) = r \leq k$ we have*

(A) if $r < k - 1$ then, $\mathsf{SH}_{k-1}(f) = \cdots = \mathsf{SH}_{r+1}(f) = 0$,
(B) if $r < k$ then, $\mathrm{lc}(\mathsf{SH}_{k+1}(f))^{k-r}\mathsf{SH}_r(f) = \delta_{k-r}\mathrm{lc}(\mathsf{SH}_k(f))^{k-r}\mathsf{SH}_k(f)$
(C) $\mathrm{lc}(\mathsf{SH}_{k+1}(f))^{k-r+2}\mathsf{SH}_{r-1}(f) = \delta_{k-r+2}\mathrm{Prem}(\mathsf{SH}_{k+1}(f), \mathsf{SH}_k(f))$,

where $\mathrm{lc}(g)$ is the leading coefficient of the polynomial g and $\mathrm{Prem}(g, h)$ is a pseudo remainder of the polynomial g by the polynomial h defined as

$$
\mathrm{Prem}(g, h) = \mathrm{remainder}(\mathrm{lc}(h)^{\deg(g) - \deg(h) + 1}g, h).
$$

2.2   A Specialized QE Algorithm for SDC

In this subsection, we describe an implementation for a specialized QE algorithm for the SDCs [11]. The flow of the algorithm for the n-th SDC problem is as follows:

1. consider all the 3^{2n+1} (at most) possible sign conditions over s_k and c_k,
2. choose all sign conditions φ_n which satisfy $V_0(\mathsf{SH}(f)) - V_\infty(\mathsf{SH}(f)) = 0$,
3. compute the Sturm-Habicht sequence associated to f,
4. construct semi-algebraic sets generated by coefficients of polynomials in $\mathsf{SH}(f)$ for each selected sign conditions and combine them as a union.

Since steps 1 and 2 are independent of an input polynomial, we can execute these steps beforehand and store the results in a database. This greatly improves the total efficiency of the algorithm. In fact, a QE computation for the fifth SDC problem $\forall x (x \geq 0 \rightarrow x^5 + \sum_{i=0}^4 a_i x^i > 0)$ by cylindrical algebraic decomposition (CAD) [7], which is a general QE algorithm, did not terminate after an hour. On the other hand, a real root classification algorithm [17], implemented as a Maple command, can solve the SDC problems. However, it also did not terminate after an hour. In contrast, the specialized algorithm took less than a second.

The result obtained from the above procedure obviously tends to be large and complicated, and hence we should reduce the admissible sign conditions φ_n as simple as possible. The simplification of φ_n makes the algorithm and post-processing, for example drawing of feasible regions, more efficient. For example, formulae can be simplified by using the well-known rules:

$$\begin{cases} < \cup > & \leftrightarrow & \neq, \\ < \cup = & \leftrightarrow & \leq, \\ > \cup = & \leftrightarrow & \geq. \end{cases} \tag{2}$$

The goal of this paper is to obtain an output formula equivalent to the SDC by simplifying the possible sign conditions φ_n.

Example 2. Let f be a quadratic polynomial in $\mathbb{R}[x]$. We consider the second SDC problem $\forall x \ (x \geq 0 \rightarrow f(x) > 0)$. Table 1 shows sign conditions which satisfy $V_0(\mathsf{SH}(f)) - V_\infty(\mathsf{SH}(f)) = 0$. Each row shows the signs of s_k and c_k when f has no real root in $x \geq 0$. We note that $s_2 = s_1 > 0$ and $s_0 = c_0$ from Remark 2, and that $c_2 > 0$ because $f(0) > 0$ implies $c > 0$. From Table 1 we obtain the following quantifier-free formula:

$$\begin{aligned}
& (s_0 < 0 \wedge c_2 > 0 \wedge c_1 < 0) \ \vee \\
& (s_0 < 0 \wedge c_2 > 0 \wedge c_1 = 0) \ \vee \\
& (s_0 < 0 \wedge c_2 > 0 \wedge c_1 > 0) \ \vee \\
& (s_0 = 0 \wedge c_2 > 0 \wedge c_1 = 0) \ \vee \\
& (s_0 = 0 \wedge c_2 > 0 \wedge c_1 > 0) \ \vee \\
& (s_0 > 0 \wedge c_2 > 0 \wedge c_1 > 0).
\end{aligned} \tag{3}$$

Table 1. φ_2: sign conditions for the 2nd SDC problem

s_2 s_1 s_0	c_2 c_1 c_0	$V_0(\mathsf{SH})$
$+$ $+$ $-$	$+$ $-$ $-$	1
$+$ $+$ $-$	$+$ 0 $-$	1
$+$ $+$ $-$	$+$ $+$ $-$	1
$+$ $+$ 0	$+$ 0 0	0
$+$ $+$ 0	$+$ $+$ 0	0
$+$ $+$ $+$	$+$ $+$ $+$	0

The formula (3) can be simplified as follows by using (2):

$$
\begin{aligned}
&(s_0 < 0 \wedge c_2 > 0) & \vee \\
&(s_0 = 0 \wedge c_2 > 0 \wedge c_1 \geq 0) \; \vee \\
&(s_0 > 0 \wedge c_2 > 0 \wedge c_1 > 0).
\end{aligned}
\tag{4}
$$

By constructing the formula (4) beforehand, the QE computation is only done by computing the Sturm-Habicht sequence and substitution for s_k and c_k. Moreover, simplification of φ_2 reduces the number of substitutions.

3   Necessary Condition for SDC

Now we again consider the formula (4) in Example 2. For any quadratic polynomial $f \in R[x]$, $s_0 = 0 \wedge c_2 > 0 \wedge c_1 = 0$ is not held. In fact, the Sturm-Habicht sequence associated to $f(x) = x^2 + p_1 x + p_0$ is $\mathsf{SH}(f) = \{x^2 + p_1 x + p_0, 2x + p_1, p_1^2 - 4p_0\}$. In this case, if $s_0 = p_1^2 - 4p_0 = 0$ and $c_1 = p_1 = 0$, we have $p_1 = p_0 = 0$. Thus, we obtain $c_2 = p_1 = 0$. From this fact, we can reduce the formula (4) to the following:

$$
\begin{aligned}
&(s_0 < 0 \wedge c_2 > 0) & \vee \\
&(s_0 \geq 0 \wedge c_2 > 0 \wedge c_1 > 0).
\end{aligned}
$$

In this section to simplify the formula derived from the possible sign conditions φ_n, we give a necessary condition for a sign sequence of the Sturm-Habicht sequence associated to a polynomial which satisfies the SDC. We use the notations introduced in the previous section.

The following theorem provides a necessary condition.

Theorem 3. *Let f be a polynomial in $\mathbb{R}[x]$ where the leading coefficient is nonzero and its degree is n, and let u be the smallest nonnegative integer k such that $s_k \neq 0$. We define s_k and c_k are zero when $k < 0$ or $k > n$. When f satisfies the SDC, the following condition holds.*

$$V_0(\mathsf{SH}(f)) - V_\infty(\mathsf{SH}(f)) = 0,$$
$$s_n > 0, s_{n-1} > 0, c_n > 0, s_0 = c_0,$$
$$s_k = 0 \to c_k = 0, \ (\forall k \in \{0, \dots, n-2\}),$$
$$c_u \neq 0,$$
$$c_{n-1} = 0 \to c_{n-2} < 0,$$
$$s_{n-2} = 0 \to s_{n-3} = \cdots = s_0 = 0,$$
$$c_{k+2} \neq 0 \wedge c_{k+1} = 0 \to c_k \neq c_{k+2}, \ (\forall k \in \mathcal{N} = \{u, \dots, n-2\}),$$
$$c_k = c_{k+1} = 0 \wedge c_{k-1}c_{k+2}s_k s_{k+1} \neq 0 \to s_k s_{k+2} < 0, \ (\forall k \in \mathcal{N}),$$
$$c_k = \cdots = c_{k+m} = 0 \to s_{k+1} = \cdots = s_{k+m-1} = 0, \ (\forall k \in \mathcal{N}, m > 1),$$
$$s_{k+2} = 0 \wedge s_{k+1} \neq 0 \to s_k \neq 0, \ (\forall k \in \mathcal{N}),$$
$$s_{k-1} \neq 0 \wedge s_k = \cdots = s_{k+m} = 0 \wedge s_{k+m+1} \neq 0 \to s_{k+m+2}^m s_{k-1} = \delta_{m+2} s_{k+m+1}^{m+1}$$
$$\wedge s_{k+m+2}^m c_{k-1} = \delta_{m+2} s_{k+m+1}^m c_{k+m+1}, \ (\forall k \in \mathcal{N}, m \geq 0).$$

The proof of the theorem is obtained by proving the following 10 lemmas. Lemma 3 and the part of Lemma 6 are already mentioned in [10].

Lemma 1. *Let $f = \sum_{i=0}^{n} p_i x^i$ be a polynomial in $\mathbb{R}[x]$ where $p_n \neq 0$. The conditions $s_n = s_{n-1} = c_n > 0$ and $s_0 = c_0$ are necessary conditions for the SDC.*

Proof. When f satisfies the SDC, $c_n > 0$ because of $f(0) = p_0 > 0$. In order to satisfy $f(x) > 0$ for sufficiently large $x > 0$, $p_n > 0$ is needed. Then $s_n = s_{n-1} > 0$. Since the degree of $\mathsf{SH}_0(f)$ with respect to x is zero, we have $s_0 = c_0$.   □

The rest of this section we assume that the leading coefficient and the constant term of a given polynomial are positive from Lemma 1. Thus $s_n = s_{n-1} = c_n > 0$. For sake of simplicity, we denote $\mathsf{SH}_k(f)$ by SH_k.

Lemma 2. *For $k \in \{0, \dots, n\}$ $s_k = 0$ implies that $c_k = 0$.*

Proof. From Remark 1, $s_k = 0$ implies that $\mathsf{SH}_k = 0$. Then $c_k = 0$.   □

Lemma 3. *$c_u \neq 0$ where u is the smallest nonnegative integer k such that $s_k \neq 0$.*

Proof. This lemma is mentioned in [10]. From the definition of u, SH_u is the greatest common factor of f and df/dx. Since the assumption $f(0) \neq 0$, $\mathsf{SH}_u(0) \neq 0$. Then $c_u \neq 0$.   □

Lemma 4. *$c_{n-1} = 0$ implies $c_{n-2} < 0$.*

Proof. By the definition we have $\mathsf{SH}_{n-1}(0) = p_1$ and $\mathsf{SH}_{n-2}(0) = p_1 p_{n-1} p_n - n^2 p_0 p_n^2$. Since p_n and p_0 are positive by the assumption and c_{n-1} is the sign of p_1, $\mathsf{SH}_{n-2}(0) = -n^2 p_0 p_n^2 < 0$.   □

Lemma 5. *$s_{n-2} = 0$ implies $s_{n-3} = \cdots = s_1 = s_0 = 0$.*

Proof. Suppose that there exists some integer $k \leq n - 3$ such that $s_k \neq 0$. From Theorem 2 the degree of SH_{n-1} must be less than $n - 2$. Since the leading coefficient of f is positive and $\mathsf{SH}_{n-1} = df/dx$, this is a contradiction.   □

Lemma 6. *For $k \in \{1, \ldots, n-1\}$, $c_{k-1} \neq c_{k+1}$ if $c_{k+1} \neq 0$ and $c_k = 0$.*

Proof. We only need to consider five cases. The case (1) is proved in [10].

(1) $\deg(\mathsf{SH}_k) = r, \deg(\mathsf{SH}_{k+1}) = k + 1, \mathsf{SH}_k(0) = c_k = 0$:
    (a) $r = k$: From Theorem 2 (C), $\mathrm{lc}(\mathsf{SH}_{k+1})^2 \mathsf{SH}_{k-1} = -\mathrm{Prem}(\mathsf{SH}_{k+1}, \mathsf{SH}_k)$.
        Then $c_{k-1} = -\mathrm{lc}(\mathsf{SH}_{k+1})^{-2} c_{k+1}$. Since $f(0) \neq 0$, x is not a common
        divisor of SH_k and SH_{k+1}. Then $c_{k+1} \neq 0$. Hence we obtain $c_{k+1} \neq c_{k-1}$.
    (b) $r < k$: From Theorem 2 (A) and (C), $c_{k-1} = 0$.
(2) $\deg(\mathsf{SH}_{k+1}) = r < k + 1, \deg(\mathsf{SH}_{k+2}) = k + 2, \mathsf{SH}_k(0) = c_k = 0$:
    (a) $r < k - 1$: From Theorem 2 (A), $c_{k-1} = 0$.
    (b) $r = k - 1$: From Theorem 2 (B), $\mathrm{lc}(\mathsf{SH}_{k+2})^2 \mathsf{SH}_{k-1} = -\mathrm{lc}(\mathsf{SH}_{k+1})^2 \mathsf{SH}_{k+1}$.
        Then $\mathsf{SH}_{k-1}\mathsf{SH}_{k+1} \leq 0$.
    (c) $r = k$: From Theorem 2 (B), $c_{k+1} = 0$.   □

Lemma 7. *For $k \in \{u + 1, \ldots, n - 2\}$, $c_{k-1} \neq 0 \wedge c_k = c_{k+1} = 0 \wedge c_{k+2} \neq 0 \wedge s_k \neq 0 \wedge s_{k+1} \neq 0 \rightarrow s_{k+2}s_k < 0$.*

Proof. From Lemma 2, $\mathsf{SH}_{k+2} \neq 0$ because of $c_{k+2} \neq 0$. We first consider the case where SH_{k+2} is not regular. From Theorem 2 (A), SH_{k+1} is regular. In addition, from Theorem 2 (B) SH_{k+1} is a constant multiple of SH_{k+2}. Thus when SH_{k+2} is not regular, $c_{k+2} \neq 0$ and $c_{k+1} = 0$ are not satisfied simultaneously.

We next consider the case where SH_{k+2} is regular. We assume that SH_{k+1} is regular. From Theorem 2 (C) $\mathrm{lc}(\mathsf{SH}_{k+2})^2 \mathsf{SH}_k = \delta_2 \mathrm{Prem}(\mathsf{SH}_{k+2}, \mathsf{SH}_{k+1})$. Since $c_k = c_{k+1} = 0$, x must be a common divisor of SH_k and SH_{k+1}. This is a contradiction to $f(0) \neq 0$. Therefore SH_{k+1} is not regular and the degree of SH_{k+1} is k from Theorem 2 (A). From Theorem 2 (B), we obtain $s_{k+2}s_k < 0$.   □

Lemma 8. *For $k \in \{u + 1, \ldots, n - 2\}$ and $m \in \{2, \ldots, n - k - 1\}$, $c_{k+m+1} \neq 0 \wedge c_{k+m} = \cdots = c_k = 0 \rightarrow s_{k+m-1} = \cdots = s_{k+1} = 0$.*

Proof. (i) Case $s_{k+m} \neq 0$. By the proof of Lemma 7, SH_{k+m+1} is regular and SH_{k+m} is not. We denote the degree of SH_{k+m} by d_1. When $d_1 > k$, by Theorem 2 (C), SH_{d_1-1} is a constant multiple of a pseudo remainder of SH_{k+m+1} by SH_{k+m}. Since $\mathsf{SH}_{d_1-1}(0) = \mathsf{SH}_{k+m}(0) = 0$, x is a common divisor of them. This is a contradiction for $f(0) \neq 0$. Thus $d_1 \leq k$. Therefore by Theorem 2 (A) we have $s_{d_1+1} = \cdots = s_{k+m-1} = 0$.
(ii) Case $s_{k+m} = 0$. We denote the degree of SH_{k+m+1} by d_2. When $d_2 \geq k$, by Theorem 2 (B), SH_{d_2} is a constant multiple of SH_{k+m+1}. This is a contradiction for $\mathsf{SH}_{d_2-1}(0) = 0$ and $\mathsf{SH}_{k+m+1}(0) \neq 0$. Thus $d_2 < k$. Therefore by Theorem 2 (A) we have $s_{d_2+1} = \cdots = s_{k+m-1} = 0$.   □

x	y	$x \cdot y$
0	0	0
0	1	0
1	0	0
1	1	1

x	y	$x + y$
0	0	0
0	1	1
1	0	1
1	1	1

x	x'
0	1
1	0

Fig. 1. Boolean operations (AND, OR, and NOT)

Lemma 9. *For $k \in \{u, \ldots, n-3\}$, $s_{k+2} = 0 \wedge s_{k+1} \neq 0$ implies $s_k \neq 0$.*

Proof. The polynomial SH_{k+1} is regular from Theorem 2 (B). From Theorem 2 (C) SH_k is a pseudo remainder of SH_r by SH_{k+1} for some $r > k+1$. Since $k \geq u$, we obtain $\mathsf{SH}_k \neq 0$.   □

Lemma 10. *For $k \in \{u+1, \ldots, n-3\}$ and $m \in \{0, \ldots, n-k-1\}$, $s_{k+m+1} \neq 0 \wedge s_{k+m} = \cdots = s_k = 0 \wedge s_{k-1} \neq 0$ implies that $s_{k+m+2}^m s_{k-1} = \delta_{m+2} s_{k+m+1}^{m+1} \wedge s_{k+m+2}^m c_{k-1} = \delta_{m+2} s_{k+m+1}^m c_{k+m+1}$.*

Proof. This is obtained from Theorem 2 (B).   □

4   Simplification of Boolean Expressions

In this section we explain an approach to simplify logical formulae by using a simplification method of Boolean expressions based on Boolean function manipulation [3].

4.1   Boolean Algebra and Simplification of Boolean Expressions

In this subsection we define a Boolean algebra and a Boolean function.

Definition 5. *A Boolean algebra is an algebraic system consisting of the set $B = \{0, 1\}$, two binary operations called AND and OR and denoted by the symbols $\cdot$ and $+$, respectively, and one unary operation called NOT and denoted by a prime, $'$. The definitions of the AND, OR, and NOT operations are as in Fig. 1.*

Definition 6. *A Boolean variable is a two-valued variable which can take any of the two distinct values 0 and 1. A literal is a Boolean variable or its complement. A product term is a literal or a conjunction of literals where no literal appears more than once. A sum of products is a product term or a disjunction of product terms.*

Definition 7. *A Boolean expression is the combination of a finite number of Boolean variables and Boolean constants by means of the Boolean operations defined in Definition 5. A (completely specified) Boolean function with n variables is a mapping $f : B^n \to B$.*

Definition 8. *An* incompletely specified Boolean function *of* n *variables is a Boolean function which is defined over a subset of* B^n. *An input combination for which the function is not specified is called a* don't care.

In general, there are a number of Boolean expressions to represent a Boolean function. For example, Boolean expressions $(x + y)'$ and $x' + y'$ represent a same Boolean function. In this paper, finding a Boolean expression which has relatively a small number of product terms is called *simplification* of Boolean expressions.

Example 3. Consider an incompletely specified Boolean function f whose truth table is defined as follows. Entry d in the column "f" means that the corresponding function value is unspecified.

$x \; y \; z$	f
0 0 0	1
0 0 1	d
0 1 0	0
0 1 1	0
1 0 0	d
1 0 1	1
1 1 0	1
1 1 1	1

For incompletely specified Boolean functions, we can expand the notion of simplification of Boolean expressions, because an incompletely specified Boolean function represents a set of completely specified Boolean functions. We can choose a completely specified Boolean function by assigning 0 or 1 to each *don't care* entry d. By considering both of assignments to *don't cares* and expressions of the functions, we may obtain better expressions. In the above example, when we assign 1 to every d, we obtain a simplified Boolean expression $f = x + y'$.

In the integrated circuit design, simplification of Boolean expressions directly corresponds to minimization of area of the designed circuit. Hence many efficient techniques have been proposed, such as a heuristic method called ESPRESSO [3] and several exact methods based on binary decision diagrams (BDDs) [8,14].

4.2 Simplification of φ_n Based on Boolean Expression Minimization

In this subsection, we consider simplification of φ_n presented in Subsection 2.2 by using simplification techniques for Boolean expressions.

First, since signs of polynomials handled in this paper is three-valued, we need two Boolean variables to represent them. In this paper, using variables x and y, we represent zero, positive, and negative as $x'y'$, xy', and $x'y$, respectively. Then we obtain simplified representation of φ_n by applying simplification techniques for Boolean expressions.

Example 4. Consider formula (3) again. Let us use x_1y_1, x_2y_2, x_3y_3 to represent the signs of s_0, c_2, c_1, respectively. Then we can represent formula (3) as the following Boolean expression.

$$x_1'y_1x_2y_2'x_3'y_3 + x_1'y_1x_2y_2'x_3'y_3' + x_1'y_1x_2y_2'x_3y_3' +$$
$$x_1'y_1'x_2y_2'x_3'y_3 + x_1'y_1'x_2y_2'x_3y_3' + x_1y_1'x_2y_2'x_3y_3'.$$

As mentioned in Subsection 4.1, introduction of *don't cares* is likely to simplify Boolean expressions further. For the specialized QE algorithm for the SDC, we can introduce *don't cares* as follows.

1. As mentioned above, we use two Boolean variables x and y to represent a sign of a polynomial. While the sign is three-valued, the two Boolean variables can represent four values. Since we do not use xy here, we can consider xy as a *don't care*.
2. Since we do not need to consider sign conditions which do not satisfy Lemmas 2–10, we consider them as *don't cares*.
3. In the same way, since it never happens that $V_0(\mathsf{SH}(f)) < V_\infty(\mathsf{SH}(f))$, we can introduce *don't cares* in this case.

5   Computational Results

In this section we present computational results.

Table 2 shows the results of simplification by our approach. All the computational experiments were executed on a personal computer with an Intel(R) Core(TM) i7-3540M CPU 3.0 GHz and 2.0 GByte memory. We used the ESPRESSO logic minimizer [1] to simplify formulae. The first column "deg" gives the degree of an input polynomial. The second column "var" gives the number of Boolean variables which are inputs to the ESPRESSO logic minimizer. When the degree of an input polynomial is n, the Sturm-Habicht sequence contains $n+1$ elements. Since $s_n = s_{n-1} = c_n > 0$ and $s_0 = c_0$ by Lemma 1, we only consider a sign sequence of $2(n+1) - 4 = 2n - 2$ polynomials. Hence the number of Boolean variables is $2(2n - 2) = 4n - 4$. The column "SyN" gives the number of product terms by the previous SyNRAC implementation in [11]. The column "DC" gives the number of product terms by the implementation with an approximate ESPRESSO method in which we introduce a *don't care* only when a sign condition satisfies the condition 1 presented in Subsection 4.2 or Lemma 2. Both conditions can be obtained without using Theorem 2. The columns "ESP app" and "ESP ex" give the number of product terms by the implementation presented in this paper with an approximate ESPRESSO method and with an exact ESPRESSO method, respectively. The seventh column "terms" gives the number of sign conditions which satisfy Theorem 2. The last two columns "time app" and "time ex" give the computing times by an approximate and by an exact minimization algorithm, respectively. Computing times are given in units of seconds. We remark that since this simplification step is executed beforehand, the comparison of the computing times is not an essential problem in QE computation.

We see that our improvements greatly reduced the number of product terms, comparing "SyN" and "ESP app", and *don't cares* are important to simplify a formula from comparing "DC" and "ESP app". We have not obtained the solutions of the seventh and the eighth SDC problems by an exact ESPRESSO method. However, we see that "ESP app" outputs good approximate solutions of up to the sixth SDC problem.

Table 2. Computational result of the SDC problems

deg	var	SyN	DC	ESP app	ESP ex	term	time app	time ex
2	4	5	2	2	2	5	0.01	0.01
3	8	17	7	4	4	21	0.01	0.01
4	12	64	24	10	10	99	0.01	0.04
5	16	302	85	18	18	480	0.05	1.12
6	20	1229	299	57	57	2352	0.72	61.92
7	24	5238	1096	121	-	11656	21.40	>350h
8	28	20468	4037	353	-	58284	757.59	>350h

Fig. 2 presents the input and output files to ESPRESSO software [1] for the third SDC problem. The rows from 5 to 42 of the input file indicate each output value of sign conditions by Boolean expressions. The rows from 6 to 9 of the output file present the simplified formula by Boolean expressions. Each row shows truth table row consisting of 8 inputs and 1 output. Each position in the input plane corresponds to an input variable where "0" implies the corresponding input literal appears complemented in the product term, "1" implies the input literal appears uncomplemented in the product term, and "-" implies the input literal does not appear in the product term. The numbers 1 and 2 at the end of row indicate that a sign condition is true and a *don't care*, respectively. The mark "-" of the input file implies the input literal does not appear. The products of the Boolean variables $x_0 y_0$, $x_1 y_1$, $x_2 y_2$ and $x_3 y_3$ show the signs of s_1, s_0, c_2 and c_1, respectively. For example, the row 18 shows $s_1 < 0 \wedge s_0 = c_0 = 0 \wedge c_2 > 0 \wedge c_1 > 0$ and this sign condition is a *don't care* because $V_0(\mathsf{SH}(f)) = 0 < 1 = V_\infty(\mathsf{SH}(f))$ which satisfies condition 3 in Section 4.2. The rows 28 to 31 show that xy is a *don't care*. The rows 32 to 33, 34 to 35, 36 to 37, 38 to 39 and 40 to 42 are from Lemmas 5, 2, 6, 4 and 3, respectively.

By our approach 22 sign conditions for the third SDC problem was reduced to 4 conditions and we obtained the following formula:

$$(s_1 < 0 \wedge s_0 > 0) \vee (s_1 < 0 \wedge c_1 < 0) \vee (s_0 < 0 \wedge c_1 < 0) \vee (c_2 \geq 0 \wedge c_1 \geq 0).$$

Let $f(x) = x^3 + ax^2 + bx + c$. The Sturm-Habicht sequence associated to f is $\{\mathsf{SH}_3 = f,\ \mathsf{SH}_2 = df/dx = 3x^2 + 2ax + b,\ \mathsf{SH}_1 = (2a^2 - 6b)x + ab - 9c,\ \mathsf{SH}_0 = -4b^3 + a^2 b^2 - 4a^3 c + 18bac - 27c^2\}$ and then we obtain a simple quantifier-free formula

$$
\begin{aligned}
c > 0 \wedge (&(2a^2 - 6b < 0 \vee 2a^2 - 6b = 0 \wedge ab - 9c < 0) \wedge \mathsf{SH}_0 > 0 \vee \\
&(2a^2 - 6b < 0 \vee 2a^2 - 6b = 0 \wedge ab - 9c < 0) \wedge ab - 9c < 0 \vee \\
&\mathsf{SH}_0 < 0 \wedge ab - 9c < 0 \ \vee \ b \geq 0 \wedge ab - 9c \geq 0).
\end{aligned}
\tag{5}
$$

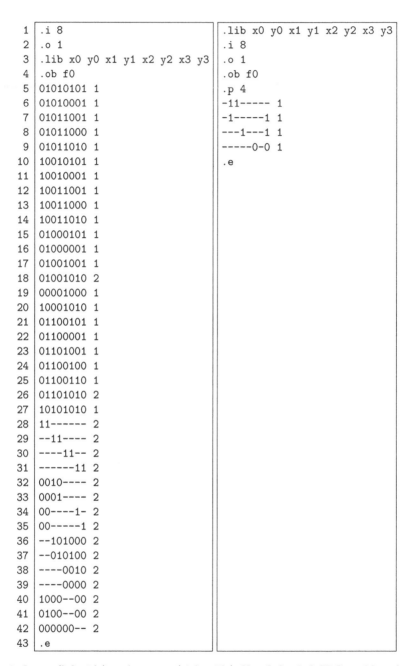

1	.i 8		.lib x0 y0 x1 y1 x2 y2 x3 y3
2	.o 1		.i 8
3	.lib x0 y0 x1 y1 x2 y2 x3 y3		.o 1
4	.ob f0		.ob f0
5	01010101 1		.p 4
6	01010001 1		-11----- 1
7	01011001 1		-1-----1 1
8	01011000 1		---1---1 1
9	01011010 1		-----0-0 1
10	10010101 1		.e
11	10010001 1		
12	10011001 1		
13	10011000 1		
14	10011010 1		
15	01000101 1		
16	01000001 1		
17	01001001 1		
18	01001010 2		
19	00001000 1		
20	10001010 1		
21	01100101 1		
22	01100001 1		
23	01101001 1		
24	01100100 1		
25	01100110 1		
26	01101010 2		
27	10101010 1		
28	11------ 2		
29	--11---- 2		
30	----11-- 2		
31	------11 2		
32	0010---- 2		
33	0001---- 2		
34	00----1- 2		
35	00-----1 2		
36	--101000 2		
37	--010100 2		
38	----0010 2		
39	----0000 2		
40	1000--00 2		
41	0100--00 2		
42	000000-- 2		
43	.e		

Fig. 2. Input (left side) and output (right side) file of the 3rd SDC problem for the ESPRESSO command

We obtained the significant simpler formula by our approach than that by [11] which has 17 product terms. However, the formula can be simplified more. For example, there do not exist real numbers a, b, and c such that

$$c > 0 \wedge ((2a^2 - 6b < 0 \vee 2a^2 - 6b = 0 \wedge ab - 9c < 0) \wedge \mathsf{SH}_0 > 0),$$

which is the first product term in (5). Thus we can reduce the formula more. The necessary condition for the SDC presented in Section 3 considers only the case that s_k or c_k is zero. To obtain the simpler formula φ_n finding the necessary and sufficient condition is a promising direction and this would be one of our future work.

Remark 3. CAD is known as an algorithm which constructs simpler quantifier-free formulae because CAD uses sign information of many projection factors. In fact, in the third SDC problem we obtain the following simpler formula by CAD [4,5,13] which used 6 projection factors:

$$c > 0 \wedge (b \geq 0 \wedge a \geq 0 \vee \mathsf{SH}_0 < 0).$$

6   Conclusion

This paper has considered the SDC problem which is the QE problem $\forall x (x \geq 0 \rightarrow f(x) > 0)$ where f is a polynomial in $\mathbb{R}[x]$. For improvement of the algorithm, simplification of formulae is important. To simplify a formula, we have shown the necessary condition for the SDC and the usage of a logic minimization method. Finally we have shown the effect of our approach by computational results.

We expect to make further reduction of φ_n and would like to find the necessary and sufficient condition for the SDC in our future work. Meanwhile we did not obtain an exact solution of the seventh and subsequent SDC problems. To obtain a simpler formula we would like to try to simplify formulae by an exact method, for example, based on BDDs.

References

1. Espresso, `http://embedded.eecs.berkeley.edu/pubs/downloads/espresso/`
2. Anai, H., Hara, S.: A parameter space approach to fixed-order robust controller synthesis by quantifier elimination. International Journal of Control 79(11), 1321–1330 (2006)
3. Brayton, R.K., Sangiovanni-Vincentelli, A.L., McMullen, C.T., Hachtel, G.D.: Logic Minimization Algorithms for VLSI Synthesis. Kluwer Academic Publishers, Norwell (1984)
4. Brown, C.W.: Solution formula construction for truth invariant cad's. PhD thesis, University of Delaware Newark (1999)
5. Brown, C.W.: QEPCAD B: A program for computing with semi-algebraic sets using CADs. SIGSAM Bulletin 37, 97–108 (2003)

6. Caviness, B.F., Johnson, J.R. (eds.): Quantifier elimination and cylindrical algebraic decomposition. Texts and Monographs in Symbolic Computation. Springer (1998)
7. Collins, G.E.: Quantifier elimination and cylindrical algebraic decomposition. In: Caviness, Johnson (eds.) [6], pp. 8–23 (1998)
8. Coudert, O., Madre, J.C., Fraisse, H.: A new viewpoint on two-level logic minimization. In: Proceedings of the Design Automation Conference, pp. 625–630 (1993)
9. González-Vega, L.: A combinatorial algorithm solving some quantifier elimination problems. In: Caviness, Johnson (eds.) [6], pp. 365–375 (1998)
10. González-Vega, L., Recio, T., Lombardi, H., Roy, M.-F.: Sturm-Habicht sequences determinants and real roots of univariate polynomials. In: Caviness, B.F., Johnson, J.R. (eds.) [6], pp. 300–316 (1998)
11. Hyodo, N., Hong, M., Yanami, H., Hara, S., Anai, H.: Solving and visualizing nonlinear parametric constraints in control based on quantifier elimination. Applicable Algebra in Engineering, Communication and Computing 18(6), 497–512 (2007)
12. Iwane, H., Yanami, H., Anai, H.: A symbolic-numeric approach to multi-objective optimization in manufacturing design. Mathematics in Computer Science 5(3), 315–334 (2011)
13. Iwane, H., Yanami, H., Anai, H., Yokoyama, K.: An effective implementation of symbolic–numeric cylindrical algebraic decomposition for quantifier elimination. Theoretical Computer Science 479, 43–69 (2013)
14. McGeer, P., Sanghavi, J., Brayton, R., Vincentelli, A.S.: ESPRESSO-SIGNATURE: A new exact minimizer for logic functions. In: Proceedings of the Design Automation Conference, pp. 618–624 (1993)
15. Sturm, T.: New domains for applied quantifier elimination. In: Ganzha, V.G., Mayr, E.W., Vorozhtsov, E.V. (eds.) CASC 2006. LNCS, vol. 4194, pp. 295–301. Springer, Heidelberg (2006)
16. Sturm, T., Tiwari, A.: Verification and synthesis using real quantifier elimination. In: Proceedings of the 36th International Symposium on Symbolic and Algebraic Computation, ISSAC 2011, pp. 329–336. ACM, New York (2011)
17. Yang, L., Xia, B.: Real solution classification for parametric semi-algebraic systems. In: Dolzmann, A., Seidl, A., Sturm, T. (eds.) Algorithmic Algebra and Logic, pp. 281–289. Books on Demand (2005)

Categories as Type Classes
in the Scala Algebra System

Raphaël Jolly

Databeans, Vélizy-Villacoublay, France
`raphael.jolly@free.fr`

Abstract. A characterization of the categorical view of computer algebra is proposed. Some requirements on the ability for abstraction that programming languages must have in order to allow a categorical approach is given. Object-oriented inheritance is presented as a suitable abstraction scheme and exemplified by the Java Algebra System. Type classes are then introduced as an alternate abstraction scheme and shown to be eventually better suited for modeling categories. Pro and cons of the two approaches are discussed and a hybrid solution is exhibited.

Keywords: categories, type classes, bounded polymorphism.

1 Introduction

The Scala Algebra System [1] is a computer algebra library build on the model of JAS [2], following a type-safe, generic and object-oriented paradigm, but written in Scala, with the aim to address a number of concerns of the Java approach. It implements polynomial arithmetic over various base rings (integer, rational, complex, modular), rational and algebraic functions, ring modules, polynomial GCD and Gröbner bases computations.

In the present work, we explain how we refactored our library using Scala's type classes [3], a feature that the language certainly owes to its functional programming origins, and that turns out a perfect match to the computer algebraic notion of categories.

2 Outline

The paper is organized as follows. Section 3 tries to extract defining features of the categorical view of computer algebra, as implemented in some historical and current computer algebra systems. Section 4 shows how categories are modeled in the Java Algebra System using bounded polymorphism. Section 5 introduce Scala's type classes as an alternate abstraction scheme, which is also suited to modeling categories. Section 6 shows that the two schemes can coexist in a hybrid solution that retains the benefits of both. Section 7 compares our design to existing solutions.

V.P. Gerdt et al. (Eds.): CASC 2013, LNCS 8136, pp. 209–218, 2013.
© Springer International Publishing Switzerland 2013

3 Categories

In the categorical view of computer algebra, mathematical objects are grouped into well defined domains, like `Integer` or `Rational`, which restrict the kind of arithmetic operations that can be applied to their elements. Domains can be assembled into towers by means of domain constructors, like `Matrix` or `Polynomial`, allowing to work with objects of arbitrary complexity, still in a precise way. This means for instance that domains `Matrix(Rational)` and `Matrix(ModInteger(2))` are incompatible, and that distinct operations must be devised for each of them. To recover flexibility, categories are introduced. A category is meant to abstract characteristics common to several domains : if in the above example, matrix elements are specified to be in category `Field`, then domain operations need only be defined once. Likewise, polynomial coefficients can be specified to be in category `Ring`, which allows the construction of recursive polynomial rings like `Polynomial(Polynomial(Integer))`, etc. Categories are tied by (possibly multiple) inheritance relations : as every field is also a ring, `Field` inherits from `Ring` and shares all its characteristics. They can optionally supplement the specification of their operations by default definitions.

Thus we can summarize the categorical approach by the following requirements:

1. arithmetic operations between elements are restricted based on their *domain*
2. domains must support abstraction by means of *categories*
3. categories must support multiple inheritance
4. (optional) categories may support default definition of operations

Over time, several attempts were made following this approach, the most prominent being undoubtedly Axiom [4], but there was also notably Gauss [5] and more recently JAS [2], DoCon [6], Mathemagix [7]. What is interesting with JAS and DoCon is that they are written in general purpose languages (Java and Haskell, respectively). Categorical programming somewhat pushes the limits of such languages and is only possible thanks to the ability of modern object-oriented or functional programming languages for abstraction and polymorphism.

4 Categorical Programming in Java: The JAS Example

In the Java Algebra System (JAS), a domain is modeled by a pair of Java classes, called *factory* and *element*, respectively. The factory is named after the Factory pattern, namely it has methods responsible for the creation of new elements [8]. The element contains the representation of a mathematical value, together with a reference to its parent factory. Operations involving one or more (usually two) element(s) are performed through the element's methods. Hence the factory is responsible for nullary operations, and unary and binary operations are handled by the element. Similarly, a category is modeled by a pair of Java interfaces, to be implemented by the factory and element of each domain that the category abstracts. These interfaces are parametrized by the type of the element, thus allowing restrictions of permitted operations on the element, in fulfillment of our requirement 1, as well as 2. As Java interfaces support multiple inheritance, but not default implementations, requirement 3 is also met, but not 4.

Below we give a code example, in the case of domain `BigInteger`, abstracted by category `Ring`. (The example is simplified, and given in Scala for easier comparison with subsequent examples, but it is strictly equivalent to the corresponding Java code.)

```scala
trait Ring[T <: Ring.Element[T]] {
  def zero: T
}
object Ring {
  trait Element[T <: Element[T]] {
    val factory: Ring[T]
    def +(that: T): T
  }
}
class BigInteger(val value: Int) extends
    Ring.Element[BigInteger] {
  val factory = BigInteger
  def +(that: BigInteger) = factory(this.value + that.value)
  override def toString = value.toString
}
object BigInteger extends Ring[BigInteger] {
  def apply(value: Int) = new BigInteger(value)
  def zero = apply(0)
}
```

A use case is given below.

```scala
BigInteger(1) + BigInteger(1)
// 2
```

This scheme, where abstraction is obtained through inheritance/implementation, is called bounded polymorphism [9]. Domain constructors are implemented as follows, in the case of domain `Polynomial[C]`:

```scala
class Polynomial[C <: Ring.Element[C]](val ring: Ring[C], val
    variable: String) extends Ring[Polynomial.Element[C]] {
  def generator = new Polynomial.Element(...)(this)
}
object Polynomial {
  def apply[C <: Ring.Element[C]](ring: Ring[C],
      variable: String) = new Polynomial(ring, variable)
  class Element[C <: Ring.Element[C]](val value: Repr[C])(val
      factory: Polynomial[C]) extends Ring.Element[Element[C]]
}
```

Polynomial coefficients are specified to be in category `Ring` through a bound on their element type `C` which is required to be a subtype of `Ring.Element[C]`. A use case is given below.

```
val r = Polynomial(BigInteger, "x")
val x = r.generator
x + x
// 2*x
```

5   Type Classes in ScAS

In the Scala Algebra system (ScAS), a domain is modeled by a single class, which is responsible for both nullary (creational) and $(n > 0)$-ary operations. Elements can be of any type. Categories are modeled by type classes. In Scala these are represented by traits, with the element type as type parameter, like in Section 4. The binding between the element type and the domain is made by an implicit definition. Operations among elements are realized in big part by the type classes machinery, and only require the manual import of some suitable implicit value. As Scala traits support multiple inheritance and default implementations, so do the obtained categories, and requirements 3 and 4 are fulfilled, in addition to 1 and 2.

```
trait Ring[T] { outer =>
  def zero: T
  def plus(x: T, y: T): T
  implicit def mkOps(value: T): Ring.Ops[T] = new Ring.Ops[T] {
    val lhs = value
    val factory = outer
  }
}
object Ring {
  trait ExtraImplicits {
    implicit def infixRingOps[T: Ring](lhs: T) =
        implicitly[Ring[T]].mkOps(lhs)
  }
  trait Ops[T] {
    val lhs: T
    val factory: Ring[T]
    def +(rhs: T) = factory.plus(lhs, rhs)
  }
}
type BigInteger = java.math.BigInteger
object BigInteger extends Ring[BigInteger] {
  def apply(i: Int) = java.math.BigInteger.valueOf(i)
  def zero = apply(0)
  def plus(x: BigInteger, y: BigInteger) = x.add(y)
```

```
}
trait ExtraImplicits {
  implicit val ZZ = BigInteger
}
object Implicits extends ExtraImplicits with Ring.ExtraImplicits
```

A use case is given below, with the required imports.

```
import Implicits.{ZZ, infixRingOps}
BigInteger(1) + BigInteger(1)
// 2
```

Domain constructors are implemented as follows, in the case of domain Polynomial[C]:

```
class Polynomial[C: Ring](val variable: String)
    extends Ring[Repr[C]] {
  val ring = implicitly[Ring[C]]
  def generator: Repr[C] = ...
}
object Polynomial {
  def apply[C: Ring](variable: String) = new Polynomial(variable)
}
```

Polynomial coefficients are specified to be in category Ring through a *context bound* on their element type C, which means that an instance of type Ring[C] must be available for the domain to be constructed. A use case is given below, with the required imports and implicit decalaration.

```
import Implicits.{ZZ, infixRingOps}
implicit val r = Polynomial(ZZ, "x")
val x = r.generator
x + x
// 2*x
```

The category hierachy of ScAS is given in Figure 1. All categories are in package scas.structure except for Equiv and Ordering which come from the Scala standard library. These are drawn in the same box to reflect the fact that the hierarchy is duplicated in unordered and ordered versions.

6   Discussion

The first advantage of type classes is, that they provide a nice, if radical, solution to the dependent type problem [2]. In Java, the type system makes no distinction between the types of elements from e.g. ModInteger(2) and ModInteger(3), and thus does not forbid operations between these domains. With type classes, such mismatches are prevented, by the fact that domains are specified through

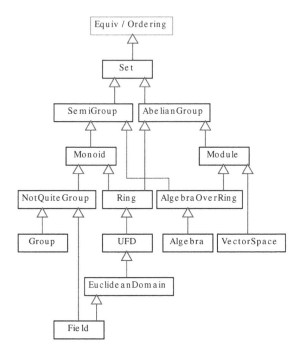

Fig. 1. Category hierarchy of ScAS

implicits, and only one such implicit definition can exist at a time for a given element type.

Compared to bounded polymorphism, the merit of type classes is also, not to impose constraints on the type of elements. It allows application of new abstractions to existing objects ("post-facto extensions", [10]). As shown in Section 5, the `Ring` abstraction can be applied to already existing domain `java.math.Big-Integer`, with no need to redevelop it or wrap it with an adapter class. This also allows unboxed primitive element types, like `Int` or `Double`, which, together with type specialization, means better efficiency, and opens the way to generic numeric-symbolic implementations, for example in linear algebra.

There are cases, however, where interface implementations are desirable. For instance, in ScAS, coercions are made through implicit conversions, and these work better (one should say only) when operations are implemented directly by the element types. To remedy this problem, and bearing in mind that operating directly on pre-existing types is mostly interesting for base types like `Integer` or `Rational`, we have investigated if it would be possible to re-introduce inheritance in a limited way, namely just for elaborate domains (polynomials, etc). There is no problem with implementing their elements as wrappers, since their representation is generally a reference type (`Array`, ...) and it is just one more indirection. This turned out to be compatible with the type classes scheme, as shown below, in the case of domain `Polynomial[C]`:

```
class Polynomial[C: Ring](val variable: String)
    extends Ring[Polynomial.Element[C]] {
  ...
  def generator = apply(...)
  def apply(value: Repr[C]) = new Polynomial.Element(value)(this)
}
object Polynomial {
  ...
  class Element[C](val value: Repr[C])(val factory: Polynomial[C])
      extends Ring.Element[Element[C]]
}
```

Conversely to domain `Polynomial[C]` in Section 5, whose elements could be of any type `Repr[C]`, here elements of domain `Polynomial[C]` are required to be subtypes of `Ring` elements. This reminiscence of bounded polymorphism is accommodated to the type classes machinery through the following addition:

```
object Ring {
  ...
  trait Element[T <: Element[T]] extends Ops[T] { this: T =>
    val lhs = this
  }
}
```

, which bridges the gap between the two abstraction schemes, forming an hybrid solution. The class layout is summarized in Figure 2.

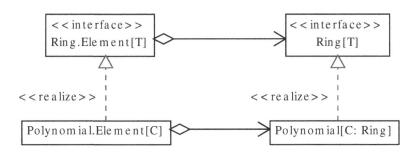

Fig. 2. Hybrid abstraction scheme

7 Related Work

Type classes were first introduced in Haskell [9]. To the best of our knowledge, they were first mentionned as possible abstractions for computer algebra structures in [11] and [12]. The computer algebra system DoCon actually uses them.

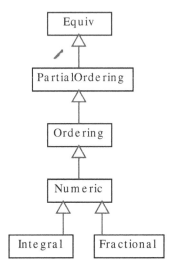

Fig. 3. Categories in the Scala standard library

Regarding Scala, one must note that type classes have been in front of computer algebraists' eyes for quite a long time, since a code example in the language documentation explicitly involves abstract algebraic constructs [13]. The computer algebra system Mathemagix uses a concept of categories that is completely equivalent to type classes, without naming them [7]. It is written in a specialized language and does not (yet) support default definitions. The Scala language itself uses types classes for algebra but only for numerical domains. Its categories are defined in package `scala.math`. Their hierachy is given in Figure 3.

The Spire project aims to redefine this hierarchy in terms of a set of more fundamental type classes, corresponding to concepts from abstract algebra [14]. Its category hierarchy is given in Figure 4 (all categories are in package `spire.algebra` except for `Numeric`, `Fractional`, `Integral`, `Order`, `Eq` which come from package `spire.math`).

Elements of the multiprecision integer domain in Scala are still implemented as an adapter class `scala.math.BigInt` but, as we have seen and thanks to type classes, the original Java class `java.math.BigInteger` could be used instead (and likewise for big decimals). This would save an indirection, but would impede coercions, which may be the reason why the adapter is kept. Thus, the adopted solution is very similar to our hybrid scheme in Section 6.

We have long thought that computer algebra in a general purpose language would not be successful until some abstract algebraic categories are included in a standard libray. But, as we can see, each projet still has its own idea of what category hierarchies should be. Furtunately, thanks to post-facto extensions, this is no longer a problem.

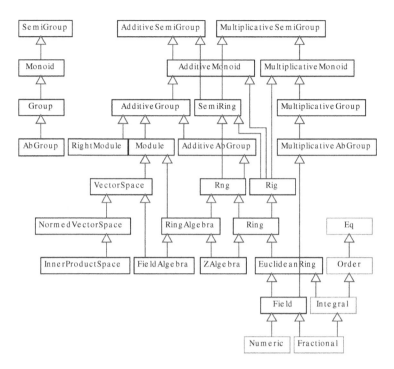

Fig. 4. Categories in Spire

8 Conclusion

Are general purpose programming languages ready for computer algebra ? This is a recurring question, that we think today may be answered positively. We have shown that the categorical approach can be tackled by modern object-oriented languages such as Java or Scala. The type classes scheme available in the latter enables a very natural implementation of categories, that in addition reveals itself compatible with the former, object-oriented scheme. Type classes based categories are already used for numerical purposes in the Scala standard library and the third-party project Spire. Computer algebra in Haskell uses them since the beginning. Computer algebra on the JVM can and should use them as well.

References

1. Jolly, R.: ScAS - Scala algebra system. Technical report (2010-2012),
   https://github.com/rjolly/scas
2. Kredel, H.: On a Java Computer Algebra System, its performance and applications.
   Science of Computer Programming 70(2-3), 185–207 (2008)
3. Odersky, M.: Poor man's type classes. Technical report (Presentation at the meeting of IFIP WG) (2006),
   http://lampwww.epfl.ch/~odersky/talks/wg2.8-boston06.pdf

4. Davenport, J.: The axiom system. In: Proceedings of NAGUA 1991 (1992)
5. Gruntz, D., Monagan, M.: Introduction to GAUSS. SIGSAM Bulletin 28(2), 3–19 (1994)
6. Mechveliani, S.: Computer algebra with haskell: applying functional-categorial-"lazy" programming. In: Gerdt, V. (ed.) Proceedings of International Workshop CAAP, Joint Institute for Nuclear Research, pp. 203–211 (2001)
7. van der Hoeven, J., Lecerf, G., Mourrain, B.: Mathemagix. Technical report (2002-2012), http://www.mathemagix.org/
8. Niculescu, V.: A design proposal for an object oriented algebraic library. Technical report, Studia Universitatis "Babes-Bolyai" (2003)
9. Oliveira, B.C., Moors, A., Odersky, M.: Type classes as objects and implicits. SIGPLAN Not. 45(10), 341–360 (2010)
10. Watt, S.: Post facto type extensions for mathematical programming. In: Proc. Domain-Specific Aspect languages (SIGPLAN/SIGSOFT DSAL 2006), October 23. ACM (2006)
11. Weber, A., Klaeren, H.: Type systems for computer algebra. Relation 10(1.54), 2615 (1993)
12. Santas, P.S.: A type system for computer algebra. J. Symb. Comput. 19(1-3), 79–109 (1995)
13. Scala developpers: A tour of scala: Implicit parameters. Technical report (2008-2010), http://www.scala-lang.org/node/114
14. Osheim, E., Switzer, T.: Powerful new number types and numeric abstractions for scala. Technical report (2011-2012), https://github.com/non/spire

Enumeration of Schur Rings over the Group A_5

Mikhail Klin and Matan Ziv-Av

Ben-Gurion University of the Negev
klin@cs.bgu.ac.il
matan@svgalib.org

Abstract. With the aid of the computer algebra system GAP (together with the extension package COCO-II) all S-rings over the smallest non-abelian simple group A_5 of order 60 are enumerated. It turns out that there are 2848 S-rings over A_5, in 163 orbits under the action of S_5. Among them there are 505 non-Schurian S-rings, in 19 orbits. We discuss the entire picture of this rich computer data and try to present an initial outline of a human explanation for most of the results obtained.

1 Introduction

The concept of a Schur ring (briefly S-ring) was coined by I. Schur in 1933. For a few decades, it was used exclusively by a rather narrow population of Schur's disciples and followers for purely group theoretical purposes. With the development of algebraic graph theory (briefly AGT), this concept was revitalized and used more frequently for the better understanding of Cayley graphs and their automorphism groups. Nowadays tools from scientific computing, in conjunction with theoretical background from permutation groups and AGT, create a serious stimulus for more extensive investigations of S-rings.

In this paper we report on the results of a computer-aided classification of S-rings over the alternating group A_5 of order 60, the smallest non-abelian simple group. The main principles of computer classification are discussed, together with a presentation of the entire picture of the lattice of S-rings, their automorphism groups and their combinatorial properties.

A number of interesting Cayley graphs over A_5 are considered, with the corresponding S-rings. Recall that A_5 is the group of rotational symmetries of the icosahedron. Since the time of Felix Klein, icosahedral symmetry has appeared as an interdisciplinary crossroads of science (see e.g. [20]). In the concluding section, we present a very brief discussion of this extra context.

2 Schur Rings, Related Concepts, and Computer-Aided Activities

In what follows we refer to such texts as [3,5,6,15,16] for more details regarding the discussed background.

V.P. Gerdt et al. (Eds.): CASC 2013, LNCS 8136, pp. 219–230, 2013.
© Springer International Publishing Switzerland 2013

Schur rings were introduced by I. Schur in 1933 ([29]), and were later developed by H. Wielandt ([31]).

Recall that the group ring $\mathbb{C}[H]$ consists of all formal linear combinations of elements of the group H with coefficients from the field $\mathbb{C}$.

A Schur ring over the group H is a subring $\mathcal{A}$ of the group ring $\mathbb{C}[H]$ such that there exists a partition P of H satisfying:

1. $\underline{P}$ is a basis of $\mathcal{A}$ (as a vector space over $\mathbb{C}$).
2. $\{e\} \in P$, where e is the identity element of H.
3. $X^{-1} \in P$ for all $X \in P$.

Here, for a subset X of H we define $X^{-1} = \{g^{-1}|g \in X\}$ and $\underline{X} = \sum_{x \in X} 1 \cdot x$, while for a set of subsets T we define $\underline{T} = \{\underline{X}|X \in T\}$.

Let (G, Ω) be a permutation group and H a regular subgroup of G. Then Ω may be identified with H. The stabilizer G_e of the identity element $e \in H$ defines an S-ring over H (see [31]). We denote this S-ring by $V(G, H)$.

An S-ring $\mathcal{A}$ is called *Schurian* if it is equal to $V(G, H)$ for a suitable overgroup (G, H) of a regular group (H, H). A group H is called a *Schur group* if all S-rings over H are Schurian. Schur [29] conjectured that all groups are Schur groups, or in other words, all S-rings are Schurian. The first examples of non-Schurian S-rings were presented by Wielandt in [31].

Let H be a group and S a subset of H. The Cayley graph $Cay(H, S) = (H, R)$ is a graph with vertex set H and with arc set $R = \{\langle x, sx\rangle|x \in H, s \in S\}$. A Cayley graph $Cay(H, S)$ is undirected if $S = S^{-1}$ and is connected if $H = \langle S\rangle$.

A *colour graph* is a pair $(\Omega, \mathcal{R})$, where $\mathcal{R} = \{R_i|i \in I\}$ is a partition of Ω^2.

Let $\mathcal{A}$ be an S-ring over group H, $\mathcal{A} = \{T_0, T_1, \ldots, T_s\}$, where $T_0 = \{e\}$, $T_1, \ldots, T_s$ are the basic sets of $\mathcal{A}$. It follows from the definitions that $\underline{T_i} \cdot \underline{T_j} = \sum_{k=0}^{s} p_{ij}^k \underline{T_k}$ for suitable non-negative integers p_{ij}^k, $0 \leq i, j, k \leq s$. The numbers p_{ij}^k are called *structure constants* of $\mathcal{A}$. We also associate with $\mathcal{A}$ the colour graph $\mathfrak{m} = (H, R_i)$, where for $0 \leq i \leq s$, R_i is the arc set of the Cayley graph $Cay(H, T_i)$. This forms a structure which is called a translation association scheme, a particular case of the more general concept of an association scheme (briefly AS). We say that the rank of $\mathfrak{m}$ is equal to $s + 1$. An AS is called *symmetric* if each basic relation is symmetric. It is called *commutative* if $p_{ij}^k = p_{ji}^k$ for all $i, j, k \in [1, r]$. The structure constants of the AS (also called *intersection numbers*) coincide with the structure constants of the corresponding S-ring.

Let $\mathfrak{m} = (X, \mathcal{R})$ be an AS. A permutation $g \in Sym(X)$ is an *automorphism* (or a strong automorphism) of $\mathfrak{m}$ if for each $R \in \mathcal{R}$, $R^g = R$. The permutation g is a *colour automorphism* of $\mathfrak{m}$ if $R^g \in \mathcal{R}$ for all $R \in \mathcal{R}$.

The group of all automorphisms of $\mathfrak{m}$ is denoted by $Aut(\mathfrak{m})$, and the group of all colour automorphisms is denoted by $CAut(\mathfrak{m})$.

Proposition 1. $Aut(\mathfrak{m})$ *is a normal subgroup of* $CAut(\mathfrak{m})$.

Let $\mathfrak{m}$ be a rank-r AS. A permutation $g \in Sym([0, r-1])$ is an *algebraic automorphism* of $\mathfrak{m}$ if $p_{ij}^k = p_{i^g j^g}^{k^g}$ for all structure constants p_{ij}^k, where $i, j, k \in [0, r-1]$. The group of all algebraic automorphisms of $\mathfrak{m}$ is denoted by $AAut(\mathfrak{m})$.

Proposition 2. *The quotient group $CAut(\mathfrak{m})/Aut(\mathfrak{m})$ is a subgroup of $AAut(\mathfrak{m})$.*

Algebraic automorphisms which are not in $CAut(\mathfrak{m})/Aut(\mathfrak{m})$ are called *proper* algebraic automorphisms.

The goal of enumeration of the S-rings over a given group is a special case of the enumeration of mergings of an AS and even more generally of a coherent configuration. We have two tools for handling this task, COCO and GAP with COCO-II. COCO is a set of programs for dealing with coherent configurations. It was developed in 1990–2, in Moscow, USSR, mainly by Faradžev and Klin ([5], [6]). The programs included in COCO allow for construction of the centralizer algebra of a permutation group, and calculation of the homogeneous mergings of this algebra, as well as their automorphism groups.

GAP ([7], [28]), an acronym for "Groups, Algorithms and Programming", is a system for computation in discrete abstract algebra. It supports easy addition of extensions (packages, in gap nomenclature), that are written in the GAP programming language which can add new features to the GAP system.

COCO-II, an extension package for GAP which is still in its initial development phase, re-implements the algorithms in COCO. In addition, COCO-II extends the abilities of COCO based on new theoretical results obtained since the original COCO package was written. For the specific task of enumeration of mergings, COCO-II uses knowledge of $AAut$ to reduce the amount of work.

Since the main function of both packages is enumeration of mergings, the programming part is simple. We need to construct the required group in COCO or GAP terms, and call a single function to enumerate all mergings.

The S-rings over cyclic groups were classified by Leung and Man in [18,19]. In 2004 Muzychuk, using the classification by Leung and Man, offered a complete solution of the isomorphism problem for circulant graphs ([25]).

Hanaki and Miyamoto ([9]) classified all association schemes of small order, in particular, all S-rings over groups of order up to 35.

Sven Reichard classified all S-rings over groups of order up to 44.

The small but challenging open cases appear to be the non-abelian group of order 55, the group $AGL(1,8)$ of order 56 and the group A_5 of order 60. We ran all three calculations, but for the group of order 55, the calculation appears to take too long. For the two other groups we achieved a classification of all S-rings. This text provides a detailed report about the S-rings over A_5.

3   Computer Results for the Group A_5

All S-rings over the group A_5 were enumerated with the aid of COCO-II; the job took about 1 month of computer time on a 3GHz CPU.

A summary of the results: there are 2848 S-rings, in 163 orbits under the action of the group $S_5 = AAut(A_5)$. Among them 505 S-rings (in 19 orbits) are non-Schurian. A complete list of orbit representatives, along with some information about each S-ring, is presented in the catalogue available from the home page of author MZA `http://www.math.bgu.ac.il/~zivav/math`. Here the full group ring $\mathbb{C}[A_5]$ has number 0, and the rank 2 S-ring has number 162. From now on, we refer to the enumeration of the orbits in this catalogue.

The results obtained appear as a massive list of computer generated data; we wish to transform this into a form more suitable for a human being. Schurian S-rings may in principle be explained in group theoretical terms. Non-Schurian S-rings are a subject of particular interest in AGT. Each such object requires special attention and analysis.

The general distribution of S-rings with respect to rank is provided in Table 1. Eight of the 19 orbits of non-Schurian S-rings are non-commutative (with ranks

Table 1. Distribution of S-rings over A_5 with respect to rank

Rank	60	33	32	22	20	19	18	17	16	15	14	13	12	11	10	9	8	7	6	5	4	3	2
Total	1	1	1	1	1	2	3	2	1	1	2	7	5	10	11	17	15	18	21	19	16	7	1
Non-Schurian	0	0	0	0	0	0	0	0	0	0	1	1	0	1	4	3	3	3	2	0	1	0	0

between 9 and 14). Up to isomorphism there are 9 different automorphism groups of non-Schurian S-rings with orders from 720 to $2^{34} \cdot 3 \cdot 5$. Altogether among the automorphism groups of the 163 orbits of S-rings there are 139 different groups (including the trivial cases of A_5 and S_{60}): too many to pay special attention to each group.

For groups of relatively small orders, GAP allows us to get one or a few "names" of abstract groups in clear algebraic terms. In addition, in each case we need to understand the (transitive) permutation representation of each group.

It was convenient for us to classify all groups into a few classes according to their order. Defining artificial borders, we distinguish:

- "small" groups, those of order up to 7680; all of them were identified with the aid of GAP;
- "large" groups, those of order 14400 up to one million. Their orders are: 14400, 14580, 24360, 29160, 46080, 61440, 122880, 230400, 466560, 933120;
- "very large" groups, of order between 1875000 and 60!.

The list of orders of small groups is presented in Table 2. Identification of the structure of large and very large groups, in general, is not possible using GAP in an automatic fashion. Instead we apply special tricks.

For example, we discuss below those groups which may be described as *wreath products* of two groups of smaller degree: $G = G_1 \wr G_2$, so that $|G| = |G_1| \cdot |G_2|^m$, where m is the degree of G_1 (we refer to [15] for definition and notation). The following simple trick was programmed in GAP:

- We start from the order $|G|$ and represent it in the form $|G| = |G_1| \cdot |G_2|^m$, where $m \cdot n = 60$.
- In addition, we require that m divides $|G_1|$, and n divides $|G_2|$.
- If these conditions are fulfilled, then we examine G in more detail, looking for transitive groups G_1, G_2 with orders $|G_1|$, $|G_2|$ and degrees m, n.

Table 2. List of orders of small groups

0	60	A_5	20	360	$C_3 \times S_5$
1	120	S_5	22	360	$A_5 \times S_3$
2	120	$C_2 \times A_5$	30	480	$(C_2 \times C_2 \times A_5) : C_2$
3	180	$GL(2,4)$	43	600	$D_{10} \times A_5$
6	240	$C_2 \times S_5$	60	720	$S_5 \times S_3$
9	240	$C_2 \times C_2 \times A_5$	67, 95, 98	720	$A_5 \times A_4$
10	240	$A_5 : C_4$	107	1200	$(C_5 \times A_5) : C_4$
12	300	$C_5 \times A_5$	42, 112, 139	1320	$PSL(2,11) : C_2$
14	360	$GL(2,4) : C_2$	109	1440	$(A_5 \times A_4) : C_2$

4, 15, 21, 35, 47, 115	1920	$C_2 \times ((C_2 \times C_2 \times C_2 \times C_2) : A_5)$
5	3840	$((C_2 \times C_2 \times C_2 \times C_2 \times C_2) : A_5) : C_2$
18, 51, 77	3840	$C_2 \times (((C_2 \times C_2 \times C_2 \times C_2) : A_5) : C_2)$
36, 45, 52, 80	3840	$((C_2 \times C_2 \times C_2 \times C_2) : A_5) : C_4$
137	7200	$(A_5 \times A_5) : C_2$
33, 44	7680	$(C_2.(((C_2 \times C_2 \times C_2 \times C_2) : A_5) : C_2) = ((C_2 \times C_2 \times C_2 \times C_2 \times C_2) : A_5).C_2) : C_2$

In fact, there are 108 very large groups: 106 of them pass the numerical test, but only 79 of them are actually wreath decomposable.

Example 1. S-ring #19 has rank 13 with $|G| = 29859840$ and valencies $48, 1^{12}$.

- $|G| = 120 \cdot 12^5$, $m = 5$, $n = 12$.
- G_1 is S_5 acting naturally on 5 points (rank 2). $|G_2|$ has to be a group of order 12 acting regularly, for example we take $G_2 = A_4$.
- We conclude that $G_1 \wr G_2$ has the required order, rank and valencies.
- We construct $G_1 \wr G_2$ using GAP and check that it and G are similar permutation groups.
- Thus $G = S_5 \wr A_4$.

Example 2. S-ring #133 has rank 5, $|G| = 137594142720000000$ and valencies 1, 50, 4, 4, 1. Here, acting in a similar fashion we recognize that $|G| = 720 \cdot 240^6$ and identify G as $S_6 \wr (S_5 \times S_2)$, where S_6 acts naturally on 6 points, while $S_5 \times S_2$ acts transitively on 10 points with rank 4 and valencies 1, 1, 4, 4.

Finally, there remain 29 very large groups which are not wreath decomposable. We expect to be able to explain these groups with the aid of more sophisticated operations over permutation groups which sometimes are called generalized wreath products, cf. [13].

Recall that an S-ring $\mathcal{A}$ is called *primitive* if all of the basic graphs of $\mathcal{A}$ are connected. In particular, a Schurian S-ring is primitive if and only if its automorphism group is a primitive permutation group. The primitive S-rings over A_5 were classified purely theoretically by M. Muzychuk ([24]):

- There are just two (non-trivial) primitive S-rings of ranks 5 and 4 over A_5 with valencies 1, 12, 12, 15, 20 and 1, 15, 20, 24 respectively.

- Both appear as centralizer algebras of the holomorph of A_5 and its subgroup of index 2.

This result is now confirmed with the aid of a computer, as one of the consequences of the project presented here.

4 Rational S-rings over A_5

The concept of a rational S-ring over an abelian group H goes back to Schur and Wielandt, see [31], where this concept, under the original name "S-ring of traces", is defined and investigated. It seems that first usage of the term "rational" can be attributed to Bridges and Mena ([2]), who at that time were not aware of the language of S-rings and were working with equivalent terminology.

Nowadays this concept may be formulated (in a more or less classical manner) for a wider class of commutative association schemes. There are several possibilities to generalize it to the case of arbitrary association schemes (also including S-rings over finite groups). We use the following definition (cf. [13]):

Definition 1. *A graph Γ is called* rational *if the spectrum of its adjacency matrix is rational (in fact, integer). An association scheme (S-ring) is rational if all its basic graphs are rational.*

There are 54 orbits of rational S-rings over the group A_5, and 49 of them have large groups. Those with large groups are in a sense less interesting, because they can be described via suitable decompositions into association schemes with a smaller (than 60) number of points.

Those with small groups are listed in Table 3. Note that S-rings #51 and #77

Table 3. Rational S-rings over A_5 having a small automorphism group

#	Rank	Valencies	$\|Aut\|$	Name
51	9	1, 16, 8, 8, 16, 4, 2, 1, 4	3840	$C_2 \times (((C_2 \times C_2 \times C_2 \times C_2) : A_5) : C_2)$
60	9	1, 12, 6, 12, 6, 2, 12, 6, 3	720	$S_5 \times S_3$
77	8	1, 16, 16, 16, 4, 2, 4, 1	3840	$C_2 \times (((C_2 \times C_2 \times C_2 \times C_2) : A_5) : C_2)$
107	6	1, 20, 5, 20, 10, 4	1200	$(C_5 \times A_5) : C_4$
109	6	1, 24, 12, 8, 12, 3	1440	$(A_5 \times A_4) : C_2$

are non-Schurian; they will be considered in more detail in Section 5. Each of the remaining 3 S-rings needs to be investigated separately using suitable ad hoc tools. The simplest pattern is presented below.

Example 3. S-ring #60. Our goal was to explain the significant properties of this S-ring. For this purpose we used the computer package COCO. It is possible, however, to perform all the necessary calculations without the use of a computer.

Let us start from the direct sum $G = S_5 + S_3$, acting intransitively on the set $[0, 7]$ with two orbits $[0, 4]$ and $[5, 7]$. Clearly, $|G| = 720$. We wish to consider the transitive faithful action of G on the cosets of a suitable subgroup K of order 12. As abstract group, K is isomorphic to the dihedral group D_6 of order 12.

We consider the combinatorial object $P = \{(0,5),(1,6),(2,7)\}$. The stabilizer $Aut(P)$ of P in G has the structure $S_3 + S_2$, and is isomorphic (as an abstract group) to D_6. Two essential extra facts are that D_6 does not contain a non-trivial normal subgroup of G, and that there exists a subgroup $A_5 \leq G$ which has a trivial intersection with D_6.

Thus, indeed we may consider the transitive action of G on the set $\Omega = \{P^g | g \in G\}$ of all images of P, and its centralizer algebra $V = V(G,\Omega)$. Naive combinatorial counting shows that $|\Omega| = \binom{5}{2} \cdot 3! = 60$. Using the orbit counting lemma (also known as the CFB lemma, see [15]), we confirm that $rank(V) = 9$ and V is a symmetric association scheme. We also calculate the valencies of the classes of V, see Table 3. This provides a reasonably elementary description of S-ring #60. The fact that it is rational was confirmed with the aid of GAP. It might be interesting to try to get an independent computer-free justification of this fact (avoiding specific calculation of the spectrum of V).

5   General Outline of Non-Schurian S-rings over A_5

The information about all 19 non-Schurian S-rings is gathered together in Table 4. The content of the first three columns is clear, and the remaining two columns are explained below. We use a few ad hoc approaches in order to explain (or better, to interpret, see [16]) the computer results obtained.

An essential initial ingredient of the desired explanation is the concept of a root group. Namely, we have a set $\mathcal{R}$ of four *root* groups of orders 720, 1320, 1920, 7680, such that each of the 19 (non-Schurian) S-rings appears as a subring (merging, in other terminology) of a suitable corresponding transitivity module.

Note that S-rings #49 and #91 originate in this way from two roots.

In particular, of the 19 S-rings, 9 appear as so-called algebraic mergings. Below we discuss each of the 4 roots separately.

Root 1: the group $\mathcal{R}_1 = PGL(2,11)$ of rank 10 and order 1320, non-commutative S-ring #42 with subdegrees 1^5, 11^5. This is a particular case of a general construction which was described by R. Mathon and his followers ([22]), a pseudocyclic association scheme in the sense of [3]. In general, these schemes allow proper algebraic automorphisms. In our case $q = 11$, $d = 2$, so we obtain such a scheme on $\frac{q^2-1}{d}$ points, which has two algebraic mergings of rank 4 and 6; the first one is generated by an antipodal distance regular graph of diameter 3 and valency 11, see also [21], [8]. This explains #112 and #139.

Root 2: the group $\mathcal{R}_2$ of order 720 and rank 8, commutative S-ring #67 with valencies 1, 3, 4^2, 12^4. The group is isomorphic to $A_5 \times A_4$ and is a subgroup of the holomorph of A_5. This allows us to justify the existence of the group $AAut(\mathcal{R}_2)$ of order 4 and the two algebraic mergings that appear in reasonably clear terms. Both mergings have rank 7. This explains #95 and #98.

Root 3: the group $\mathcal{R}_3$ of order 7680 and rank 11, non-commutative S-ring #33 with valencies 1^2, 2, 4^2, 8^6. In its natural action of degree 12 this is the

Table 4. General information about non-Schurian S-rings

#	rank	$|Aut|$	root	extra features
15	14	1920	1920	algebraic
21	13	1920	1920	algebraic
35	11	1920	1920	algebraic
39	10	3932160	1920	elementary merging #36
44	10	7680	7680	elementary merging #32
45	10	3840	1920	elementary merging #35
47	10	1920	1920	elementary merging #34
49	9	128849018880	1920,7680	wreath product, elementary merging #38, #43, #44, #46, #47
51	9	3840	1920	algebraic
52	9	3840	1920	algebraic
69	8	7864320	1920	elementary merging #49
77	8	3840	1920	elementary merging #50
80	8	3840	1920	elementary merging #51
91	7	257698037760	1920,7680	wreath product, elementary merging #68, #71, #76, #79
95	7	720	720	algebraic, elementary merging #66
98	7	720	720	algebraic, elementary merging #66
112	6	1320	1320	algebraic
115	6	1920	1920	
139	4	1320	1320	algebraic

wreath product $G = PGL(2,5) \wr S_2$ of order $120 \cdot 2^6$, where $PGL(2,5)$ acts 3-transitively on the projective line of size 6 (the entire group G is isomorphic to group #270 of degree 12 in the catalogue [4]). To describe the action of G of degree 60 combinatorially, one may start from a BIBD D with 6 points and 10 blocks, then "blow up" each point to a 2-element subset, blowing up also all blocks. For the resulting incidence structure D' we get $G = Aut(D')$. Now an appropriate substructure S of D' should be found so that the stabilizer of S in G is a suitable subgroup of order 128 in G. All possible embeddings of S into D' provide the 60 points of the S-ring #33.

Unfortunately we do not have enough space in this text to provide full details. There is one elementary merging of rank 10 (see the discussion below). The other two non-Schurian mergings are explained as wreath products of a suitable non-Schurian scheme on 30 points with $\mathbb{Z}_2$ (this is why the automorphism groups have huge sizes). This explains #44, #49, #91.

Root 4: the group $\mathcal{R}_4$ of order 1920 and rank 20, non-commutative S-ring #4 with valencies 1^4, 2^4, 4^{12}. This group acts naturally on 10 points and appears as a wreath product $A_5 \wr S_2$ (of order $60 \cdot 2^5$). Again, its transitive action of degree 60 (as S-ring #4) may be described, starting from the action of degree 10, using a suitable ad hoc explanation. It turns out that 14 of the 19 non-Schurian S-rings are mergings of transitivity module of $\mathcal{R}_4$. Of these, 5 are algebraic mergings, namely #15, #21, #35, #51, and #52.

Another very helpful trick is due to M. Muzychuk ([26]). Let us start from an S-ring $\mathcal{A}$ of rank r and its subring $\mathcal{A}'$ of rank $r - 1$. This means that $\mathcal{A}'$ is obtained from $\mathcal{A}$ by a merging of just two basic sets, while all other basic sets are unchanged. An efficient sufficient criterion, formulated in algebraic terms, guarantees the existence of such a merging $\mathcal{A}'$ of rank $r - 1$, which we call an *elementary merging* (precise details to be provided elsewhere). It turns out that altogether 11 non-Schurian S-rings may be explained (in the role of $\mathcal{A}'$) by selecting a suitable S-ring $\mathcal{A}$. The corresponding information is given in Table 4.

By now, 18 of the 19 S-rings have been explained by at least one of the two approaches described above (some having multiple explanations).

Finally, there remains one (commutative) S-ring of rank 6 from root $\mathcal{R}_4$. None of the previous tricks provides an explanation for this S-ring. This is S-ring #115, which posed us the most challenging task for a suitable computer-free interpretation. We will face this challenge in the next section.

6   The Exceptional Non-schurian S-ring #115

We are interestied in S-ring $\mathcal{A}_{115}$. This is in fact a representative of an orbit of 30 isomorphic S-rings. This non-Schurian S-ring is a symmetric (and, therefore, commutative) S-ring of rank 6, with valencies 1, 1, 5, 5, 8, 40. Its automorphism group, $Aut(\mathcal{A}_{115}) = \mathcal{R}_4$, is a group of order 1920 and rank 20. It has three imprimitivity systems, with 10, 12 and 30 cells. The basic graph of valency 40 is the only connected basic graph.

For a presentation of the relations of $\mathcal{A}_{115}$ we need two subgroups of A_5: a group K isomorphic to D_5 (the stabilizer of a pentagon), and a group L isomorphic to A_4 (the stabilizer of a point). With 6 ways to select K and 5 ways to select L, we note that the orbit is indeed of size 30.

The group $K \cap L$ is of order 2; let i be its non-identity element. The starting groups for the construction of $\mathcal{A}_{115}$ in this specific example are $K = \langle (1,4)(2,3),$ $(0,1,2,3,4) \rangle$ and $L = A_{\{1,2,3,4\}}$, so that $i = (1,4)(2,3)$.

Now we describe the basic sets as explicit subsets of A_5:
$X_0 = \{e\}$, $X_1 = \{i\}$, $X_2 = K \setminus L$ of size 8.

The icosahedron graph is a Cayley graph over L; let X_3, and X_4 be two complementing (with respect to $K \setminus L$) connection sets of the icosahedron:
$X_3 = \{(1,3)(2,4),(1,4,3),(1,2,4),(1,4,2),(1,3,4)\}$,
$X_4 = \{(1,2)(3,4),(1,2,3),(1,3,2),(2,3,4),(2,4,3)\}$.
$X_5 = A_5 \setminus (K \cup L)$ is of size 40.

The imprimitivity systems arise from:
$K \cap L = X_0 \cup X_1$: This is the connection set of the graph $30 \circ K_2$.
$K = X_0 \cup X_1 \cup X_2$: $6 \circ K_{10}$.
$L = X_0 \cup X_1 \cup X_3 \cup X_4$: $5 \circ K_{12}$.

A significant feature of this description is that we use the Cayley representation of the icosahedron graph (over the subgroup A_4) twice. Conceptually, this description is close to the concept of orthogonal block structures ([1]), though it does differ slightly.

To the best of our knowledge, $\mathcal{A}_{115}$ is new, in the sense that the automorphism group and the non-Schurian property have not appeared in the literature (though this also seems to be the case for some of the other S-rings).

7   S-rings over $AGL(1, 8)$

Another challenging problem was the enumeration of S-rings over the affine group $H = AGL(1, 8)$ over the field $\mathbb{F}_8$ with 8 elements. Recall that H is a group of order 56, which has a decomposition $H \cong (\mathbb{Z}_2)^3 \rtimes \mathbb{Z}_7$, where $(\mathbb{Z}_2)^3$ stands for the additive group of $\mathbb{F}_8$ while $\mathbb{Z}_7$ is the multiplicative group of $\mathbb{F}_8$. This problem was solved in a similar manner to the A_5 case. The same computer took about two months.

There are 2349 S-rings, in 129 orbits under $Aut(H)$ (which is isomorphic to the group $A\Gamma L(1, 8)$ of order 168). Of the 2349 S-rings, 427 are non-Schurian (forming 20 of the 129 orbits).

Here the set $\mathcal{R}$ of roots consists of two groups $\mathcal{R}_1$ and $\mathcal{R}_2$ of orders 168 and 224, both acting transitively on H. There are 5 minimal S-rings, of ranks 32, 20, 20, 20, and 14. Eight of the non-Schurian S-rings appear as mergings of the transitivity module of the root $\mathcal{R}_1$. The group $\mathcal{R}_1 = A\Gamma L(1, 8)$ is 2-transitive of degree 8 and order 168, of the form $E_8 \rtimes F_{21}$, where $F_{21} = \mathbb{Z}_7 : \mathbb{Z}_3$ is the Frobenius group of order 21. The mergings of the root $\mathcal{R}_1$, together with other interesting structures, were carefully investigated jointly with Josef Lauri. Some of the results were recently published in [14]. The second root $\mathcal{R}_2$ has order 224 and rank 20; as an abstract group we have $\mathcal{R}_2 \cong E_4 \times AGL(1, 8)$.

The full report about the enumeration of S-rings over $AGL(1, 8)$ is in preparation and will be published elsewhere.

8   Concluding Remarks

In this text, a number of results in the area of AGT, which were mainly obtained with the aid of a computer, are presented in a concentrated form.

The use of the computer system GAP, in conjunction with some of its packages (in particular, GRAPE [30] and nauty [23]), dramatically extends the capacity for research. Indeed, without the use of modern tools of scientific computing, the outlined panorama of all S-rings over A_5 could never have been achieved.

Of course, a human being still has to use some theoretical knowledge to reach a clear understanding of the extensive computer data obtained. Some initial steps in this direction are also outlined. The results of our interpretation are presented in a form suitable for an expert in AGT, as we were not trying to prepare a fully self-contained text. Of course, the terminology and notation in use also depend on the existing customs in the relevant research area. (As a simple example, the same group of order 2 is denoted in this text by C_2, $\mathbb{Z}_2$ or S_2, depending on the context.)

There still remain a number of issues within AGT, which are related to our results and which should be carefully explained in the future, e.g., relative difference sets in A_5 and maps over linear fractional groups (see [10], [11] respectively).

In addition, we are pleased to mention the very famous structure of the fullerene C_{60}. In terms of S-rings, this is simply a coloured Cayley graph which corresponds to two basic sets of sizes 2 and 1 (for example, $\{(0,1,2,3,4), (1,4,3,2,1)\}$ and $\{(1,2)(3,4)\}$). Its automorphism group $\mathbb{Z}_2 \times A_5$ of order 120 and rank 32 generates S-ring #2. Recall that this structure is nowadays at the intersection of many paths of research, from crystallography to nanotechnology and virology. Paper [17] shows the potential of a better explanation of all S-rings related to the root $\mathcal{R}_1$, while such a text as [12] opens new horizons to exploit techniques of combinatorial enumeration for the classification of icosahedral virus capsids (see [27] for an introduction to this rapidly growing new area of modern theoretical biology).

We regret not being able to provide more details here, and hope in future to exploit all these attractive links between AGT and other branches of modern science.

Personal Remarks and Acknowledgements. Issai Schur, a scientific grandfather of author MK, was born in Russia, established his mathematical career in Germany, and in 1939 was forced to escape to Palestine. MK recalls a strong wave of inspiration which we felt, together with German students, at the time of a visit to Schur's grave in the old cemetery in Tel Aviv (1995).

We are very grateful to Christian Pech and Sven Reichard for sharing with us their beta versions of COCO-II. We thank Patrick Fowler and Misha Muzychuk for helpful discussions, as well as Josef Lauri for a fruitful collaboration.

The authors acknowledge constructive criticism and helpful suggestions provided by three anonymous referees. Finally, we thank Gareth Jones for his kind attention to our text.

References

1. Bailey, R.A.: Association schemes. Designed experiments, algebra and combinatorics. Cambridge Studies in Advanced Mathematics, vol. 84. Cambridge University Press, Cambridge (2004)
2. Bridges, W.G., Mena, R.A.: Rational circulants with rational spectra and cyclic strongly regular graphs. Ars Combin. 8, 143–161 (1979)
3. Brouwer, A.E., Cohen, A.M., Neumaier, A.: Distance Regular Graphs. Springer, Berlin (1989)
4. Conway, J.H., Hulpke, A., McKay, J.: On transitive permutation groups. LMS J. Comput. Math. 1, 1–8 (1998) (electronic)
5. Faradžev, I.A., Klin, M.H.: Computer package for computations with coherent configurations. In: Proc. ISSAC 1991, pp. 219–223. ACM Press, Bonn (1991)
6. Faradžev, I.A., Klin, M.H., Muzichuk, M.E.: Cellular rings and groups of automorphisms of graphs. In: Faradžev, I.A., et al. (eds.) Investigations in Algebraic Theory of Combinatorial Objects, pp. 1–152. Kluwer Acad. Publ., Dordrecht (1994)
7. http://www.gap-system.org
8. Godsil, C.D., Hensel, A.D.: Distance regular covers of the complete graph. J. Combin. Theory Ser. B 56(2), 205–238 (1992)

9. http://math.shinshu-u.ac.jp/~hanaki/as/
10. Hiramine, Y.: Relative difference sets in Alt(5). J. Combin. Theory Ser. A 110(2), 179–191 (2005)
11. Jones, G.A., Mačaj, M., Širáň, J.: Nonorientable regular maps over linear fractional groups. Ars Math. Contemp. 6, 25–35 (2013)
12. Kerner, R.: Classification and evolutionary trends of icosahedral viral capsids. Comput. Math. Methods Med. 9(3-4), 175–181 (2008)
13. Klin, M., Kovács, I.: Automorphism groups of rational circulant graphs. Electron. J. Combin. 19(1), Paper 35, 1–52 (2012)
14. Klin, M., Lauri, J., Ziv-Av, M.: Links between two semisymmetric graphs on 112 vertices via association schemes. J. Symbolic Comput. 47, 1175–1191 (2012)
15. Klin, M., Pöschel, R., Rosenbaum, K.: Angewandte Algebra für Mathematiker und Informatiker. Einfuhrung in gruppentheoretisch-kombinatorische Methoden. Friedr. Vieweg & Sohn, Braunschweig (1988)
16. Klin, M., Pech, C., Reichard, S., Woldar, A., Ziv-Av, M.: Examples of computer experimentation in algebraic combinatorics. Ars Math. Contemp. 3, 237–258 (2010)
17. Kostant, B.: Structure of the truncated icosahedron (e.g. Fullerene or C_{60}, viral coatings) and a 60-element conjugacy class in PSL(2, 11). Selecta Math (N.S.) 1, 163–195 (1995)
18. Leung, K.H., Man, S.H.: On Schur rings over cyclic groups. Israel J. Math. 106, 251–267 (1998)
19. Leung, K.H., Man, S.H.: On Schur rings over cyclic groups II. J. Algebra 183, 273–285 (1996)
20. Levy, S. (ed.): The Eightfold Way. The Beauty of Klein's Quartic Curve. Mathematical Sciences Research Institute Publications, vol. 35. Cambridge University Press, Cambridge (1999)
21. Mathon, R.: 3-class association schemes. In: Proc. Conference on Algebraic Aspects of Combinatorics, pp. 123–155. Univ. Toronto, Toronto (1975)
22. Mathon, R.: Lower bounds for Ramsey numbers and association schemes. J. Combin. Theory Ser. B 42, 122–127 (1987)
23. McKay, B.D.: Nauty User's Guide (Version 1.5). Technical Report TR-CS-90-02, Computer Science Department, Australian National University (1990)
24. Muzychuk, M.E.: Structure of primitive S-rings over group A_5. In: VIII All-Union Symposium on Group Theory, pp. 83–84. Kiev (1982)
25. Muzychuk, M.: A solution of the isomorphism problem for circulant graphs. Proc. London Math. 88, 1–41 (2004)
26. Muzychuk, M.: Private communication (December 2012)
27. Peeters, K., Taormina, A.: Group theory of icosahedral virus capsid vibrations: A top-down approach. J. Theoret. Biol. 256, 607–624 (2009)
28. Schönert, M., et al.: GAP - Groups, Algorithms, and Programming, 5th edn. Lehrstuhl D für Mathematik. Rheinisch-Westfälische Technische Hochschule, Aachen (1995)
29. Schur, I.: Zur Theorie der einfach transitiven Permutationsgruppen. Sitzungsber. Preuss. Akad. Wiss., Phys.-Math. Kl., 598–623 (1933)
30. Soicher, L.H.: GRAPE: a system for computing with graphs and groups. In: Finkelstein, L., Kantor, W.M. (eds.) Groups and Computation. DIMACS Series in Discrete Mathematics and Theoretical Computer Science, vol. 11, pp. 287–291. A.M.S (1993)
31. Wielandt, H.: Finite Permutation Groups. Acad. Press, New York (1964)

Generalized Bruhat Decomposition
in Commutative Domains

Gennadi Malaschonok*

Tambov State University,
Internatsionalnaya 33, 392622 Tambov, Russia
malaschonok@gmail.com

Abstract. Deterministic recursive algorithms for the computation of generalized Bruhat decomposition of the matrix in commutative domain are presented. This method has the same complexity as the algorithm of matrix multiplication.

1 Introduction

A matrix decomposition of a form $A = VwU$ is called the Bruhat decomposition of the matrix A, if V and U are nonsingular upper triangular matrices and w is a matrix of permutation. It is usually assumed that the matrix A is defined in a certain field. Bruhat decomposition plays an important role in the theory of algebraic groups. The generalized Bruhat decomposition was introduced and developed by D.Grigoriev[1],[2].

In [3] there was constructed a pivot-free matrix decomposition method in a common case of singular matrices over a field of arbitrary characteristic. This algorithm has the same complexity as matrix multiplication and does not require pivoting. For singular matrices it allows us to obtain a nonsingular block of the biggest size.

Let R be a commutative domain, F be the field of fractions over R. We want to obtain a decomposition of matrix A over domain R in the form $A = VwU$, where V and U are upper triangular matrices over R, and w is a matrix of permutation, which is multiplied by some diagonal matrix in the field of fractions F. Moreover each nonzero element of w has the form $(a^i a^{i-1})^{-1}$, where a^i is some minor of order i of matrix A.

We call such triangular decomposition the Bruhat decomposition in the commutative domain R.

In [6], a fast algorithm for adjoint matrix computation was proposed. On the basis of this algorithm for computing the adjoint matrix, a fast algorithm was proposed in [8] for LDU decomposition. However, this algorithm requires the calculation of the adjoint matrix to calculate the LDU decomposition.

In this paper, we propose another algorithm that does not rely on the calculation of the adjoint matrix and which costs less number of operations. We

---

* Supported by the Russian Foundation for Basic Research, grant No. 12-07-00755a.

V.P. Gerdt et al. (Eds.): CASC 2013, LNCS 8136, pp. 231–242, 2013.
© Springer International Publishing Switzerland 2013

construct the decomposition in the form $A = LDU$, where L and U are lower and upper triangular matrices, D is a matrix of permutation, which is multiplied by some diagonal matrix in the field of fractions F and has the same rank as the matrix A. Then the Bruhat decomposition VwU in the domain R may easily be obtained using the matrices L, D and U.

2   Triangular Decomposition in Domain

Let R be a commutative domain, $A = (a_{i,j}) \in R^{n \times n}$ be a matrix of order n, $\alpha_{i,j}^k$ be $k \times k$ minor of matrix A which disposed in the rows $1, 2, \ldots, k - 1, i$ and columns $1, 2, \ldots, k - 1, j$ for all integers $i, j, k \in \{1, \ldots, n\}$. We suppose that the row i of the matrix A is situated at the last row of the minor, and the column j of the matrix A is situated at the last column of the minor. We denote $\alpha^0 = 1$ and $\alpha^k = \alpha_{k,k}^k$ for all diagonal minors ($1 \le k \le n$). And we use the notation δ_{ij} for Kronecker delta.

Let k and s be integers in the interval $0 \le k < s \le n$, $\mathcal{A}_s^k = (\alpha_{i,j}^{k+1})$ be the matrix of minors with size $(s - k) \times (s - k)$ which has elements $\alpha_{i,j}^{k+1}$, $i, j = k + 1, \ldots, s - 1, s$, and $\mathcal{A}_n^0 = (\alpha_{i,j}^1) = A$.

We shall use the following identity (see [4], [5]):

Theorem 1 (Sylvester determinant identity).
Let k and s be the integers in the interval $0 \le k < s \le n$. Then it is true that

$$\det(\mathcal{A}_s^k) = \alpha^s (\alpha^k)^{s-k-1}. \tag{1}$$

Theorem 2 (LDU decomposition of the minors matrix).
Let $A = (a_{i,j}) \in R^{n \times n}$ be the matrix of rank r, $\alpha^i \ne 0$ for $i = k, k + 1, \ldots, r$, $r \le s \le n$, then the matrix of minors $\mathcal{A}_s^k$ is equal to the following product of three matrices:

$$\mathcal{A}_s^k = L_s^k D_s^k U_s^k = (a_{i,j}^j)(\delta_{ij}\alpha^k(\alpha^{i-1}\alpha^i)^{-1})(a_{i,j}^i). \tag{2}$$

The matrix $L_s^k = (a_{i,j}^j)$, $i = k + 1 \ldots s$, $j = k + 1 \ldots r$, is a low triangular matrix of size $(s - k) \times (r - k)$, the matrix $U_s^k = (a_{i,j}^i)$, $i = k + 1 \ldots r$, $j = k + 1 \ldots s$, is an upper triangular matrix of size $(r - k) \times (s - k)$ and $D_s^k = (\delta_{ij}\alpha^k(\alpha^{i-1}\alpha^i)^{-1})$, $i = k + 1 \ldots r$, $j = k + 1 \ldots r$, is a diagonal matrix of size $(r - k) \times (r - k)$.

Proof. Let us write the matrix equation (2) for $k + 1 = r$

$$(a_{i,j}^{k+1}) = (a_{i,k+1}^{k+1})(\delta_{k+1,k+1}a^k(a^k a^{k+1})^{-1})(a_{k+1,j}^{k+1}) \tag{3}$$

This equation is correct due to Sylvester determinant identity

$$a_{i,j}^{k+1}a^{k+1} - a_{i,k+1}^{k+1}a_{k+1,j}^{k+1} = a_{i,j}^{k+2}a^k, \tag{4}$$

and the equality $a_{i,j}^{k+2} = 0$. This equality is a consequence of the fact that minors $a_{i,j}^{k+2}$ have the order greater then the rank of the matrix A.

Let for all h, $k < h < r$, the statement (1) be correct for matrices $\mathcal{A}_s^h = (a_{i,j}^{h+1})$. We have to prove it for $h = k$. Let us write one matrix element in (2) for the matrix $\mathcal{A}_s^{k+1} = (a_{i,j}^{k+2})$:

$$a_{i,j}^{k+2} = \sum_{t=k+2}^{min(i,j,r)} a_{i,t}^t \alpha^{k+1} (\alpha^{t-1}\alpha^t)^{-1} a_{t,j}^t.$$

We have to prove the corresponding expression for the elements of the matrix $\mathcal{A}_s^k$. Due to the Sylvester determinant identity (3) we obtain

$$a_{i,j}^{k+1} = a_{i,k+1}^{k+1}(\alpha^{k+1})^{-1}a_{k+1,j}^{k+1} + a^k(\alpha^{k+1})^{-1}a_{i,j}^{k+2} =$$

$$a_{i,k+1}^{k+1}\alpha^k(\alpha^k\alpha^{k+1})^{-1}a_{k+1,j}^{k+1} + a^k(\alpha^{k+1})^{-1}\sum_{t=k+2}^{min(i,j,r)} a_{i,t}^t \alpha^{k+1}(\alpha^{t-1}\alpha^t)^{-1}a_{t,j}^t =$$

$$\sum_{t=k+1}^{min(i,j)} a_{i,t}^t \alpha^k(\alpha^{t-1}\alpha^t)^{-1}a_{t,j}^t.$$

Consequence 1 (LDU decomposition of matrix A). *Let $A = (a_{i,j}) \in R^{n \times n}$, be the matrix of rank r, $r \leq n$, $\alpha^i \neq 0$ for $i = 1, 2, \ldots, r$, then matrix A is equal to the following product of three matrices:*

$$A = L_n^0 D_n^0 U_n^0 = (a_{i,j}^j)(\delta_{ij}(\alpha^{i-1}\alpha^i)^{-1})(a_{i,j}^i). \tag{4}$$

The matrix $L_n^0 = (a_{i,j}^j)$, $i = 1 \ldots n$, $j = 1 \ldots r$, is a low triangular matrix of size $n \times r$, the matrix $U_n^0 = (a_{i,j}^i)$, $i = 1 \ldots r$, $j = 1 \ldots n$, is an upper triangular matrix of size $r \times n$, and $D_n^0 = (\delta_{ij}(\alpha^{i-1}\alpha^i)^{-1})$, $i = 1 \ldots r$, $j = 1 \ldots r$, is a diagonal matrix of size $r \times r$.

Let I_n be the identity matrix and P_n be the matrix with second unit diagonal.

Consequence 2 (Bruhat decomposition of matrix A). *Let matrix $A = (a_{i,j})$ have the rank r, $r \leq n$, and $B = P_n A$. Let $B = LDU$ be the LDU-decomposition of matrix B. Then $V = P_n L P_r$ and U are upper triangular matrices of size $n \times r$ and $r \times n$ correspondingly and*

$$A = V(P_r D)U \tag{5}$$

is the Bruhat decomposition of matrix A.

We are interested in the block form of decomposition algorithms for LDU and Bruhat decompositions. Let us use some block matrix notations.

For any matrix A (or A_q^p) we denote by $A_{j_1,j_2}^{i_1,i_2}$ (or $A_{q;j_1,j_2}^{p;i_1,i_2}$) the block which stands at the intersection of rows $i_1 + 1, \ldots, i_2$ and columns $j_1 + 1, \ldots, j_2$ of the matrix. We denote by $A_{i_2}^{i_1}$ the diagonal block $A_{i_1,i_2}^{i_1,i_2}$.

3 LDU Algorithm

Input: $(\mathcal{A}_n^k, \alpha^k)$, $0 \le k < n$.
Output: $\{L_n^k, \{\alpha^{k+1}, \alpha^{k+2}, \dots, \alpha^n\}, U_n^k, M_n^k, W_n^k\}$,
where $\quad D_n^k = \alpha^k \mathrm{diag}\{\alpha^k \alpha^{k+1}, \dots, \alpha^{n-1} \alpha^n\}^{-1}$, $M_n^k = \alpha^k (L_n^k D_n^k)^{-1}$,
$\quad W_n^k = \alpha^k (D_n^k U_n^k)^{-1}$.

1. If $k = n - 1$, $\mathcal{A}_n^{n-1} = (a^n)$ is a matrix of the first order, then we obtain

$$\{a^n, \{a^n\}, a^n, a^{n-1}, a^{n-1}\}, \quad D_n^{n-1} = (\alpha^n)^{-1}.$$

2. If $k = n - 2$, $\mathcal{A}_n^{n-2} = \begin{pmatrix} \alpha^{n-1} & \beta \\ \gamma & \delta \end{pmatrix}$ is a matrix of second order, then we obtain

$$\left\{ \begin{pmatrix} \alpha^{n-1} & 0 \\ \gamma & \alpha^n \end{pmatrix}, \{\alpha^{n-1}, \alpha^n\}, \begin{pmatrix} \alpha^{n-1} & \beta \\ 0 & \alpha^n \end{pmatrix}, \begin{pmatrix} \alpha^{n-2} & 0 \\ -\gamma & \alpha^{n-1} \end{pmatrix}, \begin{pmatrix} \alpha^{n-2} & -\beta \\ 0 & \alpha^{n-1} \end{pmatrix} \right\}$$

where $\alpha^n = (\alpha^{n-2})^{-1} \begin{vmatrix} \alpha^{n-1} & \beta \\ \gamma & \delta \end{vmatrix}$, $D_n^{n-2} = \alpha^{n-2} \mathrm{diag}\{\alpha^{n-2}\alpha^{n-1}, \alpha^{n-1}\alpha^n\}^{-1}$.

3. If the order of the matrix $\mathcal{A}_n^k$ is more than two ($0 \le k < n - 2$), then we choose an integer s in the interval ($k < s < n$) and divide the matrix into blocks

$$\mathcal{A}_n^k = \begin{pmatrix} \mathcal{A}_s^k & \mathbf{B} \\ \mathbf{C} & \mathbf{D} \end{pmatrix}. \tag{6}$$

3.1. Recursive step

$$\{L_s^k, \{\alpha^{k+1}, \alpha^{k+2}, \dots, \alpha^s\}, U_s^k, M_s^k, W_s^k\} = \mathbf{LDU}(\mathcal{A}_s^k, \alpha^k)$$

3.2. We compute

$$\widetilde{U} = (\alpha^k)^{-1} M_s^k \mathbf{B}, \quad \widetilde{L} = (\alpha^k)^{-1} \mathbf{C} W_s^k, \tag{7}$$

$$\mathcal{A}_n^s = (\alpha^k)^{-1} \alpha^s (\mathbf{D} - \widetilde{L} D_s^k \widetilde{U}). \tag{8}$$

3.3. Recursive step

$$\{L_n^s, \{\alpha^{s+1}, \alpha^{s+2}, \dots, \alpha^n\}, U_n^s, M_n^s, W_n^s\} = \mathbf{LDU}(\mathcal{A}_n^s, \alpha^s)$$

3.4 *Result:*

$$\{L_n^k, \{\alpha^{k+1}, \alpha^{k+2}, \dots, \alpha^n\}, U_n^k, M_n^k, W_n^k\},$$

where

$$L_n^k = \begin{pmatrix} L_s^k & 0 \\ \widetilde{L} & L_n^s \end{pmatrix}, \quad U_n^k = \begin{pmatrix} U_s^k & \widetilde{U} \\ 0 & U_n^s \end{pmatrix}, \tag{9}$$

$$M_n^k = \begin{pmatrix} M_s^k & 0 \\ -M_n^s \widetilde{L} D_s^k M_s^k / \alpha^k & M_n^s \end{pmatrix}, \tag{10}$$

$$W_n^k = \begin{pmatrix} W_s^k & -W_s^k D_s^k \widetilde{U} W_n^s / \alpha^k \\ 0 & W_n^s \end{pmatrix}. \tag{11}$$

4   Proof of the Correctness of the LDU Algorithm

Proof of the correctness of this algorithm is based on several determinant identities.

Definition 1 ($\delta_{i,j}^k$ minors and $\mathcal{G}^k$ matrices).
    Let $A \in R^{n \times n}$ be a matrix. The determinant of the matrix obtained from the upper left block $A_{0,k}^{0,k}$ of matrix A by the replacement in matrix A of the column i by the column j is denoted by $\delta_{i,j}^k$. The matrix of such minors is denoted by

$$\mathcal{G}_s^k = (\delta_{i,j}^{k+1}) \tag{12}$$

We need the following theorem (see [4] and [5]):

Theorem 3 (Base minor's identity).
    Let $A \in R^{n \times n}$ be a matrix and i, j, s, k, be integers in the intervals: $0 \le k < s \le n$, $0 < i, j \le n$. Then the following identity is true

$$a^s \alpha_{ij}^{k+1} - \alpha^k a_{ij}^{s+1} = \sum_{p=k+1}^s \alpha_{ip}^{k+1} \delta_{pj}^s. \tag{13}$$

The minors a_{ij}^{s+1} in the left-hand side of this identity equal zero if $i < s + 1$. Therefore, this theorem gives the following

Consequence 3. Let $A \in R^{n \times n}$ be a matrix and s, k be integers in the intervals: $0 \le k < s \le n$. Then the following identities are true

$$\alpha^s U_{n;s+1,n}^{k;k+1,s} = U_s^k \mathcal{G}_{n;s+1,n}^{k;k+1,s}. \tag{14}$$

$$\alpha^s \mathcal{A}_{n;s+1,n}^{k;k+1,s} = A_s^k \mathcal{G}_{n;s+1,n}^{k;k+1,s}. \tag{15}$$

    The block $\mathcal{A}_{n;s+1,n}^{k;k+1,s}$ of the matrix $\mathcal{A}_n^k$ was denoted by **B**. Due to Sylvester identity we can write the equation for the adjoint matrix

$$(A_s^k)^* = (A_s^k)^{-1} (\alpha^s)(\alpha^k)^{s-k-1} \tag{16}$$

Let us multiply both sides of equation (15) by adjoint matrix $(A_s^k)^*$ and use equation (16). Then we get

Consequence 4

$$(A_s^k)^* \mathbf{B} = (A_s^k)^* \mathcal{A}_{n;s+1,n}^{k;k+1,s} = (\alpha^k)^{s-k-1} \mathcal{G}_{n;s+1,n}^{k;k+1,s}. \tag{17}$$

As well as $L_s^k D_s^k U_s^k = A_s^k$,

$$M_s^k = \alpha^k (L_s^k D_s^k)^{-1} = \alpha^k U_s^k (A_s^k)^{-1} \text{ and } W_s^k = \alpha^k (D_s^k U_s^k)^{-1}. \tag{18}$$

Therefore,

$$\widetilde{U} = (\alpha^k)^{-1} M_s^k \mathbf{B} = (\alpha^k)^{-1} U_s^k (A_s^k)^{-1} \mathbf{B} = (\alpha^s)^{-1}(\alpha^k)^{-s+k} U_s^k (A_s^k)^* \mathbf{B}. \tag{19}$$

Equations (19), (17), and (14) give

Consequence 5

$$\widetilde{U} = U^{k;k+1,s}_{n;s+1,n} \tag{20}$$

In the same way we can prove

Consequence 6

$$\widetilde{L} = L^{k;s+1,n}_{n;k+1,s}. \tag{21}$$

Now we have to prove identity (8). Due to equations (14)-(19) we obtain

$$\widetilde{L}D^k_s\widetilde{U} = (\alpha^k)^{-1}\mathbf{C}W^k_s D^k_s(\alpha^k)^{-1}M^k_s\mathbf{B} =$$

$$(\alpha^k)^{-2}\mathbf{C}(A^k_s)^{-1}\mathbf{B} = (\alpha^k)^{-s+k-1}(\alpha^s)^{-1}\mathbf{C}(A^k_s)^*\mathbf{B} \tag{22}$$

The identity

$$\mathcal{A}^s_n = (\alpha^k)^{-1}(\alpha^s\mathbf{D} - (\alpha^k)^{-s+k+1}\mathbf{C}(A^k_s)^*\mathbf{B}) \tag{23}$$

was proved in [4] and [5]. Due to (20) and (21) we obtain identity (8).

To prove formulas (10) and (11) it is sufficient to verify the identities $M^k_n = \alpha^k(L^k_n D^k_n)^{-1}$ and $W^k_n = \alpha^k(D^k_n U^k_n)^{-1}$ using (9),(10), (11) and definition $D^k_n = \alpha^k\mathrm{diag}\{\alpha^k\alpha^{k+1},\ldots,\alpha^{n-1}\alpha^n\}^{-1}$.

5   Complexity

Theorem 4. *The algorithm has the same complexity as matrix multiplication.*

Proof. The total amount of matrix multiplications in (7)–(15) is equal to 7, and the total amount of recursive calls is equal to 2. We do not consider multiplications of the diagonal matrices.

We can compute the decomposition of the second order matrix by means of 7 multiplicative operations. Therefore, we get the following recurrent equality for complexity

$$t(n) = 2t(n/2) + 7M(n/2), \quad t(2) = 7.$$

Let γ and β be constants, $3 \geq \beta > 2$, and let $M(n) = \gamma n^\beta + o(n^\beta)$ be the number of multiplication operations in one $n \times n$ matrix multiplication.

After summation from $n = 2^k$ to 2^1 we obtain

$$7\gamma(2^0 2^{\beta\cdot(k-1)} + \ldots + 2^{k-2}2^{\beta\cdot 1}) + 2^{k-2}7 = 7\gamma\frac{n^\beta - n2^{\beta-1}}{2^\beta - 2} + \frac{7}{4}n.$$

Therefore, the complexity of the decomposition is

$$\sim \frac{7\gamma n^\beta}{2^\beta - 2}$$

6   The Exact Triangular Decomposition

Definition 2. *A decomposition of the matrix A of rank r over a commutative domain R in the product of five matrices*

$$A = PLDUQ \qquad (24)$$

is called exact triangular decomposition *if P and Q are permutation matrces, L and PLP^T are nonsingular lower triangular matrices, U and $Q^T UQ$ are nonsingular upper triangular matrices over R, $D = \operatorname{diag}(d_1^{-1}, d_2^{-1}, .., d_r^{-1}, 0, .., 0)$ is a diagonal matrix of rank r, $d_i \in R \backslash \{0\}$, $i = 1, .. r$.*

Designation: $\mathbf{ETD}(A) = (P, L, D, U, Q)$.

Theorem 5 (Main theorem). *Any matrix over a commutative domain has an exact triangular decomposition.*

Before proceeding to the proof, we note that the exact triangular decomposition relates the LU decomposition and the Bruhat decomposition in the field of fractions.

If D matrix is combined with L or U, we get the expression $A = PLUQ$. This is the LU-decomposition with permutations of rows and columns. If the factors are grouped in the following way:

$$A = (PLP^T)(PDQ)(Q^T UQ),$$

then we obtain **LDU**-decomposition. If S is a permutation matrix in which the unit elements are placed on the secondary diagonal, then $(SLS)(S^T\mathbf{D})\mathbf{U}$ is the Bruhat decomposition of the matrix (SA).

Bruhat decomposition can be obtained from those $PLUQ$-decomposition that satisfy the additional conditions: matrices PLP^T and $Q^T UQ$ are triangular. Conversely, LU-decomposition can be obtained from the Bruhat decomposition $V'D'U'$. This can be done if the permutation matrix D can be decomposed into a product of permutation matrices $D' = PQ$ so that the $P^T L'P$ and $QU'Q^T$ are triangular matrices.

If matrix A is a zero matrix, then $\mathbf{ETD}(A) = (I, I, 0, I, I)$.

If A is a nonzero matrix of the first order, then $\mathbf{ETD}(A) = (I, a, a^{-1}, a, I)$.

Let us consider a non-zero matrix of order two. We denote

$$\mathcal{A} = \begin{pmatrix} \alpha & \beta \\ \gamma & \delta \end{pmatrix}, \quad \Delta = \begin{vmatrix} \alpha & \beta \\ \gamma & \delta \end{vmatrix}, \quad \varepsilon = \begin{cases} \Delta, & \Delta \neq 0 \\ 1, & \Delta = 0 \end{cases}, \quad \Delta^{-1} = \begin{cases} 1/\Delta, & \Delta \neq 0 \\ 0, & \Delta = 0. \end{cases}$$

Depending on the location of zero elements, we consider four possible cases. For each case, we give the exact triangular decomposition:

If $\alpha \neq 0$, then $\mathcal{A} = \begin{pmatrix} \alpha & 0 \\ \gamma & \varepsilon \end{pmatrix} \begin{pmatrix} \alpha^{-1} & 0 \\ 0 & \Delta^{-1}\alpha^{-1} \end{pmatrix} \begin{pmatrix} \alpha & \beta \\ 0 & \varepsilon \end{pmatrix}.$

If $\alpha = 0$, $\beta \neq 0$, then $\mathcal{A} = \begin{pmatrix} \beta & 0 \\ \delta & \varepsilon \end{pmatrix} \begin{pmatrix} \beta^{-1} & 0 \\ 0 & -\Delta^{-1}\beta^{-1} \end{pmatrix} \begin{pmatrix} \beta & 0 \\ 0 & \varepsilon \end{pmatrix} \begin{pmatrix} 0 & 1 \\ 1 & 0 \end{pmatrix}.$

If $\alpha = 0,\ \gamma \neq 0,$ then $\mathcal{A} = \begin{pmatrix} 0 & 1 \\ 1 & 0 \end{pmatrix} \begin{pmatrix} \gamma & 0 \\ 0 & \varepsilon \end{pmatrix} \begin{pmatrix} \gamma^{-1} & 0 \\ 0 & -\Delta^{-1}\gamma^{-1} \end{pmatrix} \begin{pmatrix} \gamma & \delta \\ 0 & \varepsilon \end{pmatrix}.$

If $\alpha = \beta = \gamma = 0, \delta \neq 0,$ then $\mathcal{A} = \begin{pmatrix} 0 & 1 \\ 1 & 0 \end{pmatrix} \begin{pmatrix} \delta & 0 \\ 0 & 1 \end{pmatrix} \begin{pmatrix} \delta^{-1} & 0 \\ 0 & 0 \end{pmatrix} \begin{pmatrix} \delta & 0 \\ 0 & 1 \end{pmatrix} \begin{pmatrix} 0 & 1 \\ 1 & 0 \end{pmatrix}.$

There are only two different cases for matrices of size 1×2:

If $\alpha \neq 0,$ then $(\alpha\ \beta) = (\alpha)(\alpha^{-1}\ 0) \begin{pmatrix} \alpha & \beta \\ 0 & 1 \end{pmatrix}.$

If $\alpha = 0,\ \beta \neq 0,$ then $(0\ \beta) = (\beta)(\beta^{-1}\ 0) \begin{pmatrix} \beta & 0 \\ 0 & 1 \end{pmatrix} \begin{pmatrix} 0 & 1 \\ 1 & 0 \end{pmatrix}.$

Two cases for matrices of size 2×1 can easily be obtained by a simple transposition.

These examples allow us to formulate

Sentence 1. *For all matrices $\mathcal{A}$ of size $n \times m$, $n, m < 3$ there exists an exact triangular decomposition.*

In addition, we can formulate the following property, which holds for triangular matrices and permutation matrices in the exact triangular decomposition.

We denote by I_s the identity matrix of order s.

Property 1 (Property of the factors). For a matrix $A \in R^{n \times m}$ of rank r, $r < n, r < m$ over a commutative domain R there exists the exact triangular decomposition (24) in which

(α) the matrices L and U are of the form

$$L = \begin{pmatrix} L_1 & 0 \\ L_2 & I_{n-r} \end{pmatrix} U = \begin{pmatrix} U_1 & U_2 \\ 0 & I_{m-r} \end{pmatrix}, \tag{25}$$

(β) the matrices PLP^T and Q^TUQ remain triangular after replacing in the matrices L and Q of unit blocks I_{n-r} and I_{m-r} by arbitrary triangular blocks.

Without loss of generality of the main theorem, we shall prove it for the exact triangular decompositions with property 1. We prove it by induction. The theorem is true for matrices of sizes smaller than three.

We consider a matrix $\mathcal{A}$ of size $N \times M$. Assume that all matrices of size less than $n \times m$ have the exact triangular decomposition. We split the matrix $\mathcal{A}$ into blocks: $\mathcal{A} = \begin{pmatrix} \mathbf{A} & \mathbf{B} \\ \mathbf{C} & \mathbf{D} \end{pmatrix}$, where $\mathbf{A} \in R^{n \times n}$, $n < N$, $n < M$.

(1). Let the block $\mathbf{A}$ have the full rank. There exists exact triangular decomposition of this block: $\mathbf{A} = P_1 L_1 D_1 U_1 Q_1$. Here the diagonal matrix D_1 has full rank, and the matrix $\mathcal{A}$ is decomposed into the factors:

$$\begin{pmatrix} P_1 & 0 \\ 0 & I \end{pmatrix} \begin{pmatrix} L_1 & 0 \\ \mathbf{C}Q_1^T U_1^{-1} D_1^{-1} & I \end{pmatrix} \begin{pmatrix} D_1 & 0 \\ 0 & \mathbf{D}^* \end{pmatrix} \begin{pmatrix} U_1 & D_1^{-1}L_1^{-1}P_1^T\mathbf{B} \\ 0 & I \end{pmatrix} \begin{pmatrix} Q_1 & 0 \\ 0 & I \end{pmatrix}.$$

Here $\mathbf{D}^* = \mathbf{D} - \mathbf{C}Q^T U^{-1} D_1^{-1} L^{-1} P^T \mathbf{B}$. The matrix $\mathbf{D}^*$ also has the exact triangular decomposition $\mathbf{D}^* = P_2 L_2 D_2 U_2 Q_2$. Substituting it in this decomposition, we obtain a new decomposition of the matrix $\mathcal{A}$:

$$\begin{pmatrix} P_1 & 0 \\ 0 & P_2 \end{pmatrix} \begin{pmatrix} L_1 & 0 \\ P_2^T C Q_1^T U_1^{-1} D_1^{-1} L_2 \end{pmatrix} \begin{pmatrix} D_1 & 0 \\ 0 & D_2 \end{pmatrix} \begin{pmatrix} U_1 & D_1^{-1} L_1^{-1} P_1^T B Q_2^T \\ 0 & U_2 \end{pmatrix} \begin{pmatrix} Q_1 & 0 \\ 0 & Q_2 \end{pmatrix}.$$

It is easy to see that this decomposition is exact triangular if both block decompositions were exact triangular.

(2) Let the block $\mathbf{A}$ has rank r, $r < n$. There exists exact triangular decomposition of this block:

$$\mathbf{A} = P_1 L_1 D_1 U_1 Q_1.$$

Here $U_1 = \begin{pmatrix} U_0 & V_0 \\ 0 & I \end{pmatrix}$, $L_1 = \begin{pmatrix} L_0 & 0 \\ M_0 & I \end{pmatrix}$ and the diagonal matrix $D_1 = \begin{pmatrix} d_1 & 0 \\ 0 & 0 \end{pmatrix}$ has a block d_1 of rank r.

Let us denote $\mathbf{C} = (\mathbf{C}', \mathbf{C}'')$ and $(\mathbf{C}_0, \mathbf{C}_1) = (\mathbf{C}' U_0, \mathbf{C}' V_0 + \mathbf{C}'') Q_1$, $\mathbf{B} = \begin{pmatrix} \mathbf{B}' \\ \mathbf{B}'' \end{pmatrix}$, $\begin{pmatrix} \mathbf{B}_0 \\ \mathbf{B}_1 \end{pmatrix} = P_1 \begin{pmatrix} L_0 \mathbf{B}' \\ M_0 \mathbf{B}' + \mathbf{B}'' \end{pmatrix}$. Then for the matrix $\mathcal{A}$ we obtain the decomposition:

$$\mathcal{A} = \begin{pmatrix} P_1 & 0 \\ 0 & I \end{pmatrix} \begin{pmatrix} L_0 & 0 & 0 \\ M_0 & I & 0 \\ C_0 d_1^{-1} & 0 & I \end{pmatrix} \begin{pmatrix} d_1 & 0 & 0 \\ 0 & 0 & \mathbf{B}_1 \\ 0 & \mathbf{C}_1 & \mathbf{D} \end{pmatrix} \begin{pmatrix} U_0 & V_0 & d_1^{-1} \mathbf{B}_0 \\ 0 & I & 0 \\ 0 & 0 & I \end{pmatrix} \begin{pmatrix} Q_1 & 0 \\ 0 & I \end{pmatrix}. \quad (26)$$

(2.1) Let $\mathbf{B}_1 = 0$ and $\mathbf{C}_1 = 0$. We can rearrange the block $\mathbf{D}$ in the upper left corner

$$\begin{pmatrix} 0 & \mathbf{B}_1 \\ \mathbf{C}_1 & \mathbf{D} \end{pmatrix} = \begin{pmatrix} 0 & I \\ I & 0 \end{pmatrix} \begin{pmatrix} \mathbf{D} & 0 \\ 0 & 0 \end{pmatrix} \begin{pmatrix} 0 & I \\ I & 0 \end{pmatrix}.$$

Let us find the exact triangular decomposition of $\mathbf{D}$:

$$\mathbf{D} = P_2 L_2 D_2 U_2 Q_2.$$

We denote

$$\mathbf{P}_3 = \begin{pmatrix} P_1 & 0 \\ 0 & P_2 \end{pmatrix} \begin{pmatrix} I & 0 & 0 \\ 0 & 0 & I \\ 0 & I & 0 \end{pmatrix}, \quad \mathbf{Q}_3 = \begin{pmatrix} I & 0 & 0 \\ 0 & 0 & I \\ 0 & I & 0 \end{pmatrix} \begin{pmatrix} Q_1 & 0 \\ 0 & Q_2 \end{pmatrix}.$$

Then for the matrix $\mathcal{A}$ we obtain the following decomposition:

$$\mathcal{A} = \mathbf{P}_3 \begin{pmatrix} L_0 & 0 & 0 \\ P_2^T C_0 d_1^{-1} & L_2 & 0 \\ M_0 & 0 & I \end{pmatrix} \begin{pmatrix} d_1 & 0 & 0 \\ 0 & D_2 & 0 \\ 0 & 0 & 0 \end{pmatrix} \begin{pmatrix} U_0 & d_1^{-1} \mathbf{B}_0 Q_2^T & V_0 \\ 0 & U_2 & 0 \\ 0 & 0 & I \end{pmatrix} \mathbf{Q}_3.$$

It is easy to check that the decomposition is exact triangular.

(2.2) Suppose that at least one of the two blocks of $\mathbf{B}_1$ or $\mathbf{C}_1$ is not zero. Let the exact triangular decomposition exist for these blocks:

$$\mathbf{C} = P_2 L_2 D_2 U_2 Q_2, \quad \mathbf{B} = P_3 L_3 D_3 U_3 Q_3.$$

We denote

$$\mathbf{P}_1 = \begin{pmatrix} P_1 & 0 \\ 0 & I \end{pmatrix}, \mathbf{P}_2 = \begin{pmatrix} I & 0 & 0 \\ 0 & P_3 & 0 \\ 0 & 0 & P_2 \end{pmatrix}, \quad \mathbf{Q}_2 = \begin{pmatrix} I & 0 & 0 \\ 0 & Q_2 & 0 \\ 0 & 0 & Q_3 \end{pmatrix}, \quad \mathbf{Q}_1 = \begin{pmatrix} Q_1 & 0 \\ 0 & I \end{pmatrix},$$

$\mathbf{P}_3 = \mathbf{P}_1 \mathbf{P}_2$, $\mathbf{Q}_3 = \mathbf{Q}_2 \mathbf{Q}_1$, $\mathbf{D}' = P_2^T L_2^{-1} \mathbf{D} U_3^{-1} Q_3^T$.

Then, basing on expansion (26) we obtain for the matrix $\mathcal{A}$ the decomposition of the form:

$$\mathcal{A} = \mathbf{P}_3 \begin{pmatrix} L_0 & 0 & 0 \\ P_3^T M_0 & L_3 & 0 \\ P_2^T \mathbf{C}_0 d_1^{-1} & 0 & L_2 \end{pmatrix} \begin{pmatrix} d_1 & 0 & 0 \\ 0 & 0 & D_3 \\ 0 & D_2 & \mathbf{D}' \end{pmatrix} \begin{pmatrix} U_0 & V_0 Q_2^T & d_1^{-1} \mathbf{B}_0 Q_3^T \\ 0 & U_2 & 0 \\ 0 & 0 & U_3 \end{pmatrix} \mathbf{Q}_3. \quad (27)$$

We denote d_2 and d_3 nondegenerate blocks of the matrices D_2 and D_3, respectively,

$$(V_1, V_4) = V_0 Q_2^T, (V_5, V_6) = d_1^{-1} \mathbf{B}_0 Q_3^T, \begin{pmatrix} M_1 \\ M_4 \end{pmatrix} = P_3^T M_0, \begin{pmatrix} M_5 \\ M_6 \end{pmatrix} = P_2^T \mathbf{C}_0 d_1^{-1}$$

$$L_2 = \begin{pmatrix} L_2' & 0 \\ M_2 & I \end{pmatrix}, L_3 = \begin{pmatrix} L_3' & 0 \\ M_3 & I \end{pmatrix}, U_2 = \begin{pmatrix} U_2' & V_2 \\ 0 & I \end{pmatrix}, U_3 = \begin{pmatrix} U_3' & V_3 \\ 0 & I \end{pmatrix}, \mathbf{D}' = \begin{pmatrix} \mathbf{D}_1' & \mathbf{D}_3' \\ \mathbf{D}_2' & \mathbf{D}_4' \end{pmatrix}.$$

$$M_7 = \mathbf{D}_2' d_3^{-1}, \quad V_7 = d_2^{-1} \mathbf{D}_1' U_3', \quad V_8 = d_2^{-1}(\mathbf{D}_1' V_3 + \mathbf{D}_3').$$

Then (27) can be written as

$$\mathcal{A} = \mathbf{P}_3 \begin{pmatrix} L_0 & 0 & 0 & 0 & 0 \\ M_1 & L_3' & 0 & 0 & 0 \\ M_4 & M_3 & I & 0 & 0 \\ M_5 & 0 & 0 & L_2' & 0 \\ M_6 & 0 & 0 & M_2 & I \end{pmatrix} \begin{pmatrix} d_1 & 0 & 0 & 0 & 0 \\ 0 & 0 & 0 & d_3 & 0 \\ 0 & 0 & 0 & 0 & 0 \\ 0 & d_2 & 0 & \mathbf{D}_1' & \mathbf{D}_3' \\ 0 & 0 & 0 & \mathbf{D}_2' & \mathbf{D}_4' \end{pmatrix} \begin{pmatrix} U_0 & V_1 & V_4 & V_5 & V_6 \\ 0 & U_2' & V_2 & 0 & 0 \\ 0 & 0 & I & 0 & 0 \\ 0 & 0 & 0 & U_3' & V_3 \\ 0 & 0 & 0 & 0 & I \end{pmatrix} \mathbf{Q}_3 =$$

$$\mathbf{P}_3 \begin{pmatrix} L_0 & 0 & 0 & 0 & 0 \\ M_1 & L_3' & 0 & 0 & 0 \\ M_4 & M_3 & I & 0 & 0 \\ M_5 & 0 & 0 & L_2' & 0 \\ M_6 & M_7 & 0 & M_2 & I \end{pmatrix} \begin{pmatrix} d_1 & 0 & 0 & 0 & 0 \\ 0 & 0 & 0 & d_3 & 0 \\ 0 & 0 & 0 & 0 & 0 \\ 0 & d_2 & 0 & 0 & 0 \\ 0 & 0 & 0 & 0 & \mathbf{D}_4' \end{pmatrix} \begin{pmatrix} U_0 & V_1 & V_4 & V_5 & V_6 \\ 0 & U_2' & V_2 & V_7 & V_8 \\ 0 & 0 & I & 0 & 0 \\ 0 & 0 & 0 & U_3' & V_3 \\ 0 & 0 & 0 & 0 & I \end{pmatrix} \mathbf{Q}_3. \quad (28)$$

Find the exact triangular decomposition $\mathbf{D}_4'$:

$$\mathbf{D}_4' = P_4 L_4 D_4 U_4 Q_4, \quad (29)$$

Let us denote the matrices $\mathbf{P}_4 = \text{diag}(I, I, I, I, P_4)$, $\mathbf{Q}_4 = \text{diag}(I, I, I, I, Q_4)$, $\mathbf{P}_5 = \mathbf{P}_3 \mathbf{P}_4$, $\mathbf{Q}_5 = \mathbf{Q}_4 \mathbf{Q}_3$, $(M_6', M_7', M_2') = P_4^T(M_6, M_7, M_2)$ $(V_6', V_8', V_3') = (V_6, V_8, V_3)Q_4^T$.

After substituting (29) into (28) we obtain the decomposition of the matrix $\mathcal{A}$ as

$$\mathcal{A} = \mathbf{P}_5 \begin{pmatrix} L_0 & 0 & 0 & 0 & 0 \\ M_1 & L_3' & 0 & 0 & 0 \\ M_4 & M_3 & I & 0 & 0 \\ M_5 & 0 & 0 & L_2' & 0 \\ M_6' & M_7' & 0 & M_2' & L_4 \end{pmatrix} \begin{pmatrix} d_1 & 0 & 0 & 0 & 0 \\ 0 & 0 & 0 & d_3 & 0 \\ 0 & 0 & 0 & 0 & 0 \\ 0 & d_2 & 0 & 0 & 0 \\ 0 & 0 & 0 & 0 & D_4 \end{pmatrix} \begin{pmatrix} U_0 & V_1 & V_4 & V_5 & V_6' \\ 0 & U_2' & V_2 & V_7 & V_8' \\ 0 & 0 & I & 0 & 0 \\ 0 & 0 & 0 & U_3' & V_3' \\ 0 & 0 & 0 & 0 & U_4 \end{pmatrix} \mathbf{Q}_5. \quad (30)$$

We rearrange the blocks d_2, d_3, and D_4 to obtain the diagonal matrix $\mathbf{d} = \mathrm{diag}(d_1, d_3, d_2, D_4, 0)$. To do it we use permutation matrices P_6 and Q_6:

$$P_6 = \begin{pmatrix} 1\,0\,0\,0\,0 \\ 0\,0\,0\,1\,0 \\ 0\,1\,0\,0\,0 \\ 0\,0\,0\,0\,1 \\ 0\,0\,1\,0\,0 \end{pmatrix}, Q_6 = \begin{pmatrix} 1\,0\,0\,0\,0 \\ 0\,1\,0\,0\,0 \\ 0\,0\,0\,0\,1 \\ 0\,0\,1\,0\,0 \\ 0\,0\,0\,1\,0 \end{pmatrix}, P_6 \begin{pmatrix} d_1\,0\,0\,0\,\,0 \\ 0\,\,0\,0\,d_3\,0 \\ 0\,\,0\,0\,0\,\,0 \\ 0\,d_2\,0\,0\,\,0 \\ 0\,\,0\,0\,0\,D_4 \end{pmatrix} Q_6 = \mathbf{d}.$$

As a result, we obtain the decomposition:

$$\mathcal{A} = \mathbf{P}_6 \mathbf{L}\mathbf{d}\mathbf{U}\mathbf{Q}_6, \qquad (31)$$

with permutation matrices $\mathbf{P}_6 = \mathbf{P}_5 P_6^T$ and $\mathbf{Q}_6 = Q_6^T \mathbf{Q}_5$, diagonal matrix $\mathbf{d}$ and triangular matrices

$$\mathbf{L} = P_6 \begin{pmatrix} L_0 & 0 & 0 & 0 & 0 \\ M_1 & L_3' & 0 & 0 & 0 \\ M_4 & M_3 & I & 0 & 0 \\ M_5 & 0 & 0 & L_2' & 0 \\ M_6' & M_7' & 0 & M_2' & L_4 \end{pmatrix} P_6^T = \begin{pmatrix} L_0 & 0 & 0 & 0 & 0 \\ M_5 & L_2' & 0 & 0 & 0 \\ M_1 & 0 & L_3' & 0 & 0 \\ M_6' & M_7' & M_2' & L_4 & 0 \\ M_4 & 0 & M_3 & 0 & I \end{pmatrix}$$

$$\mathbf{U} = Q_6^T \begin{pmatrix} U_0 & V_1 & V_4 & V_5 & V_6' \\ 0 & U_2' & V_2 & V_7 & V_8' \\ 0 & 0 & I & 0 & 0 \\ 0 & 0 & 0 & U_3' & V_3' \\ 0 & 0 & 0 & 0 & U_4 \end{pmatrix} Q_6 = \begin{pmatrix} U_0 & V_1 & V_5 & V_6' & V_4 \\ 0 & U_2' & V_7 & V_8' & V_2 \\ 0 & 0 & U_3' & V_3' & 0 \\ 0 & 0 & 0 & U_4 & 0 \\ 0 & 0 & 0 & 0 & I \end{pmatrix}$$

We show that expansion (31) is an exact triangular decomposition. To do this, we must verify that the matrices $\mathcal{L} = \mathbf{P}_6 \mathbf{L} \mathbf{P}_6^T$ and $\mathcal{Q} = \mathbf{Q}_6^T \mathbf{U}\mathbf{Q}_6$ are triangular, and the matrices $\mathbf{P}, \mathbf{L}, \mathbf{U}, \mathbf{Q}$ satisfy the properties (α) and (β).

It is easy to see that all matrices in sequence

$$\mathcal{L}_1 = P_6 \mathbf{L} P_6^T, \mathcal{L}_2 = \mathbf{P}_4 \mathcal{L}_1 \mathbf{P}_4^T, \mathcal{L}_3 = \mathbf{P}_2 \mathcal{L}_2 \mathbf{P}_2^T, \mathcal{L}_4 = \mathbf{P}_1 \mathcal{L}_3 \mathbf{P}_1^T \qquad (32)$$

are triangular and $\mathcal{L}_4 = \mathcal{L}$.

Similarly, all of the matrices in the sequence

$$\mathcal{U}_1 = Q_6^T \mathbf{L}\mathbf{Q}_6, \mathcal{U}_2 = \mathbf{Q}_4^T \mathcal{U}_1 \mathbf{Q}_4, \mathcal{U}_3 = \mathbf{Q}_2^T \mathcal{U}_2 \mathbf{Q}_2, \mathcal{U}_4 = \mathbf{Q}_1^T \mathcal{U}_3 \mathbf{Q}_1 \qquad (33)$$

are triangular and $\mathcal{U}_4 = \mathcal{U}$.

For the matrices $\mathbf{L}$ and $\mathbf{U}$ Property 1 (α) is satisfied. To verify the properties (β), the unit block in the lower right corner of the matrix $\mathbf{L}$ and $\mathbf{U}$ should be replaced by an arbitrary triangular block, respectively, the lower triangle for $\mathbf{L}$ and the upper triangular for $\mathbf{U}$. We check that all the matrices in (32) and (33) will be still triangular. This is based on the fact that the exact triangular decompositions for matrices $\mathbf{A}, \mathbf{B}, \mathbf{C}, \mathbf{D}'$ have the property (β).

7 Conclusion

Algorithms for finding the LDU and Bruhat decomposition in commutative domain are described. These algorithms have the same complexity as matrix multiplication.

8 Example

$$\begin{bmatrix} 1 & -4 & 0 & 1 \\ 4 & 5 & 5 & 3 \\ 1 & 2 & 2 & 2 \\ 3 & 0 & 0 & 1 \end{bmatrix} = \begin{bmatrix} -24 & 0 & 12 & 1 \\ 0 & 60 & 15 & 4 \\ 0 & 0 & 6 & 1 \\ 0 & 0 & 0 & 3 \end{bmatrix} \begin{bmatrix} 0 & 0 & 1/(-144) & 0 \\ 0 & 0 & 0 & 1/(-1440) \\ 0 & 1/18 & 0 & 0 \\ 1/3 & 0 & 0 & 0 \end{bmatrix} \begin{bmatrix} 3 & 0 & 0 & 1 \\ 0 & 6 & 6 & 5 \\ 0 & 0 & -24 & -16 \\ 0 & 0 & 0 & 60 \end{bmatrix}$$

References

1. Grigoriev, D.: Analogy of Bruhat decomposition for the closure of a cone of Chevalley group of a classical series. Soviet Math. Dokl. 23, 393–397 (1981)
2. Grigoriev, D.: Additive complexity in directed computations. Theoretical Computer Science 19, 39–67 (1982)
3. Malaschonok, G.: Fast Generalized Bruhat Decomposition. In: Gerdt, V.P., Koepf, W., Mayr, E.W., Vorozhtsov, E.V. (eds.) CASC 2010. LNCS, vol. 6244, pp. 194–202. Springer, Heidelberg (2010)
4. Malaschonok, G.I.: Matrix Computational Methods in Commutative Rings. Tambov University Publishing House, Tambov (2002)
5. Malaschonok, G.I.: Effective matrix methods in commutative domains. In: Krob, D., Mikhalev, A.A., Mikhalev, A.V. (eds.) Formal Power Series and Algebraic Combinatorics, pp. 506–517. Springer, Berlin (2000)
6. Malaschonok, G.I.: A fast algorithm for adjoint matrix computation. Tambov University Reports 5(1), 142–146 (2000)
7. Malaschonok, G.I.: Fast matrix decomposition in parallel computer algebra. Tambov University Reports 15(4), 1372–1385 (2010)
8. Malaschonok, G.I.: On the fast generalized Bruhat decomposition in domains. Tambov University Reports 17(2), 544–550 (2012)

Automatic Parallel Library Generation for General-Size Modular FFT Algorithms

Lingchuan Meng and Jeremy Johnson

Drexel University, Philadelphia PA 19104, USA
{lm433,jjohnson}@cs.drexel.edu

Abstract. This paper presents the automatic library generation for modular FFT algorithms with arbitrary input sizes. We show how to represent the transform and its algorithms at a high abstraction level. Symbolic manipulations and code optimizations that use rewriting systems can then be systematically applied to generate a library with recursive function closure. The generated library is automatically optimized for the target computing platforms, and is intended to support modular algorithms for multivariate polynomial computations in the modpn library used by MAPLE. The resulting scalar and vector codes provide comparable speedup to the fixed-size code presented in [LJF10], which is an order of magnitude faster over the hand-tuned modpn library. Thread-level parallelism has also been utilized by the generated library and delivers additional speedup.

Keywords: FFT, modular arithmetic, library generation, parallelization, autotuning.

1   Introduction

Fast Fourier Transforms (FFTs) are at the core of many operations in scientific computing. In computer algebra, FFTs are used for fast polynomial and integer arithmetic and modular methods (i.e. computation by homomorphic images). In recent years, the use of fast arithmetic has become prevalent and has stimulated the development of software libraries, such as modpn [FLMS06, LM06, LMP09] providing hand-optimized low-level routines implementing fast algorithms for multivariate polynomial computations over finite fields, in support of higher-level code. The modpn library has been integrated into the computer algebra system MAPLE and runs on all computer platforms supported by MAPLE. The implementation techniques employed in modpn are often platform-dependent, since cache size, associativity properties and register sets have a significant impact on performance. In order to take advantage of platform-dependent optimizations, in the context of quickly evolving hardware acceleration technologies, automated performance tuning that supports general input sizes has become necessary and should be incorporated into the modpn library.

SPIRAL [www.spiral.net] is a library generation system that automatically generates platform-tuned implementations of digital signal processing algorithms

V.P. Gerdt et al. (Eds.): CASC 2013, LNCS 8136, pp. 243–256, 2013.
© Springer International Publishing Switzerland 2013

with an emphasis on fast transforms. Currently, SPIRAL can generate highly optimized fixed-point and floating-point FFTs for a variety of platforms with automatic tuning, and has support for vectorization, threading, and shared memory parallelization. The code produced is competitive with the best available code for these platforms and SPIRAL is used by Intel for its IPP (Integrated Performance Primitives) and MKL (Math Kernel Library) libraries.

In this paper, we extend our previous work [LJF10] that requires all parameters of the modular FFT to be known at generation time to the automatic library generation for modular FFT that accepts general sizes. By incorporating and extending the new library generation mechanism in SPIRAL, we have generated general size library with comparable speedup to the fixed-size code, and have exploited additional thread level parallelization for further performance improvement. A new transform definition and associated breakdown rules and parameterization have been implemented in the library generation framework in order to generate recursive function closure. The backend has also been extended to enable the generation of high performance scalar and vectorized code for modular arithmetic. With these enhancements, we present our results showing that the parallel library generated by SPIRAL is ten to twenty times faster than the modular FFT implementation in `modpn`.

2   Background

The SPIRAL system [PMJ05] uses a mathematical framework for representing and deriving algorithms. Fig.1 shows the classic program generation framework in SPIRAL.

- **Internal languages.** SPIRAL uses internally a domain-specific language called SPL with which the algorithms are expressed symbolically as sparse matrix factorizations and are derived using rewrite rules. Although substructures produced in the factorizations are well-suited for good locality, the conquer step in these algorithms is iterative, meaning that extra passes through data are required. As a result, it is important to *merge* the iterative steps to improve data locality and reuse. Σ-SPL [FVP05], an extension to SPL, is developed to support formal loop merging by making explicit the description of loops, index mappings, and most recently parametrization and recursive calls.
- **Rewriting systems.** Algorithms in Σ-SPL representation are in turn processed by rewriting systems to perform difficult optimizations such as loop merging, vectorization [FVP062], and parallelization [FVP061] automatically. The optimizations are performed at a high abstraction level in order to overcome known compiler limitations. The sequence of applications of breakdown rules is encoded as a *ruletree* which can be translated into a formula and compiled with a special-purpose compiler into efficient code [XJJP01].
- **Search engine.** A search engine using the dynamic programing technique with a feedback loop is used to tune fixed-size implementations to particular platforms.

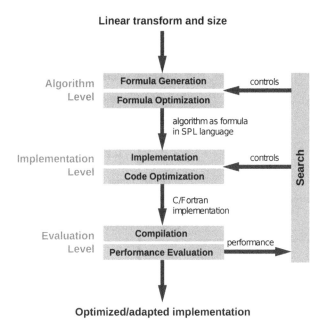

Fig. 1. Fixed-size program generation in SPIRAL

The fixed-size program generation framework was used and extended in our previous work to generate fixed-size modular FFT algorithms. However, the framework is limited by the type of code it could produce, in that transform sizes and other parameters must be known at generation time. SPIRAL has recently been extended to support general size library generation [V08].

Library Generation Extension. To generate a library for a given transform, the system has to be able to generate code for transforms of *symbolic size*, i.e., the size becomes an additional parameter. While the fixed-size transforms can be decomposed by the classic generation framework, the breakdown rules can only be applied at *runtime*, since the applicabilities of different rules are dependent on the transform size. This change leads to a new framework shown in Fig.2, with the addition of two new modules, Library Structure and Library Implementation.

- **Library Structure.** The generation of library structure operates at the algorithm level, and is independent of the library implementation. The library structure module compiles the breakdown rules of the transform into a *recursion step closure*, which corresponds to a set of recursive functions in the resulting library.
- **Library Implementation.** The goal of the library implementation module is to translate or compile the recursive step closure expressed in Σ-SPL into a program in some lower level target language. The particular set of requirements, such as adaptive or fixed, floating-point or integer, for the generated library together with target language are encapsulated in the *library target*.

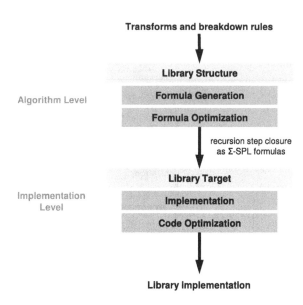

Fig. 2. Library generation in SPIRAL

The module does not require any transform specific code, which makes reusing library targets easy for new transforms.

The new framework separates the transform algorithm generation from the library implementation details in order to maximize reusability. For instance, a new transform and its algorithms can be mapped to a closure, and the *unmodified* library implementation module can immediately generate several types of libraries.

The library generation procedure is completely deterministic, and does not involve feedback-driven search at *library generation time* except for small fixed-size base cases. Instead, it moves the platform adaptation to *runtime*, by generating libraries with feedback-driven adaptation mechanisms, similar to FFTW [FJ05].

In both frameworks, new transforms can be added by introducing new symbols and their definitions, and new algorithms can be generated by adding new rules. SPIRAL was developed for floating point and fixed point computation; however, many of the transforms and algorithms carry over to finite fields. For example, the DFT of size n is defined when there is a primitive nth root of unity and many factorizations of the DFT matrix depend only on properties of primitive nth roots. In this case, the same machinery in SPIRAL can be used for generating and optimizing modular transforms.

3   Modular FFT

Let the n-point modular DFT matrix be

$$\text{ModDFT}_{n,p,\omega} = \left[\omega_n^{k\ell}\right]_{0 \le k, \ell < n}, \tag{1}$$

where ω_n is a primitive nth root of unity in $\mathbb{Z}_p$.

Let $n = rs$, then the divide and conquer step in the Cooley-Tukey algorithm [CT65] can be represented as a matrix factorization with the *tensor product*:

$$\text{ModDFT}_{n,p,\omega} = (\text{ModDFT}_{r,p,\omega_r} \otimes I_s) \, T_s^n (I_r \otimes \text{ModDFT}_{s,p,\omega_s}) \, L_r^n. \tag{2}$$

In (2), T_s^n is a diagonal matrix containing *twiddle factors*. I_s is the $s \times s$ identity matrix. Furthermore, the *stride permutation* matrix L_r^n permutes the input vector as

$$is + j \mapsto jr + i, 0 \le i < r, 0 \le j < s.$$

If the input vector is viewed as a row-major 2-D matrix of size $s \times r$, then L_r^n transposes the matrix. For instance, L_2^6 can be viewed as matrix vector computation:

$$L_2^6 x = \begin{bmatrix} 1\,0\,0\,0\,0\,0 \\ 0\,0\,1\,0\,0\,0 \\ 0\,0\,0\,0\,1\,0 \\ 0\,1\,0\,0\,0\,0 \\ 0\,0\,0\,1\,0\,0 \\ 0\,0\,0\,0\,0\,1 \end{bmatrix} \cdot \begin{bmatrix} x_0 \\ x_1 \\ x_2 \\ x_3 \\ x_4 \\ x_5 \end{bmatrix} = \begin{bmatrix} x_0 \\ x_2 \\ x_4 \\ x_1 \\ x_3 \\ x_5 \end{bmatrix} \tag{3}$$

The *tensor product*, the most important operator in this paper, is defined as

$$A \otimes B = [a_{k,l}B] = \begin{bmatrix} a_{0,0}B & \cdots & a_{0,l-1}B \\ \vdots & \ddots & \vdots \\ a_{k-1,0}B & \cdots & a_{k-1,l-1}B \end{bmatrix} \tag{4}$$

The tensor product serves as the key construct in SPIRAL and its many fast algorithms, in that it captures loops, data independence, vectorization and parallelism concisely.

Example 1. An example of the Cooley-Tukey algorithm where $r = s = 2$ is

$$\text{ModDFT}_{4,p,\omega} = (\text{ModDFT}_{2,p,\omega_2} \otimes I_2) \cdot T_2^4 \cdot (I_2 \otimes \text{ModDFT}_{2,p,\omega_2}) \cdot L_2^4,$$

which is equivalent to the matrix factorization below:

$$\begin{bmatrix} 1 & 1 & 1 & 1 \\ 1 & \omega_4 & \omega_4^2 & \omega_4^3 \\ 1 & \omega_4^2 & 1 & \omega_4^2 \\ 1 & \omega_4^3 & \omega_4^2 & \omega_4 \end{bmatrix} = \begin{bmatrix} 1\,0 & 1 & 0 \\ 0\,1 & 0 & 1 \\ 1\,0 & -1 & 0 \\ 0\,1 & 0 & -1 \end{bmatrix} \cdot \begin{bmatrix} 1\,0\,0\,0 \\ 0\,1\,0\,0 \\ 0\,0\,1\,0 \\ 0\,0\,0\,\omega_4 \end{bmatrix} \cdot \begin{bmatrix} 1 & 1 & 0 & 0 \\ 1 & -1 & 0 & 0 \\ 0 & 0 & 1 & 1 \\ 0 & 0 & 1 & -1 \end{bmatrix} \cdot \begin{bmatrix} 1\,0\,0\,0 \\ 0\,0\,1\,0 \\ 0\,1\,0\,0 \\ 0\,0\,0\,1 \end{bmatrix}$$

Mathematical operations such as the tensor product in the Cooley-Tukey algorithm produces substructures that can be interpreted as vector and parallel operations, and formulae can be transformed to adapt to a given vector length and number of cores, and permutations can be manipulated to obtain desired data access patterns [JJRT90]. Such techniques have been incorporated into SPIRAL to obtain highly optimized vector and parallel code [FVP05, FVP061, FVP062].

Fig. 3 shows an example of the substructure. The four sparse matrices from applying Cooley-Tukey algorithm to a 16-point modular DFT are represented by four frames, where only the non-zero values are plotted. In the first frame of $\text{ModDFT}_4 \otimes I_4$, for example, scalar operations can be replaced by 4-way vector operations, as each value in ModDFT_4 is duplicated four times along the block diagonals. Further, in the third frame of $I_4 \otimes \text{ModDFT}_4$, it is obvious that the same computing kernel ModDFT_4 is applied to four non-overlapping sub-vectors of the permuted input vector, which can be interpreted as a parallelizable loop with no loop-carried dependencies.

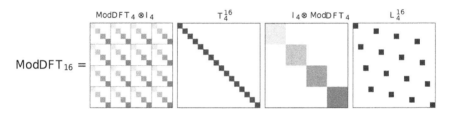

Fig. 3. Representation of the matrix factorization based on the Cooley-Tukey algorithm. Shades of gray represent values that belong to the same tensor substructure.

In fact, the general forms of the first and the third frames, $A \otimes I$ and $I \otimes A$ where A is an arbitrary matrix, are the two typical substructures for automatic vectorization and parallelization. In Sect. 4, we will discuss how SPIRAL uses *hardware tags* and *rewriting rules* to fully exploit vector level parallelism and thread level parallelism and how the library structure is generated as recursion step closure.

Modular arithmetic needs to be implemented efficiently in SPIRAL for a high performance modular FFT library. In our implementation, the Montgomery reduction proposed in [M85] is used. The algorithm effectively replaces integer division, which can be expensive on many platforms, with bitwise operations.

For arithmetic mod N, the algorithm selects a radix R coprime to N, usually the word size of the target platform, such that $R > N$ and computations modulo R are inexpensive to process. Let R^{-1} and N' be integers satisfying $0 < R^{-1} < N$ and $0 < N' < R$ and $RR^{-1} - NN' = 1$. For $i \in \mathbb{Z}_N$, let i represent the residue class containing $iR^{-1} \bmod N$. With this residue system, we can quickly compute $TR^{-1} \bmod N$ from T if $0 \leq T < RN$, as shown in the algorithm below:

With the release of the Streaming SIMD Extensions (SSE) 4.2 by Intel, the Montgomery reduction can be fully vectorized in SPIRAL by defining a library

Algorithm 1. Montgomery Reduction

---

function REDC(T)

    $m \leftarrow (T \bmod R)N' \bmod R$ $\triangleright\ m \equiv TN' \bmod R$

    $t \leftarrow (T + mN)/R$ $\triangleright\ mN \equiv TN'N \equiv -T \bmod R \Rightarrow t \equiv TR^{-1} \bmod \text{N}$

    if $t \geq N$ **then** $\triangleright\ 0 \leq T + mn < RN + RN \Rightarrow 0 \leq t < 2N$

        return $t - N$

    else

        return t

    end if

end function

---

target for modular arithmetic, adding rewriting rules based on pattern matching, and using the newly introduced integer vector instructions in the code generator.

Since the Cooley-Tukey factorization holds for any field with a primitive nth root of unity, the same machinery in SPIRAL can be used for generating and optimizing modular FFTs, provided the infrastructure is extended to support new data types and the code generation and compiler optimizations are similarly enhanced.

4   General Size Parallel Library Generation for Modular FFT

As previously mentioned, the substructures from applying recursive algorithms to transforms can be mapped to vector and parallel operations, and formulae can be transformed to adapt to hardware parameters, such as vector length and number of computing cores. In this section, we show how the rewriting systems and hardware tags are used to fully exploit vector parallelism and thread parallelism for modular FFT library. Then we show how the library structure is generated by constructing the recursion step closure as the algorithms are applied to the given transform.

4.1   Vectorization and Parallelization

Rewriting systems and *hardware tags* have been developed in SPIRAL to fully exploit two levels of parallelism: vector parallelism and thread parallelism. To generate vectorized and parallelized implementations, one first needs to obtain "fully optimized" SPL formulae, which means they can be mapped to efficient vector and parallel code. The mapping utilizes the natural interpretation of the tensor product, as shown in Fig.3, as vector and parallel operations.

Vectorization. The rewriting system for vector code uses the following four components:

– *Vectorization tags* which mark the formula as "to be vectorized", and introduce the vector length.

- *Vector formula constructs* which express subformulae that can be fully mapped to SIMD code.
- *Rewriting rules* which transform SPL formulae into vector formulae.
- *Vector backend* that generates SIMD code from vector formulae.

During the rewriting, vectorization information is propagated through the formulae via a set of *vectorization tags*. For example, $Vec_\nu(A)$ means the formula construct A is to be translated into vector code with vector length ν.

The central formula construct in *vector formula constructs* is $A \otimes I_\nu$, where A is an arbitrary matrix, and I is an *identity matrix* of size ν. Vector code is obtained on ν-way short vector extensions by generating scalar code for A and replacing scalar operations by respective ν-way vector operations. $A \overline{\otimes} I_\nu$ stipulates that the tensor product is to be mapped to vector code without any further manipulations. In addition, a set of efficient base cases for vectors of size ν are provided.

$$L_2^{2\nu}, L_\nu^{2\nu}, L_\nu^{\nu^2}, (I_{n/\nu} \otimes L_2^{2\nu}) D_n (I_{n/\nu} \otimes L_2^{2\nu}),$$

where D_n is any diagonal matrix. Both constructs marked with $\overline{\otimes}$ and base cases are *final*, i.e., they will not be changed by rewriting rules.

Rewriting rules are used for non-final constructs in the framework to generate vectorized form, and are extensively covered in [FVP062]. For example, $Vec_\nu(I_m \otimes A)$ can be rewritten into $I_{m/\nu} \otimes Vec_\nu(I_\nu \otimes A)$ based on the rewriting rules. SPIRAL derived rules for permutations and tensor products. These rules are applied with pattern matching to rewrite an SPL formula into a final vectorized formula. Similar rewriting mechanism can also be applied to Σ-SPL vectorization.

Example 2. The tensor product $I_r \overline{\otimes} \mathrm{ModDFT}_s$ produced in the Cooley-Tukey factorization of ModDFT_{rs} can rewritten for vectorization as follows

$$I_r \overline{\otimes} \mathrm{ModDFT}_s = I_{r/\nu} \otimes I_\nu \overline{\otimes} \mathrm{ModDFT}_s \tag{5}$$

$$= I_{r/\nu} \otimes (L_\nu^{s\nu}(\mathrm{ModDFT}_s \otimes I_\nu) L_s^{s\nu}) \tag{6}$$

Note that in this formula $\mathrm{ModDFT}_s \otimes I_\nu$ is already vectorized. Then the stride permutation can also be rewritten into the fully vectorizable form as follows:

$$I_{r/\nu} \otimes ((L_\nu^s \otimes I_\nu)(I_{s/\nu} \otimes L_\nu^{\nu^2})(\mathrm{ModDFT}_s \otimes I_\nu)(I_{s/\nu} \otimes L_\nu^{\nu^2})(L_{s/\nu}^s \otimes I_\nu)). \tag{7}$$

Parallelization. The same approach is followed by *shared memory parallelization* (smp) as with vectorization, starting with the formula tags and tagged SPL constructs, and finally obtaining parallel code. During parallelization, there are several issues one must address:

- *Load balancing*: All processors should have an equal workload.
- *Synchronization overhead*: Synchronization should involve as little overhead as possible and unnecessary wait should be eliminated at synchronization points.

- *Avoiding false sharing*: Private data of different processors should not be in the same cache line at the same time, otherwise cache thrashing may occur, which leads to severe performance degradation.

To generate parallel code by rewriting SPL formulae, we first notice that SPL constructs have a direct interpretation in terms of parallel code. A SPL formula fully determines the memory access pattern of the generated program, and thus one can statically schedule the loop iterations across p processors through rewriting to ensure load balancing and eliminate false sharing. The components required in the parallelization extension to SPIRAL include:

- *Parallelism tags* which introduces hardware parameters into the rewriting system.
- *Parallel formula constructs* which denote the subformulae that can be perfectly mapped to *smp* platforms.
- *Rewriting rules* which transform general formulae into parallel formulae.
- *Parallel backend* than maps parallel formulae into parallel executable code.

Parallelism tags introduces the important parameters of *smp* machines: the number of processors p, and the cache line length μ, e.g. $\mathbf{Par}_{p,\mu}(A)$.

Parallel formula constructs. For arbitrary A and A_i, the expressions

$$y = (I_p \otimes A)x, y = \left(\bigoplus_{i=0}^{p-1} A_i\right)x,$$

are embarrassingly parallel on p processors as they express block diagonal matrices with p blocks. If the matrix dimensions are a multiple of μ, then each cache line is owned by exactly one processor, thus preventing false sharing. If all A_i have the same computational cost, the resulting program is load balanced. The tagged operators declare that a construct is fully optimized for *smp* machines and does not require further manipulation.

The general formulae are transformed into fully optimized formulae with *rewriting rules*. Example of rules include:

- $\mathbf{Par}_{p,\mu}(AB) \to \mathbf{Par}_{p,\mu}(A)\mathbf{Par}_{p,\mu}(A)$, which expresses that in matrix multiplication each factor will be rewritten separately, and
- $\mathbf{Par}_{p,\mu}(A_m \otimes I_n) \to \mathbf{Par}_{p,\mu}((L_m^{mp} \otimes I_{n/p})(I_p \otimes (A_m \otimes I_{n/p}))(L_p^{mp} \otimes I_{n/p}))$, which handles tensor product with identity matrices. It distributes the workload evenly among the p processors and execute as many consecutive iterations as possible on the same processor.

4.2   Library Generation

To understand how a general size library is generated, recall the Cooley-Tukey factorization of ModDFT_{rs}, where the modulus p and the root of unity ω are omitted for simplicity.

$$\mathrm{ModDFT}_{rs} = (\mathrm{ModDFT}_r \otimes \mathrm{I}_s)\, \mathrm{T}_s^{rs}(\mathrm{I}_r \otimes \mathrm{ModDFT}_s)\, \mathrm{L}_r^{rs}\,.$$

With Σ-SPL, the twiddle factor T_s^{rs} and the permutation matrix L_r^{rs} can be fused into the nearby tensor products. As a result, the Cooley-Tukey FFT is computed in two loops corresponding to the two tensor products decorated with the twiddles and the stride permutation.

$$\mathrm{ModDFT}_{rs} = \underbrace{(\mathrm{ModDFT}_r \otimes \mathrm{I}_s)\, T_s^{rs}}_{loop}\, \underbrace{(\mathrm{I}_r \otimes \mathrm{ModDFT}_s)\, L_r^{rs}}_{loop}.$$

The fusion of the stride permutation requires a ModDFT function (*moddft_str*) with different input and output stride, and the fusion of the twiddles requires a ModDFT function (*moddft_scaled*) with an extra array parameter holding the scaling factors. These functions are implemented recursively based on the Cooley-Tukey algorithm, and do not require any new functions. Eventually, the modular DFT size becomes sufficiently small, and the recursion terminates by using codelet functions that compute small fixed-size transforms.

We say the functions *moddft*, *moddft_str* and *moddft_scaled* form a so-called *recursion step closure*, which is a minimal set of functions sufficient to compute the desired transform. The recursion step closure is the central concept in the library generation framework. The framework can compute the recursion step closure for a given transform and multiple algorithms via parametrization and descending, and generate corresponding base cases. Extensive details and examples can be found in [V08].

With the vectorization and parallelization techniques, the code generator works like a human expert for modular FFT in both algorithms and code tuning to fully exploit vector and thread level parallelism. Furthermore, the library generation techniques enable the autonomous exploration of algorithm and implementation space at runtime with arbitrary input parameters. In Sect. 5, we present the performance data that shows an order-of-magnitude speedup over the hand-tuned implementation as a result of utilizing the aforementioned techniques.

5   Performance Results

This section reports the performance data comparing the modular FFT in the hand-optimized **modpn** library with the fixed-size code and the general size library automatically generated by SPIRAL.

Test Setup. All experiments were performed on an Intel Core i7 965 quad-core processor running at 3.2 GHz with 12 GB of RAM. Generated code was compiled with gcc version 4.3.4-1 with optimization set to O3. Vector code used SSE 4.2 with 4-way 32 bit integer vectors. Initial experiments were performed using 32 bit integers and 16 bit primes. Performance is reported in Gops (giga-ops) or billions of operations per second (higher is better), which is calculated assuming that modular DFT of size N takes a total of $(3/2)N \lg(N)$ additions, subtractions and nontrivial multiplications.

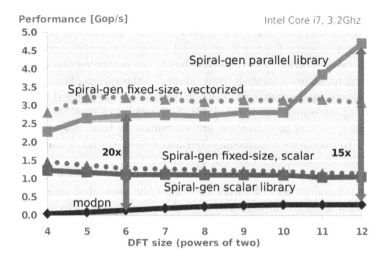

Fig. 4. Performance Comparison

Baseline Implementation. The `modpn` library is mainly written in C, which supports general multivariate polynomial computations over finite fields based on fast algorithms, and targets efficient implementation of modular algorithms. `modpn` takes advantage of certain hardware features such as SSE instructions and thus performs better on some platforms than others. Montgomery's trick is used to avoid integer division, which can be costly on many machines. In this experiment, `modpn` does not utilize the thread-level parallelism. The solid black line at the bottom of Fig. 4 represents the performance of modular FFT in `modpn` as the baseline implementation.

SPIRAL-Generated Code. The four lines labeled with *Spiral-gen* represent the performance of the generated fixed-size codes and the library. The two dotted lines labeled with *Spiral-gen fixed-size* refer to the previously reported results in [LJF10], which do not utilize thread-level parallelism. SPIRAL generated algorithms currently use the Cooley-Tukey algorithm (other algorithms like the Prime factor algorithm and Rader's algorithm have also been implemented) with a dynamic programming search engine at generation time for the fixed-size codes and at runtime for the general size library to select an "optimal" recursive breakdown strategy. More details on the DP search can be found in [PMJ05, V08]. We summarize the speedup and explain how it is achieved as a combined effort from various optimization techniques.

- **Speedup.** All SPIRAL generated codes are faster than the baseline implementation. The fixed-size vector code delivers a speedup of 26 times for medium sizes and 11 times for large sizes. The general size parallel library provides comparable speedups as the fixed-size code.

- **Search mechanism.** The nature of the recursive algorithms and complex architectural features lead to a large space of candidate implementations, which is challenging for the hand-tuned library to provide an highly optimized implementation. On the other hand, the search engine can select an optimized breakdown strategy that adapts to the target platform.
- **Scalar code optimizations.** With Σ-SPL, loops are merged to reduce the number of passes on the input data. A collection of small size transforms supporting general factorization are generated as base cases and are fully unrolled to improve cache locality and avoid loop overhead. Small functions are inlined to reduce function call overhead.
- **Vectorization and parallelization.** The rewriting systems and hardware tags introduced in Sect. 4 further optimize the generated code to fully exploit vector level parallelism and thread level parallelism by automatically adapting to the the vector length and the number of cores on the target platform.

General Size Parallel Library and Fixed-Size Code. The performance of the scalar and vector algorithms in the general size library are within 81% to 91% of the performance of corresponding fixed-size algorithms. For large sizes, thread level parallelism in the parallel library leads to roughly 1.5 time speedup over the fixed-size algorithm. The combined speedup for the general size parallel library over baseline implementation is roughly 15 times for large sizes, compared to 11 times of the fixed-size code.

Both fixed-size code generation and general size library use DP search engine to search for an optimized algorithm for a given size. For fixed-size code, all parameters including the transform size are known at generation time, which enables computation of constants like the twiddle factors at generation time. The constants are then used in fully unrolled small size transforms. On the other hand, in general size library, precomputation (e.g., computing the twiddle factors) is performed with runtime parameters and the results are kept in buffers of recursive function steps and base cases. Initial accesses to the buffers may lead to cache misses, which explains the small performance loss of the scalar and vector algorithm in the general size library.

6   Conclusions

In this paper we extend the SPIRAL system to generate general input size parallel library for modular FFT algorithms. The new transform and algorithms are represented and optimized at high abstraction levels. By applying and extending the library generation framework, we have generated a library that supports efficient modular arithmetic, fully exploits vector and thread parallelisms of the target platforms, and offers comparable performance improvement as the fixed-size code compared to the hand-tuned implementations.

The general size library is advantageous over the fixed-size approach, in that it compiles the breakdown rules to recursive function closure and search for an optimized strategy with runtime parameters. The performance results have proved the usefulness of the library generation and autotuning techniques and suggest that they should be incorporated into high-level libraries to take advantage of platform-dependent optimizations, in the context of quickly evolving hardware acceleration technologies.

The goal of the automatic generation and performance tuning of a modular FFT library is to eventually generate an optimized parallel library for polynomial multiplication. We have developed additional algorithms for modular FFT, including the Prime factor algorithm and Raders algorithm, which serve as good candidate algorithms for non power-of-two size transforms. We have also experimented with the usages of vector registers to support larger primes and hence larger transform sizes. In the future, we will systematically explore the tradeoffs and cutoffs between the 1-D transform candidates, non power-of-2 transform sizes and various padding strategies to support the automatic generation of a high performance library for polynomial multiplication.

References

[LJF10]   Meng, L., Johnson, J., Franchetti, F., Voronenko, Y., Moreno Maza, M., Xie, Y.: Spiral-Generated Modular FFT Algorithms. In: Proc. International Workshop on Parallel and Symbolic Computation (PASCO), pp. 169–170 (2010)

[V08]   Voronenko, Y.: Library Generation for Linear Transforms. PhD. thesis, Electrical and Computer Engineering, Carnegie Mellon University (2008)

[CT65]   Cooley, J.W., Tukey, J.W.: An algorithm for the machine calculation of complex Fourier series. Math. of Computation 19, 297–301 (1965)

[FLMS06]   Filatei, A., Li, X., Moreno Maza, M., Schost, É.: Implementation techniques for fast polynomial arithmetic in a high-level programming environment. In: Proc. ISSAC 2006, pp. 93–100. ACM Press, New York (2006)

[FVP05]   Franchetti, F., Voronenko, Y., Püschel, M.: Formal Loop Merging for Signal Transforms. In: Proc. Programming Languages Design and Implementation (PLDI), pp. 315–326 (2005)

[FVP061]   Franchetti, F., Voronenko, Y., Püschel, M.: FFT Program Generation for Shared Memory: SMP and Multicore. In: Proc. Supercomputing, SC (2006)

[FVP062]   Franchetti, F., Voronenko, Y., Püschel, M.: A Rewriting System for the Vectorization of Signal Transforms. In: Daydé, M., Palma, J.M.L.M., Coutinho, Á.L.G.A., Pacitti, E., Lopes, J.C. (eds.) VECPAR 2006. LNCS, vol. 4395, pp. 363–377. Springer, Heidelberg (2007)

[JJRT90]   Johnson, J., Johnson, R.W., Rodriguez, D., Tolimieri, R.: A Methodology for Designing, Modifying, and Implementing Fourier Transform Algorithms on Various Architectures. IEEE Trans. Circuits Sys. 9 (1990)

[LM06]   Li, X., Maza, M.M.: Efficient implementation of polynomial arithmetic in a multiple-level programming environment. In: Iglesias, A., Takayama, N. (eds.) ICMS 2006. LNCS, vol. 4151, pp. 12–23. Springer, Heidelberg (2006)

[LMP09]   Li, X., Moreno Maza, M., Pan, W.: Computations modulo regular
               chains. In: Proc. ISSAC 2009, pp. 239–246. ACM, New York (2009)
[LMRS08]   Li, X., Moreno Maza, M., Rasheed, R., Schost, É.: High-Performance
               Symbolic Computation in a Hybrid Compiled-Interpreted Program-
               ming Environment. In: Proc. CASA 2008. LNCS. Springer (2008)
[LMS07]   Li, X., Moreno Maza, M., Schos, É.: Fast arithmetic for triangular
               sets: From theory to practice. In: Proc. ISSAC 2007, pp. 269–276.
               ACM Press (2007)
[M85]   Montgomery, P.L.: Modular Multiplication Without Trial Division.
               Mathematics of Computation 44(170), 519–521 (1985)
[PMJ05]   Püschel, M., Moura, J., Johnson, J., Padua, D., Veloso, M., Singer, B.,
               Xiong, J., Franchetti, F., Gacic, A., Voronenko, Y., Chen, K., John-
               son, R., Rizzolo, N.: SPIRAL: Code Generation for DSP Transforms.
               Proc. IEEE Special Issue on "Program Generation, Optimization,
               and Adaptation" 93(2), 232–275 (2005)
[www.spiral.net] Spiral project website, http://www.spiral.net
[XJJP01]   Xiong, J., Johnson, J., Johnson, R., Padua, D.: SPL: A Language and
               Compiler for DSP Algorithms. In: Proc. PLDI, pp. 298–308 (2001)
[FJ05]   Frigo, M., Johnson, S.G.: The Design and Implementation of FFTW3.
               Proc. IEEE Special Issue on "Program Generation, Optimization, and
               Adaptation" 93(2), 216–231 (2005)

Extended QRGCD Algorithm*

Kosaku Nagasaka[1] and Takaaki Masui[2]

[1] Kobe University, 3-11 Tsurukabuto, Nada-ku, Kobe 657-8501, Japan
[2] Kobe-Takatsuka High School, 9-1 Mikatadai, Nishi-ku, Kobe 651-2277, Japan

Abstract. For computing the greatest common divisor of two univariate polynomials with a priori numerical errors on their coefficients, we use several approximate polynomial GCD algorithms: QRGCD, UVGCD, STLN-based, Fastgcd, GPGCD and so on. Among them, QRGCD is the most common algorithm since it has been distributed as a part of Maple and there are many papers including their comparisons of efficiency and effectiveness against QRGCD. In this paper, we give an improved QRGCD algorithm (ExQRGCD) which is unfortunately not faster than the original but more accurate and the resulting perturbation is able to satisfy the given tolerance.

1   Introduction

Computing the greatest common divisor (GCD) of polynomials is one of the most primitive computations in symbolic algebraic computations hence several algorithms are known: the Euclidean algorithm, the half GCD algorithm and so on (see also the recent result[1] and the text book[2]). However, in the practical situations (e.g. control theory, image processing and so on), the input polynomials are represented by the floating-point numbers or derived from the result of numerical computations or experiments hence in general they have a priori errors on their coefficients. For such polynomials, any conventional algorithms cannot compute their GCD since it easily becomes coprime due to a priori errors. This problem is called "approximate polynomial GCD" and there are many known studies (see Boito[3]).

Recently, most of modern algorithms for this problem use some optimization techniques and matrix decompositions. UVGCD[4], STLN-based[5], Fastgcd[6] and GPGCD[7] are typical algorithms using optimization techniques. The Gauss-Newton algorithm is used in UVGCD and Fastgcd to refine a tentative approximate GCD computed by some matrix decomposition and solving a linear system. Structured total least norm (STLN) techniques and the gradient projection algorithm are used in STLN-based and GPGCD, respectively, for seeking the approximate GCD directly.

QRGCD[8] is one of algorithms based on matrix decompositions and is also implemented as a part of the SNAP package of Maple. It uses the QR decomposition of the Sylvester matrix of the input polynomials and constructs approximate

---

* This work was supported in part by JSPS KAKENHI Grant Number 22700011.

V.P. Gerdt et al. (Eds.): CASC 2013, LNCS 8136, pp. 257–272, 2013.
© Springer International Publishing Switzerland 2013

GCD directly from the upper triangular matrix (i.e. QRGCD does not refine the resulting approximate factor). It is notable that QRGCD has been used as the benchmark algorithm for newly proposed algorithms. However, since QRGCD was proposed in the early stage of approximate GCD, its theoretical background is not enough analyzed (e.g. detectability of approximate GCD by QR factoring).

In this paper, we show some relative closeness and property of the upper triangular matrix, which clarify some weak points of QRGCD and for which we give an improved QRGCD algorithm (we call it ExQRGCD). Unfortunately our algorithm is not faster than the original but more accurate and the resulting perturbation is able to satisfy the given tolerance (note that QRGCD is not able to do this). We introduce the notations and the framework of the QRGCD algorithm in the rest of this section, and give our improved framework based on the original QRGCD in the section **2**. We give an improved QRGCD algorithm (ExQRGCD) with theoretical considerations in the section **3**. In the sections **4** and **5**, we give some numerical experiments and concluding remarks, respectively.

1.1   Notations

Let input polynomials $f(x), g(x) \in \mathbb{R}[x]$ of degree m, n be

$$f(x) = \sum_{i=0}^{m} f_i x^i, \ g(x) = \sum_{i=0}^{n} g_i x^i.$$

We assume that the input polynomials have the unit Euclidean norm (2-norm) (i.e. $\|f(x)\|_2 = \|g(x)\|_2 = 1$). Hence we scale the polynomials if they do not have the unit norm at the input. In this paper, we use the following (descending term order and row major) Sylvester matrix of $f(x)$ and $g(x)$.

$$Syl(f,g) = \begin{pmatrix} f_m & f_{m-1} & \cdots & f_1 & f_0 & & & \\ & f_m & f_{m-1} & \cdots & f_1 & f_0 & & \\ & & \ddots & \ddots & \cdots & \ddots & \ddots & \\ & & & f_m & f_{m-1} & \cdots & f_1 & f_0 \\ g_n & g_{n-1} & \cdots & g_1 & g_0 & & & \\ & g_n & g_{n-1} & \cdots & g_1 & g_0 & & \\ & & \ddots & \ddots & \cdots & \ddots & \ddots & \\ & & & g_n & g_{n-1} & \cdots & g_1 & g_0 \end{pmatrix}.$$

For vectors and matrices, $\|\cdot\|_2$ and $\|\cdot\|_F$ denote the Euclidean norm (2-norm) and the Frobenius norm, respectively. A^T and A^{-1} are the transpose[1] and the inverse of matrix A, respectively. To specify the submatrix of matrix A, we use the MATLAB like colon notation: the submatrix consisting of elements in the i_1 through i_2-th rows and j_1 through j_2-th columns is denoted by $A_{(i_1:i_2, j_1:j_2)}$.

---

[1] For the complex case, we just use the conjugate transpose (Hermitian transpose) instead of the transpose.

The coefficient vector (in the descending term order) of $p(x)$ of degree k is denoted by $\vec{p} \in \mathbb{R}^{k+1}$. We represent the evaluation of $p(x)$ at the point $\omega \in \mathbb{C}$ as $p(\omega) = \vec{p}^T \vec{\omega}_*$ where $\vec{\omega}_* = (\omega^k \ \cdots \ \omega^1 \ \omega^0)^T$. We denote the reversal polynomial of $p(x)$ by $\text{rev}(p)$ (i.e. $\text{rev}(p) = x^{\deg(p)} p(1/x)$). Moreover, we abbreviate "factor whose roots are outside the unit circle in the complex plane" to "outside-root factor" and also use "inside-root factor" similarly.

We note that approximate polynomial GCD has been defined from the several point of view in the literatures. In this paper, we use the following coefficient-wise definition without any constraint on the Bézout coefficients, which is compatible with most of known results.

Definition 1 (Approximate Polynomial GCD).
For the input polynomials[2] $f(x), g(x) \in \mathbb{R}[x]$, we call the polynomial $d(x) \in \mathbb{R}[x]$ "approximate polynomial GCD" of tolerance $\varepsilon \in \mathbb{R}_{\geq 0}$ if it satisfies

$$f(x) + \Delta_f(x) = f_1(x)d(x), \ g(x) + \Delta_g(x) = g_1(x)d(x)$$

for some polynomials $\Delta_f(x), \Delta_g(x), f_1(x), g_1(x) \in \mathbb{R}[x]$ such that $\deg(\Delta_f) \leq \deg(f)$, $\deg(\Delta_g) \leq \deg(g)$, $\|\Delta_f\|_2 < \varepsilon \|f\|_2$ and $\|\Delta_g\|_2 < \varepsilon \|g\|_2$ where $\|\cdot\|_2$ denotes the 2-norm. ◁

Remark 1. In the recent studies of approximate GCD, one may consider the extra conditions that $\deg(d)$ is maximized w.r.t. the tolerance ε or the tolerance ε is minimized w.r.t. $\deg(d)$. However, in this paper, we do not consider these conditions since it is difficult that the QRGCD or similar algorithms guarantee these properties without any refinement (optimization) step. ◁

1.2 Framework and Algorithm

The framework of the QRGCD algorithm is as follows.

1. Compute the QR decomposition of $Syl(f, g)$: $Syl(f, g) = QR$.
2. Find the gap between the k-th and $(k + 1)$-th row vectors $\vec{r}_k, \vec{r}_{k+1}$ of R and form the polynomial with coefficients $\vec{r}_k$, which is an approximate polynomial GCD (or its factor).
3. Apply the same procedures to the reversal polynomials of cofactors since R may not have the approximate outside-root factor.

This is basically based on the following well known properties of the QR decomposition of the Sylvester matrix.

Lemma 1. *The last non-zero row vector of R gives the (mathematically exact) coefficients of the polynomial GCD of $f(x)$ and $g(x)$.* (see e.g. [9]) ◁

Lemma 2. *The QR factoring is numerically backward stable. Thus, $\|Syl(f, g) - QR\|_2$ is enough small even if we compute the QR decomposition numerically.* (see e.g. [10]) ◁

---

[2] We focus on polynomials over $\mathbb{R}$ for easy understanding though QRGCD and our algorithm can work also over $\mathbb{C}$.

We briefly review the original QRGCD algorithm below.

Algorithm 1 (QRGCD in the original paper).
Input: $f(x), g(x) \in \mathbb{R}[x]$ and tolerance $\varepsilon \in \mathbb{R}_{\geq 0}$.
Output: $u(x), v(x), d(x) \in \mathbb{R}[x]$ ($d(x)$ is an approx. GCD)

1. Compute the QR decomposition of $Syl(f,g)$: $Syl(f,g) = QR$.
2. Suppose that $R^{(k)}$ is the last $(k+1) \times (k+1)$ submatrix of R such that $\|R^{(k)}\|_2 > \varepsilon$ and $\|R^{(k-1)}\|_2 < \varepsilon$.
   Case 1: $\|R^{(0)}\|_2 > \varepsilon$ (Approximately Coprime)

   (a) $d_1(x) := 1$, $u(x)$ and $v(x)$ formed by rows of Q^T.
   Case 2: $\|R^{(k-1)}\|_2 < 10\varepsilon \, \|R^{(k)}\|_2$ (Big Gap Found)

   (a) $d_1(x) :=$ the last k-th row vector of R.
   Case 3: $\exists k_1$(biggest), $\|R^{(k_1-1)}\|_2 < 10\varepsilon \, \|R^{(k_1)}\|_2$ (Gap Found)

   (a) $d_1(x) :=$ the last k_1-th row vector of R.
   Case 4: Otherwise   (Difficult Case)

   (a) find the inside-root factors of $f(x)$ and $g(x)$ by the algorithm "Split" and compute an approximate divisor $d_1(x)$ of the inside-root factors.
3. Apply the above steps to reversal polynomials of cofactors of $f(x)$ and $g(x)$ w.r.t. $d_1(x)$, to obtain $d_2(x)$.
4. Apply the above steps to cofactors $f(x)$ and $g(x)$ w.r.t. $d(x) = d_1(x)d_2(x)$, to obtain $u(x)$ and $v(x)$.
5. Output $u(x)$, $v(x)$ and $d(x)$.   ◁

Here, the algorithm "Split" was proposed in the original paper[8] which finds the inside-root factor, and we also use it in our algorithm. This is done by the Graeffe's root-squaring, contour integration, Newton's formula, Newton's iteration and lifting steps. This algorithm is not the purpose of this paper and there are similar methods in the literatures hence we omit this in detail.

2 Improved QRGCD Framework

In this section, we give new lemmas and theorems that guarantee our framework similar to that of QRGCD above. At first, we cite the following lemma given by Stetter[11] (see also [12]), which plays an essential role in the frameworks of QRGCD and our improvement.

Lemma 3 (Corollary 7 in [11]).
For $\varepsilon \in \mathbb{R}_{\geq 0}$, let $\mathcal{P}_\varepsilon$ be the polynomial set s.t. $\mathcal{P}_\varepsilon = \{\tilde{p}(x) \in \mathbb{C}[x] \mid \deg(\tilde{p}) \leq \deg(p),\ \|p(x) - \tilde{p}(x)\|_2 \leq \varepsilon\}$. For $\omega \in \mathbb{C}$, we have

$$\exists \tilde{p}(x) \in \mathcal{P}_\varepsilon(p),\ \tilde{p}(\omega) = 0 \iff |p(\omega)| \leq \varepsilon \, \|\vec{\omega}_*\|_2 .$$

(note that the original is intended for the dual norm) ◁

2.1   Detectability of Approximate GCD by QR Factoring

Let $d(x)$ be the approximate polynomial GCD of $f(x)$ and $g(x)$, of degree k and having the unit 2-norm hence $d(x)$ is the exact GCD of $f(x) + \Delta_f(x)$ and $g(x) + \Delta_g(x)$. Similar to $f(x)$ and $g(x)$, we denote the Sylvester matrix of $f(x) + \Delta_f(x)$ and $g(x) + \Delta_g(x)$ by $Syl(f + \Delta_f, g + \Delta_g)$. For simplicity, we abbreviate $Syl(f, g)$ and $Syl(f + \Delta_f, g + \Delta_g)$ to S and $S + \Delta_S$, respectively, where $\Delta_S = Syl(\Delta_f, \Delta_g)$. For S and $S + \Delta_S$, let their QR decompositions be $S = QR$ and $S + \Delta_S = \hat{Q}\hat{R}$ where $Q, \hat{Q}$ are orthogonal and $R, \hat{R}$ are upper triangular matrices. We note that the notations in this paper are different from the notations used in the original paper[8] as the result of simplifications. Moreover, we denote the matrices whose column vectors generate the null spaces of S and $S + \Delta_S$ by N and $\hat{N}$, respectively, thus we have $SN = \mathbf{0}$ and $(S + \Delta_S)\hat{N} = \mathbf{0}$ and substituting $S = QR$ gives

$$- R\hat{N} = Q^T \Delta_S \hat{N}. \qquad (2.1)$$

Lemma 4. *Let $r(x)$ be a polynomial with coefficients $\vec{r}$ which is a row vector of R, and ω be any root of $d(x)$. Then we have $|r(\omega)| \leq \|\Delta_S\|_2 \|\vec{\omega}_*\|_2$.* ◁

Proof. By the equality (2.1), we have $-R\vec{\omega}_* = Q^T\Delta_S\vec{\omega}_*$. Therefore, we have $|r(\omega)| \leq \|R\vec{\omega}_*\|_2 = \|Q^T\Delta_S\vec{\omega}_*\|_2 \leq \|\Delta_S\|_2\|\vec{\omega}_*\|_2$ since $r(\omega)$ is an element of $R\vec{\omega}_*$.

□

Note that any factor of $d(x)$ can be an approximate polynomial GCD by the definition. By $d_r(x)$ we denote such a factor of $d(x)$, and let $\Omega(d_r) = \{\omega_1, \ldots, \omega_k\}$ be the set of distinct roots of $d_r(x)$ and $\Omega_*(d_r)$ be the $k \times (m+n)$ matrix whose i-th row vector is $\vec{\omega}_{i*}^T = (\omega_i^{m+n-1} \cdots \omega_i^1 \omega_i^0)$, the transpose of the evaluation vector of ω_i. We denote the condition number of $\Omega_*(d_r)$ by $\kappa_2(\Omega_*(d_r))$.

Theorem 1. *Let $r(x)$ be a polynomial with coefficients $\vec{r}$ which is a row vector of R. Then, an upper bound of relative distance of $r(x)$ from $d_r(x)$, an approximate polynomial GCD of $f(x)$ and $g(x)$, is given by $\sqrt{k+1}\kappa_2(\Omega_*(d_r))\frac{\|\Delta_S\|_2}{\|r(x)\|_2}$. Moreover, there may exist another approximate polynomial GCD of higher degree other than $r(x)$ if $\frac{\|\Delta_S\|_2}{\|r(x)\|_2} > 1$.* ◁

Proof. First claim. Let $\delta_r(x) = d_r(x) - r(x)$ be the residual polynomial, and all the coefficient vectors are treated as polynomials of degree $m + n - 1$ by padding zeros. By Lemma 4, for any $\omega \in \Omega(d_r)$ we have

$$|r(\omega)| \leq \|\Delta_S\|_2\|\vec{\omega}_*\|_2 = \frac{\|\Delta_S\|_2}{\|r(x)\|_2}\|r(x)\|_2\|\vec{\omega}_*\|_2,$$

hence the square root of the sum of $|r(\omega)|^2$ over all roots $\omega \in \Omega(d_r)$ gives

$$\|\Omega_*(d_r)\vec{r}\|_2 \leq \frac{\|\Delta_S\|_2}{\|r(x)\|_2}\|r(x)\|_2\|\Omega_*(d_r)\|_F \leq \sqrt{k+1}\frac{\|\Delta_S\|_2}{\|r(x)\|_2}\|r(x)\|_2\|\Omega_*(d_r)\|_2.$$

Therefore, we have the following relative upper bound of $\|\delta_r(x)\|_2$ since $\Omega_*(d_r)$ is of full row rank and $\Omega_*(d_r)\vec{\delta}_r = \Omega_*(d_r)(\vec{d}_r - \vec{r}) = -\Omega_*(d_r)\vec{r}$ can be solved by the least square w.r.t. $\vec{\delta}_r$.

$$\frac{\|\vec{\delta}_r\|_2}{\|r(x)\|_2} \leq \sqrt{k+1}\kappa_2(\Omega_*(d_r))\frac{\|\Delta_S\|_2}{\|r(x)\|_2}.$$

Second claim. We note that Δ_S is the required perturbation of the $Syl(f,g)$ so that $f(x) + \Delta_f(x)$ and $g(x) + \Delta_g(x)$ may have an exact polynomial GCD (i.e. $d(x)$). Therefore, the row vector $\vec{r}$ can become the zero vector within the given tolerance if $\|\Delta_S\|_2 > \|r(x)\|_2$ since Q is orthogonal hence the perturbation of S equals to the perturbation of R in the 2-norm. This means that the degree of $d(x)$ may be larger than that of $r(x)$.   □

If $\deg(d(x)) = 1$ then the first claim is proved easily and directly by Lemma 3. Note that Theorem 1 gives us only the necessary condition for that a computed row vector of R is close to $d(x)$. It does not give us any property of each row vector of R, which will be discussed in the next subsection. Moreover, we cannot know the magnitude $\|\Delta_S\|_2$ hence we have to estimate it in advance or in the algorithm. We will discuss this later in Section 3.

Remark 2. The condition number of $\Omega_*(d_r)$ becomes large if the approximate polynomial GCD has close roots. This means that QRGCD and our algorithm may be weak for polynomials having mutually close roots.   ◁

Remark 3. A similar theorem is proved by Corless et al. in [8]. Their theorem is only for absolute closeness hence we extend it for the relative closeness. In general, lower row vectors of matrix R have smaller 2-norms. For example, we may have the case that the expected approximate polynomial GCD is $10^{-6}(x - 1)(x - 2)$ (before normalized) and the detected polynomial is $10^{-6}(x+1)(x+2)$. This result is not suitable but may happen since we compute the QR decomposition numerically. In order to prepare for such cases, we extend their theorem as above, though this is just a theoretical possibility (we didn't find any simple small example which has this behavior explicitly).   ◁

2.2   Roots Outside the Unit Circle

As reported by Corless et al. in [8], QR factoring of the Sylvester matrix may not detect any outside-root factor of approximate GCD, which has the (approximately[3]) common roots outside the unit circle. Therefore, the framework and the algorithm compute an inside-root factor and then compute an outside-root factor. In this subsection, we prove this from the different approach.

Let $p_1(x) = f(x), p_2(x) = g(x), p_3(x), \ldots \in \mathbb{R}[x]$ be the polynomial remainder sequence (PRS) of $f(x)$ and $g(x)$. We denote the k-th subresultant of $f(x)$ and $g(x)$ by $\mathrm{res}_k(f,g)$ which is defined by the determinant as follows.

---

[3] By "approximately common root" we denote the roots of approximate GCD.

$$\mathrm{res}_k(f,g) = \begin{vmatrix} Syl(f,x^n)_{(1:n-k,\,1:m+n-2k-1)} & \vec{x}_* f(x) \\ Syl(x^m,g)_{(n+1:m+n-k,\,1:m+n-2k-1)} & \vec{x}_* g(x) \end{vmatrix}$$

where $\vec{x}_* f(x) = \{x^{n-k-1}f(x) \ \cdots \ x^1 f(x) \ x^0 f(x)\}^T$, $\vec{x}_* g(x) = \{x^{m-k-1}g(x) \ \cdots$ $x^1 g(x) \ x^0 g(x)\}^T$ and $f_i = 0$ and $g_i = 0$ for any negative index $(i < 0)$. It is well known that $p_i(x)$ is a multiple of $\mathrm{res}_{\deg(p_{i-1})-1}(f,g)$ (see [13] for example) and the subresultant is related to the Sylvester's single sum ([14] and references therein) as follows.

Lemma 5 (Sylvester's single sum).
Let A and B be the sets of all the roots of $f(x)$ and $g(x)$, respectively. Then we have

$$\mathrm{res}_k(f,g) = \sum_{A' \subset A,\ \#A' = k} R(x,A') \frac{R(A \setminus A', B)}{R(A \setminus A', A')}$$

where $R(A,B) = \prod_{a \in A,\ b \in B}(a-b)$, $R(x,A) = \prod_{a \in A}(x-a)$ and $\#A$ denotes the cardinality of A. ◁

To describe some quality of PRS, we estimate the norm of each summand in the single sum for the following specific polynomials (similar estimation method used in [15] for clusters of close roots) where ω_{in_i} and $\hat{\omega}_{\mathrm{in}_j}$ are roots inside the unit circle and ω_{out_i} and $\hat{\omega}_{\mathrm{out}_j}$ are roots outside the unit circle. One may think that these polynomials are odd, however, this situation easily happens (see Example 1 for such actual polynomials).

$$f(x) = \prod_{i=1}^{m_{\mathrm{in}}}(x - \omega_{\mathrm{in}_i}) \prod_{i=1}^{m_{\mathrm{out}}}(x - \omega_{\mathrm{out}_i}), \ \deg(f) = m,$$
$$g(x) = \prod_{j=1}^{n_{\mathrm{in}}}(x - \hat{\omega}_{\mathrm{in}_j}) \prod_{j=1}^{n_{\mathrm{out}}}(x - \hat{\omega}_{\mathrm{out}_j}), \ \deg(g) = n$$

where the magnitudes of roots satisfy $|\omega_{\mathrm{in}_i}| = O(\alpha)$, $|\hat{\omega}_{\mathrm{in}_j}| = O(\alpha)$, $|\omega_{\mathrm{out}_i}| = O(\alpha\beta)$, and $|\hat{\omega}_{\mathrm{out}_j}| = O(\alpha\beta)$ with the common big O notation, for $\alpha, \beta \in \mathbb{R}_{\geq 0}$ satisfying $\alpha < 1$ and $\alpha\beta > 1$. We assume that the roots of $f(x)$ and $g(x)$ are well isolated respectively, the approximate GCD of $f(x)$ and $g(x)$ is of degree k, the corresponding (approximately common) roots satisfy $|\omega_{\mathrm{in}_i} - \hat{\omega}_{\mathrm{in}_j}| = O(\alpha\gamma_{\mathrm{in}})$ and $|\omega_{\mathrm{out}_i} - \hat{\omega}_{\mathrm{out}_j}| = O(\alpha\beta\gamma_{\mathrm{out}})$ for some $\gamma_{\mathrm{in}}, \gamma_{\mathrm{out}} \in \mathbb{R}_{\geq 0}$, and the other pairs of roots satisfy $|\omega_{\mathrm{in}_i} - \hat{\omega}_{\mathrm{in}_j}| = O(\alpha)$ and $|\omega_{\mathrm{out}_i} - \hat{\omega}_{\mathrm{out}_j}| = O(\alpha\beta)$. Moreover, c_{in} and c_{out} denote the numbers of common inside roots of $f(x)$ and $g(x)$, and common outside roots of $f(x)$ and $g(x)$, respectively, hence we have $k = c_{\mathrm{in}} + c_{\mathrm{out}}$.

Lemma 6. *With the above notations and assumptions, for the coefficient of each summand of single sum representation of $\mathrm{res}_k(f,g)$, we have*

$$\frac{R(A \setminus A', B)}{R(A \setminus A', A')} = O(\alpha^{(m-k)(n-k)} \beta^{(m-k)(n-k)-(m_{\mathrm{in}}-k_{\mathrm{in}})(n_{\mathrm{in}}-k_{\mathrm{in}})} \gamma_{\mathrm{in}}^{c_{\mathrm{in}}-k_{\mathrm{in}_c}} \gamma_{\mathrm{out}}^{c_{\mathrm{out}}-k_{\mathrm{out}_c}})$$

where k_{in}, k_{in_c} and k_{out_c} denote the numbers of inside roots included in A', common inside roots included in A' and common outside roots included in A', respectively. ◁

Proof. At first, we estimate the numerator by multiplying each part (inside non-common roots: B_{in_s}, inside common roots: B_{in_c}, outside non-common roots: B_{out_s} and outside common roots: B_{out_c}) of B as follows.

$$R(A \setminus A', B) = R(A \setminus A', B_{\mathrm{in}_s}) \times R(A \setminus A', B_{\mathrm{in}_c})$$
$$\times R(A \setminus A', B_{\mathrm{out}_s}) \times R(A \setminus A', B_{\mathrm{out}_c}),$$
$$R(A\backslash A', B_{\mathrm{in}_s}) = O(\alpha^{(m_{\mathrm{in}}-k_{\mathrm{in}})(n_{\mathrm{in}}-c_{\mathrm{in}})}(\alpha\beta)^{(m_{\mathrm{out}}-k_{\mathrm{out}})(n_{\mathrm{in}}-c_{\mathrm{in}})}),$$
$$R(A\backslash A', B_{\mathrm{in}_c}) = O(\alpha^{(m_{\mathrm{in}}-k_{\mathrm{in}})c_{\mathrm{in}}-(c_{\mathrm{in}}-k_{\mathrm{in}_c})}(\alpha\gamma_{\mathrm{in}})^{c_{\mathrm{in}}-k_{\mathrm{in}_c}}(\alpha\beta)^{(m_{\mathrm{out}}-k_{\mathrm{out}})c_{\mathrm{in}}}),$$
$$R(A\backslash A', B_{\mathrm{out}_s}) = O((\alpha\beta)^{(m-k)(n_{\mathrm{out}}-c_{\mathrm{out}})}),$$
$$R(A\backslash A', B_{\mathrm{out}_c}) = O((\alpha\beta)^{(m_{\mathrm{in}}-k_{\mathrm{in}})c_{\mathrm{out}}}(\alpha\beta\gamma_{\mathrm{out}})^{c_{\mathrm{out}}-k_{\mathrm{out}_c}}$$
$$\times (\alpha\beta)^{(m_{\mathrm{out}}-k_{\mathrm{out}})c_{\mathrm{out}}-(c_{\mathrm{out}}-k_{\mathrm{out}_c})}).$$

Therefore, we have

$$R(A\backslash A', B) = O(\alpha^{(m-k)n}\beta^{(m-k)n_{\mathrm{out}}+(m_{\mathrm{out}}-k_{\mathrm{out}})n_{\mathrm{in}}}\gamma_{\mathrm{in}}^{c_{\mathrm{in}}-k_{\mathrm{in}_c}}\gamma_{\mathrm{out}}^{c_{\mathrm{out}}-k_{\mathrm{out}_c}}).$$

Next, we estimate the denominator in the same way.

$$R(A \setminus A', A') = R(A \setminus A', A'_{\mathrm{in}}) \times R(A \setminus A', A'_{\mathrm{out}}),$$
$$= O(\alpha^{(m_{\mathrm{in}}-k_{\mathrm{in}})k_{\mathrm{in}}}(\alpha\beta)^{(m_{\mathrm{out}}-k_{\mathrm{out}})k_{\mathrm{in}}}) \times O((\alpha\beta)^{(m-k)k_{\mathrm{out}}})$$
$$= O(\alpha^{(m-k)k}\beta^{(m-k)k_{\mathrm{out}}+(m_{\mathrm{out}}-k_{\mathrm{out}})k_{\mathrm{in}}}).$$

By substituting $m_{\mathrm{out}} - k_{\mathrm{out}} = (m - k) - (m_{\mathrm{in}} - k_{\mathrm{in}})$ for the power of β, the lemma follows from the above directly.   □

Theorem 2. *There exists a pair of polynomials $f(x)$ and $g(x)$ such that the QR decomposition of $\mathrm{Syl}(f, g)$ cannot detect any outside-root factor of the approximate GCD. Moreover, we need to detect such factors from $\mathrm{rev}(f)$ and $\mathrm{rev}(g)$ or their cofactors several times and combine them.*   ◁

Proof. With the assumptions of Lemma 6 for a fixed $k = \deg(d)$, let A'_0 be the set of all the (approximately) common roots of $f(x)$ and $g(x)$ and A'_s be a set of A'_0 replaced s common outside roots with s (not common) inside roots. We focus on the roots of $r(x)$ with coefficients $\vec{r}$ which is the last $(k + 1)$-th row vector of R.

By Lemmas 5 and 6, the coefficient $R(A\backslash A'_s, B)/R(A\backslash A'_s, A'_s)$ of $R(x, A'_s)$, is $O(\beta^{(m_{\mathrm{in}}+n_{\mathrm{in}}-2k_{\mathrm{in}}-s)s}\gamma_{\mathrm{out}}^s)$ times larger than $R(A \setminus A'_0, B)/R(A \setminus A'_0, A'_0)$ of $R(x, A'_0)$. Therefore, $r(x)$ easily becomes to be without common outside roots, regardless of their closeness, hence the QR decomposition cannot detect any outside-root factor of the approximate GCD. For such factors, we have to compute the QR decomposition of matrix of $\mathrm{rev}(f)$ and $\mathrm{rev}(g)$ (i.e. transform the outside roots into inside).   □

One may think that Theorems 1 and 2 seem like a contradiction since Theorem 1 does not restrict roots to the inside. However, there is no contradiction. Theorem 2 states only one of properties of PRS which may not have common outside roots. As the result of the computation of PRS, Theorem 1 states only a necessary condition that the resulting PRS is an approximate GCD (or its approximate factor).

Moreover, the above theorem indicates a concrete example below that we have to compute approximate GCD from the both of input and reversal polynomials. In the rest of paper, by "normal side" and "reversal side" we denote the computations from the input and reversal polynomials (finding inside and outside roots), respectively.

Example 1 (Fake Common Inside Roots).
We compute the QR decomposition of $Syl(f, g)$ of the following $f(x)$ and $g(x)$.

$$f(x) = (x+0.001)(x-0.001)(x+1000)(x-1000),$$
$$g(x) = (x+0.0010000001)(x-0.009)(x+1000.0001)(x-2000).$$

They have the following approximate GCD of tolerance 10^{-12}.

$$d(x) = 9.99998 \times 10^{-4}x^2 + 0.999999x + 9.99998 \times 10^{-4}$$
$$\approx 9.99998 \times 10^{-4}(x + 0.00100000)(x + 1000.00).$$

The 2-nd, 3-rd and 4-th last row vectors of R are as follows.

4-th last row: $0.00143003x^3 + 1.39855x^2 - 0.00559397x - 6.99252 \times 10^{-6}$
$$\approx 0.00143003(x-0.00499981)(x+0.001)(x+977.993),$$
3-rd last row: $0.20978x^2 - 0.000839074x - 1.04885 \times 10^{-6}$
$$\approx 0.20978(x - 0.00499978)(x + 0.001),$$
2-nd last row: $0.00565685x + 5.65685 \times 10^{-6}$
$$\approx 0.00565685(x + 0.001).$$

As in the proof of Theorem 2, the 3-rd last row is not related to the above $d(x)$ and has a non-common inside root (i.e. 0.00499978 instead of approximate value of 1000). To detect the correct common outside root (≈ 1000), we have to transform the outside roots into inside. This is a reason that we have to compute with reversal polynomials. In the rest of this paper, these non-common inside roots detected instead of the common outside roots are called "fake common inside roots".   ◁

We note that the QRGCD algorithm also works for this example since there is only one fake common root. However, for polynomials having more than one fake common roots, QRGCD may not detect the expected degree of GCD since QRGCD computes from the normal and reversal sides only once. For example, suppose that the row vectors of R consist of roots in this order from bottom to top: common, fake, common, fake, common roots. In this case, we need to compute from the normal, reversal and normal sides (the last normal side is not done by QRGCD). This fact leads us the improved QRGCD algorithm in the next section.

3   Improved QRGCD Algorithm

We propose the following algorithm based on the discussions, which is unfortunately not faster than the original but more accurate and the resulting perturbation is able to satisfy the given tolerance. We will explain the algorithm in subsections 3.1–3.5.

Algorithm 2 (Extended QRGCD).

Input: $f(x), g(x) \in \mathbb{R}[x]$ and tolerance $\varepsilon \in \mathbb{R}_{\geq 0}$.

Output: $f_1(x), g_1(x), d(x) \in \mathbb{R}[x]$ s.t.

$$\|f(x) - d(x)f_1(x)\|_2 < \varepsilon, \quad \|g(x) - d(x)g_1(x)\|_2 < \varepsilon.$$

1. Determine the first side (normal or reversal sides), put $i := 1$, $f_1(x) := f(x)$ and $g_1(x) := g(x)$ (or $\mathrm{rev}(f)$ and $\mathrm{rev}(g)$, respectively, if the first side is reversal).
2. Compute the QR decomposition of $Syl(f_1, g_1)$: $Syl(f_1, g_1) = QR$.
3. Suppose that $R^{(k)}$ is the bottom-rightmost $(k+1) \times (k+1)$ submatrix of R and $\vec{r}_k$ is the top row vector of $R^{(k)}$.
   Case 1: $\|R^{(0)}\|_F > \varepsilon\sqrt{m+n}$ (Approx. Coprime)
   (a) $d_i(x) := 1$.
   Case 2: $\|R^{(0)}\|_F \leq \varepsilon\sqrt{m+n}$ (Trial Divisions)
   (a) for all $k \geq 1$ s.t. $\|R^{(k-1)}\|_F \leq \varepsilon\sqrt{m+n}$, compute $er_k := \|R^{(k-1)}\|_F / \|\vec{r}_k\|_2$, and sort er_k in ascending order s.t. $er_{k_1} \leq er_{k_2} \leq er_{k_3} \leq \cdots$.
   (b) for $k = k_1, k_2, k_3$ (i.e. up to the 3-rd smallest er_k at most), do
       (i) $d_i(x) := r_k(x)$ and if $\exists f_1, g_1$, $\|f(x) - d(x)f_1(x)\|_2 < \varepsilon$ and $\|g(x) - d(x)g_1(x)\|_2 < \varepsilon$ for $d(x) := \prod d_i(x)$, then goto step 4.
       (if no factor found in the recursive call by step (e) below, then put $d_i(x) := 1$ and goto step 4.)
   (c) find the last two rows of R whose norm is not less than 1.0 and let $p_1(x)$ and $p_2(x)$ be their polynomial representations s.t. $\deg(p_1) > \deg(p_2)$.
   (d) find the inside-root factors of $p_1(x), p_2(x)$ by the algorithm "Split" and let them be $p_{1_{\mathrm{in}}}(x), p_{2_{\mathrm{in}}}(x)$.
   (e) apply steps 2 and 3 to $p_{1_{\mathrm{in}}}(x)$ and $p_{2_{\mathrm{in}}}(x)$ and form the divisor $d_i(x)$.
4. $i := i + 1$, change the side (normal $\leftrightarrow$ reversal) and apply the above steps 2 and 3 to (if in the reversal side, reversal polynomials of) cofactors $f_1(x)$ and $g_1(x)$ of $f(x)$ and $g(x)$ w.r.t. $d(x) := \prod d_i(x)$, to obtain $d_{i+1}(x)$ until $d_i(x) = d_{i+1}(x) = 1$ for some i.
5. Output $f_1(x), g_1(x)$ and $d(x)$.   ◁

3.1   The Matrix Norm Used

In the algorithm, we use the Frobenius norm instead of the 2-norm since it is easy to compute. It should be the 2-norm if we can compute it fastly (but usually not fast).

3.2   Approximate Coprime Condition

The QRGCD and our algorithms are based on the QR decomposition of $Syl(f, g)$ hence they cannot detect $d(x)$ directly but may be able to detect $r(x)$ close to $d(x)$ as the result of the QR factoring. Therefore, any (approximately) co-prime detection must not be against the expected $d(x)$ but be against $r(x)$ with coefficients $\vec{r}$ which is the last $(k+1)$-th row of R. From this point of view, $\|R^{(k)}\|_F$ represents a sufficient "unstructured" magnitude to make $Syl(f, g)$ to be rank deficient and $f(x)$ and $g(x)$ may have $r(x)$ as an approximate GCD. Note that "unstructured" perturbation does not preserve a Toeplitz-block structure of Sylvester matrix $Syl(f, g)$. Let $f(x) + \delta_f$ and $g(x) + \delta_g$ be polynomials whose (exact) GCD is $r(x)$. In general, $Syl(\delta_f, \delta_g)$ which makes $Syl(f + \delta_f, g + \delta_g)$ to be rank deficient is the "structured" perturbation hence its norm is larger than $\|R^{(k)}\|_F$ of "unstructured" perturbation in most cases. This means that $r(x)$ is not any approximate GCD if we have $\|R^{(k)}\|_F > \varepsilon\sqrt{m+n}$ hence $\|Syl(\delta_f, \delta_g)\|_F > \varepsilon\sqrt{m+n}$.

3.3 Trial Divisions

By the same reason of the above subsection and Theorem 1, the QR decomposition might not detect all the inside common roots but detect some $r(x)$ close to an approximate factor of $d(x)$ and its closeness can be estimated by $er_k := \|R^{(k-1)}\|_F / \|\vec{r}_k\|_2$ though this is depending on the condition number $\Omega_*(d_r)$ and is not the sufficient condition. If we take $r(x)$ of the maximal degree, $r(x)$ may have some fake inside roots and it becomes difficult to detect the common outside roots from the reversals of cofactors. Therefore, we try to find $r(x)$ for which er_k is as small as possible (since we cannot estimate the condition number $\Omega_*(d_r)$). According to our experiments, the smallest er_k gives enough close factors in most cases but to make sure we try to do trial divisions up to the 3-rd smallest er_k at most.

3.4 The Polynomials to be Applied to "Split"

Let $p_i(x)$ and $p_{i+1}(x)$ be successive two elements of PRS of $f(x)$ and $g(x)$, and $r(x)$ be an approximate GCD of $f(x)$ and $g(x)$. We have

$$f(x) + \delta_f(x) = f_1(x)r(x), \quad g(x) + \delta_g(x) = g_1(x)r(x),$$
$$u_i(x)f(x)+v_i(x)g(x)=p_i(x), \quad u_{i+1}(x)f(x)+v_{i+1}(x)g(x)=p_{i+1}(x)$$

where the coefficients of $u_i(x), v_i(x), u_{i+1}(x), v_{i+1}(x)$ are some column vectors of the orthogonal matrix Q of the QR decomposition of $Syl(f, g)$ hence they have the unit 2-norm by the orthogonality of Q. Then $r(x)$ is an approximate GCD of $p_i(x)$ and $p_{i+1}(x)$ of tolerance $(\|\delta_f\|_2 + \|\delta_g\|_2)/ \min\{\|p_i\|_2, \|p_{i+1}\|_2\}$ since we have

$$p_i(x)=(u_i(x)f_1(x)+v_i(x)g_1(x))r(x)-(u_i(x)\delta_f(x)+v_i(x)\delta_g(x)),$$
$$p_{i+1}(x)=(u_{i+1}(x)f_1(x)+v_{i+1}(x)g_1(x))r(x)-(u_{i+1}(x)\delta_f(x)+v_{i+1}(x)\delta_g(x)).$$

Therefore, we should split $p_i(x)$ and $p_{i+1}(x)$ whose norms are not small (e.g. $\|p_i\|_2, \|p_{i+1}\|_2 \geq 1.0$) since there is a chance that $p_i(x)$ and $p_{i+1}(x)$ have (approximately) common roots other than that of $f(x)$ and $g(x)$ if their norms are small (i.e. tolerance for $p_i(x)$ and $p_{i+1}(x)$ becomes large).

3.5   The Fail-Safe Retry Loop

By the same reason of the section 3.3, we should try to detect a piece of approximate factors which is enough close to the ideal factors. Therefore, our algorithm seeks such a factor by computing from the normal and reversal sides several times. At the worst case, this makes the computational cost k (the degree of approximate GCD) times larger than that of the original, hence our algorithm unfortunately is not faster than the original. However, this is an inevitable cost proven by Theorem 2 and ExQRGCD becomes more stable than the original as in the following examples. Note that in our implementation, the first side of this retry loop is the normal side if $\min\{|f_m|, |g_n|\} \geq \min\{|f_0|, |g_0|\}$.

4   Numerical Experiments

All the computations in this section are done by Maple 16 with $Digits := 16$ on Linux (Intel Core i7 3.30GHz and 64GB memory).

4.1   Against SNAP and QRGCD

At first, we compare ExQRGCD with QRGCD: our algorithm (Algorithm 2) and the SNAP implementation of QRGCD (Algorithm 1), respectively. The vertical axes of Figure 1 and 2 denote the sum of detected degrees of approximate GCD (larger is better) and the magnitude of required perturbation in the common logarithmic scale (smaller is better), respectively. We note that the official SNAP implementation increases the precision in case of the difficult case. However in our experiments we omit this functionality[4] to make the condition equal.

Example 2 (Random Polynomials).
For $i = 1, \ldots, 10$, we have generated 100 pairs of (f, g) such that

$$f(x) = f_1(x)d(x), \ g(x) = g_1(x)d(x), \ d(x) = \sum_{j=0}^{5i} d_j x^j,$$
$$f_1(x) = \sum_{j=0}^{5i} f_{1,j} x^j, \ g_1(x) = \sum_{j=0}^{5i} g_{1,j} x^j$$

where $f_{1,j}$, $g_{1,j}$, $d_j \in [-99, 99] \subset \mathbb{Z}$ is randomly chosen, $f(x), g(x)$ are normalized ($\|f(x)\|_2 = \|g(x)\|_2 = 1$) and rounded with $Digits := 10$. We computed with tolerance 10^{-5}.

Figure 1 shows the result that ExQRGCD and QRGCD are not so different. However, we note that QRGCD outputs "failure" or GCDs that do not satisfy the given tolerance as in Table 1 (ExQRGCD does not have this problem and always can work well). As for computing time, ExQRGCD is 1.59 times slower than QRGCD. The average of resulting perturbations of QRGCD is better than ExQRGCD except failure cases.   ◁

---

[4] With this functionality, the QRGCD and ExQRGCD algorithms become to be much better since the QR decomposition becomes to be more stable.

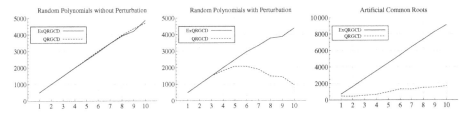

Fig. 1. Sum of Detected Degrees (failure is counted as 0)

Example 3 (Random Poly. with Perturbations).
For $i = 1, \ldots, 10$, we have generated 100 pairs of (f, g) such that

$$f(x) = f_1(x)d(x)/\|f_1d\|_2 + 10^{-8}\Delta_f(x)/\|\Delta_f\|_2,$$
$$g(x) = g_1(x)d(x)/\|g_1d\|_2 + 10^{-8}\Delta_g(x)/\|\Delta_g\|_2,$$
$$\Delta_f(x) = \sum_{j=0}^{10i} \Delta_{f_j}x^j, \ \Delta_g(x) = \sum_{j=0}^{10i} \Delta_{g_j}x^j$$

where $\Delta_{f_j}, \Delta_{g_j} \in [-99, 99] \subset \mathbb{Z}$ is randomly chosen, $f_1(x)$, $g_1(x)$, $d(x)$ are the polynomials of Example 2 and rounded with $Digits := 10$. We computed with tolerance 10^{-5}.

Figure 1 and Table 1 show the result that ExQRGCD is explicitly better than QRGCD. As for computing time, ExQRGCD is 1.99 times slower than QRGCD. The average of resulting perturbations of QRGCD is better than ExQRGCD except failure cases. ◁

Example 4 (Fake Common Roots).
For $i = 1, \ldots, 10$, we have generated 100 pairs of (f, g) such that

$$f(x) = d(x) \prod_{j=1}^{2i}(x - \omega_{f,j}) \prod_{j=1}^{2i}(x - \hat{\omega}_{f,j}),$$
$$g(x) = d(x) \prod_{j=1}^{2i}(x - \omega_{g,j}) \prod_{j=1}^{2i}(x - \hat{\omega}_{g,j}),$$
$$d(x) = \prod_{j=1}^{3i}(x - \omega_{d,j}) \prod_{j=1}^{3i}(x - \hat{\omega}_{d,j})$$

where $\omega_{\cdot,j} = O(10^{-2}), \hat{\omega}_{\cdot,j} = O(10^2)$ is randomly chosen, $f(x), g(x)$ are normalized (i.e. $\|f(x)\|_2 = \|g(x)\|_2 = 1$) and rounded with $Digits := 10$. We computed with tolerance 10^{-5}. We note that the degree of approximate GCD should be larger than or equal to $6i$.

Figure 1 and Table 1 show the result that ExQRGCD is explicitly better than QRGCD. As for computing time, ExQRGCD is 39.8 times slower than QRGCD (note that QRGCD outputs failure for 62% pairs so computing time is very fast for the rest easy cases). The average of resulting perturbations of ExQRGCD is better than QRGCD. ◁

Other than the above, we have tested with several examples of polynomials of higher degree (up to 1020) which are not listed here due to the page restriction. Our algorithm also works for such polynomials. For some of examples we have tested by the native version (C with ATLAS, LAPACK and LAPACKE) since it is about 100 times faster or more.

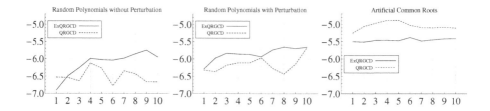

Fig. 2. Resulting Perturbations (i.e. $(\|\Delta_f\|_2 + \|\Delta_g\|_2)/2$) (failure is excepted)

Table 1. Number of fail events or wrong perturbations by QRGCD
(The numbers by ExQRGCD are 0)

	$i = 1$	2	3	4	5	6	7	8	9	10	Δ_f	Δ_g
Random Polys. w/o Perturbation	0	0	0	0	0	0	1	1	3	6	6	6
Random Polys. with Perturbation	0	0	0	9	17	31	46	63	68	81	6	6
Fake Common Roots	29	63	68	71	64	60	66	66	69	68	110	119

4.2   Against Fastgcd and UVGCD

In this subsection, we compare ExQRGCD with Fastgcd and UVGCD by examples given by Bini and Boito[6,3]. We note that the results of Fastgcd and UVGCD are quoted from there hence the following comparisons are just subsidiary data since the conditions may not be equal.

Example 5 (Mignotte-like polynomials).
Let the input polynomials be $f(x) = x^{100} + (x - 1/2)^{17}$ and $g(x) = f'(x)$. We compute with tolerance $\varepsilon = 10^{-1}, \ldots, 10^{-12}$. Table 2 shows the detected degrees. ExQRGCD is almost better than others. We note that this is Example 8.2.2 [3] and ExQRGCD is not better than Fastgcd for the polynomials in Example 8.2.1 [3] but same as UVGCD.   ◁

Example 6 (An ill-conditioned case).
Let n be an even positive integer and $k = n/2$; define $f(x) = f_1(x)d(x)$ and

Table 2. Detected degrees (Mignotte-like polys.)

ε	ExQRGCD	ε	Fastgcd	ε	UVGCD
10^{-1}	99	10^{-1}	99	$10^{-1} \ldots 10^{-2}$	99
10^{-2}	25	$10^{-2} \ldots 10^{-5}$	17	$10^{-3} \ldots 10^{-5}$	17
10^{-3}	23	$10^{-6} \ldots 10^{-7}$	6	$10^{-6} \ldots 10^{-11}$	16
10^{-4}	20	$10^{-8} \ldots 10^{-9}$	4	10^{-12}	0
$10^{-5} \ldots 10^{-10}$	16	10^{-10}	3		
10^{-11}	0	10^{-11}	0		

Table 3. Resulting Perturbations (ill-conditioned case)

n	ExQRGCD	Fastgcd	UVGCD
12	2.86×10^{-12}	1.65×10^{-14}	9.99×10^{-15}
14	2.67×10^{-11}	4.81×10^{-14}	3.66×10^{-14}
16	9.31×10^{-10}	2.27×10^{-13}	1.54×10^{-13}
18	4.38×10^{-9}	1.08×10^{-12}	5.21×10^{-13}
20	1.22×10^{-7}	(detected degree fails)	1.59×10^{-12}

$g(x) = g_1(x)d(x)$ where

$$d(x) = \prod_{j=1}^{k}((x - r_1\alpha_j)^2 + r_1^2\beta_j^2), \ r_1 = 0.5, \ r_2 = 1.5,$$
$$f_1(x) = \prod_{j=1}^{k}((x - r_2\alpha_j)^2 + r_2^2\beta_j^2), \ \alpha_j = \cos(j\pi/n),$$
$$g_1(x) = \prod_{j=k+1}^{h}((x - r_1\alpha_j)^2 + r_1^2\beta_j^2), \ \beta_j = \sin(j\pi/n).$$

Table 3 shows the resulting perturbations in the 2-norm (i.e. $\sqrt{\|\Delta_f\|_2^2 + \|\Delta_g\|_2^2}$). ExQRGCD is not good though it detected the correct degree of approximate GCD.   ◁

Other than the above, we have tested several examples in Boito[3]. Although ExQRGCD is more competitive than the original QRGCD algorithm, it is not better than Fastgcd and UVGCD for most of those examples. Moreover, ExQRGCD requires higher precision (e.g. $Digits := 24$ or 32) to make it competitive against them for those examples. ExQRGCD and QRGCD do not refine the resulting factor hence the resulting perturbation may become larger than those algorithms with refinement steps.

5   Concluding Remarks

In this paper, we improved the QRGCD algorithm from the different approach and our algorithm works as same as the original for polynomials without perturbations and much better than the original for polynomials with perturbations or having fake common roots. We again note that ExQRGCD does not refine the output hence it is notable that ExQRGCD is almost better than Fastgcd and UVGCD for Mignotte-like polynomials in Example 5.

We note that our preliminary implementations on Maple and written in C, and generated polynomial data are available at our website:

http://wwwmain.h.kobe-u.ac.jp/~nagasaka/research/snap/exqrgcd/

though some routines derived from the SNAP package are not included.

Acknowledgments. The authors would like to thank an anonymous reviewer for his/her many constructive comments that are very helpful to improve the manuscript.

References

1. Roy, M.F., Sedjelmaci, S.M.: New fast euclidean algorithms. J. Symbolic Comput. 50, 208–226 (2013)
2. von zur Gathen, J., Gerhard, J.: Modern computer algebra, 2nd edn. Cambridge University Press, Cambridge (2003)
3. Boito, P.: Structured Matrix Based Methods for Approximate GCD. Ph.D. Thesis. Department of Mathematics, University of Pisa, Italia (2007)
4. Zeng, Z.: The numerical greatest common divisor of univariate polynomials. In: Randomization, relaxation, and complexity in polynomial equation solving, Providence, RI. Contemp. Math. Amer. Math. Soc, vol. 556, pp. 187–217 (2011)
5. Kaltofen, E., Yang, Z., Zhi, L.: Approximate greatest common divisors of several polynomials with linearly constrained coefficients and singular polynomials. In: ISSAC 2006: Proceedings of the 2006 International Symposium on Symbolic and Algebraic Computation, pp. 169–176 (2006)
6. Bini, D.A., Boito, P.: A fast algorithm for approximate polynomial GCD based on structured matrix computations. In: Numerical Methods for Structured Matrices and Applications. Oper. Theory Adv. Appl, vol. 199, pp. 155–173. Birkhäuser Verlag, Basel (2010)
7. Terui, A.: An iterative method for calculating approximate GCD of univariate polynomials. In: ISSAC 2009: Proceedings of the 2009 International Symposium on Symbolic and Algebraic Computation, pp. 351–358 (2009)
8. Corless, R.M., Watt, S.M., Zhi, L.: QR factoring to compute the GCD of univariate approximate polynomials. IEEE Trans. Signal Process. 52(12), 3394–3402 (2004)
9. Laidacker, M.A.: Another theorem relating Sylvester's matrix and the greatest common divisor. Math. Mag. 42, 126–128 (1969)
10. Golub, G.H., Van Loan, C.F.: Matrix computations, 3rd edn. Johns Hopkins Studies in the Mathematical Sciences. Johns Hopkins University Press, Baltimore (1996)
11. Stetter, H.J.: The nearest polynomial with a given zero, and similar problems. SIGSAM Bull. 33(4), 2–4 (1999)
12. Rezvani, N., Corless, R.M.: The nearest polynomial with a given zero, revisited. SIGSAM Bull. 39(3), 73–79 (2005)
13. Brown, W.S., Traub, J.F.: On Euclid's algorithm and the theory of subresultants. J. Assoc. Comput. Mach. 18, 505–514 (1971)
14. D'Andrea, C., Hong, H., Krick, T., Szanto, A.: An elementary proof of Sylvester's double sums for subresultants. J. Symbolic Comput. 42(3), 290–297 (2007)
15. Sasaki, T.: The subresultant and clusters of close roots. In: ISSAC 2003: Proceedings of the 2003 International Symposium on Symbolic and Algebraic Computation, pp. 232–239 (2003)

Polynomial Evaluation and Interpolation and Transformations of Matrix Structures

Victor Y. Pan

Departments of Mathematics and Computer Science
Lehman College and the Graduate Center of the City University of New York
Bronx, NY 10468 USA
victor.pan@lehman.cuny.edu
http://comet.lehman.cuny.edu/vpan/

Abstract. Multipoint polynomial evaluation and interpolation are fundamental for modern numerical and symbolic computing. The known algorithms solve both problems over any field of constants in nearly linear arithmetic time, but the cost grows to quadratic for numerical solution. By combining a variant of the Multipole celebrated numerical techniques with transformations of matrix structures of [10] we achieve dramatic speedup and for a large class of inputs yield solution algorithms running in nearly linear time as well. The algorithms support similar speedup of approximation of the products of a Vandermonde matrix, its transpose, inverse, and the transpose of the inverse by a vector.

1 Introduction

Multipoint polynomial evaluation and interpolation are fundamental for modern numerical and symbolic computations. Both problems can be solved in nearly linear arithmetic time over any field [12]. For numerical solution, in the presence of rounding errors, however, these algorithms are prone to error propagation, and the users employ quadratic time algorithms [2], [1]. We propose novel reduction of these tasks to computations with HSS matrices, followed by application of the efficient numerically stable algorithms of papers [9], [3], and [18], based on the Multipole celebrated techniques. ("HSS" is the acronym for "hierarchically semiseparable".) This enables dramatic acceleration of the known numerical algorithms, and for a large class of inputs our new algorithms use nearly linear arithmetic time for approximate multipoint polynomial evaluation and interpolation as well as for multiplication by a vector of a Vandermonde matrix and its transpose and inverse (see Theorem 11). The Multipole techniques could possibly help to extend these results to a larger class of inputs. Paper [13] (revisiting the approach of paper [10]) extends our algorithms to a large class of structured matrices (as well as to rational interpolation) via transformation of matrix structures. Due to the size limitation we leave many details, proofs, figures, and references to paper [14].

V.P. Gerdt et al. (Eds.): CASC 2013, LNCS 8136, pp. 273–287, 2013.
© Springer International Publishing Switzerland 2013

2   Definitions and Auxiliary Results

Hereafter "flop" stands for "arithmetic operation performed in the field $\mathbb{C}$ of complex numbers with no error". $|\mathcal{S}|$ denotes the cardinality of a set $\mathcal{S}$.

$M = (m_{i,j})_{i,j=0}^{m-1,n-1}$ is an $m \times n$ matrix. M^T is its transpose, M^H is its Hermitian transpose. For two sets $\mathcal{I} \subseteq \{1, \ldots, m\}$ and $\mathcal{J} \subseteq \{1, \ldots, n\}$ define the submatrix $M(\mathcal{I}, \mathcal{J}) = (m_{i,j})_{i \in \mathcal{I}, j \in \mathcal{J}}$. Write $M(\mathcal{I}, .) = M(\mathcal{I}, \mathcal{J})$ for $\mathcal{J} = \{1, \ldots, n\}$. Write $M(., \mathcal{J}) = M(\mathcal{I}, \mathcal{J})$ for $\mathcal{I} = \{1, \ldots, m\}$. $\mathcal{C}(B)$ and $\mathcal{R}(B)$ are the sets of indices of the rows and columns of a submatrix B of $M = (m_{i,j})_{i,j=1}^{m,n}$, respectively. $\mathcal{R}(B) = \mathcal{I}$ and $\mathcal{C}(B) = \mathcal{J}$ if and only if $B = M(\mathcal{I}, \mathcal{J})$. An $m \times n$ matrix M has a non-unique *generating pair* (F, G^T) *of a length* ρ if $M = FG^T$ for two matrices $F \in \mathbb{C}^{m \times \rho}$ and $G \in \mathbb{C}^{n \times \rho}$. The rank of a matrix is the minimum length of its generating pairs. A matrix M has a rank ρ if and only if it has a nonsingular $\rho \times \rho$ submatrix $M(\mathcal{I}, \mathcal{J})$, and if so, then $M = M(., \mathcal{J})M(\mathcal{I}, \mathcal{J})^{-1}M(\mathcal{I}, .)$, that is $(M(., \mathcal{J}), M(\mathcal{I}, \mathcal{J})^{-1}M(\mathcal{I}, .)$ and $(M(., \mathcal{J})M(\mathcal{I}, \mathcal{J})^{-1}, M(\mathcal{I}, .))$ are two generating pairs of M of length ρ. We refer to generating pairs and *triples* such as $(M(., \mathcal{J}), M(\mathcal{I}, \mathcal{J})^{-1}, M(\mathcal{I}, .))$ as *generators*.

$(B_0 \mid \ldots \mid B_{k-1})$ is a $1 \times k$ block matrix with k blocks $B_0, \ldots, B_{k-1}$, whereas $D_B = \mathrm{diag}(B_0, \ldots, B_{k-1}) = \mathrm{diag}(B_j)_{j=0}^{k-1}$ is a $k \times k$ block diagonal matrix with k diagonal blocks $B_0, \ldots, B_{k-1}$, possibly rectangular. For 1×1 blocks $b_j = B_j$ we arrive at a vector $\mathbf{b}^T = (b_0 \mid \ldots \mid b_{k-1})$ and a $k \times k$ diagonal matrix $D_{\mathbf{b}} = \mathrm{diag}(b_j)_{j=0}^{k-1}$, respectively. $O = O_{m,n}$ is the $m \times n$ matrix filled with zeros. $I = I_n = \mathrm{diag}(1)_{j=0}^{n-1}$ is the $n \times n$ identity matrix. An $n \times n$ matrix M is nonsingular if $\mathrm{rank}(M) = n$ or equivalently if it has the inverse $X = M^{-1}$ such that $XM = MX = I$. M is a $k \times l$ *unitary* matrix if $M^H M = I_l$ or $MM^H = I_k$. These two equations imply one another and imply that $M^H = M^{-1}$ if $k = l$. If a unitary matrix is a vector we call it a *unit vector*.

$\|M\| = \|M\|_2$ denotes the 2-norm of a matrix M. It holds that $\|U\| = 1$ and $\|MU\| = \|UM\| = \|M\|$ for a unitary matrix U. For an $m \times n$ matrix M of a rank ρ define its *SVD* or *full SVD*, $M = S_M \Sigma_M T_M^H$ where S_M and T_M are square unitary matrices, $S_M S_M^H = S_M^T S_M = I_m$, $T_M T_M^H = T_M^H T_M = I_n$, $\Sigma_M = \mathrm{diag}(\mathrm{diag}(\sigma_j(M))_{j=1}^{\rho}, O_{m-\rho,n-\rho})$, $\sigma_j = \sigma_j(M)$ is the jth largest *singular value* of a matrix M for $j = 1, \ldots, \rho$, $\sigma_j = 0$ for $j > \rho$, $\sigma_\rho > 0$, $\sigma_1 = \max_{\|\mathbf{x}\|=1} \|M\mathbf{x}\| = \|M\|$, and

$$\min_{\mathrm{rank}(B) \le s-1} \|A - B\| = \sigma_s(A), \quad s = 1, 2, \ldots \tag{1}$$

Delete the last $n - \rho$ columns of the SVD matrices S_M and Σ_M and the last $m - \rho$ rows of the matrices Σ_M and T_M^H and obtain a *compact SVD* $M = \bar{S}_M \bar{\Sigma}_M \bar{T}_M^H$, defining a generating triple $(\bar{S}_M, \bar{\Sigma}_M, \bar{T}_M^H)$ of the minimum length ρ for the matrix M. Rank revealing factorizations (such as ULV factorization in [3] and [18] for two unitary matrices U and V and a triangular matrix L) are less costly and also supply generating triples of the minimum length.

$\alpha(M)$ and $\beta(M)$ denote the numbers of flops for computing the vectors $M\mathbf{u}$ and $M^{-1}\mathbf{u}$, respectively, maximized over all unit vectors $\mathbf{u}$ and minimized over all algorithms. We write $\beta(M) = \infty$ where the matrix M is singular.

Theorem 1. $\alpha(M) \leq (2m + 2n - 1)\rho - m < 2(m + n)\rho$ *for an* $m \times n$ *matrix* M *given with its generating pair of a length* ρ.

A matrix $\tilde{M}$ is an ϵ-*approximation* of a matrix M if $||\tilde{M} - M|| \leq \epsilon$. We similarly define other ϵ-concepts such as the ϵ-*rank* of a matrix M, denoting the integer $\min_{||\tilde{M}-M|| \leq \epsilon} \text{rank}(\tilde{M})$ (so that $\sigma_\rho(M) > \epsilon \geq \sigma_{\rho+1}(M)$), an ϵ-*basis* for a linear space $\mathbb{S}$ of dimension k, denoting a set of vectors that ϵ-approximate the k vectors of a basis for this space, and an ϵ-generator of a matrix, which is a generator of its ϵ-approximation. $\alpha_\epsilon(M)$ and $\beta_\epsilon(M)$ replace the bounds $\alpha(M)$ and $\beta(M)$ where we ϵ-approximate the vectors $M\mathbf{u}$ and $M^{-1}\mathbf{u}$ instead of evaluating them.

Equations (1) imply the following result.

Theorem 2. *Assume block matrices* $M = \begin{pmatrix} V & B\,Y \\ W & O\,Z \end{pmatrix}$ *and* $M_- = (W \mid Z)$. *If the matrix* M *has* ϵ-*rank at most* ρ *and if* $\sigma_\rho(B) > \epsilon$, *then* $||M_-|| \leq \epsilon$.

In Section 5 we deal with block tridiagonal matrices extended into the southeastern and northwestern corners as in the following 5×5 example,

$$T = \begin{pmatrix} \Sigma_0 & B_0 & O & O & A_0 \\ A_1 & \Sigma_1 & B_1 & O & O \\ O & A_2 & \Sigma_2 & B_2 & O \\ O & O & A_3 & \Sigma_3 & B_3 \\ B_4 & O & O & A_4 & \Sigma_4 \end{pmatrix}. \tag{2}$$

We represent such *extended block tridiagonal* matrices as *extended block diagonal* matrices in two dual ways, $T = \Sigma^{(c)} = \text{diag}(\Sigma_0^{(c)}, \ldots, \Sigma_{k-1}^{(c)}) = \Sigma^{(r)} = \text{diag}(\Sigma_0^{(r)}, \ldots, \Sigma_{k-1}^{(r)})$. Here $\Sigma_q^{(c)} = \begin{pmatrix} B_{q-1 \bmod k} \\ \Sigma_q \\ A_{q+1 \bmod k} \end{pmatrix}$ and $\Sigma_q^{(r)} = (A_q \mid \Sigma_q \mid B_q)$ for $q = 0, \ldots, k-1$ denote the *extended diagonal blocks*, each made up of a triple of the blocks of the matrix T. The central block Σ_q is adjacent to both of its neighbours in both triples if we glue together the lower and upper boundaries of the matrix as well as its right and left boundaries.

3   Dense Structured Matrices. Polynomial and Rational Evaluation and Interpolation. DFT, IDFT, FFT, and Some Transformations of Matrix Structures

$T = (t_{i-j})_{i,j=0}^{n-1}$, $H = (h_{i+j})_{i,j=0}^{n-1}$, $V = V_\mathbf{s} = (s_i^j)_{i,j=0}^{n-1}$, and $C = C_{\mathbf{s},\mathbf{t}} = \left(\frac{1}{s_i-t_j}\right)_{i,j=0}^{n-1}$ denote *Toeplitz, Hankel, Vandermonde, and Cauchy* matrices, respectively, all linked to computations with polynomials [12, Chapters 2 and 3]. It holds that $\alpha(M) = O(n(\log(n))^d)$ for $d \leq 2$, and if $\det(M) \neq 0$ then $\beta(M) = O(n(\log(n))^2)$ for the matrices M of these classes or more generally for the matrices having structures of Toeplitz, Hankel, Vandermonde, or Cauchy type.

Problem 1. Multipoint polynomial evaluation or Vandermonde-by-vector multiplication. INPUT: $2n$ complex scalars $p_0, \ldots, p_{n-1}; s_0, \ldots, s_{n-1}$. OUTPUT: n complex scalars $v_0, \ldots, v_{n-1}$ satisfying $v_i = p(s_i)$ for $p(x) = p_0 + p_1 x + \cdots + p_{n-1} x^{n-1}$ and $i = 0, \ldots, n - 1$ or equivalently

$$V\mathbf{p} = \mathbf{v} \text{ for } V = V_{\mathbf{s}} = (s_i^j)_{i,j=0}^{n-1}, \ \mathbf{p} = (p_j)_{j=0}^{n-1}, \text{ and } \mathbf{v} = (v_i)_{i=0}^{n-1}. \quad (3)$$

Problem 2. Polynomial interpolation or the solution of a Vandermonde linear system of equations.
INPUT: $2n$ complex scalars $v_0, \ldots, v_{n-1}; s_0, \ldots, s_{n-1}$, the last n of them distinct.
OUTPUT: n complex scalars $p_0, \ldots, p_{n-1}$ satisfying equation (3).

Likewise the equations $C\mathbf{u} = \mathbf{v}$, $C = C_{\mathbf{s},\mathbf{t}} = \left(\frac{1}{s_i - t_j}\right)_{i,j=0}^{n-1}$, $\mathbf{u} = (u_j)_{j=0}^{n-1}$, and $\mathbf{v} = (v_i)_{i=0}^{n-1} = (\sum_{j=0}^{n-1} \frac{u_j}{s_i - t_j})_{i=0}^{n-1}$ link multipoint rational evaluation and interpolation to Cauchy matrix computations. One can solve all these polynomial and rational computational problems numerically by using $O(n \log(n))$ flops [12] provided the knots are the roots of 1, $s_i = w^i$ for $i = 0, \ldots, n - 1$, $w = w_n = \exp(2\pi\sqrt{-1}/l)$ denotes a primitive nth root of 1, $V_{\mathbf{s}} = (w^{ij})_{i,j=0}^{n-1}$, and $\Omega = \frac{1}{\sqrt{n}} V_{\mathbf{s}}$. In this case Problems 1 and 2 turn into the tasks of computing the forward and inverse discrete Fourier transforms, *DFT* and *IDFT*. The *FFT* (Fast Fourier transform) and Inverse FFT perform DFT and IDFT by using $1.5n \log_2(n)$ and $1.5n \log_2(n) + n$ numerically stable flops, respectively, if n is a power of 2, whereas $O(n \log(n))$ flops are sufficient for any n [12, Problem 2.4.2]. Note that $\Omega^H \Omega = I_n$, $\Omega^T = \Omega$, Ω and $\Omega^H = \Omega^{-1} = \frac{1}{\sqrt{n}}(w^{-ij})_{i,j=0}^{n-1}$ are unitary matrices.

In spite of some common properties the four matrix structures have quite distinct features. The matrix structure of Cauchy type is invariant in row and column interchange (in contrast to the structures of Toeplitz and Hankel types) and enables expansion of the matrix entries into Laurent series (unlike the structures of the three other types). The paper [10], however, links the four structures to each other by means of structured matrix multiplication and exploits *this link to extend any successful matrix inversion algorithm for the matrices of any of the four classes to the matrices of the three other classes*. The present paper shows a new specialization of this general approach. We first employ the Vandermonde–Cauchy links defined by the following equation [12, Section 3.6],

$$C_{\mathbf{s},\mathbf{t}} = \text{diag}(t(s_i)^{-1})_{i=0}^{n-1} V_{\mathbf{s}} V_{\mathbf{t}}^{-1} \text{diag}(t'(t_j))_{j=0}^{n-1} \quad (4)$$

where $\mathbf{s} = (s_i)_{i=0}^{n-1}$, $\mathbf{t} = (t_i)_{i=0}^{n-1}$, and $t(x) = \prod_{i=0}^{n-1}(x - t_i)$. One can first compute the values $v(t_0) = -t_0^n, \ldots, v(t_{n-1}) = -t_{n-1}^n$ of the polynomial $v(x) = t(x) - x^n$, by using $O(n \log n)$ flops, and then compute the coefficients of this polynomial as the solution of Problem 2 of polynomial interpolation in a special case. For $\mathbf{t} = (fw^j)_{j=0}^{n-1}$ the knots t_i are the roots of 1 scaled by f, $t(x) = x^n - f^n$, $t'(x) = nx^{n-1}$, $V_{\mathbf{t}} = \sqrt{n}\Omega \, \text{diag}(f^j)_{j=0}^{n-1}$, and $V_{\mathbf{t}}^{-1} = \text{diag}(f^{-j})_{j=0}^{n-1} \Omega^H/\sqrt{n}$, and equation (4) turns into the equation

$$C_{\mathbf{s},f} = \sqrt{n} \, \text{diag}\left(\frac{1}{s_i^n - f^n}\right)_{i=0}^{n-1} V_{\mathbf{s}} \, \text{diag}(f^{-j})_{j=0}^{n-1} \Omega^H \, \text{diag}(w^{-j})_{j=0}^{n-1}, \quad (5)$$

which links Vandermonde matrix $V_\mathbf{s}$ and its inverse to the Cauchy matrix $C_{\mathbf{s},f} = C_{\mathbf{s},\mathbf{t}}$, having a knot set $\mathcal{T} = \{t_j = f\omega^j\}_{j=0}^{n-1}$ and said to be a *CV matrix*. $C_{\mathbf{s},f} = -C_{f,\mathbf{s}}^T$ is said to be a *CV^T matrix*.

4   HSS Matrices and Neutered Blocks

Next we link CV matrices to the class of HSS matrices, which extend banded matrices and their inverses. We refer the reader to [5], [9], [3], [17], and the bibliography therein on such extensions, long studied by many authors under the names of matrices with low Hankel rank, rank structured matrices, quasiseparable, and weakly, recursively, or sequentially semiseparable matrices.

Definition 1. *(Cf. [14, Figures 1–3].) Assume an $m \times n$ matrix M with a block diagonal $\widehat{\Sigma}' = (\Sigma_0', \ldots, \Sigma_{k-1}')$ and the two dual extended block diagonals $\widehat{\Sigma}^{(c)} = (\Sigma_0^{(c)}, \ldots, \Sigma_{k-1}^{(c)})$ and $\widehat{\Sigma}^{(r)} = (\Sigma_0^{(r)}, \ldots, \Sigma_{k-1}^{(r)})$ (cf. the end of Section 2). For simplicity we also write $\widehat{\Sigma} = (\Sigma_0, \ldots, \Sigma_{k-1})$ instead of $\widehat{\Sigma}^{(c)} = (\Sigma_0^{(c)}, \ldots, \Sigma_{k-1}^{(c)})$, dropping the superscript $^{(c)}$. (i) (See [9, Section 1].) A block of the matrix M is neutered unless it overlaps the extended block diagonal, and so such block is either subdiagonal or superdiagonal. For a set of consecutive indices $\mathcal{J}$ remove all rows of the block column $M(., \mathcal{J})$ that overlap the extended block diagonal and obtain the neutered block column $N(\mathcal{J}) = N^{(c)}(\mathcal{J})$. (ii) It is a basic neutered block column, dual to an extended diagonal block Σ_q and denoted N_q if $\mathcal{J} = \mathcal{C}(\Sigma_q)$, that is if this neutered block column shares its column indices with the submatrix Σ_q. (iii) The neutered union $N(N(\mathcal{J}), N(\mathcal{K}))$ of two neutered block columns $N(\mathcal{J})$ and $N(\mathcal{K})$ is the neutered block column $N(\mathcal{J} \cup \mathcal{K})$. (iv) Similarly define neutered block rows $N^{(r)}(\mathcal{I})$, their unions, and the basic neutered block rows $N_p^{(r)}$ dual to the extended diagonal blocks $\Sigma_p^{(r)}$.*

Definition 2. *A matrix M, given with its extended block diagonal, is an (l, u)-HSS block matrix if l is the maximum rank of its subdiagonal neutered blocks and if u is the maximum rank of its superdiagonal neutered blocks. A matrix M is ρ-neutered if it is a (ρ, ρ)-HSS block matrix or equivalently if ρ is the maximum rank of its neutered blocks. A matrix M is basically ρ-neutered if all its basic neutered block columns have ranks at most ρ. By replacing ranks with ϵ-ranks we define (ϵ, l, u)-HSS block matrices, basically (ϵ, ρ)-neutered matrices, and (ϵ, ρ)-neutered matrices.*

We immediately verify the following results.

Theorem 3. *Every neutered block is a block submatrix of the neutered union of some basic neutered block columns as well as of some basic neutered block rows.*

Theorem 4. *Assume a matrix $M = (B_0 \mid \ldots \mid B_{k-1})$ having k block columns $B_0, \ldots, B_{k-1}$ and given with its extended block diagonal $\widehat{\Sigma} = (\Sigma_0, \ldots, \Sigma_{k-1})$ and with k generating pairs of lengths at most ρ defining the k basic neutered block* columns $N_0, \ldots, N_{k-1}$. Then one can modify the extended diagonal blocks Σ_q to obtain generating pairs of lengths at most ρ defining the k block columns B'_q of the resulting matrix $M' = (B'_0 \mid \ldots \mid B'_{k-1})$.

Corollary 1. *(Cf. Theorem 1.) Under the assumptions of Theorem 4 let $\Sigma = \mathrm{diag}(M)$ be generic matrix. Then $\alpha(M) \leq \alpha(\Sigma) + (2m + 2n - 1)k\rho$.*

One can yield superior estimates in the case of ρ-neutered matrices.

Theorem 5. *(See [4, Section 3], [3, Sections 3 and 4], and our Remark 1.) Assume an (l, u)-HSS block matrix M of size $m \times n$ having an extended block diagonal $\widehat{\Sigma} = (\Sigma_0, \ldots, \Sigma_{k-1})$ with $m_q \times n_q$ extended diagonal blocks Σ_q, $q = 0, \ldots, k - 1$ such that $\sum_{q=0}^{k-1} m_q n_q = O((l + u)(m + n))$. Then it holds that*
(i) $\alpha(M) \leq 2\sum_{q=0}^{k-1}((m_q + n_q)(l + u) + m_q n_q) + 2l^2 k + 2u^2 k = O((l+u)(m+n))$, and (ii) if $m = n$ and if the matrix M is nonsingular, then $\beta(M) = O((l+u)^2 n)$.

Part (i) is supported by the algorithms of [4, Section 3]. They separately multiply by a vector the extended block diagonal matrix $\Sigma = \mathrm{diag}(M)$, the sub- and superdiagonal parts of the matrix M. Part (ii) is supported by the algorithms of [3]. Both papers as well as the study in [17], [18], and [5] rely on the representation of an (l, u)-HSS matrix M with *HSS generators*, which we define in Theorem 6 below, but first demonstrate by the following 4×4 example,

$$M = \begin{pmatrix} \Sigma_0 & S_0 T_1 & S_0 B_1 T_2 & S_0 B_1 B_2 T_3 \\ P_1 Q_0 & \Sigma_1 & S_1 T_2 & S_1 B_2 T_3 \\ P_2 A_1 Q_0 & P_2 Q_1 & \Sigma_2 & S_2 T_3 \\ P_3 A_2 A_1 Q_0 & P_3 A_2 Q_1 & P_3 Q_2 & \Sigma_2 \end{pmatrix}.$$

Theorem 6. *(Cf. [5], [17], the bibliography therein, and our Remark 1.) Assume a $k \times k$ matrix M with an extended block diagonal $\widehat{\Sigma} = (\Sigma_0, \ldots, \Sigma_{k-1})$, where $\Sigma_q = M(I_q, J_q)$, $q = 0, \ldots, k - 1$. Then M is an (l, u)-HSS block matrix if and only if there exists a nonunique family of HSS generators $\{P_i, Q_h, S_h, T_i, A_g, B_g\}$ such that $M(I_i, J_h) = P_i A_{i-1} \cdots A_{h+1} Q_h$ and $M(I_h, J_i) = S_h B_{h+1} \cdots B_{i-1} T_i$ for $0 \leq h < i < k$. Here P_i, Q_h and A_g are $|I_i| \times l_i$, $l_{h+1} \times |J_h|$, and $l_{g+1} \times l_g$ matrices, respectively, whereas S_h, T_i and B_g are $|I_h| \times u_{h+1}$, $u_i \times |J_i|$, and $u_g \times u_{g+1}$ matrices, respectively, $g = 1, \ldots, k - 2$, $h = 0, \ldots, k - 2$, $i = 1, \ldots, k - 1$, $l = \max_g\{l_g\}$, and $u = \max_h\{u_h\}$.*

Now one can redefine the (l, u)-HSS block matrices as those allowing representation with some HSS generator families $\{P_h, Q_i, A_g\}$ and $\{S_h, T_i, B_g\}$ above, where the integers l and u called *lower and upper lengths* or *orders* of the HSS generators, respectively.

Remark 1. The cited bibliography covers matrices with block diagonals, but the study presented there, including Theorems 5 and 6, is readily extended to the case of matrices with extended block diagonals as well as (ϵ, l, u)-HSS block matrices and (ϵ, ρ)-neutered matrices, which we define by replacing ranks with ϵ-ranks, generators with ϵ-generators, computation with ϵ-approximation, and the bounds $\alpha(M)$ and $\beta(M)$ with $\alpha_\epsilon(M)$ and $\beta_\epsilon(M)$, respectively.

5 ϵ-approximation of CV Matrices by HSS Matrices

[9, Section 4] and [3, Section 2.4] ϵ-approximate the matrix $C_{1,\omega_{2n}}$ by an (l, u)-HSS block matrix for $l = u = O(\log(n))$, ϵ of order $c'/n^{c''}$, and two positive constants c' and c'', and then solve a linear system of equations with this matrix numerically in nearly linear time by applying the Multipole techniques. We yield similar results for multiplication by a vector of CV and CV^T matrices as well as of their inverses where there exist the inverses. Namely, in this section, given a vector $\bar{\mathbf{s}} = (\bar{s}_i)_{i=0}^{n-1}$ and a complex point f on the unit circle $\{f : |f| = 1\}$, we seek a permutation matrix P and a ρ-neutered matrix that approximates a CV matrix $C = C_{\mathbf{s},f} = (\frac{1}{s_i - f\omega^{j-1}})_{i,j=0}^{n-1}$ within a norm bound ϵ for $\mathbf{s} = (s_i)_{i=0}^{n-1} = P\bar{\mathbf{s}}$ and $\rho = O(\log(n/\sqrt{\epsilon}))$.

5.1   Small-rank Approximation of Cauchy Matrices Where the Knot Sets $\mathcal{S}$ and $\mathcal{T}$ Are Separated from One Another (cf. [3])

Definition 3. *(See [3, page 1254].) For a separation bound $\theta < 1$ and a complex separation center c, two complex points s and t are (θ, c)-separated from one another if $|\frac{t-c}{s-c}| \le \theta$. Two sets of complex numbers $\mathcal{S}$ and $\mathcal{T}$ are (θ, c)-separated from one another if every two points $s \in \mathcal{S}$ and $t \in \mathcal{T}$ are (θ, c)-separated from one another. $\delta_{c,\mathcal{S}} = \min_{s \in \mathcal{S}} |s - c|$ and $\delta_{c,\mathcal{T}} = \min_{t \in \mathcal{T}} |t - c|$ denote the distances from the center c to the sets $\mathcal{S}$ and $\mathcal{T}$, respectively.*

Lemma 1. *(See [3, equation (2.8)].) Suppose two complex values s and t are (θ, c)-separated from one another for a positive $\theta < 1$ and a complex c and write $q = \frac{t-c}{s-c}$, $|q| \le \theta$. Then for every positive integer ρ it holds that*

$$\frac{1}{s-t} = \frac{1}{s-c} \sum_{i=0}^{\rho-1} \frac{(t-c)^i}{(s-c)^i} + \frac{q_\rho}{s-c} \text{ where } |q_\rho| = \frac{|q|^\rho}{1-|q|} \le \frac{\theta^\rho}{1-\theta}.$$

Corollary 2. *(Cf. [3, Section 2.2].) Suppose two sets of $2n$ distinct complex numbers $\mathcal{S} = \{s_0, \ldots, s_{n-1}\}$ and $\mathcal{T} = \{t_0, \ldots, t_{n-1}\}$ are (θ, c)-separated from one another for $0 < \theta < 1$ and a complex c. Write $C = (\frac{1}{s_i - t_j})_{i,j=0}^{n-1}$ and $\delta = \delta_{c,\mathcal{S}} = \min_{i=0}^{n-1} |s_i - c|$. Then for every positive integer ρ it holds that $C = FG^T + E$, where $F = (1/(s_i - c)^{\nu+1})_{i,\nu=0}^{n-1,\rho-1}$, $G = ((t_j - c)^\nu)_{j,\nu=0}^{n-1,\rho-1}$, and $\|E\| \le \frac{n\theta^\rho}{(1-\theta)\delta}$. Furthermore we can compute the $n \times \rho$ matrices F and G by applying $2\rho n$ flops.*

In the corollary and throughout, we can replace $\delta = \delta_{c,\mathcal{S}} = \min_{i=0}^{n-1} |s_i - c|$ by $\delta = \delta_{c,\mathcal{T}} = \min_{j=0}^{n-1} |t_j - c|$ because of the symmetric roles of the sets $\mathcal{S}$ and $\mathcal{T}$. Corollary 2 defines a $\|E\|$-generating pair of a length at most ρ for the Cauchy matrix C. Unless any of the values $1 - \theta$ and δ is small, the norm $\|E\|$ is small already for moderately large integers ρ.

5.2   An Extended Block Diagonal of a CV Matrix (Cf. (2), our Figures 1 and 2, and [14, Algorithm 1].)

Assume the polar coordinates for the knots $\bar{s}_i = |\bar{s}_i| \exp(2\pi\bar{\phi}_i\sqrt{-1})$ where $0 \le \bar{\phi}_i < 2\pi$, $\bar{\phi}_i = 0$ if $\bar{s}_i = 0$, and $i = 0, \ldots, n-1$. Write $\phi_0 = \min_{i=0}^{n-1} \bar{\phi}_i$, reorder the angles $\bar{\phi}_i$ in the nondecreasing order breaking ties arbitrarily, let P denote the permutation matrix that defines this reordering, and write $(\phi_i)_{i=0}^{n-1} = P(\bar{\phi}_i)_{i=0}^{n-1}$ and $(s_i)_{i=0}^{n-1} = P(\bar{s}_i)_{i=0}^{n-1}$. To simplify the notation assume that $f = 1$ and $n = hk$ for two positive integers h and k. Let $\mathcal{S}'_q$ and $\mathcal{T}_q = \{\omega^j\}_{j=qh}^{(q+1)h-1}$ denote the subsets of the sets $\mathcal{S}$ and $\mathcal{T}$, respectively, consisting of the knots s_i and $t_j = \omega^j$ such that $qh \le \phi_i < (q+1)h - 1$, $qh \le j < (q+1)h - 1$, $q = 0, \ldots, k - 1$. The two subsets are the intersections of the sets $\mathcal{S}$ and $\mathcal{T}$ with the semi-open sector Γ_q of the complex plane bounded by the pairs of the rays from the origin to the points ω^{hq} and $\omega^{(q+1)h}$. Write $\mathcal{S}_q = \mathcal{S}'_{q-1 \bmod k} \cup \mathcal{S}'_q \cup \mathcal{S}'_{q+1 \bmod k}$, $q = 0, \ldots, k - 1$. Now for the matrix $C = \left(\frac{1}{s_i - \omega^j}\right)_{i,j=0}^{n-1}$ define the k diagonal blocks $\Sigma'_q = \left(\frac{1}{s_i - \omega^j}\right)_{i \in \mathcal{S}'_q, \, j \in \mathcal{T}_q}$, $q = 0, \ldots, k - 1$, which have hn entries overall, and the k extended diagonal blocks $\Sigma_q = \left(\frac{1}{s_i - \omega^j}\right)_{i \in \mathcal{S}_q, j \in \mathcal{T}_q}$, $q = 0, \ldots, k - 1$, which have $3hn$ entries overall, and then for every q, $q = 0, \ldots, k - 1$, partition the block column $\left(\frac{1}{s_i - \omega^j}\right)_{0 \le i < n, j \in \mathcal{T}_q}$ into the block Σ_q and the dual basic neutered block column $N_q = N_q^{(c)}$.

5.3   The ϵ-ranks of Basic Neutered Block Columns

Theorem 7. *(Cf. Remarks 2–5 and Figure 2.) Assume a CV matrix C of the previous subsection. Then (i) the extended diagonal blocks $\Sigma_0, \ldots, \Sigma_{k-1}$ together have exactly $3hn$ entries and (ii) for $|p - q \bmod k| > 1$ the row index sets $\mathcal{R}(\Sigma_p)$ and $\mathcal{R}(\Sigma_q)$ have no overlap. Furthermore (iii) there are points $c_0, \ldots, c_{k-1}$ on the unit circle $\{z : |z| = 1\}$ and at the distance of at least $0.5/(kn)$ from the set $\mathcal{S}$ such that the sets $\mathcal{S}_p$ and $\mathcal{T}_q$ are (θ, c_q)-separated from one another for $\theta = (1.5\pi/k)/\sin(3\pi/k) \approx 0.5/\cos(1.5\pi/k)$ as long as $|p - q \bmod k| > 1$.*

Proof. One can readily verify parts (i) and (ii). Let us prove part (iii). Let $\mathcal{A}(s, t)$ denote the arc of the unit circle $\{z : |z| = 1\}$ with the end points s and t. For every q, $q = 0, \ldots, k - 1$, choose a center c_q on the arc $\mathcal{A}(\omega_{4n}^{(4q+1)h}, \omega_{4n}^{(4q+3)h})$ at the distance at least $\frac{2}{kn}$ from the set $\mathcal{S}$ (as we require). This is possible because the set has exactly n elements. The arc has length π/k and shares the midpoint $\omega_{2n}^{(2q+1)h}$ with the arc $\mathcal{A}(\omega^{qh}, \omega^{(q+1)h})$ of length $2\pi/k$, on which all points of the set $\mathcal{T}_q$ lie. Therefore, these points lie at the distance less than $\frac{3\pi}{2k}$ from the center c_q. Furthermore, unless $|p - q| \le 1$ or $|p - q| = k - 1$, all points of the set $\mathcal{S}_p$ lie outside the sector of the complex plane bounded by the pairs of the rays from the origin to the points $\omega^{(q-1)h}$ and $\omega^{(q+2)h}$, and then $\text{Distance}(c_q, \mathcal{S}_p) \ge \sin(3\pi/k)$. Therefore, the sets $\mathcal{S}_p$ and $\mathcal{T}_q = \{\omega^j\}_{j=(q-1)h}^{hq-1}$ are (θ, c_q)-separated from one another for $\theta = (1.5\pi/k)/\sin(3\pi/k)$ unless $|p - q| \le 1$ or $|p - q| = k - 1$. It holds that $1.5\pi/k \approx \sin(1.5\pi/k)$ for sufficiently large values $k = n/h$, whereas $\sin(1.5\pi/k)/\sin(3\pi/k) = 0.5/\cos(1.5\pi/k)$. Summarizing obtain part (iii).

Corollary 3. *Assume a CV matrix C of Section 5.1 and its extended block diagonal. Assume a positive integer ρ and write $\theta = \frac{1.5\pi/k}{\sin(3\pi/k)}$ and $\epsilon = \frac{n\theta^\rho}{(1-\theta)\sin(3\pi/k)}$. Then C is a basically (ϵ, ρ)-neutered matrix.*

Proof. Theorem 7 implies that every basic neutered block column is (θ, c_q)-separated for some center c_q and for the above θ. Apply Corollary 2 for $\delta = \sin(3\pi/k)$ and deduce that the ϵ-ranks are at most ρ.

Remark 2. To bound ρ from above in terms of k, n, and ϵ we rewrite the expression of the corollary to obtain $\frac{1}{\theta^\rho} = \frac{2kn^2}{\epsilon}\frac{1}{1-\theta}$, and so $\rho = (\log_2(\frac{2kn^2}{\epsilon}) + \log_2(\frac{1}{1-\theta}))/\log_2(\frac{1}{\theta})$. We can assume that n is reasonably large, then choose, say, $k \geq 12$, and obtain from Theorem 7 that $\theta < 0.5554$ (one can specify such bounds for other choices of k, noting that $\theta \to 1/2$ as $k = n/h \to \infty$). Now for $k \geq 12$ deduce that $\rho < 1.2\log_2(kn^2/\epsilon) + 2.6 < 1.2\log_2(n^3/\epsilon) + 2.6$ and $\epsilon < kn^2/2^{0.83\rho-2.17} \leq n^3/2^{0.83\rho-2.17}$.

Remark 3. In the proof of Theorem 7, we can rotate all sectors Γ_q and centers c_q by a fixed angle ϕ, $0 \leq \phi < 2\pi$, redefine the block diagonal matrix Σ accordingly, and then readily extend Theorem 7 and Corollary 3. In particular, rotation by the angle π/k produces new basic neutered block columns, each overlapping a pair of old adjacent basic neutered block columns.

Remark 4. Among the basic neutered block columns N_q only N_0 and N_{k-1} are blocks, whereas for $q = 1, \ldots, k-2$ the matrices N_q are made up of the pairs of super- and subdiagonal blocks. By gluing the upper and lower boundaries of the matrix C, however, we can turn all these pairs into single blocks. Furthermore our study is invariant under block row and block column permutations, with which we can move any diagonal block into the northwestern position $(0,0)$ or the southeastern position $(k-1, k-1)$.

Remark 5. One can extend the results of this section in various ways. To simplify the notation we have assumed that $f = 1$ and $n = hk$ throughout, but the same arguments and proofs can be applied to any complex f and any triple of integers (h, k, n) satisfying $|f| = 1$ and $k = \lceil n/h \rceil$. Since $-C_{s,t}^T = C_{t,s}$ one can replace the CV matrix of Corollary 3 by a CV^T matrix $(\frac{1}{f\omega^i - t_j})_{i,j=0}^{n-1}$ for $|f| = 1$. The proof techniques enable extension to rectangular CV and CV^T matrices as well as to a Cauchy matrix $C_{s,t}$ with the set $\mathcal{T}$ or $\mathcal{S}$ more or less equally spaced about a segment of a line or a smooth curve on the complex plane, the unit circle $\{x : |x| = 1|\}$ being an example of such a curve.

5.4   The $((2k-1)\epsilon)$-ranks of Neutered Blocks

Next, under an additional assumption on the CV matrices, we extend the bound ρ of Corollary 3 on the ϵ-rank from the basic neutered block columns at first to the neutered unions of the pairs and the chains of such adjacent block columns and then (by virtue of Theorem 3) to all neutered blocks of the matrix.

Definition 4. *(Cf. [14, Figures 5 and 6].) Assume four positive integers h, k, $n = hk$, and $\rho \leq \lfloor n/2 \rfloor$ and a basically (ϵ, ρ)-neutered $n \times n$ matrix C for ϵ of Remark 2. In addition to the extended block diagonal $\widehat{\Sigma}$ define another one, $\widehat{\Sigma}'$, by applying Remark 3 for $\phi = \pi/k$. For $q = 0, \ldots, k - 1$ let $N_q = N_c(\Sigma_q)$ and $N'_q = N_c(\Sigma'_q)$ denote the basic neutered block columns dual to the extended diagonal blocks Σ_q and Σ'_q, respectively, and let $\cap_q$ and $\cap'_q$ denote the two neutered blocks defined by the common entries of the matrix pairs (N_q, N'_q) and $(N_{q+1 \mod k}, N'_q)$, respectively. We call the matrix C weakly (ϵ, ρ)-neutered if the blocks $\cap_q$ and $\cap'_q$ have ϵ_+-ranks at least ρ for some $\epsilon_+ > \epsilon$ and all q.*

Remark 6. For a weakly (ϵ, ρ)-neutered matrix we must have $\lfloor h/2 \rfloor \geq \rho$ because each matrix $\cap_q$ and $\cap'_q$ has at least $\lfloor h/2 \rfloor$ columns and has ϵ_+-rank at least ρ.

Theorem 8. *For a weakly (ϵ, ρ)-neutered $n \times n$ CV matrix C the $((2k - 1)\epsilon)$-ranks of the neutered unions of the blocks of any chain of adjacent basic neutered block columns cannot exceed ρ.*

Proof. (Cf. [14, Figure 6].) By virtue of Corollary 3 the basic neutered block columns N_q and N'_q and therefore also the blocks $\cap_q$ and $\cap'_q$ have ϵ-ranks at most ρ for all q, but actually all these blocks have ϵ-ranks exactly ρ for all q because the matrix C is weakly (ϵ, ρ)-neutered. Let $\cap_q = \bar{S}_{\cap_q} \bar{\Sigma}_{\cap_q} \bar{T}_{\cap_q}^H$ be a compact SVD for $0 \leq q < k$. Then $\bar{S}_{\cap_q}^H \cap_q = \bar{\Sigma}_{\cap_q} \bar{T}_{\cap_q}^H = \begin{pmatrix} M_\rho \\ E_q \end{pmatrix}$ where M_ρ is a $\rho \times \rho$ matrix, $||E_q|| \leq \epsilon$, and $\sigma_\rho(M_\rho) = \sigma_\rho(\cap_q) > \epsilon$ because the matrix C is weakly (ϵ, ρ)-neutered. Represent the matrix $U_q = \bar{S}_{\cap_q}^H N(N_q, N'_q)$ as $\begin{pmatrix} V & M_\rho & Y \\ W & E_q & Z \end{pmatrix}$, ϵ-approximate the matrix E_q by the null matrix O, apply Theorem 2 for $B = M_\rho$, and obtain that $||(W \mid Z)|| \leq \epsilon$. Consequently only the first ρ rows of the matrix $U_q - E'_q$ are nonzero for some matrix E'_q satisfying $||E'_q|| \leq 2\epsilon$. Extend this argument to prove that for any pair of integers l and q, $0 < l < k$, $0 \leq q < k$, and the neutered union U of the chain of the $2l - 1$ overlapping neutered blocks $N_q, N'_q, N_{q+1 \mod k}, N'_{q+1 \mod k}, \ldots, N_{q+l-1 \mod k}, N'_{q+l-1 \mod k}$ there is a matrix $E_{q,l}$ such that the matrix $U - E_{q,l}$ has only ρ nonzero rows, and so $\mathrm{rank}(U - E_{q,l}) \leq \rho$, whereas $||E_{q,l}|| \leq (2l - 1)\epsilon$. This implies the theorem. □

Theorems 3 and 8 and Remark 1 together imply the following corollary.

Corollary 4. *A CV $n \times n$ matrix is $((2k - 1)\epsilon, \rho)$-neutered if it is weakly (ϵ, ρ)-neutered.*

Remark 7. Clearly a basically (ϵ, ρ)-neutered matrix is weakly (ϵ, ρ)-neutered unless ρ exceeds the ϵ_+-rank of some block $\cap_q$ or $\cap'_q$ for $\epsilon_+ > \epsilon$. For CV matrices $C_{1,f}$ and more generally CV matrices C with the sets $\mathcal{S}$ more or less uniformly distributed over the unit circle $\{x : |x| = 1\}$, all blocks $\cap_q$ and $\cap'_q$ have the same ϵ-rank. This is also the ϵ_+-rank of the blocks $\cap_q$ and $\cap'_q$ for all q if we choose ϵ such that the ϵ-rank does not change under a sufficiently small increase of ϵ. Therefore the CV matrix C is weakly (ϵ, ρ)-neutered and consequently, by

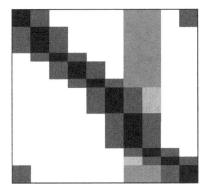

Fig. 1. The neutered union of two basic neutered block columns is shown in green. The rest of them is shown in blue and pink.

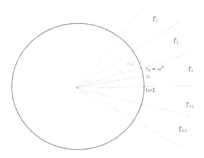

Fig. 2. Blue lines bound the sectors Γ_0, Γ_1, Γ_3, Γ_{k-2}, and Γ_{k-1}

virtue of Corollary 4, is $((2k-1)\epsilon, \rho)$-neutered for ϵ and ρ of Remark 2 provided $12 \le k < n$, $h = n/k \ge 2\rho$ (cf. Remark 6), and so $\rho = O(\log(n))$ where $1/\epsilon = O(\log(n))$. Such bound was stated in papers [9] and [3].

6 Multiplication of CV and Vandermonde Matrices and Their Transposes and Inverses by a Vector

Given an $n \times n$ CV matrix, we seek an $(n\epsilon)$-approximation of its product by a vector. Our algorithm supporting the following theorem accelerates by a factor of $\sqrt{n/\log(n)}$ the known algorithms provided that $\log(1/\epsilon) = O(\log(n))$.

Theorem 9. *Assume positive integers k and n, $n > k$, complex $s'_0, \ldots, s'_{n-1}$, and f, $|f| = 1$, positive ϵ, and a CV matrix $C' = (\frac{1}{s'_i - \omega^j})_{i,j=0}^{n-1}$. Then it holds that (i) $\alpha_{(k+1)\epsilon}(C') < 4n\sqrt{3n\rho}$ for ϵ and ρ of Remark 2. (ii) Consequently it holds that $\alpha_{n\epsilon}(C') = O(n\sqrt{n\log(n)})$ where $\log(1/\epsilon) = O(\log(n))$.*

Proof. As in Section 5.1, define a permutation matrix P, an extended block diagonal matrix Σ, and the basic neutered block columns $N_0, \ldots, N_{k-1}$ of the matrix $C = PC'$. Clearly $\alpha_{(k+1)\epsilon}(C') = \alpha_{(k+1)\epsilon}(C) \le \alpha_\epsilon(\Sigma) + \sum_{q=0}^{k-1} \alpha_\epsilon(N_q) + kn$. By virtue of Theorem 7, every basic neutered block column N_q has rank at most ρ and the matrix Σ has at most $3hn$ nonzero entries. Therefore, $\alpha(\Sigma) \le 6hn - n$, whereas by virtue of Theorem 1, $\alpha(N_q) \le 2(h + n - n_q)\rho - \rho - (n - n_q)$ for the $(n - n_q) \times h$ basic neutered block column N_q. Note that $\sum_{q=0}^{k-1} n_q = 3n$, and so $\sum_{q=0}^{k-1} \alpha(N_q) \le 2(h + n)k\rho - 6n\rho - k\rho + (3 - k)n = 2kn\rho - (4n + k)\rho + (3 - k)n$ because $hk = n$. Combine the above bounds and deduce that $\alpha_{(k+1)\epsilon}(C) \le 6hn + 2kn\rho - (4n + k)\rho + (2 - k)n < 6hn + 2kn\rho$. Choose $k\rho = 3h = 3n/k$, and deduce that $k = \sqrt{3n/\rho}$, $k\rho = \sqrt{3n\rho}$, and therefore $\alpha_{(k+1)\epsilon}(C) < 4n\sqrt{3n\rho}$. This proves part (i), which implies part (ii). $\square$

Remark 8. The theorem can be readily extended to the case of rectangular CV matrices as well as CV^T matrices $C = (\frac{1}{f\omega^i - s_j})_{i,j=0}^{m-1,n-1}$ for $|f| = 1$.

For $((2k-1)\epsilon)$-approximate multiplication by a vector of a weakly (ϵ, ρ)-neutered CV matrix C we yield nearly linear upper bounds on $\alpha_{(2k-1)\epsilon}(C)$, $\alpha_{(2k-1)\epsilon}(C^T)$, $\beta_{(2k-1)\epsilon}(C)$, and $\beta_{(2k-1)\epsilon}(C^T)$ provided that $\log(1/\epsilon) = O(\log(n))$.

Theorem 10. *Assume a weakly (ϵ, ρ)-neutered $n \times n$ CV matrix C. Then it holds that (i) $\alpha_{(2k-1)\epsilon}(C) \le 16n\rho + 4k\rho^2 + 6hn \le 30n\rho + 4n\lfloor h/2 - \rho \rfloor = O(n\rho)$ for $k = n/h$ and $\rho = \lfloor h/2 \rfloor$ (cf. Remark 6). Therefore, $\alpha_{(2k-1)\epsilon}(C) = O(n\log(n/\sqrt{\epsilon}))$ for $\rho = O(\log(n/\sqrt{\epsilon}))$ of Remark 2. (ii) Furthermore if the matrix C is nonsingular and if $(2k - 1)\epsilon \|C^{-1}\| < 1/3$, then $\beta_{(2k-1)\epsilon}(C) = O(n\rho^2)$, and so $\beta_{(2k-1)\epsilon}(C) = O(n(\log(n))^2)$ where $1/\epsilon = O(\log(n))$.*

Proof. Combine Corollary 4 with part (i) of Theorem 5 for $l = u = \rho$, note that in this application of the theorem it holds that $\sum_{q=0}^{k-1} m_q = 3n$, $\sum_{q=0}^{k-1} n_q = n$, and $\sum_{q=0}^{k-1} m_q n_q \le 3hn$, deduce that $\alpha_{(2k-1)\epsilon}(C) \le 16n\rho + 4k\rho^2 + 6hn$, and obtain the bounds of part (i) of Theorem 10. Likewise, to deduce its part (ii) combine Corollary 4 with part (ii) of Theorem 5 and [16, Corollary 1.4.19]. $\square$

Theorem 11. *Suppose equation (5) has been applied to express an $n \times n$ Vandermonde matrix V_s through a CV matrix $C = C_{s,f}$ for a complex f such that $|f| = 1$. Assume that $\mathbf{u}$ is a unit vector. Then (i) one can ν-approximate the vector $V_s\mathbf{u}$ for $\nu = (\epsilon\sqrt{n})\max_{i=0}^{n-1} |s_i^n - f|$ by using at most $4n\sqrt{7.2\log_2(n/\sqrt{\epsilon})}$ flops. Furthermore let this CV matrix C be weakly (ϵ, ρ)-neutered. Then (ii) one can ν-approximate the product $V_s\mathbf{u}$ for $\nu = ((2k - 1)\epsilon/\sqrt{n})\max_{i=0}^{n-1} |s_i^n - f|$ by using at most $30\rho n + 4n$ flops for ϵ and ρ of Remark 2, which means $O(n(\log(n)))$ flops where $1/\epsilon = O(\log(n))$. Furthermore (iii) if the matrix V_s is nonsingular and if $\nu' = 1.5(2k - 1)\epsilon\sqrt{n}\mu_f\|V_s^{-1}\| \le 1/2$, then one can ν'-approximate the vector $V_s^{-1}\mathbf{u}$ by using $O(n\rho^2)$ flops where $\mu_f = \max_{i=0}^{n-1} |s_i^n - f^n|^{-1}$, $|f| = 1$, and one can choose f such that $\mu_f \le n/\pi$. The same estimates hold where the transposes V_s^T and $-C_{f,s} = C_{s,f}^T$ replace the matrices V_s and $C = C_{s,f}$, respectively, as well as for the problems of approximate evaluation of a polynomial* *of degree* $n-1$ *at the* n *knots* $s_0, \ldots, s_{n-1}$ *and approximate interpolation to this polynomial from its* n *values at these knots.*

Proof. Deduce the claims about the matrix $V_{\mathbf{s}}$ and its transpose by combining Theorems 9 and 10 with equation (5) and note that μ_f is minimized under $|f| = 1$ for the complex points s_i equally spaced on the unit cirle $\{x : |x| = 1\}$. Extend the estimates to the case of polynomials by applying equation (3).

Remark 9. The proofs of Theorems 9–11 are constructive. One can readily devise algorithms that compute some centers c_q and c_q' (defining basic neutered block columns N_q and N_q') and then compute ULV factorizations of the neutered blocks $\cap_q$ and $\cap_q'$ for all q. One can avoid a large part of these computations by following papers [3] and [18]: they bypass the computation of the centers c_q and c_q' and compute the HSS generators for the blocks defined by HSS trees. These blocks are not as numerous as all neutered block columns.

Remark 10. In the extensive tests reported in [18], the ϵ-ranks of the matrix $C_{1,\omega_{2n}}$ for small ϵ consistently grew much slower than $\log(n)$ as n grew large. This empirical observation and formal arguments in [14, Remark 11] suggest that at least for a large class of CV matrices if not for a "typical" CV matrix our upper estimate of Remark 2 for ρ is overly pessimistic.

Remark 11. Reduction of Vandermonde to Cauchy computations was the initial step for our current progress (cf. (4)), but the reduction into the opposite direction seems to be also promising. Indeed exact or approximate polynomial evaluation and interpolation are equivalent to computing or approximating the products of a Vandermonde matrix and its inverse by a vector (cf. Problems 1 and 2). Now assume any successful algorithms for the latter tasks and employ equation (4) to extend them to the multiplication of a Cauchy matrix and its inverse by a vector and consequently to some problems of rational interpolation [12] as well as to a broad area of computations with matrices having structures of Vandermonde and Cauchy types [13].

7   Conclusions

We first proposed a novel transformation of a Vandermonde matrix into an HSS matrix, and then application of the efficient algorithms of [9], [3], and [18] enabled dramatic acceleration of the known approximation algorithms for polynomial interpolation and multipoint evaluation. At first we accelerated ϵ-approximate evaluation of an nth degree polynomial at n points by a factor of $\sqrt{n/\log n}$ provided $\log(1/\epsilon) = O(\log(n))$. Then for a large class of input matrices, which we call weakly (ρ, ϵ)-neutered CV matrices (cf. Definition 4), we obtained further speedup reaching nearly optimal arithmetic cost bounds for both ν-approximate evaluation and interpolation for $\nu = O(\epsilon)$ specified in Theorem 11. Formal and experimental study of the above matrix class can be a natural research subject. The basic computational blocks of our algorithms are the FFT and Multipole type algorithms, well studied, efficiently implemented, and numerically stable.

As a potential beneficiary of our progress we recall the classical task of univariate polynomial root-finding. The current best package of subroutines MPSolve [2] relies on Ehrlich–Aberth iterations, which amount essentially to recursive multipoint polynomial evaluation. MPSolve performs it in quadratic time by means of Horner's algorithm, versus nearly linear time of our algorithms. For a large class of inputs a number of initial iterations can be performed with the IEEE standard double precision, and at these stages, our algorithms clearly supersede Horner's.

Transformations of matrix structures can extend our progress to computations with various matrices having structures of Vandermonde and Cauchy types (cf. [13]), for example, confluent Vandermonde matrices and Loewner matrices, and to various problems of rational interpolation such as the Nevanlinna–Pick and matrix Nehari problems, although numerical stability problems may limit the power of these extensions. One can seek efficient algorithms for approximation of the products of a Cauchy matrix and its inverse by a vector by employing their link to computations with polynomials and rational functions (cf. [12]). As we pointed out in Remark 11 transformations of matrix structures should help to extend any progress in exact or approximate polynomial evaluation and interpolation to a broad area of computations with structured matrices, polynomials, and rational functions. One may employ the algorithms that are relatively little known but support *the Boolean complexity bounds of order $n \log(1/\epsilon)$ up to polylogarithmic factors for ϵ-approximate multipoint polynomial evaluation* provided that $\log(1/\epsilon)$ dominates the precision of other input parameters (cf. [8, Theorem 3.9 and Section 5.3] and [15]), and similarly for ϵ-approximate polynomial interpolation. These algorithms involve arithmetic operations with a high precision, but their small Boolean cost implies that they proceed with a lower precision of computing on the average and therefore can be competitive under sufficient support from the field of Computer Arithmetic (cf. [11]). Other natural subjects for further formal and experimental study, include the ones pointed out in Remarks 5, 10, and 11 and the estimation of the threshold input sizes for which our algorithms, running in nearly linear time, outperform the known numerical algorithms, running in quadratic time, as well as the algorithms supporting Theorem 9 and part (i) of Theorem 11.

Acknowledgements. This research has been supported by the NSF Grant CC 1116736 and the PSC CUNY Awards 64512–0042 and 65792–0043. Reviewers' comments were thoughtful and valuable.

References

1. Bella, T., Eidelman, Y., Gohberg, I., Olshevsky, V.: Computations with quasiseparable polynomials and matrices. TCS 409(2), 158–179 (2008)
2. Bini, D.A., Fiorentino, G.: Design, analysis, and implementation of a multiprecision polynomial rootfinder. Numer. Algs. 23, 127–173 (2000)
3. Chandrasekaran, S., Gu, M., Sun, X., Xia, J., Zhu, J.: A superfast algorithm for Toeplitz systems of linear equations. SIMAX 29, 1247–1266 (2007)

 4. Eidelman, Y., Gohberg, I.: A modification of the Dewilde–van der Veen method for inversion of finite structured matrices. Linear Algebra and Its Applications 343, 419–450 (2002)
 5. Eidelman, Y., Gohberg, I., Haimovici, I.: Separable Type Representations of Matrices and Fast Algorithms, vol. 1. Birkhäuser, Basel (2013)
 6. Gu, M.: Stable and efficient algorithms for structured systems of linear equations. SIAM J. Matrix Anal. Appl. 19, 279–306 (1998)
 7. Gohberg, I., Kailath, T., Olshevsky, V.: Fast Gaussian elimination with partial pivoting for matrices with displacement structure. Mathematics of Computation 64, 1557–1576 (1995)
 8. Kirrinnis, P.: Polynomial factorization and partial fraction decomposition by Newton's iteration. J. Complexity 14, 378–444 (1998)
 9. Martinsson, P.G., Rokhlin, V., Tygert, M.: A fast algorithm for the inversion of Toeplitz matrices. Comput. Math. Appl. 50, 741–752 (2005)
10. Pan, V.Y.: On computations with dense structured matrices. Math. Computation 55(191), 179–190 (1990); Proceedings version in Proc. ISSAC 1989, pp. 34–42. ACM Press, New York (1989)
11. Priest, D.: Algorithms for arbitrary precision floating point arithmetic. In: Kornerup, P., Matula, D. (eds.) Proc. 10th Symp. Computer Arithmetic, pp. 132–145. IEEE Computer Society Press, Los Angeles (1991)
12. Pan, V.Y.: Structured Matrices and Polynomials: Unified Superfast Algorithms. Birkhäuser/Springer, Boston/New York (2001)
13. Pan, V.Y.: Transformations of Matrix Structures Work Again. Tech. Report TR 2013004, PhD Program in Comp. Sci., Graduate Center, CUNY (2013), http://www.cs.gc.cuny.edu/tr/techreport.php?id=449
14. Pan, V.Y.: Polynomial Evaluation and Interpolation and Transformations of Matrix Structures 1. Tech. Report TR 2013007, PhD Program in Comp. Sci. Graduate Center, CUNY (2013), http://www.cs.gc.cuny.edu/tr/techreport.php?id=452
15. Pan, V.Y., Tsigaridas, E.P.: On the Boolean complexity of the real root refinement. In: Kauers, M. (ed.) Proc. ISSAC 2013. ACM Press, New York (2013)
16. Stewart, G.W.: Matrix Algorithms. Basic Decompositions, vol. I. SIAM, Philadelphia (1998)
17. Vandebril, R., van Barel, M., Mastronardi, N.: Matrix Computations and Semiseparable Matrices: Linear Systems, vol. 1. The Johns Hopkins University Press, Baltimore (2007)
18. Xia, J., Xi, Y., Gu, M.: A superfast structured solver for Toeplitz linear systems via randomized sampling. SIMAX 33, 837–858 (2012)

A Note on the Need for Radical Membership Checking in Mechanical Theorem Proving in Geometry

Eugenio Roanes-Lozano and Eugenio Roanes-Macías

Universidad Complutense de Madrid, Algebra Dept., 28040-Madrid, Spain
{eroanes,roanes}@mat.ucm.es,
http://www.ucm.es/info/secdealg/ERL/

Abstract. In his famous 1988 book "Mechanical Geometry Theorem Proving", Shang-Ching Chou details a method with two steps to check whether a geometric theorem is "generally true" or not using Groebner bases. The second step consists of checking the membership of the thesis polynomial to the radical (J) of a certain ideal (L). Chou mentions: "However, for all theorems we have found in practice, J=L." In his 2007 book "Selected topics in geometry with classical vs. computer proving", Pavel Pech shows a beautiful example where checking the radical membership, not only the ideal membership, is required. Using a kind of "reverse engineering" we shall show how to easily find examples of theorems where to check the radical membership is required. The idea is just to construct a thesis involving an ideal such that the ideal of its variety is not itself (therefore, the ideal is not equal to its radical).

Keywords: mechanical theorem proving in geometry, ideals theory, radical ideal.

1 Introduction

In his famous book "Mechanical Geometry Theorem Proving" [6], Shang-Ching Chou details a method with two steps (page 78) to check whether a geometric theorem is "generally true" or not, using Groebner bases [4, 8, 18, 19].

The adjective "generally" is used because there can be degenerate cases (like a triangle degenerating into a segment when its three vertices are aligned). Let us note that there is some controversy on what actually means "generally true" [1, 5, 7, 9, 15], but we shall not address this topic here, as it deals with complex issues as the distinction between parameters and free variables in special cases. Regarding this topic, we shall underline that minimal canonical comprehensive Groebner bases [11] have been successfully applied to mechanical theorem proving [12]. Finally, regarding the truthfulness of geometric theorems, in the recent [25], comprehensive Groebner systems and Rabinovitch's trick have been used to successfully distinguish the situations: true / true on component / completely false / generically false.

V.P. Gerdt et al. (Eds.): CASC 2013, LNCS 8136, pp. 288–300, 2013.
© Springer International Publishing Switzerland 2013

Mechanical theorem proving in geometry is not restricted to the 2D case (see, for instance [21–23]). Moreover, there are different attempts to link dynamic geometry systems and GUIs to computer algebra systems for mechanical theorem proving and related issues [2, 3, 17, 26], but we shall not address these topics here.

1.1 Some Preliminary Notes about Algebraic Geometry

This section can be skipped by the reader familiar with basic algebraic geometry (as it only overviews some basic concepts and results in order the paper to be self-contained).

Definition 1. *An ideal of a ring is a special subset of a ring such that it is also a ring and the product of any element of the ring by an element of the subset is again in the subset.*

Example 1. The set of even numbers is an ideal of the integers.

Definition 2. *The radical of an ideal I, $Rad(I)$, is the set of elements, α, such that some integer power of α is in I.*

Proposition 1. *$Rad(I)$ is also an ideal and I is obviously contained in $Rad(I)$.*

Example 2. In $\mathbb{Z}$, the radical of the ideal of the set of multiples of 9 is the set of multiples of 3.

Definition 3. *A basis of an ideal is a set of generators of the ideal (i.e., the elements of the ideal are the linear algebraic combinations of the elements in the basis). The ideal generated by the polynomials $p_1, ..., p_n$ is denoted $\langle p_1, ..., p_n \rangle$.*

Corollary 1. *$\langle 1 \rangle$ is the whole ring.*

Property 1. A Groebner basis of a polynomial ideal is a very special basis. Once the way the monomials have to be ordered and the order for the variables are fixed, the (reduced) Groebner basis of an ideal is unique (that is, it completely characterizes the ideal). Moreover, Buchberger's algorithm provides a method to obtain the Groebner basis of any ideal of a polynomial ring over a field in a finite number of variables.

Corollary 2. *As a consequence of the properties of Groebner bases, checking some very complex algebraic issues turns easily decidable. For instance, checking whether two ideals of a polynomial ring are equal or not can be decided by simply comparing their (reduced) Groebner bases.*

Definition 4. *An algebraic variety is the set of solutions of a system of polynomial equations (that is, the set of zeros of the ideal generated by the corresponding polynomials). We shall denote by $v(I)$ the algebraic variety of ideal I.*

Example 3. The variety of ideal $\langle x^2 + y^2 - 1 \rangle$ in the Euclidean space $\mathbb{R}^3$ is the cylinder of axis the z axis and radius 1.

Definition 5. *The ideal of an algebraic variety V, $i(V)$, is the set of all polynomials which vanish on all the points of the set V. This set turns out to be an ideal.*

Theorem 1. *Hilbert's Nullstellensatz states that, if the base field is algebraically closed, then: $i(v(I)) = Rad(I)$.*

Example 4. Let us consider $\mathbb{C}$ as the base field. $v(\langle x^3, y-1\rangle) = \{(0,1)\} \subseteq \mathbb{C}^2$. We have: $i(\{(0,1)\}) = \langle x, y-1\rangle$. Consequently: $i(v(\langle x^3, y-1\rangle)) = \langle x, y-1\rangle$, and, as expected: $\langle x, y-1\rangle = Rad(\langle x^3, y-1\rangle)$ (see Figure 1).

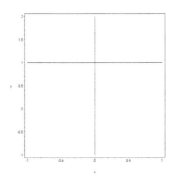

Fig. 1. The variety of ideal $\langle x^3, y-1\rangle$ is the intersection of lines $x = 0$ and $y = 1$. Observe that, although the example is considered in $\mathbb{C}^2$, we can only draw it in $\mathbb{R}^2$ (we would need a 4-dimensional space to represent $\mathbb{C}^2$).

We shall use the following nontrivial surprising result (radical membership criterion):

Proposition 2. *Let $K[x_1, ..., x_n]$ be a polynomial ring over the field K, let $p \in K[x_1, ..., x_n]$ and let $I = \langle p_1, p_2, ..., p_m\rangle$ be an ideal of $K[x_1, ..., x_n]$. Let w be another variable, independent from $x_1, x_2, ..., x_n$. Then:*

$$p \in Rad(I) \text{ (in the polynomial ring } K[x_1, ..., x_n]) \Leftrightarrow$$
$$\Leftrightarrow \text{ the ideal } \langle p_1, p_2, ..., p_m, 1 - p \cdot w\rangle \text{ of } K[x_1, ..., x_n, w] \text{ is the whole ring, } \langle 1\rangle.$$

1.2   Yet another Preliminary Note

In logic we have the following:

Definition 6. *A formula F' is a tautological consequence of a formula F iff every valuation that causes F to be true also causes F' to be true. This situation is denoted $F \models F'$.*

Proposition 3. *In classic Boolean logic, $F \models F'$ is equivalent to the formula $F \Rightarrow F'$ being a tautology (i.e., to being always true).*

But, if S is a set of polynomial equations and p is a polynomial equation (all in the same polynomial ring), what does it mean that $S \Rightarrow p$?

Example 5. If we say (in $\mathbb{C}$) that:

$$x - 3_{120^\circ} = 0 \Rightarrow x^3 - 27 = 0$$

(where 3_{120° represents the complex number of modulus 3 and argument 120 degrees), we mean that all the solutions of the LHS equation satisfy the RHS equation.

Example 6. If we say (in $\mathbb{R}^2$) that:

$$\{x + y - 2 = 0, x - y = 0\} \Rightarrow x - 1 = 0$$

we mean that all the solutions of the LHS linear system satisfy the RHS equation.

2   Chou's Remark Regarding Mechanical Theorem Proving in Geometry and the Radical Membership Checking

Roughly speaking, we can describe the underlying idea as follows. Let H be the ideal describing the hypotheses polynomials and let t be the thesis polynomial.

1st Step (ideal membership): If t belongs to H, t is a linear algebraic combination of some of the elements in H. Consequently, at the points where all the polynomials in H vanish, any linear algebraic combination of these polynomials will vanish, so t will also vanish. Consequently, $v(H)$ is a subset of $v(\langle t \rangle)$. Therefore, according to the ideas in Section 1.1, the theorem is "generally true".

However, this 1st step is only a sufficient condition. A mentioned above (Theorem 1), if the base field is algebraically closed, for any ideal I, $i(v(I)) = Rad(I)$, and it is not difficult to prove that a necessary and sufficient condition is:

2nd Step (radical membership): t belongs to $Rad(H)$ iff the theorem is "generally true".

Chou mentions at the same page 78 of [6]:

"However, for all theorems we have found in practice, $H = Rad(H)$."

(that is, to check the "1st step" was always sufficient in practice).
Other introductory sources are [10, 20, 24].

3   Pavel Pech's Counterexample Revisited

In his book "Selected topics in geometry with classical vs. computer proving" [14], Pavel Pech shows a beautiful theorem (Section 8.1) where H is not equal to $Rad(H)$, that is, checking the radical membership, not only the ideal membership, is required. This theorem is about the planarity of a (3D) regular skew (equilateral) pentagon:

> "**Theorem 8.1.** A regular skew polygon $ABCDE$ in the Euclidean space E^3 is given. Then $ABCDE$ is a planar pentagon."

This counterexample can also be found in [13].

We shall begin by revisiting Pech's example using the computer algebra system Maple 16 (nevertheless, what is done afterwards could be done with any CAS with a Groebner bases implementation and plotting capabilities, what most of them have).

We firstly load the Groebner bases package and define a "square distance" auxiliary function:

```
> with(Groebner):
> unprotect(D):
> dist2:=(P,Q)->(P[1]-Q[1])^2+(P[2]-Q[2])^2+(P[3]-Q[3])^2:
```

We give the coordinates of the vertices of the pentagon (for the sake of simplicity, we consider one of the vertices to be $(1,0,0)$ and another one $(0,0,0)$, what doesn't represent any restriction but an adequate selection of the coordinate axes and the unit).

```
> A:=[1,0,0]:
> B:=[b1,b2,0]:
> C:=[c1,c2,c3]:
> D:=[d1,d2,d3]:
> E:=[0,0,0]:
```

According to the coordinate system chosen, A, B, E lie on plane $z = 0$. Therefore, the thesis is that C, D lie on plane $z = 0$, so there are two thesis polynomials. For instance: C lies on plane $z = 0 \Leftrightarrow 1 \cdot b_2 \cdot b_3 = 0$. Let us consider this thesis polynomial (the other is analogous).

We can then use the radical membership criterion to check whether the thesis polynomial belongs to the hypothesis ideal:

```
> Basis({ dist2(A,B)-1 , dist2(B,C)-1 ,
>         dist2(C,D)-1 , dist2(D,E)-1 ,
>         dist2(B,D)-dist2(A,C) , dist2(C,E)-dist2(A,C) ,
>         dist2(D,A)-dist2(A,C) , dist2(E,B)-dist2(A,C) ,
>         1*b2*c3*w-1 } ,
>         plex(b1,b2,c1,c2,c3,d1,d2,d3,w) );
                      [1]
```

so the theorem holds.

But if we compare the hypotheses ideal with the ideal generated by the hypotheses together with the thesis (observe that w is no longer a variable):

```
> GB1:=Basis({ dist2(A,B)-1 , dist2(B,C)-1 ,
>              dist2(C,D)-1 , dist2(D,E)-1 ,
>              dist2(B,D)-dist2(A,C) , dist2(C,E)-dist2(A,C) ,
>              dist2(D,A)-dist2(A,C) , dist2(E,B)-dist2(A,C) } ,
>              plex(b1,b2,c1,c2,c3,d1,d2,d3) );
```

$$GB1 := [d3^2, 5 - 20*d2^2 + 16*d2^4, -3 + 4*d2^2 + 2*d1, c3,$$
$$2*d2 - 4*d2^3 + c2, -1 + 2*c1, b2 - d2, 1 - 4*d2^2 + 2*b1]$$

```
> GB2:=Basis({ dist2(A,B)-1 , dist2(B,C)-1 ,
>              dist2(C,D)-1 , dist2(D,E)-1 ,
>              dist2(B,D)-dist2(A,C) , dist2(C,E)-dist2(A,C) ,
>              dist2(D,A)-dist2(A,C) , dist2(E,B)-dist2(A,C) ,
>              1*b2*c3 } ,
>              plex(b1,b2,c1,c2,c3,d1,d2,d3) );
```

$$GB2 := [d3^3, 5 - 14*d3^2 - 20*d2^2 + 24*d2^2*d3^2 + 16*d2^4, -3 + 3*d3^2 + 4*d2^2 + 2*$$
$$d1, d3^2 - 2*d2^2*d3^2 + c3*d3, -d3^2 + c3^2, 10*d2 - 15*d2*d3^2 - 20*d2^3 - 4*d2^3*d3^2 +$$
$$5*c2, -1 + 2*c1, -5*d2 - 10*d2*d3^2 + 8*d2^3*d3^2 + 5*b2, 1 - 3*d3^2 - 4*d2^2 + 2*b1]$$

They are not equal!

Therefore, for this theorem, the membership of the thesis to the hypotheses ideal is not equivalent to the radical membership.

For the other thesis condition the same holds:

```
> GB3:=Basis({dist2(A,B)-1 , dist2(B,C)-1 ,
>              dist2(C,D)-1 , dist2(D,E)-1 ,
>              dist2(B,D)-dist2(A,C) , dist2(C,E)-dist2(A,C) ,
>              dist2(D,A)-dist2(A,C) , dist2(E,B)-dist2(A,C) ,
>              1*b2*d3 } , plex(b1,b2,c1,c2,c3,d1,d2,d3) );
```

$$GB3 := [d3, 5 - 20*d2^2 + 16*d2^4, -3 + 4*d2^2 + 2*d1, c3^2,$$
$$2*d2 - 4*d2^3 + c2, -1 + 2*c1, b2 - d2, 1 - 4*d2^2 + 2*b1]$$

```
> evalb(GB1=GB3);
```

$$false$$

They are not equal! (observe that the fourth element of this last Groebner basis is c_3^2, instead of c_3).

3.1 Continuing with Pavel Pech's Example

We can extend Pavel Pech's example by analysing the different solutions to the problem.

Maple includes a `solve` command that can be used with nonlinear polynomial systems (it is based on Groebner bases techniques). Now that we know that the theorem holds, we can solve the system of hypotheses conditions:

```
> solve(GB2,[b1,b2,c1,c2,c3,d1,d2,d3]);
```

$$[[b1 = -\tfrac{1}{2} + \tfrac{1}{2}RootOf(5 - 5\_Z + \_Z^2),$$
$$b2 = \tfrac{1}{2}RootOf(\_Z^2 - RootOf(5 - 5\_Z + \_Z^2)), c1 = 1/2,$$
$$c2 = \tfrac{1}{2}(-2 + RootOf(5 - 5\_Z + \_Z^2)) * RootOf(\_Z^2 - RootOf(5 - 5\_Z + \_Z^2)),$$
$$c3 = 0, d1 = \tfrac{3}{2} - \tfrac{1}{2}RootOf(5 - 5\_Z + \_Z^2),$$
$$d2 = \tfrac{1}{2}RootOf(\_Z^2 - RootOf(5 - 5\_Z + \_Z^2)), d3 = 0]]$$

If we ask for the explicit solutions we obtain four solutions:

```
> allvalues(%);
```

$$[[b1 = \tfrac{3}{4} - \tfrac{\sqrt{5}}{4}, b2 = \tfrac{\sqrt{10-2\sqrt{5}}}{4}, c1 = \tfrac{1}{2}, c2 = \tfrac{(\tfrac{1}{2}-\tfrac{\sqrt{5}}{2})\sqrt{10-2\sqrt{5}}}{4},$$
$$c3 = 0, d1 = \tfrac{\sqrt{5}}{4} + \tfrac{1}{4}, d2 = \tfrac{\sqrt{10-2\sqrt{5}}}{4}, d3 = 0]],$$
$$[[b1 = \tfrac{3}{4} - \tfrac{\sqrt{5}}{4}, b2 = -\tfrac{\sqrt{10-2\sqrt{5}}}{4}, c1 = \tfrac{1}{2}, c2 = -\tfrac{(\tfrac{1}{2}-\tfrac{\sqrt{5}}{2})\sqrt{10-2\sqrt{5}}}{4},$$
$$c3 = 0, d1 = \tfrac{\sqrt{5}}{4} + \tfrac{1}{4}, d2 = -\tfrac{\sqrt{10-2\sqrt{5}}}{4}, d3 = 0]],$$
$$[[b1 = \tfrac{3}{4} + \tfrac{\sqrt{5}}{4}, b2 = \tfrac{\sqrt{10+2\sqrt{5}}}{4}, c1 = \tfrac{1}{2}, c2 = \tfrac{(\tfrac{1}{2}+\tfrac{\sqrt{5}}{2})\sqrt{10+2\sqrt{5}}}{4},$$
$$c3 = 0, d1 = \tfrac{\sqrt{5}}{4} - \tfrac{1}{4}, d2 = \tfrac{\sqrt{10+2\sqrt{5}}}{4}, d3 = 0]],$$
$$[[b1 = \tfrac{3}{4} + \tfrac{\sqrt{5}}{4}, b2 = -\tfrac{\sqrt{10+2\sqrt{5}}}{4}, c1 = \tfrac{1}{2}, c2 = -\tfrac{(\tfrac{1}{2}+\tfrac{\sqrt{5}}{2})\sqrt{10+2\sqrt{5}}}{4},$$
$$c3 = 0, d1 = \tfrac{\sqrt{5}}{4} - \tfrac{1}{4}, d2 = -\tfrac{\sqrt{10+2\sqrt{5}}}{4}, d3 = 0]]$$

Finally, if we plot these four solutions we obtain two symmetrical convex pentagons and two symmetrical star pentagons (see Figure 2).

4   Designing Other Counterexamples

The idea is to use a kind of "reverse engineering" to easily find examples of theorems where to check the radical membership is required. It is enough to construct a hypotheses ideal such that the ideal of its variety is not itself (i.e., an ideal such that it is not equal to its radical) and an adequate thesis polynomial.

We shall include afterwards some trivial examples: one involving two circles and a cubic, another one involving three parabolae, another one involving two circumferences and a line,...

4.1   First Example

Idea: Let us consider a simple locus described by an ideal that is not radical. For example, point $(0,0)$ can be described as the intersection of two circumferences. Now an ideal which variety contains point $(0,0)$ should be contained in the ideal

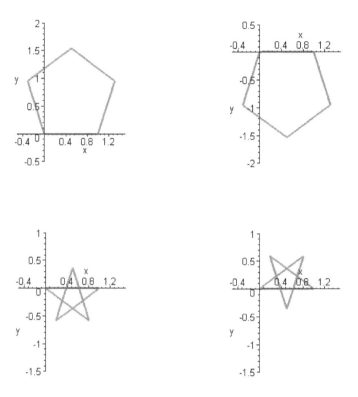

Fig. 2. The four solutions of the regular pentagon problem computed with Maple 16

generated by the two circumferences. Let us consider such ideal to be simple and reducing a power of y by means of x, for instance $y = x^3$. Summarizing, we can consider the following:

Theorem 2. *The intersection point of the circumferences of centers $(1,0)$ and $(-1,0)$ and radius 1 lie on the cubic $y = x^3$ (see Figure 3).*

The hypothesis polynomial are:

$$(x-1)^2 + y^2 - 1, (x+1)^2 + y^2 - 1$$

and the thesis polynomial is:

$$y = x^3 \ .$$

If we apply the radical membership criterion:

```
> Basis( [ (x-1)^2+y^2-1 , (x+1)^2+y^2-1, 1-w*(y-x^3) ],
                                      plex(x,y,w) );
```

$$[1]$$

we obtain that the theorem holds.

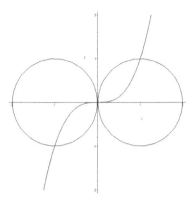

Fig. 3. A first example on the need to apply the radical membership criterion

But if we apply the ideal membership criterion:

```
> Basis( [ (x-1)^2+y^2-1 , (x+1)^2+y^2-1) ], plex(x,y) );
```

$$[y^2, x]$$

```
> Basis( [ (x-1)^2+y^2-1 , (x+1)^2+y^2-1) , y-x^3 ], plex(x,y) );
```

$$[y, x]$$

we obtain that the two Groebner bases are different!

There is a Maple package specialized in computations with ideals. Using such package the previous computations are straightforward:

```
> with(PolynomialIdeals):
> thesis:=y-x^3:
> J:=<(x-1)^2+y^2-1,(x+1)^2+y^2-1>:
> IdealMembership( thesis , J );
```

$$false$$

```
> RadicalMembership( thesis , J );
```

$$true$$

Remark 1. Let us emphasize that the intersection point of the two circumferences in the previous example is a tangency point. That is a key issue in order the corresponding ideal not to be radical (and to obtain a counterexample).

For instance, if the centre of the second circumference is moved to $(1, 1)$, the two circumferences intersect at points $(0, 0)$ and $(1, 1)$, and the two Groebner bases of the ideal membership criterion are equal.

4.2   Second Example

It is clear that we can use endlessly the idea of the first example, and the way the theorem is enunciated can be camouflaged in a "geometric style":

Theorem 3. *The locus of the points of $\mathbb{R}^2$ that are at the same distance of point $(0, 1/4)$ and line $y = -\frac{1}{4}$ and are also at the same distance of point $(0, -1/4)$ and line $y = \frac{1}{4}$ is contained in the locus of the points that are at the same distance of point $(1/4, 0)$ and line $x = -\frac{1}{4}$.*

Let us not directly recognize what these curves are, and let us compute the situation with Maple (an auxiliary function "square of the distance" is defined firstly):

```
> dist2:=(P,Q)->(P[1]-Q[1])^2+(P[2]-Q[2])^2:
> simplify( dist2([x,y],[0,1/4])-(y+1/4)^2 )=0;
> simplify( dist2([x,y],[0,-1/4])-(y-1/4)^2 )=0;
> simplify( dist2([x,y],[1/4,0])-(x+1/4)^2 )=0;
```

$$x^2 - y = 0$$
$$x^2 + y = 0$$
$$-x + y^2 = 0$$

Therefore, the situation is that of Figure 4.
Let us apply both criteria:

```
> with(PolynomialIdeals):
> thesis:=simplify( dist2([x,y],[1/4,0])-(x+1/4)^2 ):
> J:=<dist2([0,1/4],[x,y])-(y+1/4)^2,
>     dist2([0,-1/4],[x,y])-(y-1/4)^2>:
> IdealMembership( thesis , J );
```

$$false$$

```
> RadicalMembership( thesis , J );
```

$$true$$

4.3   Other Examples

More complex and interesting (from the geometric point of view) examples can be found, now that we know what to look for. The following examples are not detailed for the sake of space, but behave as the examples previously analysed.

Theorem 4. *The intersection point of two tangent circumferences lies on their centers line.*

Theorem 5. *If T and X are the tangency points of the inscribed circle and the excribed circle on the side AB of triangle ABC (respectively), then $dist(A, X) = dist(T, B)$ (Figure 5) .*

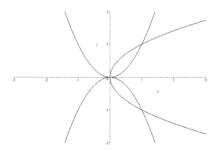

Fig. 4. A second example on the need to apply the radical membership criterion

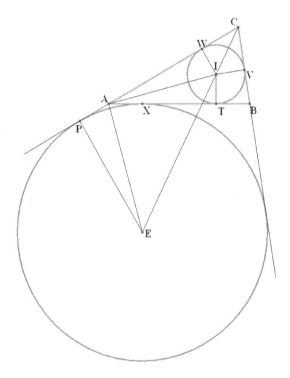

Fig. 5. A more complex example on the need to apply the radical membership criterion

5   Remark

This is an extended and very improved version of an unpublished previous talk presented at the "Applications of Computer Algebra 2012 (ACA'2012)" conference. The Extended Abstract can be found at the web page of ACA'2012.

6   Conclusions

Summarizing, Pavel Pech's regular pentagon example is not an isolated rare case of mechanical theorem proving in geometry where the radical membership is required to be checked.

Moreover, using a kind of "reverse engineering" we have shown how to easily find examples of theorems where to check the radical membership is also required.

Acknowledgments. This work was partially supported by the research project *TIN2012-32482* (Government of Spain).

The author is grateful to the anonymous referees for valuable suggestions and recommendations which improved the paper.

References

1. Bazzotti, L., Dalzotto, G., Robbiano, L.: Remarks on geometric theorem proving. In: Richter-Gebert, J., Wang, D. (eds.) ADG 2000. LNCS (LNAI), vol. 2061, pp. 104–128. Springer, Heidelberg (2001)
2. Botana, F., Valcarce, J.L.: A software tool for the investigation of plane loci. Mat. Comp. Simul. 61(2), 141–154 (2003)
3. Botana, F.: A web-based resource for automatic discovery in plane geometry. Intl. J. Comp. Mat. Learning 8(1), 109–121 (2003)
4. Buchberger, B: Applications of Gröbner bases in non-linear computational geometry, in: Rice, J.R. (ed.) Mathematical Aspects of Scientific Software, pp. 59–87. Springer (1988)
5. Bulmer, M., Fearnley-Sander, D., Stokes, T.: The kinds of truth of geometry theorems. In: Richter-Gebert, J., Wang, D. (eds.) ADG 2000. LNCS (LNAI), vol. 2061, pp. 129–142. Springer, Heidelberg (2001)
6. Chou, S.-C.: Mechanical geometry theorem proving. Reidel, Dordrecht (1988)
7. Conti, P., Traverso, C.: Algebraic and Semialgebraic Proofs: Methods and Paradoxes. In: Richter-Gebert, J., Wang, D. (eds.) ADG 2000. LNCS (LNAI), vol. 2061, pp. 83–103. Springer, Heidelberg (2001)
8. Cox, D., Little, J., O'Shea, D.: Ideals, Varieties, and Algorithms. Undergraduate Texts in Mathematics. Springer, New York (1997)
9. Kutzler, B.: Careful algebraic translations of geometry theorems. In: Gonnet, G.H. (ed.) Proc. ISSAC 1989, pp. 254–263. ACM Press, Portland (1989)
10. Kutzler, B., Stifter, S.: On the application of Buchberger's algorithm to automated geometry theorem proving. J. Symb. Comp. 2(4), 389–397 (1986)
11. Manubens, M., Montes, A.: Minimal Canomical Comprehensive Groebner system. J. Symb. Comput. 44, 463–478 (2006)
12. Montes, A., Recio, T.: Automatic discovery of geometry theorems using minimal canonical comprehensive gröbner systems. In: Botana, F., Recio, T. (eds.) ADG 2006. LNCS (LNAI), vol. 4869, pp. 113–138. Springer, Heidelberg (2007)
13. Pech, P.: On the need of radical ideals in automatic proving: A theorem about regular polygons. In: Botana, F., Recio, T. (eds.) ADG 2006. LNCS (LNAI), vol. 4869, pp. 157–170. Springer, Heidelberg (2007)
14. Pech, P.: Selected topics in geometry with classical vs. computer proving. World Scientific Publishing Co. Pte. Ltd, Singapore (2007)

15. Recio, T., Botana, F.: Where the truth lies (in automatic theorem proving in elementary geometry). In: Laganá, A., Gavrilova, M.L., Kumar, V., Mun, Y., Tan, C.J.K., Gervasi, O. (eds.) ICCSA 2004. LNCS, vol. 3044, pp. 761–770. Springer, Heidelberg (2004)

16. Recio, T., Vélez, M.P.: An Introduction to Automated Discovery in Geometry through Symbolic Computation. In: Langer, U., Paule, P. (eds.) Numerical and Symbolic Scientific Computing. Texts & Monographs in Symbolic Computation, vol. 1, pp. 257–271. Springer, Vienna (2012)

17. Roanes-Lozano, E., Roanes-Macías, E., Villar-Mena, M.: A bridge between dynamic geometry and computer algebra. Mat. Comp. Mod. 37(9-10), 1005–1028 (2003)

18. Roanes-Lozano, E., Roanes-Macías, Laita, L.M.: The geometry of algebraic systems and their exact solving using Groebner bases. Comp. Sci. Eng. 6(2), 76–79 (2004)

19. Roanes-Lozano, E., Roanes-Macías, Laita, L.M.: Some Applications of Groebner bases. Comp. Sci. Eng. 6(3), 56–60 (2004)

20. Roanes-Macías, E., Roanes-Lozano, E.: Nuevas Tecnologías en Geometría. Editorial Complutense, Madrid (1994)

21. Roanes-Macías, E., Roanes-Lozano, E.: A Maple package for automatic theorem proving and discovery in 3D-geometry. In: Botana, F., Recio, T. (eds.) ADG 2006. LNCS (LNAI), vol. 4869, pp. 171–188. Springer, Heidelberg (2007)

22. Roanes-Macías, E., Roanes-Lozano, E., Fernández-Biarge, J.: Extensión natural a 3D del teorema de Pappus y su configuración completa. Bol. Soc. "Puig Adam" 80, 38–56 (2008)

23. Roanes-Macías, E., Roanes-Lozano, E., Fernández-Biarge, J.: Obtaining a 3D extension of Pascal theorem for non-degenerated quadrics and its complete configuration with the aid of a computer algebra system. RACSAM (Rev. R. Acad. C. Exactas, Fís. Nat., Serie A, Mat.) 103(1), 93–109 (2009)

24. Roanes-Macías, E., Roanes-Lozano, E.: Un método algebraico-computacional para demostración automática en geometría euclídea. Bol. Soc. "Puig Adam" 88, 31–63 (2011)

25. Sun, Y., Wang, D., Zhou, J.: A new method of automatic geometric theorem proving and discovery by comprehensive Groebner systems. In: Proc. ADG 2012 (2012) (to appear), http://dream.inf.ed.ac.uk/events/adg2012/uploads/proceedings/ADG2012-proceedings.pdf#page=163

26. Wang, D.: GEOTHER 1.1: Handling and proving geometric theorems automatically. In: Winkler, F. (ed.) ADG 2002. LNCS (LNAI), vol. 2930, pp. 194–215. Springer, Heidelberg (2004)

A Symbolic Approach to Boundary Problems for Linear Partial Differential Equations

Applications to the Completely Reducible Case of the Cauchy Problem with Constant Coefficients

Markus Rosenkranz and Nalina Phisanbut*

University of Kent, Canterbury, Kent CT2 7NF, United Kingdom
{M.Rosenkranz,N.Phisanbut}@kent.ac.uk

Abstract. We introduce a general algebraic setting for describing linear boundary problems in a symbolic computation context, with emphasis on the case of partial differential equations. The general setting is then applied to the Cauchy problem for completely reducible partial differential equations with constant coefficients. While we concentrate on the theoretical features in this paper, the underlying operator ring is implemented and provides a sufficient basis for all methods presented here.

1 Introduction

A symbolic framework for boundary problems was built up in [11,13] for linear ordinary differential equations (LODEs); see also [15,6,7] for more recent developments. One of our long-term goals is to extend this to boundary problems for linear partial differential equations (LPDEs). Since this is a daunting task in full generality, we want to tackle it in stages of increasing generality. In the first instance, we restrict ourselves to *constant coefficients*, where the theory is quite well-developed [2]. Within this class we distinguish the following three stages:

1. The simplest is the Cauchy problem for *completely reducible* operators.
2. The next stage will be the *Cauchy problem* for general hyperbolic LPDEs.
3. After that we plan to study *boundary problems* for elliptic/parabolic LPDEs.

In this paper we treat the first case (Section 4). But before that we build up a *general algebraic framework* (Sections 2 and 3) that allows a symbolic description for all boundary problems (LPDEs/LODEs, scalar/system, homogeneous/inhomogeneous, elliptic/hyperbolic/parabolic).[1] Using these concepts and tools we develop a general solution strategy for the Cauchy problem in the case (1). See the Conclusion for some thoughts about the next two steps.

The passage from LODEs to LPDEs was addressed at two earlier occasions:

---

* The authors acknowledge support from the EPSRC First Grant EP/I037474/1.
[1] Due to space limitations we omit proofs; they can be found in arXiv:1304.7380.

V.P. Gerdt et al. (Eds.): CASC 2013, LNCS 8136, pp. 301–314, 2013.
© Springer International Publishing Switzerland 2013

- An *abstract theory of boundary problems* was developed in [9], including LODEs and LPDEs as well as linear systems of these. The concepts and results of Sections 2 and 3 are built on this foundation, adding crucial concepts whose full scope is only appreciated in the LPDE setting: boundary data, semi-homogeneous problem, state operator.
- An algebraic language for *multivariate differential and integral operators* was introduced in Section 4 of [14], with a prototype implementation described in Section 5 of the same paper. This language is generalized in the PIDOS algebra[2] of Section 4, and it is also implemented in a Mathematica package.

In this paper we will not describe the current state of the *implementation* (mainly because of space limitations). Let us thus say a few words about this here. A complete reimplementation of the PIDOS package described in [14] is under way. The new package is called OPIDO (Ordinary and Partial Integro-Differential Operators), and it is implemented as a standalone Mathematica package unlike its predecessor, which was incorporated into the THEOREMA system. In fact, our reimplementation reflects several important design principles of THEOREMA, emphasizing the use of functors and a strong support for modern two-dimensional (user-controllable) parsing rules. We have called this programming paradigm FUNPRO, first presented at the Mathematica symposium [12]. The last current stable version of the (prototype) package can be found at `http://www.kent.ac.uk/smsas/personal/mgr/index.html`.

While the interested reader will find all the relevent details (in particular the rather large rewrite system) in this package, here are a few *remarks on the current state of the implementation*:

- The FUNPRO paradigm emphasizes the a functional style of programming while at the same time encapsulating all mathematical domains similar to object-oriented languages. For example, a ring $\mathcal{R}$ would come along with its various operations like $+_{\mathcal{R}}$ and $*_{\mathcal{R}}$.
- We make heavy use of two-dimensional syntax, using parsing and formatting rules that allow us to write integro-differential operators in a notation that is close to the one we use on paper.
- At this stage we do not aim at efficiency. Whenever there was a conflict between speed and conceptual clarity, we gave preference to the latter. The reason is that we view the current package as a prototype, with the goal of writing a more efficient (and stable) MAPLE package once the content is sufficiently matured.
- Since we rely on Mathematica for computing integrals in closed form, we inherit all the usual limitations in that respect. However, this is not a real issue in the current scope (constant coefficients) since here we can stay within the class of exponential polynomials.

We plan to describe more details of the implementation at another occasion.

---

[2] PIDOS = Partial Integro-Differential Operator System.

At the time of writing, the ring of *ordinary integro-differential operators* is completed and the ring of *partial integro-differential operators* is close to completion (for two independent variables). Compared to [14], the new PIDOS ring contains several crucial new rewrite rules (instances of the substitution rule for resolving multiple integrals). Our conjecture is that the new rewrite system is noetherian and confluent but this shall also be analyzed at another occasion.

Notation. The algebra of $m \times n$ matrices over a field K is written as K_n^m, where $m = 1$ or $n = 1$ is omitted. Thus we identify $K^m = K \oplus \cdots \oplus K$ with the space of column vectors and $K_n = (K^n)^*$ with the space of row vectors. More generally, we have $K_n^m \cong K^n \to K^m$. As usual we write $C^\omega(\mathbb{R}^n)$ for the algebra of (complex-valued) analytic functions with real arguments is denoted by.

2   An Algebraic Language for Boundary Data

As mentioned in the Introduction, we follow the *abstract setting* developed in [9]. We will motivate and recapitulate some key concepts here, but for a fuller treatment of these issues we must refer the reader to [9] and its references.

Let us recall the notion of boundary problem. Fix vector spaces $\mathcal{F}$ and $\mathcal{G}$ over a common ground field K of characteristic zero (for avoiding trivialities one may assume $\mathcal{F}$ and $\mathcal{G}$ to be infinite-dimensional). Then a *boundary problem* $(T, \mathcal{B})$ consists of an epimorphism $T \colon \mathcal{F} \to \mathcal{G}$ and a subspace $\mathcal{B} \subseteq \mathcal{F}^*$ that is orthogonally closed in the sense defined below. We call T the differential operator and $\mathcal{B}$ the *boundary space*.

Similar to the correspondence of ideals/varieties in algebraic geometry, we make use of the following Galois connection [9, A.11]. If $\mathcal{A}$ is any subspace of the space $\mathcal{F}$, its *orthogonal* $\mathcal{A}^\perp \le \mathcal{F}^*$ is defined as $\{\varphi \in \mathcal{F}^* \mid \varphi(a) = 0 \text{ for all } a \in \mathcal{A}\}$. Dually, for a subspace $\mathcal{B}$ of the dual space $\mathcal{F}^*$, the orthogonal $\mathcal{B}^\perp \le \mathcal{F}$ is defined by $\{f \in \mathcal{F} \mid \beta(f) = 0 \text{ for all } \beta \in \mathcal{B}\}$. If we think of $\mathcal{F}$ as "functions" and of $\mathcal{F}^*$ as "boundary conditions", then $\mathcal{A}^\perp$ is the space of valid conditions (the boundary conditions satisfied by the given functions) while $\mathcal{B}^\perp$ is the space of admissible functions (the functions satisfying the given conditions).

Naturally, a subspace of either $\mathcal{S}$ of $\mathcal{F}$ or $\mathcal{F}^*$ is called *orthogonally closed* if $\mathcal{S}^{\perp\perp} = \mathcal{S}$. But while any subspace of $\mathcal{F}$ itself is always orthogonally closed, this is far from being the case of the subspaces of the dual $\mathcal{F}^*$. Hence the condition on boundary spaces $\mathcal{B}$ to be orthogonally closed is in general not trivial. However, if $\mathcal{B}$ is finite-dimensional as in boundary problems for LODEs (as in Example 1 below), then it is automatically orthogonally closed. For LPDEs, the condition of orthogonal closure is important; see Example 2 for an intuitive explanation.

In [11,13] and also in the abstract setting of [9] we have only considered what is sometimes called the *semi-inhomgeneous boundary problem* [16], more precisely the semi-inhomogeneous incarnation of $(T, \mathcal{B})$; see Definition 7 for the full picture. This means we are given a *forcing function* $f \in \mathcal{G}$ and we search for a solution $u \in \mathcal{F}$ with

$$\boxed{\begin{aligned} Tu &= f, \\ \beta(u) &= 0 \; (\beta \in \mathcal{B}). \end{aligned}} \tag{1}$$

In other words, u satisfies the inhomogeneous "differential equation" $Tu = f$ and the homogeneous "boundary conditions" $\beta(u) = 0$ given in $\mathcal{B}$.

A boundary problem which admits a unique solution $u \in \mathcal{F}$ for every forcing function $f \in \mathcal{G}$ is called *regular*. In terms of the spaces, this condition can be expressed equivalently by requiring that $\operatorname{Ker} T + \mathcal{B}^{\perp} = \mathcal{F}$; see [9] for further details. In this paper we shall deal exclusively with regular boundary problems. For singular boundary problems we refer the reader to [6] and [5].

For a regular boundary problem, one has a linear operator $G\colon \mathcal{G} \to \mathcal{F}$ sending f to u is known as the *Green's operator* of the boundary problem $(T, \mathcal{B})$. From the above we see that G is characterized by $TG = 1$ and $\operatorname{Im} G = \mathcal{B}^{\perp}$.

Example 1. A classical example of this notion is the two-point boundary problem. As a typical case, consider the simplified model of *stationary heat conduction* described by

$$\boxed{\begin{aligned} u'' &= f, \\ u(0) &= u(1) = 0. \end{aligned}}$$

Here we can choose $\mathcal{F} = \mathcal{G} = C^{\infty}(\mathbb{R})$ for the function space such that the differential operator is given by $T = D^2\colon C^{\infty}(\mathbb{R}) \to C^{\infty}(\mathbb{R})$ and the boundary space by the two-dimensional subspace of $C^{\infty}(\mathbb{R})^*$ spanned by the linear functionals $L\colon u \mapsto u(0)$ and $R\colon u \mapsto u(1)$ for evaluation on the left and right endpoint. In the sequel we shall write $\mathcal{B} = [L, R]$, employing an important generalization for LPDEs, described in the next eamples. We can express its Green's operator in the language of integro-differential operators as explained in [13].

Example 2. As a typical counterpart in the world of LPDEs, consider the *equation for waves in an inhomogeneous medium*, described by

$$\boxed{\begin{aligned} u_{xx} - u_{tt} &= f(x, t) \\ u(x, 0) = u_t(x, 0) &= u(0, t) = u(1, t) = 0 \end{aligned}}$$

in one space dimension. In this case we choose $\mathcal{F} = \mathcal{G} = C^{\omega}(\mathbb{R} \times \mathbb{R}^+)$; again one could choose much larger spaces of functions (or distributions) in analysis and in the applications. Here the differential operator is $T = D_{xx} - D_{tt}\colon C^{\omega}(\mathbb{R} \times \mathbb{R}^+) \to C^{\omega}(\mathbb{R} \times \mathbb{R}^+)$ while the boundary space $\mathcal{B}$ is the orthogonal closure of the linear span of the families of functionals β_x, γ_x $(x \in \mathbb{R})$ and $\kappa_t, \lambda_t; (t \in \mathbb{R}^+)$ defined by $\beta_x(u) = u(x, 0), \gamma_x(u) = u_t(x, 0)$ and $\kappa_t(u) = u(0, t), \lambda_t(u) = u(1, t)$. Using the notation $[\dots]$ for denoting the orthogonal closure of the linear span, we can thus write $\mathcal{B} = [\beta_x, \gamma_x, \kappa_t, \lambda_t \mid x \in \mathbb{R}, t \in \mathbb{R}^+]$ for the boundary space under consideration.

The point of the *orthogonal closure* is that the given conditions imply other conditions not in their span, for example $u_x(1/2, 0) = 0$ or $\int_{-3}^{5} u(0, \tau)\, d\tau = 0$. Rather than being linear consequences, these two examples are differential and integral consequences. (Of course the full boundary space also contains many functionals without a natural analytic interpretation.)

In the problems above, the differential equation is inhomogeneous while the boundary conditions are homogeneous. A *semi-homogeneous boundary problem* is the opposite, combining a homogeneous differential equation with inhomogeneous boundary conditions. While this is a simple task for LODEs (as always we assume that the fundamental system is available to us in some form!), it is usually a nontrivial problem for LPDEs (even when they have constant coefficients). We will give the formal definition of a semi-inhomogeneous boundary problem in the next section (Definition 7). Here it suffices to consider an example for developing the necessary auxiliary notions.

Example 3. The *Cauchy problem for the wave equation* in one dimension is

$$\boxed{\begin{aligned} &u_{xx} - u_{tt} = 0, \\ &u(x,0) = f(x), u_t(x,0) = g(x). \end{aligned}}$$

Being a hyperbolic problem, we could use rather general function spaces for the "boundary data" f, g. For reasons of uniformity we will nevertheless restrict ourselves here to the analytic setting, so assume $f, g \in C^\omega(\mathbb{R})$. Note that the association of u to (f, g) is again a linear operator mapping (two univariate) functions to a (bivariate) function; we will come back to this point in Definition 7.

Going back to the abstract setting, one is tempted to define the notion of boundary data as some kind of functions depending on "fewer" variables. But the problem with this approach is that—abstractly speaking—we are not dealing with any functions depending on any number of variables (but see below). Moreover, the inhomogeneous boundary conditions in the form $u(0, x) = f(x), u_t(0, x) = g(x)$ are basis-dependent while the whole point of the abstract theory is to provide a *basis-free description* (which leads to an elegant setting for describing composition and factorization of abstract boundary problems); see Proposition 9. We shall therefore develop a basis-independent notion of boundary data (we can go back to the traditional description by choosing a basis).

We define first the *trace map* trc: $\mathcal{F} \to \mathcal{B}^*$ as sending $f \in \mathcal{F}$ to the functional $\beta \mapsto \beta(f)$. In Example 3 this would map the function $u(x, t)$ to its position and velocity values on $\mathbb{R} \times \{0\}$. We call $\mathrm{trc}(f)$ the trace of f and write it as f^*. Moreover, we denote the image of the map trc by $\mathcal{B}'$ and refer to its elements as *boundary data*. Note that $\mathcal{B}'$ is usually much smaller than the full dual $\mathcal{B}^*$ since a continuous function (let alone an analytic one) cannot assume arbitrary values. (This situation is vaguely reminiscent of the algebraic and continuous dual of a topological vector space.)

Since by definition the trace map is surjective from $\mathcal{F}$ to $\mathcal{B}'$, it has some right inverse $\mathcal{B}^\diamond : \mathcal{B}' \to \mathcal{F}$. We refer to $\mathcal{B}^\diamond$ as an *interpolator* for $\mathcal{B}$ since it constructs a "function" $f = \mathcal{B}^\diamond(B) \in \mathcal{F}$ from given boundary values $B \in \mathcal{B}'$ such that $\beta(f) = B(\beta)$. Of course, the choice of f is usually far from being unique. Apart from its use for describing boundary data (see at the end of this section), the notion of interpolator will turn out to be useful for solving the semi-homogeneous boundary problem (see Proposition 10).

Let us now describe how to relate these abstract notions to the usual setting of initial and boundary values problems as they actually in analysis: essentially by *choosing a basis*. However, we have to be a bit careful since we must deal with the orthogonal closure.

Definition 4. If $\mathcal{B} \leq \mathcal{F}^*$ is any orthogonally closed subspace, we call a family $(\beta_i \mid i \in I)$ a *boundary basis* if $\mathcal{B} = [\beta_i \mid i \in I]$, meaning $\mathcal{B}$ is the orthogonal closure of the span of the β_i.

Note that a boundary basis is typically smaller than a K-linear basis of $\mathcal{B}$. All traditional boundary problems are given in terms of such a boundary basis. In Example 3, the boundary basis could be spelled out by using $I = \mathbb{R} \uplus \mathbb{R}$ with $\beta_{(x,0)}(u) = u(x,0)$ and $\beta_{(x,1)}(u) = u_t(x,0)$. Relative to a boundary basis $(\beta_i \mid i \in I)$, we call $\bar{f} = \beta_i(f)_{i \in I} \in K^I$ the *boundary values* of $f \in \mathcal{F}$. As we can see from the next proposition, we may think of the trace as a basis-free description of boundary values. Conversely, one can always extract from any given boundary data $B \in \mathcal{B}'$ the boundary values $B(\beta_i)_{i \in I}$ as its coordinates relative to the boundary basis (β_i).

Lemma 5. *Let $\mathcal{B} \leq \mathcal{F}^*$ be a boundary space with boundary basis $(\beta_i \mid \in I)$. If for any $B, \tilde{B} \in \mathcal{B}'$ one has $B(\beta_i)_{i \in I} = \tilde{B}(\beta_i)_{i \in I}$ then also $B = \tilde{B}$. In particular, for any $f \in \mathcal{F}$, the trace f^* depends only on the boundary values $f(\beta_i)_{i \in I}$.*

The *analytic interpretation* of this proposition is clear in concrete cases like Example 2: Once the values $u(x,0), u_t(x,0)$ and $u(0,t), u(1,t)$ are fixed, all differential and integral consequences, as in the above examples $u_x(1/2,0) = 0$ or $\int_{1/4}^{3/4} u(0,\tau)\, d\tau$, are likewise fixed. It is therefore natural that an interpolator need only consider the boundary values rather than the full trace information. This is the contents of the next lemma.

Lemma 6. *Let $\mathcal{B} \leq \mathcal{F}^*$ be a boundary space with boundary basis $(\beta_i \mid \in I)$ and write $f_I = f(\beta_i)_{i \in I} \in K^I$ for the boundary values of any $f \in \mathcal{F}$ and $\mathcal{B}'_I$ for the K-subspace of K^I generated by all boundary values f_I. Then any linear map $J \colon \mathcal{B}'_I \to \mathcal{F}$ with $J(f_I)_I = f_I$ induces a unique interpolator $\mathcal{B}^\diamond \colon \mathcal{B}' \to \mathcal{F}$ defined by $B \mapsto J(B(\beta_i)_{i \in I})$.*

As noted above, we can always extract the boundary values $B(\beta_i)_{i \in I} \in K^I$ of some boundary data $B \in \mathcal{B}'$ relative to fixed basis (β_i) of $\mathcal{B}$. However, since one normally has got *only* the boundary values (coming from some function), where does the corresponding $B \in \mathcal{B}'$ come from? By definition, it has to assign values to all $\beta \in \mathcal{B}$, not only to the β_i making up the boundary basis. As suggested by the above lemmata, for actual computations those additional values will be irrelevant. Nevertheless, it gives a feeling of confidence to provide these values: If $\mathcal{B}^\diamond$ is any *interpolator*, we have $B(\beta) = \beta(\mathcal{B}^\diamond(B_i)_{i \in I})$. This follows immediately from the fact that $\mathcal{B}^\diamond$ is a right inverse of the trace map and that it depends only on the boundary values $(B_i)_{i \in I}$ by Lemma 6. In the analysis setting this means we interpolate the given boundary value and then do with the resulting function whatever is desired (like derivatives and integrals in Example 2).

3   Green's Operators for Signals and States

Using the notion of boundary data developed in the previous section, we can now give the formal definition of the *semi-homogeneous boundary problem*. In fact, we can distinguish three different incarnations of a "boundary problem" (as we assume regularity, the *fully homogeneous problem* is of course trivial).

Definition 7. Let $(T, \mathcal{B})$ be a regular boundary problem with $T \colon \mathcal{F} \to \mathcal{G}$ and boundary space $\mathcal{B} \subseteq \mathcal{F}^*$. Then we distinguish the following problems:

Given $(f, B) \in \mathcal{G} \oplus \mathcal{B}'$,
find $u \in \mathcal{F}$ with

$$\begin{array}{|l|}\hline Tu = f, \\ \beta(u) = B(\beta) \ (\beta \in \mathcal{B}). \\ \hline \end{array}$$

Given $f \in \mathcal{G}$,
find $u \in \mathcal{F}$ with

$$\begin{array}{|l|}\hline Tu = f, \\ \beta(u) = 0 \ (\beta \in \mathcal{B}). \\ \hline \end{array}$$

Given $B \in \mathcal{B}'$,
find $u \in \mathcal{F}$ with

$$\begin{array}{|l|}\hline Tu = 0, \\ \beta(u) = B(\beta) \ (\beta \in \mathcal{B}). \\ \hline \end{array}$$

They are, respectively, called the *fully inhomogeneous*, the *semi-inhomogeneous* and the *semi-homogeneous* boundary problem for $(T, \mathcal{B})$. The corresponding linear operators will be written as $F \colon \mathcal{G} \oplus \mathcal{B}' \to \mathcal{F}$, $(f, B) \mapsto u$ and $G \colon \mathcal{G} \to \mathcal{F}$, $f \mapsto u$ and $H \colon \mathcal{B}' \to \mathcal{F}$, $B \mapsto u$.

Lemma 8. *Each of the three problems in Definition 7 has a unique solution for the respective input data, so the operators F, G, H are well-defined.*

The terminology for the operators F, G, H is not uniform in the literature. In the past, we have only considered G and called it the "Green's operator" acting on a "forcing function" f. While this is in good keeping with the engineering tradition and large parts of the standard mathematical culture [16], it is difficult to combine with suitable terminology for F and H. In this paper, we shall follow the systems theory jargon [8] and refer to F as the (full) *transfer operator*, to G as the (zero-state) signal transfer operator or briefly *signal operator*, and to H as the (zero-signal) state transfer operator or briefly *state operator*. This terminology reflects the common view of forcing functions $f \in \mathcal{F}$ as "signals" and boundary data $B \in \mathcal{B}'$ as (initial) "states".

One of the advantages of the abstract formulation is that it allows us to describe the *product of boundary problems* in a succinct, basis-free manner (and it includes LODEs and LPDEs as well as systems of these). The composite boundary problem can then be solved, both in its semi-inhomogneous and its semi-homogeneous incarnation (the latter is presented here for the first time).

Proposition 9. *Define the product of two boundary problems $(T, \mathcal{B})$ and $(\tilde{T}, \tilde{\mathcal{B}})$ with $\mathcal{F} \xrightarrow{T} \mathcal{G} \xrightarrow{T} \mathcal{H}$ and $\mathcal{B} \subseteq \mathcal{G}^*$, $\tilde{\mathcal{B}} \subseteq \mathcal{F}$ by*

$$(T, \mathcal{B})(\tilde{T}, \tilde{\mathcal{B}}) = (T\tilde{T}, \mathcal{B}\tilde{T} + \tilde{\mathcal{B}}).$$

Then $(T, \mathcal{B})(\tilde{T}, \tilde{\mathcal{B}})$ is regular if both factors are. In that case, if $(T, \mathcal{B})$, $(\tilde{T}, \tilde{\mathcal{B}})$ have, respectively, the signal operators G, $\tilde{G}$ and the state operators H, $\tilde{H}$, then $(T, \mathcal{B})(\tilde{T}, \tilde{\mathcal{B}})$ has the signal operator $\tilde{G}G$ and the state operator $(\mathcal{B}\tilde{T} + \tilde{\mathcal{B}})' \to \mathcal{F}$ acting by $B + \tilde{B} \mapsto \tilde{G}H(B\tilde{T}^*) + \tilde{H}(\tilde{B})$.

As detailed in [11,9], the computation of the *signal operator* G can be decomposed in two parts: (1) Finding a *right inverse* $T^\diamond$ of the differential operator T, which involves only the differential equation without boundary conditions (so we may replace the boundary by intial conditions, thus having again a unique solution: this is the so-called fundamental right inverse). (2) Determining the *projector* onto the homogeneous solution space along the space of functions admissible for the given boundary conditions—the projector "twists" the solutions coming from the right inverse into satisfying the boundary conditions. An analogous result holds for the computation of the *state operator* H if we replace the right inverse $T^\diamond$ of T by the interpolator $\mathcal{B}^\diamond$ for the boundary space $\mathcal{B}$.

Proposition 10. *Let $(T, \mathcal{B})$ be regular with operators F, G, H as in Definition 7. Then we have $G = (1 - P)\,T^\diamond$ and $H = P\mathcal{B}^\diamond$, hence $F = (1 - P)\,T^\diamond \oplus P\mathcal{B}^\diamond$ for the transfer operator. Here $T^\diamond \colon \mathcal{G} \to \mathcal{F}$ is any right inverse of the differential operator $T \colon \mathcal{F} \to \mathcal{G}$ and $\mathcal{B}^\diamond \colon \mathcal{B}' \to \mathcal{F}$ any interpolator for $\mathcal{B}$ while $P \colon \mathcal{F} \to \mathcal{F}$ is the projector determined by $\operatorname{Im} P = \operatorname{Ker} T$ and $\operatorname{Ker} P = \mathcal{B}^\perp$.*

If $T \in \mathbb{C}[D]$ is a completely reducible differential operator[3] with constant coefficients in $\mathbb{C}$, the determination of $T^\diamond$ reduces to solving an inhomogeneous first-order equation with constant coefficients—which is of course straightforward (Lemma 14). Also the determination of the interpolator $\mathcal{B}^\diamond$ turns out to be easy for a Cauchy problem since it is essentially given by the corresponding Taylor polynomial (3). Hence it remains to find some means for computing the *kernel projector* P for a boundary problem $(T, \mathcal{B})$.

In the case of a LODE of order n, the method for computing P given in the proof of Theorem 26 of [13] and in Section 6 of [9] is essentially a Gaussian elimination on the so-called *evaluation matrix* $\beta(u) = [\beta_i(u_j)]_{ij} \in K^{n \times n}$ formed by evaluating the i-th boundary condition β_i on the j-th fundamental solution u_j. So here we assume $u_1, \ldots, u_n$ is a basis of $\operatorname{Ker} T$ and $\beta_1, \ldots, \beta_n$ a basis of $\mathcal{B}$. Unfortunately, this is not a very intuitive description of P, and it is not evident how to generalize it to the LPDE case. We have to gain a more conceptual perspective at $\beta(u)$ for making the generalization transparent.

Let us write $\operatorname{Ev} \colon \mathcal{B} \oplus \operatorname{Ker} T \to K$ for the bilinear operation of evaluation $(\beta, u) \mapsto \beta(u)$. Choosing bases $\beta_1, \ldots, \beta_n$ for $\mathcal{B}$ and $u_1, \ldots, u_n$ for $\operatorname{Ker} T$, the coordinate matrix of Ev is clearly $\beta(u)$. By the usual technique of dualization, we can also think of Ev as the map $\mathcal{B} \colon \operatorname{Ker} T \to \mathcal{B}^*$ that sends $u \in \operatorname{Ker} T$ to the functional $\beta \mapsto \beta(u)$. But this map is nothing else than the *restriction of the trace map* $\operatorname{trc} \colon \mathcal{F} \to \mathcal{B}'$ to $\operatorname{Ker} T \subset \mathcal{F}$. It is easy to check that the restricted trace is bijective and that its inverse gives rise to the projector.

Proposition 11. *Let $(T, \mathcal{B})$ be a regular boundary problem with $E \colon \operatorname{Ker} T \to \mathcal{B}'$ being the restricted trace map. Then E is bijective with the state operator H as its inverse, and $P = H \circ \operatorname{trc}$ is the projector with $\operatorname{Im} P = \operatorname{Ker} T$ and $\operatorname{Ker} P = \mathcal{B}^\perp$.*

---

[3] By definition, this means its characteristic polynomial $T(\lambda) \in \mathbb{C}[\lambda] = \mathbb{C}[\lambda_1, \ldots, \lambda_n]$ splits into linear factors.

We observe that the formula $P(u) = H(u^*)$ has a very *natural interpretation*: The kernel projector picks up the boundary data u^* of an arbitrary function $u \in \mathcal{F}$ and then constructs the required kernel element $H(u^*) \in \operatorname{Ker} T$ by solving the semi-homogeneous boundary problem with boundary data u^*. In the LODE case, the relation $P = H \circ \operatorname{trc}$ reduces to the aforementioned formulae (see [9] after Proposition 6.1) after choosing bases $u_1, \ldots, u_n$ for $\operatorname{Ker} T$ and $\beta_1, \ldots, \beta_n$ for $\mathcal{B}$. Apart from its conceptual clarity, the advantage of Proposition 11 is that it can also be used in the LDPE case (see after Lemma 14).

4   The Cauchy Problem for Analytic Functions

At this point we switch from the abstract setting of Sections 2 and 3 to the concrete setting of *analytic functions*. Note that we are dealing with complex-valued functions of real arguments. This means the ground field is $K = \mathbb{C}$, and $\mathcal{F}$ is the integro-differential algebra of entire functions restricted to real arguments.

More precisely, we shall employ the following conventions for easing the burden of book-keeping: As elements of $\mathcal{F}$ we take all holomorphic functions $\mathbb{R}^n \to \mathbb{C}$ for any $n \in \mathbb{N}$, including the constant functions $u \in \mathbb{C}$ for $n = 0$. In other words, $\mathcal{F}$ is a direct limit of algebras. Moreover, we have derivations D_n and integrals A_n for all $n > 0$, namely $D_n(u) = \partial u / \partial x_n$ and

$$A_n(u) = \int_0^{x_n} u(\ldots, \xi, \ldots) \, d\xi,$$

where ξ occurs at the n-th position. Clearly, we have then integro-differential algebras $(\mathcal{F}, D_n, A_n)$ for every $n > 0$. In fact, $\mathcal{F}$ has the structure of a *hierarchical integro-differential algebra*. This notion will be made precise at another occasion; for the moment it suffices to make the following observations. If $\mathbb{N}^{\circledast}$ is the sublattice of the powerset $\mathcal{P}(\mathbb{N}^+)$ that consists of finite sets $\alpha = \{\alpha_1, \ldots, \alpha_k\}$ and the full set $\mathbb{N}^+ = \{1, 2, 3, \ldots\}$, we define for $\alpha \in \mathbb{N}^{\circledast}$ the subalgebras

$$\mathcal{F}_\alpha = \{f \in \mathcal{F} \mid D_i f = 0 \text{ for all } i \notin \alpha\},$$

consisting of the functions depending (at most) on $x_{\alpha_1}, \ldots, x_{\alpha_k}$. Then $(\mathcal{F}_\alpha, \subseteq)$ is a sublattice of $(\mathcal{F}, \subseteq)$ that is isomorphic to the lattice $(\mathbb{N}^{\circledast}, \subseteq)$. The bottom element is of course $\mathcal{F}_\emptyset = \mathbb{C}$, the top element $\mathcal{F}_{\mathbb{N}^+} = \mathcal{F}$. We write $\mathcal{F}_n$ as an abbreviation for $\mathcal{F}_{\{1,\ldots,n\}}$.

As in the earlier paper [14], we add to this algebraic structure all linear *substitution operators*. In accordance with the above hierarchical structure, we use the ring $\mathbb{C}_*^*$ of row and column finite matrices with complex entries.[4] This means

---

[4] The *usage of complex substitutions* in functions of a real argument may sound strange at first. But an analytic function on $\mathbb{R}^n$ is of course also analytic on $\mathbb{C}^n$ with values in $\mathbb{C}$, so there is no problem with this view. For example, the substitution $(1, i)^*$ sends $f(x) = e^x \in \mathcal{F}_1$ to $f(x + iy) = e^x \cos y + i e^y \sin y \in \mathcal{F}_2$. Moreover, complex substitutions are indispensible for specifying the general solution of elliptic equations like the Laplace equation.

any $M \in \mathbb{C}_*^*$ can actually be seen as a finite matrix $M \in \mathbb{C}_n^m$ with m rows and n columns, extended by zero rows and columns. As usual, we identify $M \in \mathbb{C}_n^m$ with the linear map $M \colon \mathbb{C}^n \to \mathbb{C}^m$, yielding the substitution operator $M^* \colon \mathcal{F}_m \to \mathcal{F}_n$ defined by $u(x) \mapsto u(Mx)$.

We write $\mathcal{F}[D, A]$ for the *PIDOS algebra* generated over $\mathbb{C}$ by the operators D_n, A_n ($n > 0$), the substitutions M^* induced by $M \in \mathbb{C}_*^*$ and the exponential basis polynomials $x^\alpha e^{\lambda x} \in \mathcal{F}$. Here x denotes the arguments $x = (x_1, \ldots, x_n)$ for any $n \geq 0$, with exponents $\alpha = (\alpha_1, \ldots, \alpha_n) \in \mathbb{C}$ and frequencies $\lambda = (\lambda_1, \ldots, \lambda_n) \in \mathbb{C}$. Obviously, $\mathcal{F}[D, A]$ acts on $\mathcal{F}$, with $D = (D_1, D_2, \ldots)$ denoting the differential operators and $A = (A_1, A_2, \ldots)$ the integral operators, similar to the univariate case in the older notation of [11]. Here we avoid the notation $\int$ for the integrals since the powers $\int^n$ might be mistaken as integrals with upper bound n.

The algebra $\mathcal{F}[D, A]$ can be described by a *rewrite system* (PIDOS = partial integro-differential operator system), analogous to the one given in [14]. We will present this system in more detail—in particular proofs of termination and confluence—at another occasion.

Since in this paper we restrict ourselves to the analytic setting, we can appeal to the well-known *Cauchy-Kovalevskaya theorem* [10, Thm. 2.22] for ensuring the existence and uniqueness of the solution of the Cauchy problem. While the theorem in its usual form yields only local results, there is also a global version [4, Thm. 7.4] that provides a good foundation for our current purposes.[5] Since this form of the theorem is not widely known, we repeat the statement here.

As usual, we designate one *lead variable* t, writing the other ones $x_1, x_2, \ldots$ as before. Note that in applications t is not necessarily time. The apparently special form of the differential equation $Tu = 0$ implies no loss of generality: Whenever $T \in \mathbb{C}[D]$ is a differential operator of order $\deg T = m$, the change of variables $\bar{t} = t, \bar{x}_i = x_i + t$ leads to an equation of the required form.

Theorem 12 (Global Cauchy-Kovalevskaya). *Let* $T \in \mathbb{C}[D_t, D_1, \ldots, D_n]$ *be a differential operator in Caucy-Kovalevskaya form with respect to* t, *meaning* $T = D_t^m + \tilde{T}$ *with* $\deg(\tilde{T}, t) < m$ *and* $\deg(\tilde{T}) \leq m$. *Then the Cauchy problem*

$$\left. \begin{array}{l} Tu = 0 \\ D_t^{i-1} u(0, x_1, \ldots, x_n) = f_i(x_1, \ldots, x_n) \text{ for } i = 1, \ldots, m \end{array} \right\} \qquad (2)$$

has a unique solution $u \in \mathcal{F}_{n+1}$ *for given* $(f_1, \ldots, f_m) \in \mathcal{F}_n^m$.

In the *abstract language* of Sections 2 and 3 this is the semi-homogeneous boundary problem $(T, \mathcal{B})$ with boundary space

$$\mathcal{B} = [L_{0,\xi} D_t^i \mid i = 0, \ldots, m - 1 \text{ and } \xi \in \mathbb{R}^m],$$

where the evaluation $u(t, x_1, \ldots, x_n) \mapsto u(0, \xi_1, \ldots, \xi_n)$ is written as the substitution $L_{0,\xi} = \operatorname{diag}(0, \xi_1, \ldots, \xi_m)^*$ denotes . Hence the solution of (2) is given

---

[5] Of course the problem may still be *ill-posed*; we will not treat this issue here.

by the state operator $(f_1, \ldots, f_m) \in \mathcal{F}_n^m \mapsto u$ if we identify the boundary data $B \in \mathcal{B}'$ with its coordinate representation $(f_1, \ldots, f_m) \in \mathcal{F}_n^m$ relative to the above boundary basis $(L_{0,\xi} D_t^i)$. In detail, $B \colon \mathcal{B} \to \mathbb{C}$ is the unique linear map sending $L_{0,\xi} D_t^i \in \mathcal{B}$ to $f(\xi) \in \mathbb{C}$; confer Lemma 5 for the uniqueness statement. In the sequel these identifications will be implicit.

For future reference, we mention also that the usual Taylor polynomial allows one to provide a natural *interpolator* for the initial data, namely

$$\mathcal{B}^\diamond(f_1, \ldots, f_m) = f_1(x) + t\, f_2(x) + \cdots + \tfrac{t^{m-1}}{(m-1)!}\, f_m(x), \qquad (3)$$

which we will not need here because compute the kernel projector directly from its first-order factors.

In this paper, we will study the Cauchy problem (2) for a *completely reducible operator* $T(D)$. Hence assume $T = T_1^{m_1} \cdots T_k^{m_k}$ with first-order operators $T_1, \ldots, T_k \in \mathbb{C}[D]$. By a well-known consequence of the Ehrenpreis-Palamodov theorem, the general solution of $Tu = 0$ is the sum of the general solutions of the factor equations $T_1^{m_1} u = 0, \ldots, T_k^{m_k} u = 0$; see the Corollary on [1, p. 187]. Hence it remains to consider differential operators that are powers of first-order ones (we may assume all nonconstant coefficients are nonzero since otherwise we reduce n after renaming variables).

Lemma 13. *Let $T = a + a_0 D_t + a_1 D_1 + \cdots + a_n D_n \in \mathbb{C}[D]$ be a first-order operator with all $a_i \neq 0$. Order the variables such that all cumulative sums $a_0 + a_1 + \cdots + a_{i-1}$ are nonzero. Then the general solution of $T^m u = 0$ is given by*

$$u(t, x_1, \ldots, x_n) = \sum_{i=1}^{m} c_i(\bar{x}_1, \ldots, \bar{x}_n)\, \frac{t^{i-1} e^{-at/a_0}}{(i-1)!}, \qquad (4)$$

$$\bar{x}_i = t + x_1 + \cdots + x_{i-1} - (a_0 + a_1 + \cdots + a_{i-1})\, x_i / a_i, \qquad (5)$$

where $(f_1, \ldots, f_m) \in \mathcal{F}_{n-1}^m$ are arbitrary functions of the indicated arguments.

In principle, one could now combine the general solutions (4) for each factor, substitute them into the initial conditions of (2) and then solve for the c_i in terms of the prescribed boundary data $(f_1, \ldots, f_m)$. With this choice of c_i, the general solution will become the state operator for the Cauchy problem. However, this is a very laborious procedure, and therefore we prefer to use another route. Since we assume a completely reducible operator, we can employ the product representation of Proposition 9. In that case, it remains to consider the case of a *single first-order factor*.

Lemma 14. *Let $T = a + a_0 D_t + a_1 D_1 + \cdots + a_n D_n \in \mathbb{C}[D]$ be a first-order operator with all $a_i \neq 0$. Then the Cauchy problem $Tu = 0$, $u(0, x_1, \ldots, x_n) = f(x_1, \ldots, x_n)$ has the state operator $H(f) = e^{-at/a_0}\, Z^* \tilde{Z}_x^* f$ and the signal operator $G = a_0^{-1}\, e^{at/a_0}\, Z^* A_t\, e^{-at/a_0}\, \tilde{Z}^*$, where $Z \in \mathbb{C}_{n+1}^{n+1}$ is the transformation (5) with $\bar{t} = t$, and $\tilde{Z}$ is its inverse (written as $\tilde{Z}_x$ when restricted to the arguments $x_1, \ldots, x_n$).*

By Proposition 11, we can determine the *kernel projector* for the Cauchy problem of Lemma 14 as $P = H \circ \mathrm{trc}$, where $\mathrm{trc}(u) = u(0, x_1, \ldots, x_n)$ in this simple case. Having the kernel projector and the right inverse $T^\diamond$ in Lemma 14, the *signal operator* is computed by $G = (1 - P)T^\diamond$ as usual. Now we can tackle the general Cauchy problem (2) by a simple special case of Proposition 9.

Proposition 15. *Let $T_1, T_2 \in \mathbb{C}[D]$ be two first-order operators with nonzero coefficients for D_t. If $L_{0,\xi}$ is the evaluation defined after Theorem. 12, then we have*

$$(T_1, [L_{0,\xi} \mid \xi \in \mathbb{R}]) \, (T_2, [L_{0,\xi} \mid \xi \in \mathbb{R}]) = (T_1 T_2, [L_{0,\xi}, L_{0,\xi} D_t \mid \xi \in \mathbb{R}]$$

for the product of the Cauchy problems.

This settles the completely reducible case: Using Proposition 15 we can break down the *general Cauchy problem* (2) into first-order factors with single initial conditions. For each of these we compute the state and signal operator via Lemma 14, hence the state and signal operator of (2) by Proposition 9.

Example 16. As a typical example, let us consider a *modified wave equation* as treated in [3, §5.2] within the general class of higher-order hyperbolic equations with constant coefficients. We want to solve the initial value problem given by

$$\begin{array}{l} u_{tt} - 4\, u_{tx} + 4\, u_{xx} - 9\, u_{yy} = f, \\ u(0, x, y) = f_1(x, y), \quad u_t(0, x, y) = f_2(x, y) \end{array} \tag{6}$$

for a given forcing function f and initial data f_1, f_2. Note that the differential operator $T = D_t^2 - 4\, D_t D_x + 4\, D_x^2 - 9\, D_y^2$ factors completely as $T = T^- T^+$ with the two first-order factors $T^\pm = D_t - 2\, D_x \pm 3\, D_y$. In view of Proposition 9, it suffices to consider the two first-order factor problems

$$\begin{array}{l} u_t - 2\, u_x \pm 3\, u_y = f, \\ u(0, x, y) = f^\pm(x, y). \end{array} \tag{7}$$

Using Lemma 14, we obtain $H^\pm f^\pm (t, x, y) = f^\pm(x + 2t, y \mp 3t)$ for the state operators and

$$G^\pm f (t, x, y) = \int_0^t f(\tau, x + 2t - 2\tau, y \mp 3t \pm 3\tau)\, d\tau$$

for the signal operators. By Proposition 15 the composition of the two semi-inhomogeneous boundary problems for (7) yields the semi-inhomogeneous problem for (6), hence we may compute the signal operator as $G = G^+ G^-$ with

$$Gf (t, x, y) = \int_0^t \int_0^\sigma f(\tau, x + 2t - 2\tau, y - 3t - 3\tau + 6\sigma)\, d\tau\, d\sigma.$$

Using Proposition 9 now for computing the composite state operator one obtains at first $u(t, x, y) = H(f^-, f^+) = G^+ H^- f^- + H^+ f^+$ with

$$u(t, x, y) = f^+(x + 2t, y - 3t) + \int_0^t f^-(x + 2t, y - 3t + 6\tau)\, d\tau.$$

But this result should be interpreted cautiously: It solves the differential equation of (6) for the initial conditions $u(0, x, y) = f^+(x, y)$ and $T^+u(0, x, y) = f^-(x, y)$. But since $T^+u(0, x, y) = u_t(0, x, y) - (2D_x - 3D_y)f_1(x, y)$, we just have to pick $f^+(x, y) = f_1(x, y)$ and $f^-(x, y) = f_2(x, y) - 2D_xf_1(x, y) + 3D_yf_1(x, y)$ so that $H(f_1, f_2) = u$ with

$$u(t, x, y) = f_1(x + 2t, y - 3t) + \int_0^t (f_2 - 2\,D_xf_1 + 3\,D_yf_1)(x + 2t, y - 3t + 6\tau)\,d\tau$$

is the state operator for the given initial value problem (6).

5   Conclusion

As explained in the Introduction, we see the framework developed in this paper as the first stage of a more ambitious endeavor aimed at boundary problems for general constant-coefficient (and other) LPDEs. Following the enumeration of the Introduction, the next steps are as follows:

1. Stage (1) was presented in this paper, but the *detailed implementation* for some of the methods explained here is still ongoing. The crucial feature of this stage is that it allows us to stay within the (rather narrow) confines of the PIDOS algebra. In particular, no Fourier transformations are needed in this case, so the analytic setting is entirely sufficient.
2. As we enter Stage (2), it appears to be necessary to employ stronger tools. The most popular choice is certainly the *framework of Fourier transforms* (and the related Laplace transforms). While this can be algebraized in a manner completely analogous to the PIDOS algebra, the issue of choosing the right function space becomes more pressing: Clearly one has to leave the holomorphic setting for more analysis-flavoured spaces like the Schwartz class or functions with compact support. (As of now we stop short of using distributions since that would necessitate a more radical departure, forcing us to give up rings in favor of modules.)
3. For the treatment of genuine boundary problems in Stage (3) our plan is to use a powerful generalization of the Fourier transformation—the *Ehrenpreis-Palamodov integral representation* [1], also applicable to systems of LPDEs.

Much of this is still far away. But the *general algebraic framework* for boundary problems from Sections 2 and 3 is applicable, so the main work ahead of us is to identify reasonable classes of LPDEs and boundary problems that admit a symbolic treatment of one sort or another.

References

1. Hansen, S.: On the "fundamental principle" of L. Ehrenpreis. In: Partial Differential Equations (Warsaw, 1978). Banach Center Publ., vol. 10, pp. 185–201. PWN, Warsaw (1983)
2. Hörmander, L.: Linear partial differential operators. Springer, Berlin (1976)

3. John, F.: Partial differential equations, 4th edn. Applied Mathematical Sciences, vol. 1. Springer, New York (1982)
4. Knapp, A.W.: Advanced real analysis. Cornerstones. Birkhäuser Boston Inc., Boston (2005)
5. Korporal, A.: Symbolic Methods for Generalized Green's Operators and Boundary Problems. PhD thesis, Johannes Kepler University, Linz, Austria (November 2012); Abstracted in ACM Communications in Computer Algebra 46(4(182)) (December 2012)
6. Korporal, A., Regensburger, G., Rosenkranz, M.: Regular and singular boundary problems in MAPLE. In: Gerdt, V.P., Koepf, W., Mayr, E.W., Vorozhtsov, E.V. (eds.) CASC 2011. LNCS, vol. 6885, pp. 280–293. Springer, Heidelberg (2011)
7. Korporal, A., Regensburger, G., Rosenkranz, M.: Symbolic computation for ordinary boundary problems in maple. In: Proceedings of the 37th International Symposium on Symbolic and Algebraic Computation, ISSAC 2012 (2012) (software presentation)
8. Oberst, U., Pauer, F.: The constructive solution of linear systems of partial difference and differential equations with constant coefficients. Multidimens. Systems Signal Process. 12(3-4), 253–308 (2001); Special issue: Applications of Gröbner bases to multidimensional systems and signal processing
9. Regensburger, G., Rosenkranz, M.: An algebraic foundation for factoring linear boundary problems. Ann. Mat. Pura Appl (4) 188(1), 123–151 (2009), doi:10.1007/s10231-008-0068-3
10. Renardy, M., Rogers, R.C.: An introduction to partial differential equations, 2nd edn. Texts in Applied Mathematics, vol. 13. Springer, New York (2004)
11. Rosenkranz, M.: A new symbolic method for solving linear two-point boundary value problems on the level of operators. J. Symbolic Comput. 39(2), 171–199 (2005)
12. Rosenkranz, M.: Functorial programming & integro-differential operators. In: Talk at the International Mathematica Symposium (IMS 2012), London, United Kingdom, June 13 (2012), http://www.homepages.ucl.ac.uk/~ucahwts/ims2012/ims2012announce1/IMS2012.html
13. Rosenkranz, M., Regensburger, G.: Solving and factoring boundary problems for linear ordinary differential equations in differential algebras. Journal of Symbolic Computation 43(8), 515–544 (2008)
14. Rosenkranz, M., Regensburger, G., Tec, L., Buchberger, B.: A symbolic framework for operations on linear boundary problems. In: Gerdt, V.P., Mayr, E.W., Vorozhtsov, E.V. (eds.) CASC 2009. LNCS, vol. 5743, pp. 269–283. Springer, Heidelberg (2009)
15. Rosenkranz, M., Regensburger, G., Tec, L., Buchberger, B.: Symbolic analysis of boundary problems: From rewriting to parametrized Gröbner bases. In: Langer, U., Paule, P. (eds.) Numerical and Symbolic Scientific Computing: Progress and Prospects, pp. 273–331. Springer (2012)
16. Stakgold, I.: Green's functions and boundary value problems. John Wiley & Sons, New York (1979)

Towards Industrial Application of Approximate Computer Algebra*

Tateaki Sasaki[1], Daiju Inaba[2], and Fujio Kako[3]

[1] Professor emeritus, University of Tsukuba,
Tsukuba-city, Ibaraki 305-8571, Japan
sasaki@math.tsukuba.ac.jp
[2] Japanese Association of Mathematics Certification,
Ueno 5-1-1, Tokyo 110-0005, Japan
d.inaba@su-gaku.net
[3] Dept. Info. Comp. Sci., Nara Women's University,
Nara-city, Nara 630-8506, Japan
kako@ics.nara-wu.ac.jp

Abstract. The approximate computer algebra has scarcely been used for computations in industry so far. In order to break through this situation, we consider the series expansion of multivariate eigenvalues at their critical points in an aircraft control model. We show that the approximate square-free decomposition of univariate polynomial is quite useful in finding critical points semi-numerically and that the approximate factorization is successfully used for factoring multivariate polynomials with floating-point number coefficients. Furthermore, the "effective floating-point numbers" are quite useful in eliminating fully-erroneous terms from small but meaningful terms.

Keywords: approximate computer algebra, approximate polynomial factorization, approximate square-free decomposition, Hensel series.

1 Introduction

In this paper, we abbreviate floating-point number to FLOAT, and by $\mathbb{F}$ we denote the fixed-precision FLOATs. By $\| \circ \|$, with $\circ$ a polynomial or numerical vector, we denote the norm; we employ the infinity norm in this paper.

In industry, expressions with FLOAT coefficients are widely used. However, most algorithms in computer algebra before 1990 were developed by assuming exact coefficients, such as rational numbers and algebraic numbers. In late 1980's, the first author started on studying "approximate computer algebra", that is, algebraic computation of expressions with inaccurate coefficients represented by FLOATs. With collaboration of Noda et al., he proposed algorithms of approximate GCD (greatest common divisor) and approximate square-free decomposition of univariate polynomials in 1989 [19,20] and multivariate ones

---

* Work supported by Japan Society for the Promotion of Science under Grants 23500003.

V.P. Gerdt et al. (Eds.): CASC 2013, LNCS 8136, pp. 315–330, 2013.
© Springer International Publishing Switzerland 2013

in 1991 [10,9]. The algorithms are based on Euclid's method. Soon, he and his collaborators proposed an algorithm of approximate factorization of multivariate polynomials [21,22], which is based on the multivariate Hensel construction. In 1993, Sasaki and Kako proposed a method of expanding multivariate algebraic functions in a series form at critical points (see Subsect. 3.1) [17], by extending the multivariate Hensel construction so that it works at critical points. Later, Sasaki and Inaba revealed the properties of the series and named the series *Hensel series* [13,5,15].

Study of approximate algebraic computation has spread to the world very soon. Shirayanagi proposed a stabilization technique for computing Gröbner bases using FLOATs [25]. Stetter and Thallinger investigated singular systems of polynomials having multiple zeros each of which is separated into isolated zeros under almost all perturbations of the system [27]. Corless, Giesbrecht and Jeffrey proposed "approximate polynomial decomposition": given $f(x) \in \mathbb{C}[x]$, find polynomials g and h satisfying $(f + \delta f)(x) = g(h(x))$, where $\|\delta f\|$ is much smaller than $\|f\|$ and $\deg(\delta f) \leq \deg(f)$ [3]. Many researchers tried to construct a theory of approximate Gröbner bases since ∼1995, and after various trials, the first author has constructed a theory based on a concept of "approximate ideal" and proposed an algorithm to compute them [12]. He and Inaba then proposed a concept of "approximately singular system" to be a polynomial system the dimension of whose zero manifold increases by suitable small perturbations of the system [16].

There are quite many and various studies on approximate computer algebra, especially on the approximate GCD and the approximate factorization. Furthermore, studies on the nearest polynomial and its related topics may be included into them. Although the studies include many interesting and important ones, we do not refer to each because there are so many papers on these topics.

There are a number of applications of approximate computer algebra to mathematical computations, such as irreducibility testing of multivariate polynomials and the analytic continuation of algebraic functions. However, so long as computations in industry are concerned, the applications are very few. The present authors remind only the followings: Kitamoto proposed to use the approximate Puiseux series for eigenvalues appearing in the control theory [7], Li et al. utilized the approximate GCD to deconvolute blurred images [8], and Giesbrecht and Pham employed a symbolic-numeric method to compute null space basis of a Jacobian matrix from a multibody mechanical system [4]. The current situation is completely unsatisfactory.

In order to break through this situation, the authors decided to perform a study of applying approximate computer algebra to industrial computation. As an industrial problem, we have chosen symbolic-numeric analysis of the passengers aircraft control; the aircraft is described theoretically by a small number of parameters. We learned the aircraft control by a textbook [1], and found that the aircraft motion is described usually by being separated into the longitudinal motion (rotary motion around the "wing axis") and the lateral motion (rotary motions around the "body axis" and the "yaw axis"). Hence, we developed an aircraft model which describe these motions unitedly. We formulate the aircraft motion in the scheme of linear control theory [2], by approximating the equations of aircraft motion suitably, and obtain a 9×9 matrix A containing many aircraft "constants" and free variables. An important expression in the linear control theory is the eigenpolynomial of A. Substituting suitable numbers for the aircraft constants in the eigenpolynomial, we obtain a polynomial $\bar{C}(X, \boldsymbol{u})$ with FLOATs coefficients, where $\boldsymbol{u}$ denotes three interesting free variables.

We consider searching critical points of $\bar{C}(X, \boldsymbol{u})$ near the origin and performing Hensel series expansion of some roots of $\bar{C}(X, \boldsymbol{u})$ at two critical points. Owing to this specialized but concrete computation, we will be able to find typical phenomena for which we should be careful enough and situations in which we can utilize approximate algebraic operations successfully.

In Sect. 2, we present a model of aircraft control and show an eigenpolynomial the roots of which we will compute. In Subsect. 3.1, after defining the critical point, we explain how to compute the Hensel series. In Subsect. 3.2, we consider searching for critical points near the origin, and show that the approximate square-free decomposition is quite useful in the searching. In Subsect. 3.3, we consider separating eigenpolynomial into two parts, one containing only critical eigenvalues and the other containing only non-critical eigenvalues. By this, we show that it is quite important to remove "fully-erroneous" terms. In Subsect. 3.4, we show that the approximate factorization is nicely applicable to factor the so-called Newton polynomial with FLOAT coefficients; the Newton polynomial plays a crucial role in the "extended Hensel construction". In Subsect. 3.5, we perform the Hensel series expansion of engenvalues at critical points. In Sect. 4, we give conclusions and comments on our model from an expert.

2 Eigenpolynomial in an Aircraft Model

We denote the time by t, and the derivative dX/dt by $\dot{X}$. Let (x, y, z) be a Cartesian coordinate system fixed to the ground, such that the origin is at the aircraft center of gravity at time t_0, the z-axis is vertical to the ground, and the x- and y-axes are horizontal such that if x-axis points north then y-axis points east, where the x-axis points the aircraft head at time t_0. We also consider three mutually orthogonal aircraft axes called *body axis* (or *roll axis*), *wing axis* (or *pitch axis*) and *yaw axis*, where the axes pass the aircraft center always: the body axis penetrates the aircraft body from the tail to head, the wing axis is parallel to the wings and points from left to right, and the yaw axis is vertical from the floor to the ceiling. Let θ be the angle between the body axis and the horizontal plane, with $\theta > 0$ if the body axis points upward, ϕ the angle between the wing axis and the horizontal plane, with $\phi > 0$ if the wing axis points upward, and ψ the angle between the x-axis and the orthogonal projection of the body axis on the horizontal plane ($\psi = 0$ at time t_0). We express the state of aircraft at time t by state variables $v_x, v_y, v_z, \phi, \theta, \psi, \omega_\phi, \omega_\theta, \omega_\psi$. Here, $(v_x, v_y, v_z) = (\dot{x}, \dot{y}, \dot{z})$ and $(\omega_\phi, \omega_\theta, \omega_\psi) = (\dot{\phi}, \dot{\theta}, \dot{\psi})$. We assume that the aircraft is equipped with left and right engines below the wings and left and right *flaps* in the rear of wings, where the left and the right ones are controllable independently.

Our aircraft model is the following first-order differential system for state vector U given below, where (u_x, u_y, u_z) is the dimensionless velocity defined by $(v_x, v_y, v_z) = V_0(1+u_x, u_y, u_z)$, with V_0 the aircraft speed at time t_0. Because of the page limit, we omit the derivation of the model; the reader can see the omitted part at http://kako.ics.nara-wu.ac.jp/sasaki/CASC13paper.pdf. Below, matrix A_0 is for the aircraft motion in the case of $F = 0$ and $T_L = T_R$, and A_F shows the effect of F and $T_L \neq T_R$, where $T(= T_L+T_R)$ and $F(= F_L+F_R)$ denote the "thrust" and the force due to flaps, respectively (left and right).

$$\dot{U} = (A_0 + A_F)U + (b_0 + b_F), \quad U = (u_x, u_y, u_z, \phi, \theta, \psi, \omega_\phi, \omega_\theta, \omega_\psi)^t, \quad (2.1)$$

$$A_0 = \begin{pmatrix} -2D & 0 & 0 & 0 & -\alpha_{m0}L & 0 & 0 & 0 & 0 \\ 0 & -D-P_v & 0 & -\alpha_{m0}L & 0 & T+P_v & 0 & 0 & 0 \\ 2\alpha_{m0}L & 0 & -D-L-P_h & 0 & L+T+P_h & 0 & 0 & 0 & 0 \\ 0 & 0 & 0 & 0 & 0 & 0 & 1 & 0 & 0 \\ 0 & 0 & 0 & 0 & 0 & 0 & 0 & 1 & 0 \\ 0 & 0 & 0 & 0 & 0 & 0 & 0 & 0 & 1 \\ 0 & \alpha_{m0}\Lambda+Q_{vv} & 0 & 0 & 0 & -\alpha_{m0}\Lambda-Q_{vv} & 0 & 0 & -\alpha_{m0}Q_o \\ 0 & 0 & Q_h & 0 & -Q_h & 0 & 0 & 0 & 0 \\ 0 & Q_v & 0 & 0 & 0 & -Q_v & 0 & 0 & 0 \end{pmatrix}, \quad b_0 = \begin{pmatrix} T_d \\ 0 \\ L_d \\ 0 \\ 0 \\ 0 \\ Q_{ai} \\ Q_{el} \\ Q_{ru} \end{pmatrix}.$$
$$(2.2)$$

$$A_F = \begin{pmatrix} 0 & 0 & 0 & 0 & -f_+F_{ac} & 0 & 0 & 0 & 0 \\ 0 & 0 & 0 & -f_+F_{ac} & 0 & 0 & 0 & 0 & 0 \\ 0 & 0 & 0 & 0 & -f_+F_{as} & 0 & 0 & 0 & 0 \\ 0 & 0 & 0 & 0 & 0 & 0 & 0 & 0 & 0 \\ 0 & 0 & 0 & 0 & 0 & 0 & 0 & 0 & 0 \\ 0 & 0 & 0 & 0 & 0 & 0 & 0 & 0 & 0 \\ 0 & 0 & 0 & 0 & 0 & 0 & 0 & 0 & 0 \\ 0 & 0 & 0 & 0 & 0 & 0 & 0 & 0 & 0 \\ 0 & 0 & 0 & 0 & 0 & 0 & 0 & 0 & 0 \end{pmatrix}, \quad b_F = \begin{pmatrix} -f_+F_{as} \\ 0 \\ f_+F_{ac} \\ 0 \\ 0 \\ 0 \\ -f_-H_{\phi ac} \\ -f_+H_{\theta ac} + (2+e_+)T_\theta \\ -f_-H_{\psi as} + e_-T_\psi \end{pmatrix}.$$
$$(2.3)$$

We explain the symbols briefly. In deriving a model, many approximations are made by assuming $u_x^2 + u_y^2 + u_z^2 \ll 1$ and $\theta^2 + \phi^2 + \psi^2 \ll 1$. The α, β, λ denote, respectively, the *attack angle* ($\alpha = \theta - \mathrm{atan}(v_z/v_x)$), the *side-slide angle* ($\beta = \psi - \mathrm{atan}(v_y/v_x)$), the *sweptback angle* of the wings; we neglect the *dihedral angle* of the wings. The L, D, P_h, P_v denote the *lift*, frictional *drag*, forces acting on *tail plates* and *vertical tail*, respectively. L is proportional to α approximately, and we simplify L to be $L = (\alpha_m+\alpha)\tilde{L}$, where $\alpha_m\tilde{L}$ denotes the lift in the horizontal flight. We put $\alpha_{m0} = \alpha_m+\alpha_0$, where α_0 is the attack angle at time t_0. Similarly, $P_h = \alpha\tilde{P}_h$ and $P_v = \beta\tilde{P}_v$. $T_d = T-D$ and $L_d = \alpha_m\tilde{L} - gM$, where g is the gravitational constant and M is the weight of the aircraft. All these forces are divided implicitly by MV_0 and $\tilde{\ }$ is omitted from $\tilde{L}, \tilde{P}_h$ and $\tilde{P}_v$ in (2.2). The $\Lambda, Q_o, Q_{vv}, Q_h, Q_v, Q_{ai}, Q_{el}, Q_{ru}$ are torques caused by the sweptback wings ($\Lambda \propto \sin\beta \tan\lambda$), the rotary motion around the yaw axis, P_v, the *ailerons*, the *elevators*, the *rudder*, respectively. The T_θ and T_ψ are torques due to the engines. They are divided by $I_\phi, I_\phi, I_\phi, I_\theta, I_\psi, I_\phi, I_\theta, I_\psi, I_\phi, I_\psi$, respectively, where

I_ϕ, I_θ, I_ψ are moments of inertia around the body axis, the wing axis, the yaw axis, respectively. The α_F is the angle between the wings and the flaps taken out downward; the force F is approximately proportional to $\alpha + \alpha_F$: $F = (\alpha + \alpha_F)\tilde{F}$. The F_{ac} and F_{as} (and H_{*ac} and H_{*as}, where $*$ is either ϕ, θ or ψ) denote the forces due to flaps (and the torques due to flaps, resp.); suffices ac and as mean that $\alpha_F \cos\alpha_F$ and $\alpha_F \sin\alpha_F$ are multiplied to F (and multiplied to H_*, resp.). These forces and toques are divided implicitly by MV_0 and I_*, respectively, and $\tilde{\ }$ is omitted from $\tilde{F}$ in (2.3). We have mentioned that we may have $T_L \neq T_R$ and $F_L \neq F_R$. So, we put $T_L + T_R = (2 + e_+)T_0$, $T_L - T_R = e_-T_0$, $F_L + F_R = f_+F_m$ and $F_L - F_R = f_-F_m$, where F_m is the maximum force acting on each flap when the flap is taken out maximally from the wing.

The equations of motion were determined by the x-, y- and z-components of forces T, D, L, P_h, P_v, F at time t_0, by treating $\tilde{L}, \tilde{P}_h, \tilde{P}_v, \tilde{F}$ to be constant. However, $\tilde{L}, \tilde{P}_h, \tilde{P}_v, \tilde{F}$ are proportional to V^2, hence their values also change as t. (Even M changes gradually because the fuel is being exhausted). Therefore, A_0 and A_F must be reset if the values of T, V, α are changed noticeably.

We put $A = A_0 + A_F$ and $\boldsymbol{b} = \boldsymbol{b}_0 + \boldsymbol{b}_F$. The solution of (2.1) is given by

$$\boldsymbol{U}(t) = e^{A(t-t_0)}\boldsymbol{U}(t_0) + \int_{t_0}^t e^{A(t-t')}\boldsymbol{b}(t')dt'. \tag{2.4}$$

Therefore, 9 eigenvalues of the matrix A play an important role for the long-term behavior of $\boldsymbol{U}$; the short-time behavior of $\boldsymbol{U}$ is determined mostly by the vector $\boldsymbol{b}$. The eigenpolynomial $|XI_9 - A|$ has a factor X, so we put

$$C(X) \overset{\text{def}}{=} |XI_9 - A|/X = X^8 + C_7X^7 + \cdots + C_1X + C_0. \tag{2.5}$$

We show coefficient polynomials C_0, C_1, C_2 explicitly.

$$\begin{cases} C_0 = 2 \times \alpha_{m0}Q_hQ_vQ_o \times (\alpha_{m0}L + f_+F_{ac}) \\ \qquad \times \{(2T_0 + T_0e_+ - D)D - \alpha_{m0}L(\alpha_{m0}L + f_+F_{ac}) - Df_+F_{ac}\}, \\ C_1 = -\alpha_{m0}Q_h \times [16 \text{ terms}] + 2 \times DQ_h \times (2T_0 + T_0e_+ - D - f_+F_{as}) \\ \qquad\qquad\qquad\qquad \times \{(2T_0 + T_0e_+ - D)Q_v - f_+Q_{vv}F_{ac}\}, \\ C_2 = \alpha_{m0} \times [22 \text{ terms}] \\ \qquad + Q_hQ_v \times \{T_0^2(2 + e_+)^2 - D(12T_0 + 6T_0e_+ - 5D)\} \\ \qquad + f_+F_{as}Q_hQ_v \times (-2T_0 - T_0e_+ + 3D) \\ \qquad + f_+F_{ac}Q_hQ_{vv} \times (-2T_0 - T_0e_+ + 3D + f_+F_{as}), \end{cases}$$
$$\tag{2.6}$$

The $C(X)$ contains many parameters, most of them are "constants" determined by the aircraft structure and the aircraft speed V. The parameters which are independent of $\tilde{L}, \tilde{P}_h, \tilde{P}_v, \tilde{F}$ are α_{m0}, e_+ and f_+. Therefore, below, we analyze the dependence of $C(X)$ on α_{m0}, e_+, f_+. By $\bar{C}(X, \alpha_{m0}, e_+, f_+) = \bar{C}(X, \boldsymbol{u})$ we denote the eigenpolynomial, with the other aircraft parameters being substituted by suitable numbers. Since the coefficients of α_{m0} in $\bar{C}$ are pretty large, we redefine $\bar{C}(X, \boldsymbol{u})$ by replacing α_{m0} by $\alpha_{m0}/100$.

The aircraft parameters are estimated for a middle-sized passengers aircraft of $M = 250,000\,kg$, $S_W = 350\,m^2$, $S_p = 100\,m^2$, $S_F = 50\,m^2$, $l_T = 30\,m$, $l_W = 12\,m$,

$\lambda = 30$ degrees, etc., flying horizontally at a high altitude where $\rho = 0.5\,kg/m^3$, with speed $250\,m/s$. Here, m, kg, s denote meter, kilogram, second, respectively, and S_W, S_p, S_F denote the surface areas of the wings, the tail plates or the vertical tail, and each flap, respectively; l_T, l_W etc. denote the distances between the aircraft center and the tail plates, between an effective center of one wing and the body axis, and so on. The estimated values of parameters in the matrix A are: $L = 1.10$, $T_0 = 0.20$, $D = 0.0039$, $P_h = P_v = 0.315, \Lambda = 26.5$, $Q_o = 8.83$, $Q_h = 14.8$, $Q_v = 12.3$, $Q_{vv} = 6.57$, $F_{ac} = 0.023$, $F_{as} = 0.0067$.

3   Expansion of Multivariate Algebraic Functions at Critical Points

In this section, we will expand multivariate algebraic functions, which are roots of $\bar{C}(X, \boldsymbol{u})$ w.r.t. X, into Hensel series at critical points. By this, we want to show usefulness of the approximate square-free decomposition of univariate polynomials and the approximate factorization of multivariate polynomials. Furthermore, we will also show usefulness of "eFLOATs" (effective floating-point numbers) which were designed to detect the accuracy of each FLOAT approximately but quickly [6].

3.1   Hensel Series: A Brief Survey

Let $F(x, u_1, \ldots, u_\ell)$, with $\ell \geq 2$, be a multivariate polynomial over $\mathbb{C}$; for simplicity, we denote $(u_1, \ldots, u_\ell)$ by $(\boldsymbol{u})$. Consider the generalized Hensel construction at $(\boldsymbol{u}) = (\boldsymbol{0}) \stackrel{\text{def}}{=} (0, \ldots, 0)$. The first step of the construction is to factor univariate polynomial $F(x, \boldsymbol{0})$ into two or more mutually irreducible polynomials, and the construction fails if $F(x, \boldsymbol{0})$ has no such factors. Such a case occurs if, for example, $F(x, \boldsymbol{0}) = x^n$.

Definition 1 (critical point) *Let* $F(x, u) \in \mathbb{C}[x, u]$. *If* $F(x, \boldsymbol{u}_c)$, *with* $\boldsymbol{u}_c \in \mathbb{C}^\ell$, *has multiple roots in* $\mathbb{C}$ *then* $\boldsymbol{u}_c$ *is called a critical point of* $F(x, \boldsymbol{u})$.

In this subsection, we assume that $F(x, \boldsymbol{0})$ has m multiple root at $x = 0$. Then, the generalized Hensel construction, with initial factors $G^{(0)} = x^m$ and $H^{(0)} = F(x, \boldsymbol{0})/x^m$, gives $G^{(k)}(x, \boldsymbol{u}), H^{(k)}(x, \boldsymbol{u}) \in \mathbb{C}[x, \boldsymbol{u}], \forall k \in \mathbb{N}$, satisfying

$$F(x, \boldsymbol{u}) \equiv G^{(k)}(x, \boldsymbol{u})H^{(k)}(x, \boldsymbol{u}) \pmod{\langle \boldsymbol{u} \rangle^{k+1}}, \tag{3.1}$$

where $\langle \boldsymbol{u} \rangle$ denotes an ideal generated by $u_1, \ldots, u_\ell$, and we have $G^{(k)}(x, \boldsymbol{0}) = x^m$. In the rest of this subsection, we consider $G^{(k)}(x, \boldsymbol{u})$.

As is well-known, the algebraic functions, multivariate as well as univariate, can be expanded into Taylor series at non-critical points. The univariate algebraic functions can be expanded into Puiseux series at critical points. The multivariate algebraic functions (i.e., the case of $\ell \geq 2$) can also be expanded into multivariate Puiseux series, but the series are hard to utilize practically. In

1993, Sasaki and Kako devised a new technique to expand multivariate algebraic function into a series form by extending the Hensel construction, and they called the new construction *extended Hensel construction* or *EHC* in short.

The so-called Newton polynomial defined below plays a crucial role in the EHC. We explain the Newton polynomial in the simplest case; for more general cases, see [13].

Definition 2 (Newton polynomial) *We assume that $\deg_x(F(x, \boldsymbol{u})) = \deg_x(F(x, \boldsymbol{0})) = n$. For each nonzero term $c\,x^{e_x} u_1^{e_1} \cdots u_\ell^{e_\ell}$ of $F(x, \boldsymbol{u})$, plot a dot at the point (e_x, e_t) in the two-dimensional plane, where $e_t = e_1 + \cdots + e_\ell$. Let $\mathcal{L}_{\mathrm{New}}$ be a line which passes the point $(n, 0)$ and another dot plotted, such that no dot lies below it. The $F_{\mathrm{New}}(x, \boldsymbol{u})$ is the sum of the terms plotted on $\mathcal{L}_{\mathrm{New}}$.*

Figure 1 shows a polynomial $F(x, u, v)$ and its Newton line: the Newton polynomial in this case is $F_{\mathrm{New}}(x, u, v) = x^3 - u^2$.

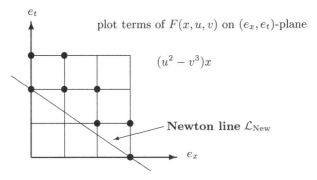

Fig. 1. Newton polynomial is $x^3 - u^2$

In the EHC, we factor $F_{\mathrm{New}}(x, \boldsymbol{u})$ and perform the Hensel construction by using the mutually irreducible factors of $F_{\mathrm{New}}(x, \boldsymbol{u})$ as initial factors; in the above example, the initial factors are $x - \omega_i \sqrt[3]{u^2}$ $(i = 1, 2, 3)$, where ω_i is the cube root of 1. The modulus of EHC is chosen in such a way that the Newton line $\mathcal{L}_{\mathrm{New}}$ is shifted upward without changing its slope so that all the lattice points above $\mathcal{L}_{\mathrm{New}}$, of (e_x, e_t)-plane are scanned.

Using the generalized and extended Hensel constructions, we factor $F(x, \boldsymbol{u})$ as $F(x, \boldsymbol{u}) \equiv C_n(\boldsymbol{u}) \prod_{i=1}^{n} (x - \chi_i(\boldsymbol{u})) \ (\langle \boldsymbol{u} \rangle^{k+1})$. Then, we obtain n algebraic functions defined by $F(x, \boldsymbol{u}) = 0$ in series forms; the series corresponding to single roots of $F(x, \boldsymbol{0})$ are Taylor series.

3.2 Determination of Critical Points and Approximate Square-Free Decomposition

From now on, we treat $\bar{C}(X, \boldsymbol{u}) = \bar{C}(X, \alpha_{m0}, e_+, f_+)$ defined in Sect. 2. We put

$$\begin{cases} \bar{C}(X, \boldsymbol{u}) = X^8 + \bar{C}_7(\boldsymbol{u})X^7 + \cdots + \bar{C}_2(\boldsymbol{u})X^2 + \bar{C}_1(\boldsymbol{u})X + \bar{C}_0(\boldsymbol{u}), \\ \bar{C}(X, \boldsymbol{0}) = X^8 + \cdots + 182.2\cdots X^4 + 1.383\cdots X^3 + 0.00028\cdots X^2 + 1.41\cdots\mathrm{e}{-8}\,X. \end{cases}$$
$$(3.2)$$

The first thing we must do is to determine a critical point $\boldsymbol{u}_c$ at which we perform the EHC. We consider mostly the case that $\bar{C}(X, \boldsymbol{u}_c)$ has multiple roots at $X = 0$. Then, $\boldsymbol{u}_c$ can be determined by solving $\bar{C}_1(\boldsymbol{u}) = \bar{C}_0(\boldsymbol{u}) = 0$. Fortunately, we have factorized expressions of C_1 and C_0. We see $C_0 \propto \alpha_{m0}$ in (2.6), so we set $\alpha_{m0} = 0$. Then, C_1 becomes 0 if we set

$$\begin{cases} \boldsymbol{u}_{c1} : (\alpha_{m0}, e_+, f_+) = (0,\ 0,\ (2T_0 - D)Q_{\mathrm{v}}/Q_{\mathrm{vv}}F_{ac}) = (0,\ 0,\ 0.0081\cdots), \\ \boldsymbol{u}_{c2} : (\alpha_{m0}, e_+, f_+) = (0,\ 0,\ (2T_0 - D)/F_{as}) \qquad = (0,\ 0,\ 0.0149\cdots), \quad (3.3) \\ \boldsymbol{u}_{c3} : (\alpha_{m0}, e_+, f_+) = (0,\ (2T_0 - D)/T_0,\ 0) \qquad = (0,\ 0.0500\cdots,\ 0). \end{cases}$$

We see that $\bar{C}_2(\boldsymbol{u}_{c2}) \neq 0$ but $\bar{C}_2(\boldsymbol{u}_{c3}) = 0$, hence $\boldsymbol{u}_{c2}$ and $\boldsymbol{u}_{c3}$ are critical points of the second and the third orders, respectively.

The above determination relies on the exact factorization crucially, which we cannot anticipate practically. If C_0 and C_1 are not factorizable then $\bar{C}_0(\boldsymbol{u})$ and $\bar{C}_1(\boldsymbol{u})$ will mostly be not factorizable, too. Then, we determine critical points by solving $\bar{C}_0(\boldsymbol{u}) = \bar{C}_1(\boldsymbol{u}) = 0$ numerically. This system of equations is often under-determined hence has infinitely many solutions, which is true in our case, then it may have solutions near the origin $\boldsymbol{0}$. If this expectation is right then $\bar{C}(X, \boldsymbol{0})$ will have approximately multiple roots, so we perform the approximate square-free decomposition of $\bar{C}(X, \boldsymbol{0})$. In our algebra system, this operation is executed by computing "normalized" remainder sequence which is cutoff if the norm of remainder is reduced to ε ($\ll 1$), compared with that of the quotient. Changing the "cutoff parameter" as $\varepsilon = 10^{-3} \to 10^{-4} \to \cdots \to 10^{-10}$, we obtained the following three decompositions.

$$\begin{cases} (X + 0.0019239\cdots)^4 \times (X^4 + 1.73\cdots X^3 + \cdots + 182.\cdots) \quad (\text{tol } 0.01793\cdots), \\ (X + 6.5961\cdots\mathrm{e}{-}5)^3 \times (X^5 + 1.74\cdots X^4 + \cdots + 1.41\cdots) \quad (\text{tol } 3.63\cdots\mathrm{e}{-}6), \quad (3.4) \\ (X - 9.9999\cdots\mathrm{e}{-}5)^2 \times (X^6 + 1.74\cdots X^5 + \cdots + 1.41\cdots) \quad (\text{tol } 0.00000\cdots). \end{cases}$$

Here, the "(tol $0.01793\cdots$)" for example means that

$$\| (X + 0.0019239\cdots)^4 \times (X^4 + 1.73\cdots X^3 + \cdots) - \bar{C}_0 \| / \|\bar{C}_0\| = 0.01793\cdots.$$

The above decompositions show that there is a very small cluster of 2 close roots, which is contained in a little bigger cluster of 3 close roots, and which is contained in a much bigger cluster of 4 close roots. We expect that two small clusters correspond to critical points.

Following this observation, we search for critical points near the origin $\boldsymbol{0}$, by taking out only terms of total-degrees ≤ 2 w.r.t. $\boldsymbol{u}$, of $\bar{C}_1(\boldsymbol{u})$ and $\bar{C}_0(\boldsymbol{u})$ of (3.2):

$$\begin{aligned} \bar{C}_1 :\ & 1.571292819390\mathrm{e}{-}12 && -\ 9.232303419296\mathrm{e}{-}11\,\alpha_{m0} \\ & +\ 6.285171277561\mathrm{e}{-}11\,e_+ - 2.983156903935\mathrm{e}{-}10\,f_+ \\ & +\ 6.285171277561\mathrm{e}{-}10\,e_+^2 - 5.966313807871\mathrm{e}{-}09\,e_+ f_+ \\ & +\ 1.293361779012\mathrm{e}{-}08\,f_+^2 + [3 \text{ terms} \propto \alpha_{m0}] && = 0, \\ \bar{C}_0 :\ & 4.154005264743\mathrm{e}{-}12\,\alpha_{m0}^2 + 8.685647371734\mathrm{e}{-}12\,\alpha_{m0} f_+ = 0. \end{aligned}$$

Solving these equations by putting $\alpha_{m0} = 0$, we obtain three special solutions $(e_+, f_+) = (-0.050000000\cdots,\ 0)$, $(0,\ 0.0149253731\cdots)$ and $(0,\ 0.00813976573\cdots)$.

Since $\bar{C}_1$ contains only $e_+^{e_e} f_+^{e_f}$-terms with $e_e + e_f \leq 2$, the former two become the same as those in (3.4). In general, we must determine e_+ and f_+ iteratively.

Remark 1. The result of square-free decomposition is very accurate. If a nonzero FLOAT f contains an error e then we say that the accuracy of f is $|e/f|$. If we compute m multiple roots or very close roots numerically, say by famous Durand-Kerner's method with FLOATs of machine-epsilon ε_M, then the accuracies of the results are $O(\sqrt[m]{\varepsilon_M})$. On the other hand, in the approximate square-free decomposition, it is assumed that $F(x)$ can be expressed as $F(x) = F_m(x)^m G(x) + D(x)$, $\|D\|/\|F\| = \varepsilon$, where ε is much smaller than 1, and $F_m(x)$ is computed as $\texttt{appgcd}(d^{m-2}F/dx^{m-2}, d^{m-1}F/dx^{m-1})$, where $\texttt{appgcd}$ is the operation of approximate GCD computation. In [20,11], it is shown that if the size of a cluster containing m close roots is δ then $\varepsilon = O(\delta^2)$. The reason is as follows: let c be the center of the cluster containing m close roots of $F_m(x)$ then the center of the cluster containing $m-1$ corresponding close roots of dF/dx is $c + O(\delta^2)$. Note that if $F(x)$ and $G(x)$ have no special relation except that they have mutually close roots of "closeness" δ then $\varepsilon = O(\delta)$.

The above computation is for the simplest case. General case is that we search for $\boldsymbol{u}_c$ such that $\bar{C}(X, \boldsymbol{u}_c)$ has multiple roots at $X = X_c \neq 0$, with unknown X_c. Even in this case, if $\|\boldsymbol{u}_c\| \ll 1$ then the approximate square-free decomposition tells us an approximate value of X_c. Hence, the decomposition is more useful in this case.

The first decomposition in (3.4) is also utilized as follows.

Empirical rule-1: For treating a cluster of m close-roots around the origin, we had better regularize $F(x)$ as $F(x) \mapsto x^n + \cdots + x^m + c_{m-1} x^{m-1} + \cdots$.

Following this rule, we regularize $\bar{C}(X, \boldsymbol{u})$ as

$$\bar{C}(X, \boldsymbol{u}) \mapsto \widetilde{C}(X, \boldsymbol{u}) \overset{\text{def}}{=} \bar{C}(\bar{s}X, \boldsymbol{u})/\bar{s}^8, \qquad \bar{s} = \{\text{coef}(\bar{C}(X, \boldsymbol{0}), X^4)\}^{1/4}. \quad (3.5)$$

Here, $\text{coef}(F(x), x^m)$ denotes the coefficient of x^m-term of $F(x)$.

In the following, we consider only critical points $\boldsymbol{u}_{c2}$ and $\boldsymbol{u}_{c3}$, and we put

$$\begin{cases} S_2(X, \boldsymbol{u}) \overset{\text{def}}{=} \widetilde{C}(X, \boldsymbol{u} - \boldsymbol{u}_{c2}), \\ \quad S_2(X, \boldsymbol{0}) = X^8 + \cdots + 0.00214 \cdots X^3 + 4.81 \cdots \text{e-}8\, X^2, \\ S_3(X, \boldsymbol{u}) \overset{\text{def}}{=} \widetilde{C}(X, \boldsymbol{u} - \boldsymbol{u}_{c3}), \\ \quad S_3(X, \boldsymbol{0}) = X^8 + \cdots + 1.000012 \cdots X^4 + 0.00212 \cdots X^3. \end{cases} \quad (3.6)$$

It is quite interesting to note that, although $\widetilde{C}(X, \boldsymbol{u})$ contains no $X^0, X^0 \alpha_{m0}$, $X^0 e_+, X^0 f_+$ terms, $S_2(X, \boldsymbol{u})$ contains $X^0 \alpha_{m0}$ term and $S_3(X, \boldsymbol{u})$ contains no $X^{e_X} \boldsymbol{u}^{e_t}$ term for $e_X + e_t \leq 2$.

3.3   Separation of Critical Factors and Effective FLOATs

In order to compute algebraic functions in Hensel series, we must separate critical factors $\widehat{S}_2(X, \boldsymbol{u})$ and $\widehat{S}_3(X, \boldsymbol{u})$ from $S_2(X, \boldsymbol{u})$ and $S_3(X, \boldsymbol{u})$, respectively. However, we must notice the following rule; see [24].

Empirical rule-2: Suppose we perform the Hensel construction of $F(x, \boldsymbol{u})$ with initial factors $G_0(x)$ and $H_0(x)$, where $G_0, H_0 \in \mathbb{F}[x]$. We should choose $G_0(x)$ and $H_0(x)$ so that they have no mutually close roots, otherwise we will encounter large cancellation errors.

If $G_0(x)$ and $H_0(x)$ have mutually close roots, the Hensel construction will often give *fully-erroneous terms* (terms with coefficients having no significant bit). We remove the fully-erroneous terms by eFLOATs. An eFLOAT is expressed as $\#e(f, e)$, where f is the conventional FLOAT and e is an error computed by the following arithmetic.

$$\begin{cases} \#e(f_a, e_a) + \#e(f_b, e_b) \Longrightarrow \#e[f_a + f_b, \; \max\{e_a, e_b\}], \\ \#e(f_a, e_a) - \#e(f_b, e_b) \Longrightarrow \#e[f_a - f_b, \; \max\{e_a, e_b\}], \\ \#e(f_a, e_a) \times \#e(f_b, e_b) \Longrightarrow \#e[f_a \times f_b, \; \max\{|f_b e_a|, |f_a e_b|\}], \\ \#e(f_a, e_a) \div \#e(f_b, e_b) \Longrightarrow \#e[f_a \div f_b, \; \max\{|e_a/f_b|, |f_a e_b/f_b^2|\}]. \end{cases} \quad (3.7)$$

Remark 2. The error part e is set initially as follows. For the aircraft constant c, we set e to be $10^{-15}|c|$. For numerically computed root r, we set e to be Smith's error bound for r [26].

Remark 3. An eFLOAT $\#e(f, e)$ is simplified to 0 if $|f| < e$.

We show an example of simple computation for $G_0(x)$ and $H_0(x)$ having mutually close roots, as follows:

$$G_0 = (x - 0.0001) \cdot (3x^2 + x - 2)/3.0, \quad H_0 = (x^2 + 0.0002x) \cdot (7x^2 - 5x + 3)/7.0.$$

In the Hensel construction (and in EHC, too), Moses-Yun's polynomials $A_i(x)$ and $B_i(x)$ satisfying $A_i(x)G_0(x) + B_i(x)H_0(x) = x^i$, $(i = 0, 1, \ldots, n-1)$, play crucial role. We check these relations for $0 \le i \le n-1$ by computing $A_i(x)$ and $B_i(x)$ with double-precision FLOATs and eFLOATs, respectively.

Table I shows comparison of the computations by FLOATs and eFLOATs, for $i = 1, 3, 5$. We see that the initial errors are magnified by $10^6 \sim 10^7$ and that many small terms are created by the computation with FLOATs. The small terms are unnecessary and fully-erroneous; they appear because terms which will cancel exactly in the exact arithmetic do not cancel in the FLOAT arithmetic. The eFLOATs over-estimate errors considerably, but they remove the fully-erroneous terms successfully due to the mechanism of Remark 3.

Table 1. Computation of $A_i(x)G_0(x) + B_i(x)H_0(x)$

i	computation by FLOATs	computation by eFLOATs
1	$+2.24\mathrm{e}{-8}\, x^5 + \cdots + 0.9999999999998\, x$	$\#e(0.9999999999998, 5.43\mathrm{e}{-12})\, x$
3	$-7.75\mathrm{e}{-9}\, x^6 + \cdots + 1.000000001092\, x^3 + \cdots$	$\#e(1.000000001092, 0.000121)\, x^3$
5	$-1.66\mathrm{e}{-9}\, x^6 + 0.9999999967943\, x^5 + \cdots$	$\#e(0.9999999967943, 0.000258)\, x^5$

Remark 4. If $G_0(x) = x^m$ then Moses-Yun's polynomials can be computed more accurately. The reason is as follows: the division by $G_0(x)$ is performed many times in the computation, and the division in this case is exact.

Following Empirical rule-2, we first separate clusters of 4 close-roots around $X = 0$ from $S_l(X, \boldsymbol{u})$ $(l = 1, 2)$.

$$
\begin{cases}
S_l(X, \boldsymbol{u}) \equiv \bar{S}_l^{(k)}(X, \boldsymbol{u}) H_l^{(k)}(X, \boldsymbol{u}) \pmod{\langle \boldsymbol{u} \rangle^{k+1}}, \\
\bar{S}_l^{(k)}(X, \boldsymbol{0}) = X^4 + \bar{s}_{l,3} X^3 + \cdots, \quad \text{contains 4 close-roots.}
\end{cases}
\tag{3.8}
$$

Since the initial factors of this Hensel construction have no mutually close roots, we can perform this separation quite accurately.

The computation of Hensel series is similar to the computation of close roots of univariate polynomials proposed by the first author and his collaborators [19,18,23]. First, perform the approximate square-free decomposition to get information on close-roots clusters: their locations, their sizes and the numbers of close-roots contained. Second, move the origin to one of the close-root cluster, separate the cluster, and enlarge the cluster size to $O(1)$. Finally, determine the roots contained in the cluster. Following this method, we perform the scaling of X. Consider $\bar{S}_3^{(k)}(X, \boldsymbol{0})$, for example, which contains three zeros and one small root. So, we perform the following scaling for $\bar{S}_l^{(k)}(X, \boldsymbol{u})$ $(l = 2, 3)$.

$$
\begin{cases}
\bar{S}_l^{(k)}(X, \boldsymbol{u}) \mapsto \widetilde{S}_l^{(k)}(X, \boldsymbol{u}) \overset{\text{def}}{=} \bar{S}_l^{(k)}(\bar{s}_l X, \boldsymbol{u}) / \bar{s}_l^4, \\
\bar{s}_l = \text{coef}(\bar{S}_l^{(k)}(X, \boldsymbol{0}), X^3) = 0.0021 \cdots .
\end{cases}
\tag{3.9}
$$

This scaling makes $\widetilde{S}_l^{(k)}(X, \boldsymbol{0})$ $(l=1, 2)$ as follows: $X^4 + X^3 + 0.01046283125455 X^2$ for $l = 2$, and $X^4 + X^3$ for $l = 3$. Finally, we separate the critical factors $\widehat{S}_l^{(k)}(X, \boldsymbol{u})$, $(l = 2, 3)$, by the Hensel construction of $\widetilde{S}_l^{(k)}(X, \boldsymbol{u})$, with initial factors $(G_2(X), H_2(X)) = (X^2, \ X^2 + X + 0.01046283125455)$ for $l = 2$, and $(G_3(X), H_3(X)) = (X^3, \ X+1)$ for $l = 3$.

3.4   Approximate Factorization of Newton Polynomials

We show low-order terms of $\widehat{S}_l^{(k)}(X, \boldsymbol{u})$ $(l = 2, 3)$ computed in the previous subsection.

$$
\begin{aligned}
\widehat{S}_2^{(k)} = {}& X^2 + X^1 \big(- 46.00 \cdots \alpha_{m0} - 0.2537 \cdots e_+ + 0.8499 \cdots f_+ \\
& + 202961. \cdots \alpha_{m0}^2 - 1106. \cdots \alpha_{m0} e_+ + 3715. \cdots \alpha_{m0} f_+ \\
& + 4.86 \cdots \text{e-5} e_+^2 - 0.00033 \cdots e_+ f_+ + 0.00055 \cdots f_+^2 + \cdots \big) \\
& + X^0 \big(- 5.665 \cdots \alpha_{m0}^2 - 11.70 \cdots \alpha_{m0} e_+ - 39.20 \cdots \alpha_{m0} f_+ \\
& + 24745. \cdots \alpha_{m0}^3 - 51624. \cdots \alpha_{m0}^2 e_+ + 172562. \cdots \alpha_{m0}^2 f_+ \\
& + 280.7 \cdots \alpha_{m0} e_+^2 - 1880. \cdots \alpha_{m0} e_+ f_+ + 3150. \cdots \alpha_{m0} f_+^2 + \cdots \big).
\end{aligned}
\tag{3.10}
$$

$$\widehat{S}_3^{(k)} = X^3 + X^2 (+\ 0.7532\cdot\cdot\,\alpha_{m0} - 0.5128\cdot\cdot\,e_+ + 2.434\cdot\cdot\,f_+$$
$$+\ 4.011\cdot\cdot\,\alpha_{m0}^2 - 7.81\cdot\cdot\,\mathrm{e}\text{-}5\,\alpha_{m0}e_+ + 8.387\cdot\cdot\,\alpha_{m0}f_+$$
$$+\ 6.24\cdot\cdot\,\mathrm{e}\text{-}5\,e_+^2 - 0.00049\cdot\cdot\,e_+f_+ + 0.00105\cdot\cdot\,f_+^2 + \cdots)$$
$$+\ X^1 (-\ 15.96\cdot\cdot\,\alpha_{m0}^2 - 0.1931\cdot\cdot\,\alpha_{m0}e_+ + 0.0657\cdot\cdot\,e_+^2$$
$$+\ 1.352\cdot\cdot\,f_+^2 + 2.993\cdot\cdot\,\alpha_{m0}^3 - 1.027\cdot\cdot\,\alpha_{m0}^2 e_+ + \cdots)$$
$$+\ X^0 (+\ 4.093\cdot\cdot\,\alpha_{m0}^2 e_+ - 13.71\cdot\cdot\,\alpha_{m0}^2 f_+ + 8.559\cdot\cdot\,\alpha_{m0}e_+f_+$$
$$-\ 28.67\cdot\cdot\,\alpha_{m0}f_+^2 - 0.0020\cdot\cdot\,\alpha_{m0}e_+^2 f_+ + 0.0152\cdot\cdot\,\alpha_{m0}e_+f_+^2$$
$$-\ 0.027\cdot\cdot\,\alpha_{m0}f_+^3 - 0.00099\cdot\cdot\,\alpha_{m0}^2 e_+^2 + \cdots).$$

$$(3.11)$$

The Newton polynomials for $\widehat{S}_2^{(k)}$ and $\widehat{S}_3^{(k)}$, let them be $\widehat{S}_{2,\mathrm{New}}$ and $\widehat{S}_{3,\mathrm{New}}$, respectively, are as follows.

$$\widehat{S}_{2,\mathrm{New}} = X^2$$
$$+ X^1 (-46.00100771767\alpha_{m0} - 0.2536988065195e_+ + 0.8498910018404f_+) \quad (3.12)$$
$$+ X^0 (-5.664934365224\alpha_{m0}^2 + 11.70156007615\alpha_{m0}e_+ - 39.20022625510\alpha_{m0}f_+),$$
$$\widehat{S}_{3,\mathrm{New}} = X^3$$
$$+ X^2 (+0.7532833020638\alpha_{m0} - 0.5128205128205e_+ + 2.434021263290f_+) \quad (3.13)$$
$$+ X^1 (-15.96482577252\alpha_{m0}^2 - 0.1931495646317\alpha_{m0}e_+ - 32.73394830102\alpha_{m0}f_+$$
$$+0.06574621959237e_+^2 - 0.6241080162281e_+f_+ + 1.352924904989f_+^2)$$
$$+ X^0 (+4.093545069876\alpha_{m0}^2 e_+ - 13.71337598409\alpha_{m0}^2 f_+$$
$$+8.559230600651\alpha_{m0}e_+f_+ - 28.67342251218\alpha_{m0}f_+^2).$$

The authors thought that $\widehat{S}_{2,\mathrm{New}}$ may be factorizable but $\widehat{S}_{3,\mathrm{New}}$ is not. However, surprisingly, the computer has factored them approximately as follows.

$$\widehat{S}_{2,\mathrm{New}} = (X - 46.12382784406\alpha_{m0}) \times$$
$$(X + 0.1228201263873\alpha_{m0} - 0.2536988065195e_+ \qquad\qquad (3.14)$$
$$+ 0.8498910018404f_+) \quad (\text{tol } 1.07\mathrm{e}{-}12),$$
$$\widehat{S}_{3,\mathrm{New}} = (X - 0.2564102564101e_+ + 0.8589743589743f_+) \times$$
$$(X^2 + 0.7532833020638X\alpha_{m0} - 0.2564102564104Xe_+ \qquad\qquad (3.15)$$
$$+ 1.575046904316Xf_+$$
$$- 15.96482577252\alpha_{m0}^2 - 33.38099934254\alpha_{m0}f_+) \quad (\text{tol } 1.82\mathrm{e}{-}13).$$

Smallness of the above tolerances indicates that the above factorizations are nearly exact, because the coefficients of $\widehat{S}_2^{(k)}$ and $\widehat{S}_3^{(k)}$ are contaminated by errors of FLOATs.

3.5   Computation of Hensel Series by Using FLOATs

With the factorization (3.14), we performed the EHC of $\widehat{S}_2$ up to order 1, without using eFLOATs. The result is as follows; very small terms will be fully-erroneous.

$$G_2^{(1)} = X + 0.1228201263873\alpha_{m0} - 0.2536988065195e_+ + 0.8498910018404f_+$$
$$+ (182.7\alpha_{m0}^3 - 2.461\alpha_{m0}^2 e_+ + 388.8\alpha_{m0}^2 f_+ - 4.918\text{e--}11\,\alpha_{m0}^2$$
$$+ 7.401\text{e--}3\,\alpha_{m0}e_+^2 - 2.156\alpha_{m0}e_+f_+ - 1.781\text{e--}12\,\alpha_{m0}e_+$$
$$+ 7.109\alpha_{m0}f_+^2 - 2.943\text{e--}16\,\alpha_{m0}f_+ - 1.234\text{e--}5\,e_+^3$$
$$+ 1.240\text{e--}4\,e_+^2 f_+ - 4.155\text{e--}4\,e_+f_+^2 + 4.640\text{e--}4\,f_+^3)$$
$$/(46.25\alpha_{m0} - 0.2537e_+ + 0.8499f_+),$$

$$H_2^{(1)} = X - 46.12382784406\alpha_{m0}$$
$$+ (9.386\text{e+}6\,\alpha_{m0}^3 - 1.027\text{e+}4\,\alpha_{m0}^2 e_+ + 3.439\text{e+}5\,\alpha_{m0}^2 f_+$$
$$+ 1.741\text{e--}10\,\alpha_{m0}^2 + 280.7\alpha_{m0}e_+^2 - 1.881\text{e+}3\,\alpha_{m0}e_+f_+$$
$$+ 1.096\text{e--}12\,\alpha_{m0}e_+ + 3.151\text{e+}3\,\alpha_{m0}f_+^2 + 2.295\text{e--}12\,\alpha_{m0}f_+)$$
$$/(46.25\alpha_{m0} - 0.2537e_+ + 0.8499f_+).$$

We have also performed the EHC of $\widehat{S}_3$; we omit the result to save the space.

Note the denominator above, which is a characteristic feature of the Hensel series. The denominator appears by the following reason. Let $F_{\text{New}}(x, \boldsymbol{u})$ be an irreducible Newton polynomial, and put $G_0 = (x - \alpha)$ and $H_0 = F_{\text{New}}(x, \boldsymbol{u})/(x - \alpha)$, where α is a root of F_{New}. Then, Moses-Yun's "polynomials" A_j and B_j satisfying $A_j G_0 + B_j H_0 = x^j$, $0 \le j \le \deg_x(F_{\text{New}}) - 1$, have the denominator $F'_{\text{New}}(\alpha_i, \boldsymbol{u})$ [14].

In the generalized Hensel construction, the initial factors are determined by factoring univariate polynomial $F(x, \boldsymbol{0})$, and the factorization is quite easy even if $F(x, \boldsymbol{0}) \in \mathbb{F}[x]$. In the extended Hensel construction, the initial factors are determined to be factors of Newton polynomial which is multivariate. Therefore, if the Newton polynomial can be factored exactly then the computation of Hensel series can be simplified largely; otherwise, we must introduce the roots of the Newton polynomial symbolically and perform the computation by using the Newton polynomial as a defining polynomial, which is quite time-consuming. Therefore, the nearly exact factorization of Newton polynomials is very useful.

4   Conclusion and Comments

In this paper, we have convinced the following three points.

Usefulness of eFLOATs. The algebraic expression handled in industry may contain terms of very different magnitudes. In fact, $\bar{C}(X, \boldsymbol{0})$ in (3.2) contains terms of magnitudes 180 and 10^{-8}. On the other hand, computation of expressions with coefficients of FLOATs creates many fully-erroneous terms usually, as Table I in Subsect. 3.3 shows. Therefore, we must distinguish small meaningful terms from fully-erroneous terms strictly. The eFLOAT is an easy and useful device, although it often over-estimates errors. Without such a device as eFLOAT, it is too dangerous to apply approximate computer algebra to industrial computations.

Usefulness of approximate square-free decomposition. The approximate square-free decomposition is based on a simple algorithm but it is extremely useful because it gives almost full information on well-separated close-root cluster; the location, the size and the number of close-roots contained.

Usefulness of approximate factorization. The factorization is a very strong operation which simplifies the analysis largely. We think the people in many application areas cannot imagine to factor multivariate polynomials with FLOAT coefficients. However, in approximate computer algebra, the approximate factorization is a standard operation.

Unfortunately, the Hensel series obtained in this paper are not useful for the aircraft control: they are quite small in value near the expansion point hence they are not influential for the aircraft motion. This fact is due to that the slope of the Newton line is negative. We are now looking for other industrial computations in which the slope of the Newton line is positive at the expansion point.

We asked an expert on aircraft design to criticize our model. His main comments are as follows. 1) Authors' linear model can also be divided into two sets. 2) The attack angle α is usually not treated as a variable; when α is treated as an input variable, the rigid-body motion of the aircraft is ignored. 3) Authors' model cannot treat large angles; large angles are treated as constants in linear models. 4) The aircraft-fixed coordinate system is used commonly, as the authors do so, but the aircraft center of gravity is expressed usually in the ground-fixed coordinate system. 5) Currently, there is no aircraft whose left and right flaps are controllable independently. 6) Except for these points, authors' model is not much strange, so long as only small angles are treated.

Currently, people in the industry express their machine models by sets of mathematical formulas containing many symbols usually and perform the computation as exactly as possible. However, the final answers required are mostly numbers. In such cases, the computation is extremely simplified by substituting actual numbers for symbols. Then, it is inevitable to handle expressions with FLOAT coefficients. Therefore, we are convinced of wide usage of approximate computer algebra in the industry in future.

Acknowledgments. The authors are very grateful to an anonymous reviewer for many comments on revising the manuscript. They also thank very much Professor Seiya Ueno of Yokohama National University for very valuable comments on our aircraft model from the viewpoint of expert.

References

1. Katayanagi, R.: Introduction to Aircraft Design. Nikkan Kohgyo Shinbun (Manufacturing Newspaper, Co.), Tokyo (2009) (in Japanese)
2. Suzuki, T., Itamiya, K.: Fundamentals of Modern Control. Morikita Publ. Co., Tokyo (2011) (in Japanese)
3. Corless, R.M., Giesbrecht, M.W., Jeffrey, D.J.: Approximate polynomial decomposition. In: Proceedings of ISSAC 1999 (Intn'l Symp. on Symbolic and Algebraic Computation), pp. 213–219. ACM Press (1999)
4. Giesbrecht, M., Pham, N.: A symbolic computation approach to the projection method. In: Proceedings of ASCM 2012 (Asian Symp. on Computer Mathematics) (2012) (to appear)

5. Inaba, D., Sasaki, T.: A numerical study of extended Hensel series. In: Proceedings of SNC 2007 (Intn'l Workshop on Symbolic-Numeric Computation), pp. 103–109. ACM Press (2007)
6. Kako, F., Sasaki, T.: Proposal of "effective" floating-point number. Preprint of Univ. Tsukuba (May 1997) (unpublished)
7. Kitamoto, T.: On Puiseux expansion of approximate eigenvalues and eigenvectors. IEICE Trans. Fundamentals E81-A, 1242–1251 (1998)
8. Li, Z., Yang, Z., Zhi, L.: Blind image deconvolution via fast approximate GCD. In: Proceedings of ISSAC 2010 (Intn'l Symp. on Symbolic and Algebraic Computation), pp. 155–162. ACM Press (2010)
9. Noda, M.-T., Sasaki, T.: Approximate GCD and its application to ill-conditioned algebraic equations. J. Comput. App. Math. 38, 335–351 (1991)
10. Ochi, M., Noda, M.-T., Sasaki, T.: Approximate GCD of multivariate polynomials and application to ill-conditioned system of algebraic equations. J. Inf. Proces. 14, 292–300 (1991)
11. Sasaki, T.: The subresultant and clusters of close roots. In: Proceedings of ISSAC 2003 (Intn'l Symp. on Symbolic and Algebraic Computation), pp. 232–239. ACM Press (2003)
12. Sasaki, T.: A theory and an algorithm of approximate Gröbner bases. In: Proceedings of SYNASC 2011 (Symbolic and Numeric Algorithms for Scientific Computing), pp. 23–30. IEEE Computer Society Press (2012)
13. Sasaki, T., Inaba, D.: Hensel construction of $F(x, u_1, \ldots, u_\ell)$, $\ell \geq 2$, at a singular point. ACM SIGSAM Bulletin 34, 9–17 (2000)
14. Sasaki, T., Inaba, D.: Convergence and many-valuedness of Hensel series near the expansion point. In: Proceedings of SNC 2009 (Intn'l Workshop on Symbolic-Numeric Computation), pp. 159–167. ACM Press (2009)
15. Sasaki, T., Inaba, D.: A study of Hensel series in general case. In: Proceedings of SNC 2011 (Intn'l Workshop on Symbolic-Numeric Computation), pp. 34–43. ACM Press (2011)
16. Sasaki, T., Inaba, D.: Approximately singular systems and ill-conditioned polynomial systems. In: Gerdt, V.P., Koepf, W., Mayr, E.W., Vorozhtsov, E.V. (eds.) CASC 2012. LNCS, vol. 7442, pp. 308–320. Springer, Heidelberg (2012)
17. Sasaki, T., Kako, F.: Solving multivariate algebraic equation by Hensel construction, pp. 257–285. Preprint of Univ., Tsukuba (1999); Japan J. Indust. Appl. Math. 16, 257–285 (1999)
18. Sasaki, T., Kako, F.: An algebraic method for separating close-root clusters and the minimum root separation. In: Proceedings of SNC 2005 (Intn'l Workshop on Symbolic-Numeric Computation), pp. 126–143. ACM Press (2005); Trends in Mathematics (Symbolic-Numeric Computation), Birkhäuser, 149–166 (2007)
19. Sasaki, T., Noda, M.-T.: Approximate square-free decomposition and root-finding of ill-conditioned algebraic equations. J. Inf. Proces. 12, 159–168 (1989)
20. Sasaki, T., Sasaki, M.: Analysis of accuracy decreasing in polynomial remainder sequence with floating-point number coefficients. J. Inf. Proces. 12, 394–403 (1989)
21. Sasaki, T., Suzuki, M., Kolář, M., Sasaki, M.: Approximate factorization of multivariate polynomials and absolute irreducibility testing. Japan J. Indust. Appl. Math. 8, 357–375 (1991)
22. Sasaki, T., Saito, T., Hirano, T.: Analysis of approximate factorization algorithm I. Japan J. Indust. Appl. Math. 9, 351–368 (1992)
23. Sasaki, T., Terui, A.: Computing clustered close-roots of univariate polynomials. In: Proceedings of SNC 2009 (Intn'l Workshop on Symbolic-Numeric Computation), pp. 177–184. ACM Press (2009)

24. Sasaki, T., Yamaguchi, T.: An analysis of cancellation error in multivariate Hensel construction with floating-point number arithmetic. In: Proceedings of ISSAC 1998 (Intn'l Symp. on Symbolic and Algebraic Computation), pp. 1–8. ACM Press (1998)
25. Shirayanagi, K.: An algorithm to compute floating-point Gröbner bases. In: Mathematical Computation with Maple V. Ideas and Applications, Birkäuser, pp. 95–106 (1993)
26. Smith, B.T.: Error bounds for zeros of a polynomial based on Gerschgorin's theorems. J. ACM 17, 661–674 (1970)
27. Stetter, H.J., Thallinger, G.H.: Singular systems of polynomials. In: Proceedings of ISSAC 1998 (Intn'l Symp. on Symbolic and Algebraic Computation), pp. 9–16. ACM Press (1998)

A Note on Sekigawa's Zero Separation Bound

Stefan Schirra

Otto von Guericke University Magdeburg,
Faculty of Computer Science,
Department of Simulation and Graphics,
39106 Magdeburg, Germany

Abstract. Regarding zero separation bounds for arithmetic expressions, we prove that the BFMS bound dominates the Sekigawa bound for division-free radical expressions with integer operands.

1 Introduction

In computational geometry software [5, 11], the exact decision geometric computation paradigm [20, 24] has been succesfully applied to avoid robustness problems [9, 16, 22] due to inconsistencies caused by numerical imprecision and to allow for the use of symbolic perturbation schemes [6, 17] in order to avoid the handling of degeneracies. Exact decision computation requires exact sign computation. However, if we compute the sign numerically using arbitrary precision floating-point arithmetic, high precision is required only if the value whose sign we want to compute is close to or equal to zero. Adaptive precision computations compute approximations starting with low precision and iteratively increasing the precision as long as the sign computation has not been verified. User-friendly implementations of this approach [1, 8, 13] record the computation history in expression trees, more precisely, in expression dags, in order to have an exact representation that allows one to repeatedly compute approximations with increasing precision.

The adaptive precision approach faces the numerical halting problem [23]. If the exact value is not zero, computing a sufficiently close approximation with an error bound smaller than its absolute value gives us the verified correct sign which we are interested in. Thus iteratively increasing the precision and computing error bounds, i.e., basically using interval arithmetic, we will finally reach our goal. If the value is zero, however, we can not verify the sign this way, since the numerical error will most likely be non-zero. This is where constructive zero separation bounds come into play.

Given an arithmetic expression E over a set of allowed operations, a constructive zero separation bound comprises rules to derive a positive real number $\mathrm{sep}(E)$ that is a lower bound on the absolute value of E, unless this value is zero. Let ξ denote the value of E. Then, $\mathrm{sep}(E)$ is a separation bound if

$$\xi \neq 0 \quad \Rightarrow \quad \mathrm{sep}(E) \leq |\xi|.$$

V.P. Gerdt et al. (Eds.): CASC 2013, LNCS 8136, pp. 331–339, 2013.
© Springer International Publishing Switzerland 2013

Here, we consider the case where all operands are integers and the basic arithmetic operations $+, -, \cdot$ and radical operations ($\sqrt[k]{\ }$) are permitted. Fortunately, these operations suffice for many geometric computations. A constructive separation bound allows us to resolve our numerical halting problem: Once we know that the sum of the absolute value $|\tilde{\xi}|$ of the current approximation $\tilde{\xi}$ and the corresponding error bound is less than the computed zero separation bound we may conclude that the actual value ξ is actually zero.

Fortunately, constructive zero separation bounds exist for arithmetic expressions whose values are real algebraic numbers. In particular, such constructive zero separation bounds exist for real algebraic integers and subsets of the set of real algebraic integers. Let $\mathscr{E}$ be the class of arithmetic expressions with operations $+, -, \cdot$ and $\sqrt[k]{\ }$ and integer operands. Several constructive zero separation bounds have been proposed for $\mathscr{E}$ and more general classes of expressions in the literature, most notably the degree-length bound [21, p. 177], the degree-measure bound [2, 10, 12], Canny's polynomial system bound [4], the BFMS bound [2], Li and Yap's conjugate bound [10], Scheinerman's eigenvalue bound [15], and the BFMSS bound [3]. Sekigawa [18, 19] proposes an improved version of the degree-measure bound. We call this the Sekigawa bound. Pion and Yap present a technique that leads to improved bounds for expressions involving k-ary rational numbers [14]. The improvement on Canny's gap theorem [4] by Emiris et al. [7] gives rise to a constructive zero separation bound better than Canny's polynomial system bound. Note that for the division-free expressions in $\mathscr{E}$, BFMS, Li and Yap's conjugate bound, and BFMSS are identical.

We say a separation bound sep dominates another bound sep$'$ for a class of expressions, if $\text{sep}(E) \geq \text{sep}'(E)$ for all expressions E in this class. We prove that BFMS dominates the Sekigawa bound for the class $\mathscr{E}$.

2   Constructive Bounds for Division-Free Radical Expressions

If an annihilating integer polynomial for the value ξ of an arithmetic expression E is known, i.e., a non-zero polynomial $P(X) \in \mathbb{Z}[X]$ such that $P(\xi) = 0$, root bounds for polynomials provide us with a zero separation bound. Usually, for a real algebraic number given by an arithmetic expression E, we do not know such a polynomial and in most cases it would be too expensive to compute one explicitly, since the adaptive approach will resolve the sign more quickly. The rules of constructive zero separation bounds typically allow us to derive bounds on certain quantities of such a polynomial based on the structure of E. These bounds then provide us with a separation bound. Most of the constructive zero separation bounds involve a bound on the degree of ξ. This degree bound is computed as

$$D(E) = \prod_{i=1}^{r} k_i,$$

where $k_1, \ldots, k_r$ are the indices of the radical operations in E.

The *length* and the *height* of a polynomial $P(X) = \sum_{i=0}^{d} a_i X^i \in \mathbb{Z}[X]$ are the sum of the magnitudes and the maximum of the magnitudes, respectively, of its coefficients:

$$\text{length}(P) = \sum_{i=1}^{d} |a_i| \quad \text{and} \quad \text{height}(P) = \max_{i=1}^{d} |a_i|.$$

The degree-length bound [21, p. 177] computes a bound on the length of an annihilating polynomial by the following rules, where the degree is computed as described above. It computes a quantity $\ell(E)$ inductively according to the structure of expression E.

E	$\ell(E)$
integer N	$|N|$
$A \cdot B$	$\ell(A)^{D(B)} \cdot \ell(B)^{D(A)}$
$A \pm B$	$\ell(A)^{D(B)} \cdot \ell(B)^{D(A)} \cdot 2^{D(A)D(B)+\min(D(A),D(B))}$
$\sqrt[k]{A}$	$\ell(A)$

Since $\ell(E)$ is an upper bound on the length of an annihilating polynomial for the value ξ of E, we know that

$$\ell(E)^{-1}$$

is a zero separation bound for E.

Similarly, the following rules [24] compute a bound $h(E)$ on the height of an annihilating polynomial for the value ξ of E.

E	$h(E)$
integer N	$|N|$
$A \cdot B$	$(h(A)\sqrt{1+D(A)})^{D(B)} \cdot (h(B)\sqrt{1+D(B)})^{D(A)}$
$A \pm B$	$(h(A)2^{1+D(A)})^{D(B)} \cdot (h(B)\sqrt{1+D(B)})^{D(A)}$
$\sqrt[k]{A}$	$h(A)$

The corresponding zero separation bound, called the degree-height bound, is

$$(1 + h(E))^{-1}$$

Both degree-length bound and degree-height bound are based on resultant calculus.

Scheinerman [15] presents a constructive zero separation bound based on matrix eigenvalues. He computes quantities $n(E)$ and $b(E)$ as shown in the table below.

E	$n(E)$	$b(E)$
integer N	1	$|N|$
$A \cdot B$	$n(A) \cdot n(B)$	$b(A) \cdot b(B)$
$A \pm B$	$n(A) \cdot n(B)$	$b(A) + b(B)$
$\sqrt[k]{A}$	$k \cdot n(A)$	$b(A)$

The corresponding zero separation bound is

$$(n(E)b(E))^{1-n(E)}$$

The *measure* of a polynomial $P(X) = \sum_{i=0}^{d} a_i X^i = a_d \prod_{i=1}^{d}(X - \alpha_i) \in \mathbb{Z}[X]$
is

$$|a_d| \prod_{i=1}^{d} \max(1, |\alpha_i|).$$

The degree-measure bound [2, 10, 12] is based on the rules

E	$\hat{M}(E)$
integer N	$|N|$
$A \cdot B$	$\hat{M}(A)^{D(B)} \cdot \hat{M}(B)^{D(A)}$
$A \pm B$	$\hat{M}(A)^{D(B)} \cdot \hat{M}(B)^{D(A)} \cdot 2^{D(E)}$
$\sqrt[k]{A}$	$\hat{M}(A)$

The corresponding zero separation bound is then

$$\hat{M}(E)^{-1}.$$

Sekigawa [18, 19] splits the measure in the leading coefficient and the remaining part and provides rules to bound both quantities. In our case, the values of all expressions are real algebraic integers. Therefore, the leading coefficient is always one. Thus there is no need to maintain it. Therefore, in the division-free case, the rules for the Sekigawa bound simplify to

E	$M(E)$
integer N	$|N|$
$A \cdot B$	$M(A)^{D(B)} \cdot M(B)^{D(A)}$
$A \pm B$	$(*)$
$\sqrt[k]{A}$	$M(A)$

where $(*)$ is the product of the $D(E)$ largest values of

$$M(A)+M(B), \underbrace{M(A) + 1, ..., M(A) + 1}_{D(B)-1}, \underbrace{M(B) + 1, ..., M(B) + 1}_{D(A)-1}, \underbrace{2, ..., 2}_{(D(A)-1)(D(B)-1)}$$

Then

$$M(E)^{-1}$$

is a separation bound.

The BFMS bound [2] computes a quantity $U(E)$ that is an upper bound on the absolute value of the conjugates of the value ξ of E. For expressions in $\mathscr{E}$ we have the following rules:

E	$U(E)$
integer N	$\lvert N \rvert$
$A \pm B$	$U(A) + U(B)$
$A \cdot B$	$U(A) \cdot U(B)$
$\sqrt[k]{A}$	$\sqrt[k]{U(A)}$

Then

$$U(E)^{-(D(E)-1)}$$

is a separation bound for E.

For division-free arithmetic expressions involving radicals, the Li and Yap's conjugate bound and the BFMSS bound use exactly the same rules.

3   Dominance Results

Remember that we consider dominance with respect to the class $\mathscr{E}$ of division-free radical expressions over the integers only.

It is easy to see that the degree-measure bound dominates both the degree-length bound and the degree-height bound.

Burnikel et al. [2] show that BFMS dominates the degree-measure bound. Li and Yap [10] show that BFMS dominates Scheinerman's eigenvalue bound.

Sekigawa's bound refines and improves the degree-measure bound. By inspecting the rules given in the previous section we see that

$$M(E) \leq \hat{M}(E)$$

for all expressions E in $\mathscr{E}$. Thus we have

Theorem 1. *The Sekigawa bound dominates the degree-measure bound for division-free radical expressions with integer operands.*

However, the BFMS separation bound still dominates Sekigawa's improved measure bound. We prove

Theorem 2. *The* BFMS *bound dominates the Sekigawa bound for division-free radical expressions with integer operands.*

Theorem 2 follows from

Lemma 1
$$U(E)^{D(E)} \leq M(E)$$

We prove Lemma 1 by structural induction. The lemma holds for the base case where E is an integer N, since the rules are identical and $D(E) = 1$.

For the induction step, we are left with three cases: $E = \sqrt[k]{A}$, $E = A \cdot B$, and $E = A \pm B$. By induction hypothesis we have

$$U(A)^{D(A)} \leq M(A) \qquad \text{and} \qquad U(B)^{D(B)} \leq M(B).$$

First, let $E = \sqrt[k]{A}$. Then $M(E) = M(A)$ and $U(E) = \sqrt[k]{U(A)}$. Furthermore $D(E) = k \cdot D(A)$. Thus

$$\begin{aligned}
U(E)^{D(E)} &= (\sqrt[k]{U(A)})^{k \cdot D(A)} \\
&= U(A)^{D(A)} \\
&\leq M(A) \qquad \text{by induction hypothesis} \\
&= M(E)
\end{aligned}$$

Next, let $E = A \cdot B$. Then $M(E) = M(A)^{D(B)} M(B)^{D(A)}$ and $U(E) = U(A) \cdot U(B)$. Then

$$\begin{aligned}
U(E)^{D(E)} &= (U(A) \cdot U(B))^{D(E)} \\
&\leq (U(A) \cdot U(B))^{D(A) \cdot D(B)} \\
&= U(A)^{D(A) \cdot D(B)} \cdot U(B)^{D(B) \cdot D(A)} \\
&\leq M(A)^{D(B)} \cdot M(B)^{D(A)} \qquad \text{by induction hypothesis} \\
&= M(E)
\end{aligned}$$

Finally, let $E = A \pm B$. This is the non-trivial case. Let us assume $D(E) = D(A) \cdot D(B)$. Then $M(E)$ is $(M(A) + M(B)) \cdot (M(A) + 1)^{D(B)-1} \cdot (M(B) + 1)^{D(A)-1} \cdot 2^{(D(A)-1)(D(B)-1)}$ and $U(E) = U(A) + U(B)$.

The proof uses the following lemma by Sekigawa [18]:

Lemma 2. *Let S be a set of pairs (i, j), $1 \leq i \leq m$, $1 \leq j \leq n$, and let*

$$F(x_1, \ldots, x_m, y_1, \ldots, y_n) = \prod_{(i,j) \in S} (x_i + y_j).$$

For constants $\mathbf{a}, \mathbf{b} \geq 1$, the maximum value of the continous function F on the compact set

$$\mathscr{D} = \left\{ (x_1, \ldots, x_m, y_1, \ldots, y_n) \in \mathbb{R}^{m+n} \;\middle|\; \begin{array}{l} \prod_{i=1}^{m} x_i = \mathbf{a}, \;\; x_i \geq 1, i = 1, \ldots, m \\ \prod_{j=1}^{m} y_j = \mathbf{b}, \;\; y_j \geq 1, j = 1, \ldots, n \end{array} \right\}$$

is not greater than the product of the $|S|$ largest numbers among the following mn numbers:

$$\mathbf{a} + \mathbf{b}, \underbrace{\mathbf{a} + 1, \ldots, \mathbf{a} + 1}_{n-1}, \underbrace{\mathbf{b} + 1, \ldots, \mathbf{b} + 1}_{m-1}, \underbrace{2, \ldots, 2}_{(n-1)(m-1)}$$

We have

$$U(E)^{D(E)} = (U(A) + U(B))^{D(E)}$$
$$\leq \left({}^{D(A)}\!\sqrt{M(A)} + {}^{D(B)}\!\sqrt{M(B)} \right)^{D(E)} \qquad \text{by induction hypothesis}$$

Now we apply Lemma 2 with

$$m = D(A)$$
$$x_i = {}^{D(A)}\!\sqrt{M(A)} \text{ for } i = 1, \ldots, m$$
$$\mathbf{a} = M(A)$$
$$n = D(B)$$
$$y_j = {}^{D(B)}\!\sqrt{M(B)} \text{ for } j = 1, \ldots, n$$
$$\mathbf{b} = M(B)$$

Then we get

$$\left({}^{D(A)}\!\sqrt{M(A)} + {}^{D(B)}\!\sqrt{M(B)} \right)^{D(E)}$$
$$= \prod_{i=1}^{m} \prod_{j=1}^{n} (x_i + y_j)$$
$$\leq (\mathbf{a} + \mathbf{b}) \cdot (\mathbf{a} + 1)^{n-1} \cdot (\mathbf{b} + 1)^{m-1} \cdot 2^{(m-1)(n-1)}$$
$$= (M(A) + M(B)) \cdot (M(A) + 1)^{D(B)-1}$$
$$\qquad \cdot (M(B) + 1)^{D(A)-1} \cdot 2^{(D(A)-1)(D(B)-1)}$$
$$= M(E)$$

If we have $D(E) < D(A)D(B)$, we apply Lemma 2 with a set S with cardinality $D(E) < nm$. The remaining part is analogous.

4   Conclusions

We have shown that BFMS also dominates Sekigawa's improved measure bound for division-free radical expressions over the integers and thus all constructive zero separation bounds currently known. Thus, the relevance of the Sekigawa bound is rather as a substitute for the degree-measure bound in combined bounds, e.g. the Li-Yap bound.

As for expressions involving division operations, there is no bound among the known ones that dominates all others. Thus, in this case, the best approach is to combine bounds not dominated by any others, i.e., Li and Yap's conjugate bound, BFMSS, and the degree-measure bound.

References

[1] Burnikel, C., Fleischer, R., Mehlhorn, K., Schirra, S.: Efficient exact geometric computation made easy. In: Proc. 15th Annu. ACM Sympos. Comput. Geom., pp. 341–350 (1999)

[2] Burnikel, C., Fleischer, R., Mehlhorn, K., Schirra, S.: A strong and easily computable separation bound for arithmetic expressions involving radicals. Algorithmica 27(1), 87–99 (2000)

[3] Burnikel, C., Funke, S., Mehlhorn, K., Schirra, S., Schmitt, S.: A separation bound for real algebraic expressions. In: Meyer auf der Heide, F. (ed.) ESA 2001. LNCS, vol. 2161, pp. 254–265. Springer, Heidelberg (2001)

[4] Canny, J.F.: The complexity of robot motion planning. MIT Press, Cambridge (1988)

[5] CGAL, Computational Geometry Algorithms Library, http://www.cgal.org

[6] Edelsbrunner, H., Mücke, E.P.: Simulation of simplicity: A technique to cope with degenerate cases in geometric algorithms. ACM Trans. Graph. 9(1), 66–104 (1990)

[7] Emiris, I.Z., Mourrain, B., Tsigaridas, E.P.: The DMM bound: multivariate (aggregate) separation bounds. In: ISSAC 2010, pp. 243–250 (2010)

[8] Karamcheti, V., Li, C., Pechtchanski, I., Yap, C.: A core library for robust numeric and geometric computation. In: Proceedings of the 15th Annual ACM Symposium on Computational Geometry, Miami, Florida, pp. 351–359 (1999)

[9] Kettner, L., Mehlhorn, K., Pion, S., Schirra, S., Yap, C.: Classroom examples of robustness problems in geometric computations. Computational Geometry: Theory and Applications 40(1), 61–78 (2008)

[10] Li, C., Yap, C.: A new constructive root bound for algebraic expressions. In: 12th ACM-SIAM Symposium on Discrete Algorithms (SODA) (January 2001)

[11] Mehlhorn, K., Näher, S.: LEDA: A Platform for Combinatorial and Geometric Computing. Cambridge University Press, Cambridge (2000)

[12] Mignotte, M.: Mathematics for computer algebra. Springer-Verlag New York, Inc., New York (1992)

[13] Mörig, M., Rössling, I., Schirra, S.: On design and implementation of a generic number type for real algebraic number computations based on expression dags. Mathematics in Computer Science 4(4), 539–556 (2010)

[14] Pion, S., Yap, C.: Constructive root bound for k-ary rational input numbers. Journal of Theoretical Computer Science (TCS) 369(1-3), 361–376 (2006)

[15] Scheinerman, E.R.: When close enough is close enough. Am. Math. Mon. 107(6), 489–499 (2000)

[16] Schirra, S.: Robustness and precision issues in geometric computation. In: Sack, J.R., Urrutia, J. (eds.) Handbook of Computational Geometry, ch. 14, pp. 597–632. Elsevier (1999)

[17] Seidel, R.: The nature and meaning of perturbations in geometric computing. Discrete & Computational Geometry 19(1), 1–17 (1998)

[18] Sekigawa, H.: Using interval computation with the Mahler measure for zero determination of algebraic numbers. Josai University Information Sciences Research 9(1), 83–99 (1998)

[19] Sekigawa, H.: Zero Determination of Algebraic Numbers using Approximate Computation and its Application to Algorithms in Computer Algebra. PhD thesis, University of Tokyo (2004)

[20] Yap, C.K.: Towards exact geometric computation. Computational Geometry: Theory and Applications 7, 3–23 (1997)

[21] Yap, C.-K.: Fundamental problems of algorithmic algebra. Oxford University Press (2000)

[22] Yap, C.K.: Robust geometric computation. In: Goodman, J.E., O'Rourke, J. (eds.) Handbook of Discrete and Computational Geometry, ch. 41, 2nd edn., pp. 927–952. Chapmen & Hall/CRC, Boca Raton (2004); revised and expanded from 1997 version

[23] Yap, C.K.: In praise of numerical computation. In: Albers, S., Alt, H., Näher, S. (eds.) Efficient Algorithms. LNCS, vol. 5760, pp. 380–407. Springer, Heidelberg (2009)

[24] Yap, C.K., Dubé, T.: The exact computation paradigm. In: Du, D.-Z., Hwang, F.K. (eds.) Computing in Euclidean Geometry, 2nd edn. Lecture Notes Series on Computing, vol. 4, pp. 452–492. World Scientific, Singapore (1995)

Applications of Symbolic Calculations and Polynomial Invariants to the Classification of Singularities of Differential Systems[*]

Dana Schlomiuk[1] and Nicolae Vulpe[2]

[1] Département de Mathématiques et de Statistiques Université de Montréal
dasch@dms.umontreal.ca
[2] Institute of Mathematics and Computer Science, Academy of Science of Moldova
nvulpe@gmail.com

Abstract. The goal of this paper is to present applications of symbolic calculations and polynomial invariants to the problem of classifying planar polynomial systems of differential equations. For these applications, we use some previously defined, and some new polynomial invariants. This is part of a much larger work by the authors together with J.C. Artés and J. Llibre which is in progress. We show here how polynomial invariants and their symbolic calculations are instrumental in obtaining the bifurcation diagram of the global configurations of singularities (finite and infinite), of quadratic differential systems having a unique simple finite singularity. This bifurcation diagram is given in the twelve-dimensional space of the coefficients of the systems, and the bifurcation points form an algebraic set. The classification of singularities is done using the notion of *geometric equivalence relation* of configurations of singularities, which is finer than the topological equivalence. The bifurcation diagram is expressed in terms of polynomial invariants. The results can, therefore, be applied to any family of quadratic systems, given in any normal form. Determining the configurations of singularities for any family of quadratic systems thus becomes a simple task using computer symbolic calculations.

1 Introduction and Statement of Main Results

We consider here differential systems of the form

$$\frac{dx}{dt} = p(x, y), \qquad \frac{dy}{dt} = q(x, y), \tag{1}$$

where $p, q \in \mathbb{R}[x, y]$, i.e. p, q are polynomials in x, y over $\mathbb{R}$. We call *degree* of a system (1) the integer $m = \max(\deg p, \deg q)$. In particular, we call *quadratic* a differential system (1) with $m = 2$. We denote here by **QS** the whole class of real quadratic differential systems.

---

[*] This work was supported by NSERC. The second author is partially supported by FP7-PEOPLE-2012-IRSES-316338 and by the grant 12.839.08.05F from SCSTD of ASM.

V.P. Gerdt et al. (Eds.): CASC 2013, LNCS 8136, pp. 340–354, 2013.
© Springer International Publishing Switzerland 2013

Quadratic differential systems occur very often in many areas of applied mathematics. Yet several problems on the class **QS**, formulated more than a century ago, are still open for this class. We do not have the topological classification of these systems in spite of consistent attempts to advance our knowledge about them. There are three reasons for this situation, the first one being the elusive nature of limit cycles. These periodic solutions, which are an essential ingredient in the construction of phase portraits, are hard to pin down. The second reason is the rather large number of parameters involved. This family of systems depends on twelve parameters but due to the group action of real affine transformations and time homotheties, the class ultimately depends on five parameters. This means that to obtain the bifurcation diagram of this class we need to work in a five dimensional space which is not $\mathbb{R}^5$ but a much more complicated topological space, quotient of $\mathbb{R}^{12}$ by this group action.

The third reason is the fact that to gain global insight into this class one needs to perform a large number of ample calculations which could not have been done before the more powerful methods of symbolic and numerical calculations emerged. This situation is, however, now changing. We have now several tools for performing such calculations, among them computer algebra systems like Mathematica, Maple, Reduce, Cocoa, Macaulay 2. With the help of these programs progress is now being made. We also have the program P4 (see [18]) which combines both symbolic and numerical calculations for drawing phase portraits of individual planar polynomial differential systems.

For example, some subclasses of **QS** depending on at most three parameters were studied globally, including global bifurcation diagrams (for example [2]) with the help of Mathematica, Maple, and P4. For the moment, the global studies of four-dimensional subclasses of **QS**, modulo the group action, remain a challenge. On the other hand, we can restrict the study of *the whole quadratic class* by focusing on specific global features of the systems in this family. We may thus focus on the global study of singularities and their bifurcation diagram. The singularities are of two kinds: finite and infinite. The infinite singularities are obtained by compactifying on the sphere or on the Poincaré disk (see, for example, [18] or [23]) the planar polynomial differential systems.

It is now possible to do a complete study of the global configurations of singularities, finite and infinite of the class **QS**. Indeed, the whole bifurcation diagram of the global configurations of singularities, finite and infinite, in quadratic vector fields and more generally in polynomial vector fields can be obtained by using only algebraic means, among them, the algebraic tool of polynomial invariants computed using symbolic computations.

The goal of this article is to show how this is done for the family of quadratic differential systems having a single finite singularity which, in addition, is simple. What we show here for this particular family can also be done in the general case, and work of the authors, together with J.C. Artés and J. Llibre, in this direction is in progress.

For the classification of configurations of singularities, finite and infinite we use the notion of *geometric equivalence relation*, a notion which was introduced in [3] (see also [4]) and which is finer than the topological equivalence relation (for these equivalence notions see the next section). Unlike the topological equivalence relation, the geometric one distinguishes between foci and nodes, between the strong and weak foci, between strong and weak saddles, between foci of different orders, between saddles of different orders, between integrable saddles and weak saddles of positive orders, and between the different kinds of nodes. Such distinctions are important in the production of limit cycles. Indeed, for example, the maximum number of limit cycles which can be produced close to the weak foci in perturbations depends on the orders of the foci. The notion of *geometric equivalence relation* is completely defined in terms of an algebraic nature.

This equivalence relation is also finer than the *qualitative equivalence relation* introduced by Jiang and Llibre in [22] since it distinguishes among the foci of different orders and among the various types of nodes.

Algebraic information may not be significant for the local (topological) phase portrait around a singularity. For example, topologically there is no distinction between a focus and a node or between a weak and a strong focus. However, as indicated before, algebraic information plays a fundamental role in the study of perturbations of systems possessing such singularities.

At this point, the following question comes to mind:

How far can we go in the global theory of quadratic (or more generally polynomial) vector fields by using mainly algebraic means?

While algebraic methods are not sufficient for attacking such difficult transcendental problems as Hilbert's 16th problem, the applications of these methods have not yet been exhausted. For example, the purely algebraic problem of constructing the global bifurcation diagram of configurations of singularities has not yet been completed.

To distinguish among the foci (or saddles) of various orders we use the algebraic concept of Poincaré–Lyapunov constants. We call *strong focus* (or *strong saddle*) a focus with non–zero trace of the linearization matrix at this point. Such a focus (or saddle) will be considered to have the order zero. A focus (or saddle) with trace zero is called a *weak focus* (or *weak saddle*). It is known that a singular point with purely imaginary eigenvalues is either a center or a focus in which case we call it weak focus. We may assume the systems to be $dx/dt = -y + \ldots$, $dy/dt = x + \ldots$. In this case, it is easily shown that there exists a formal power series with coefficients in $\mathbb{Q}[a_{20}, \ldots, b_{0,n}]$ such that $dF/dt = \Sigma_{i=1}^{\infty} V_i(x^2 + y^2)^{i=1}$. A weak focus is of the order i if for all $j < i$ we have $V_j = 0$ and $V_i \neq 0$.

In this work, we show how to merge the results for infinite singularities in [3] (see also [4]) with work on finite singularities, when we have a single finite singularity which is in addition simple.

We distinguish two cases:

1) Suppose we have a finite number of infinite singular points. In this case, we call *configuration of singularities*, finite and infinite, the set of all these singularities each endowed with its own multiplicity together with their local phase portraits endowed with additional geometric structure involving the concepts of tangent, order and blow–up equivalences defined in [3] (see also [4]).

2) If the line at infinity $Z = 0$ (see [23]) is filled up with singularities, in both charts at infinity $X \neq 0$ and $Y \neq 0$, the system is degenerate, and we need to do a rescaling of an appropriate degree of the system, so that the degeneracy be removed. The resulting systems have only a finite number of singularities on the line $Z = 0$. In this case, we call *configuration of singularities*, finite and infinite, the union of the set of all points at infinity (they are all singularities) with the set of finite singularities, taking care of singling out the singularities of the "reduced" system at infinity, taken together with the local phase portraits of finite singularities endowed with additional geometric structure as above and of the infinite singularities of the reduced system.

We obtain the following

Main Theorem. (*A*) *The configurations of singularities, finite and infinite, of all quadratic vector fields with a single finite singularity which is simple (the total multiplicity of finite singularities is $m_f = 1$) are classified in Figure 1 according to the geometric equivalence relation. We have 52 geometrically distinct global configurations of singularities. We have: only one configuration with a center but 5 configurations with a finite integrable saddle; 6 configurations with a strong focus but 7 configurations with a strong finite saddle; 4 configurations with a weak focus of order one but 2 configurations with a weak finite saddle of order one; no configurations with either weak focus or a finite weak saddle of order greater than one.*

(*B*) *Necessary and sufficient conditions for each one of the 52 different equivalence classes can be assembled from these diagram in terms of 25 invariant polynomials with respect to the action of the affine group and time rescaling, indicated by Remark.1.*

(*C*) *The Figure 1 actually contains the global bifurcation diagram in the 12-dimensional space of parameters, of the global configurations of singularities, finite and infinite, of this family of quadratic differential systems.*

Remark 1. The invariants and comitants of differential equations used for proving our main results are obtained following the theory of algebraic invariants of polynomial differential systems, developed by Sibirsky and his disciples (see, for instance, [39,42,27,8,13]). More exactly:

- the invariant polynomials $\mu_i(a, x, y), i \in \{0, 1, \ldots, 4\}$,, $\kappa(a)$, $\kappa_1(a)$, $\eta(a)$, $\widetilde{M}(a, x, y)$, $C_2(a, x, y)$, $\tilde{L}(a, x, y)$, $\widetilde{K}(a, x, y)$, $K_1(a, x, y)$ and $K_3(a, x, y)$ were constructed in [32];
- the invariant polynomials $W_4(a)$, $W_7(a)$, $W_8(a)$ and $W_{11}(a, x, y)$ were constructed in [5];
- the invariant polynomials $\mathcal{T}_4(a)$, $\mathcal{B}_1(a, x, y)$, $\mathcal{B}_3(a, x, y)$ and $\sigma(a, x, y)$ were constructed in [40];
- the invariant polynomials $U_3(a, x, y)$, $U_4(a, x, y)$, and $U_5(a, x, y)$ were constructed here in Section 4.

We point out that we keep the notations introduced in the indicated papers for all invariant polynomials.

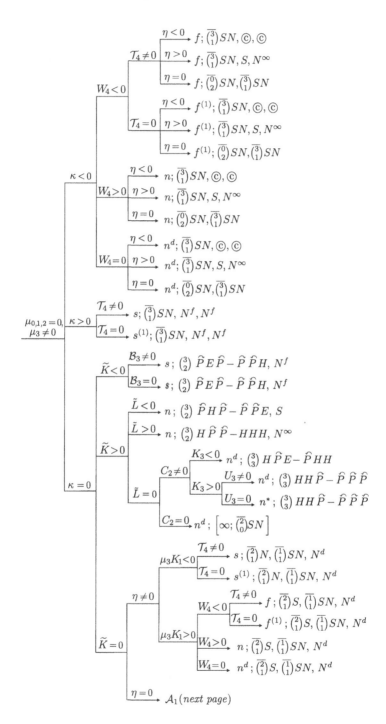

Fig. 1. Global configurations: case $\mu_0 = \mu_1 = \mu_2 = 0$, $\mu_3 \neq 0$

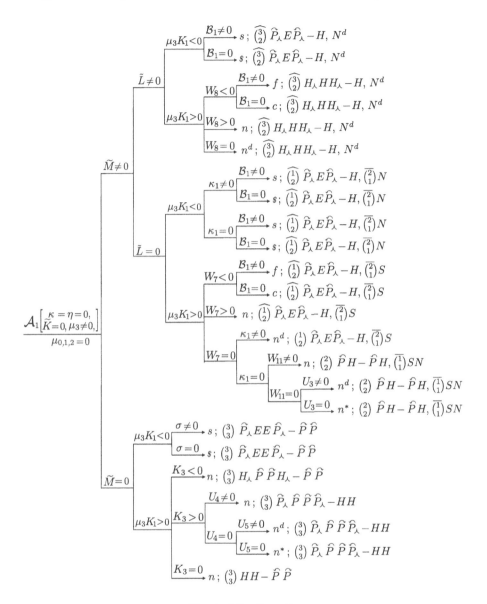

Fig. 1. (*Continued.*)

2 Equivalence Relations for Singularities of Planar Polynomial Vector Fields

We first recall the *topological equivalence relation* as it is used in most of the literature. Two singularities p_1 and p_2 are *topologically equivalent* if there exist open neighborhoods N_1 and N_2 of these points and a homeomorphism $\Psi : N_1 \to N_2$

carrying orbits to orbits and preserving their orientations. To reduce the number of cases, by *topological equivalence* we shall mean here that the homeomorphism Ψ *preserves or reverses* the orientation.

In [22], Jiang and Llibre introduced another equivalence relation for singularities which is finer than the topological equivalence: We say that p_1 and p_2 are *qualitatively equivalent* if i) they are topologically equivalent through a local homeomorphism Ψ; and ii) two orbits are tangent to the same straight line at p_1 if and only if the corresponding two orbits are also tangent to the same straight line at $p_2 = \Psi(p_1)$.

Intuitively it is clear what we mean by an orbit γ to be tangent to a line L passing through a limit point p (singular point of the system) of the orbit γ. It is an exercise to make this notion precise.

We say that two simple finite nodes, with the respective eigenvalues λ_1, λ_2 and σ_1, σ_2, are *tangent equivalent* if and only if they satisfy one of the following three conditions: a) $(\lambda_1 - \lambda_2)(\sigma_1 - \sigma_2) \neq 0$; b) $\lambda_1 - \lambda_2 = 0 = \sigma_1 - \sigma_2$ and both linearization matrices at the two singularities are diagonal; c) $\lambda_1 - \lambda_2 = 0 = \sigma_1 - \sigma_2$ and the corresponding linearization matrices are not diagonal.

We say that two infinite simple nodes P_1 and P_2 are *tangent equivalent* if and only if their corresponding singularities on the sphere are tangent equivalent and in addition, in case they are generic nodes, we have $(|\lambda_1| - |\lambda_2|)(|\sigma_1| - |\sigma_2|) > 0$ where λ_1 and σ_1 are the eigenvalues of the eigenvectors tangent to the line at infinity.

Finite and infinite singular points may either be real or complex. In case we have a complex singular point we will specify this with the symbols ⓒ and Ⓒ for finite and infinite points, respectively. We point out that the sum of the multiplicities of all singular points of a quadratic system with a finite number of singular points is always 7. (Here of course we refer to the compactification on the complex projective plane $P_2(\mathbb{C})$ of the foliation with singularities associated to the complexification of the vector field (see [23]). The sum of the multiplicities of the infinite singular points is always at least 3, more precisely it is always 3 plus the sum of the multiplicities of the finite points which disappeared at infinity.

We use here the following terminology for singularities:

We call *elemental* a singular point with its both eigenvalues not zero;
We call *semi–elemental* a singular point with exactly one of its eigenvalues equal to zero;
We call *nilpotent* a singular point with both its eigenvalues zero but with its Jacobian matrix at that point not identically zero;
We call *intricate* a singular point with its Jacobian matrix identically zero.

In the literature, *intricate* singularities are usually called *linearly zero*. We use here the term *intricate* to indicate the rather complicated behavior of phase curves around such a singularity.

Roughly speaking, a singular point p of an analytic differential system χ is a *multiple singularity of multiplicity m* if p generates m singularities, as close to p as we wish, under analytic perturbations χ_ε of this system and m is the maximal such number. In polynomial differential systems of fixed degree n, we have several possibilities for obtaining multiple singularities. i) A finite singular point splits into several finite singularities in n-degree polynomial perturbations. ii) An infinite singular point splits into some finite and some infinite singularities in n-degree polynomial perturbations. iii) An infinite singularity splits only in infinite singular points of the systems in n-degree perturbations. To all these cases, we can give a precise mathematical meaning using the notion of intersection multiplicity at a point p of two algebraic curves (see [30], [31]).

We will say that two foci (or saddles) are *order equivalent* if their corresponding orders coincide.

Semi–elemental saddle–nodes are always topologically equivalent.

The notion of *geometric equivalence relation* of singularities involves the notion of *blow–up equivalence* for nilpotent and intricate singular points. The *blow–up equivalence* is a technical notion, and we refer to [3] (see also [4]) for the technical details involved in defining this notion based on the process of desingularization (see, for example, [18]).

Definition 1. *Two singularities p_1 and p_2 of two polynomial vector fields are locally geometrically equivalent if and only if they are topologically equivalent, they have the same multiplicity, and one of the following conditions is satisfied:*

- *p_1 and p_2 are order equivalent foci (or saddles);*
- *p_1 and p_2 are tangent equivalent simple nodes;*
- *p_1 and p_2 are both centers;*
- *p_1 and p_2 are both semi–elemental singularities;*
- *p_1 and p_2 are blow–up equivalent nilpotent or intricate singularities.*

Definition 2. *Let χ_1 and χ_2 be two polynomial vector fields each having a finite number of singularities. We say that χ_1 and χ_2 have geometric equivalent configurations of singularities if and only if we have a bijection ϑ carrying the singularities of χ_1 to singularities of χ_2 and for every singularity p of χ_1, $\vartheta(p)$ is geometrically equivalent with p.*

3   Notations for Singularities of Polynomial Differential Systems

We denote the finite singularities with lower case letters and the infinite ones with capital letters placing first the finite ones, then the infinite ones, separating them by a semicolon ';'.

Elemental Points. We use the letters 's', 'S' for "saddles"; 'n', 'N' for "nodes"; 'f' for "foci"; 'c' for "centers" and ⓒ (respectively Ⓒ) for complex finite (respectively infinite) singularities. We distinguish the finite nodes as follows:

- 'n' for a node with two distinct eigenvalues (generic node);
- 'n^d' (a one–direction node) for a node with two identical eigenvalues whose Jacobian matrix is not diagonal;

– 'n^*' (a star–node) for a node with two identical eigenvalues whose Jacobian matrix is diagonal.

Moreover, in the case of an elemental infinite generic node, we want to distinguish whether the eigenvalue associated to the eigenvector directed towards the affine plane is, in absolute value, greater or lower than the eigenvalue associated to the eigenvector tangent to the line at infinity. We will denote them as 'N^∞' and 'N^f', respectively.

When the trace of the Jacobian matrix of an elemental saddle or focus is zero, in the quadratic case, one may have up to 3 finite orders. We denote them by '$s^{(i)}$' and '$f^{(i)}$' where $i = 1, 2, 3$ is the order. In addition, we have the centers which we denote by 'c' and saddles of infinite order (integrable saddles) which we denote by '$\hat{s}$'.

Non–elemental singular points are multiple points, i.e., in perturbations they can produce at least two elemental singular points as close as we wish to the multiple point. For finite singular points, we denote with a subscript their multiplicity as in '$\overline{s}_{(5)}$' or in '$\widehat{es}_{(3)}$' (the notation '$-$' indicates that the singular point is semi–elemental and '$\frown$' indicates that the singular point is nilpotent). In order to describe the various kinds of multiplicity for infinite singular points we use the concepts and notations introduced in [32]. Thus, we denote by '$\binom{a}{b}$...' the maximum number a (respectively b) of finite (respectively infinite) singularities which can be obtained by perturbation of the multiple point. For example, '$\overline{\binom{1}{1}} SN$' means a saddle–node at infinity produced by the collision of one finite singularity with an infinite one; '$\overline{\binom{0}{3}} S$' means a saddle produced by the collision of 3 infinite singularities.

Semi–elemental Points. We always denote the semi–elemental points with an overline, for example, '$\overline{sn}$', '$\overline{s}$' and '$\overline{n}$' with the corresponding multiplicity. In the case of infinite points, we will put '$-$' on top of the parenthesis with multiplicities.

Nilpotent Points. They can either be saddles, nodes, saddle–nodes, elliptic–saddles, cusps, foci or centers. The first four of these could be at infinity. We denote the nilpotent singular points with a hat '$\frown$' as in $\widehat{es}_{(3)}$ for a finite nilpotent elliptic–saddle of multiplicity 3 and $\widehat{cp}_{(2)}$ for a finite nilpotent cusp point of multiplicity 2. In the case of nilpotent infinite points, we will put the '$\frown$' on top of the parenthesis with multiplicity, for example, $\widehat{\binom{1}{2}} PEP - H$ (the meaning of $PEP - H$ will be explained in next paragraph). The relative position of the sectors of an infinite nilpotent point, with respect to the line at infinity, can produce topologically different phase portraits. This forces us to use a notation for these points similar to the notation which we will use for the intricate points.

Intricate Points. The neighborhood of any singular point of a polynomial vector field (except for foci and centers) is formed by a finite number of sectors which could be: parabolic, hyperbolic, and elliptic (see [18]). To describe intricate and nilpotent points we use a sequence formed by the types of their sectors taken in clockwise order (starting anywhere). In this article, we only have one finite singularity which is simple. For infinite intricate and nilpotent singular points, we insert a dash (hyphen) between the sectors appearing on one side or the other of the equator of the sphere. In this way, we will distinguish between $\binom{2}{2}PHP - PHP$ and $\binom{2}{2}PPH - PPH$. Whenever we have an infinite nilpotent or intricate singular point, we will always start with a sector bordering the infinity (to avoid using two dashes).

We are interested in additional geometrical features such as the number of characteristic directions which figure in the final global picture of the blow-up (or the desingularization) of the singularity.

We call *borsec* (contraction of border and sector) any orbit of the original system which carried through consecutive stages of the desingularization ends up as an orbit of the phase portrait in the final stage which is either a separatrix or a representative orbit of a characteristic angle (as defined in [3]) of a node or a of saddle–node in the final desingularized phase portrait.

Using the concept of borsec, we define a *geometric local sector* with respect to a neighborhood V as a region in V delimited by two consecutive borsecs.

If two consecutive borsecs arrive at the singular point with the same slope and direction, then the sector will be denoted by $H_\curvearrowleft$, $E_\curvearrowleft$ or $P_\curvearrowleft$, and it is called cusp-like. In the case of parabolic sectors, we want to distinguish whether the orbits arrive tangent to one or to the other borsec. We distinguish the two cases by $\overset{\curvearrowleft}{P}$ if they arrive tangent to the borsec limiting the previous sector in clock-wise sense or $\overset{\curvearrowright}{P}$ if they arrive tangent to the borsec limiting the next sector. A parabolic sector denoted by P^* is a sector in which orbits arrive with all possible slopes between those of the borsecs.

Line at Infinity Filled up with Singularities. It is known that any such system has in a sufficiently small neighborhood of infinity one of 6 topologically distinct phase portraits (see [35]). To determine these portraits we study the reduced systems on the infinite local charts after removing the degeneracy of the systems within these charts. In case a singular point still remains on the line at infinity we study such a point. If after the removal of the degeneracy in the local charts at infinity a node remains, this could either be of the type N^d, N or $N^\star$ (this last case does not occur in quadratic systems as shown in [3] or [4]). To point out the lack of an infinite singular point after removal of a degeneracy, we will use the symbol $\emptyset$. Other types of singular points at infinity of quadratic systems, after removal of the degeneracy, can be saddles, centers, semi–elemental saddle–nodes or nilpotent elliptic–saddles. To convey the way these singularities were obtained as well as their nature, we use the notation $[\infty; \emptyset]$, $[\infty; N]$, $[\infty; N^d]$, $[\infty; S]$, $[\infty; C]$, $[\infty; \overline{\binom{1}{0}}SN]$ or $[\infty; \widehat{\binom{3}{0}}ES]$.

4   Invariant Polynomials

Consider real quadratic systems of the form:

$$
\begin{aligned}
\frac{dx}{dt} &= p_0 + p_1(x, y) + p_2(x, y) \equiv P(x, y), \\
\frac{dy}{dt} &= q_0 + q_1(x, y) + q_2(x, y) \equiv Q(x, y)
\end{aligned}
\tag{2}
$$

with homogeneous polynomials p_i and q_i $(i = 0, 1, 2)$ of degree i in x, y:

$$p_0 = a_{00}, \quad p_1(x, y) = a_{10}x + a_{01}y, \quad p_2(x, y) = a_{20}x^2 + 2a_{11}xy + a_{02}y^2,$$
$$q_0 = b_{00}, \quad q_1(x, y) = b_{10}x + b_{01}y, \quad q_2(x, y) = b_{20}x^2 + 2b_{11}xy + b_{02}y^2.$$

Let $a = (a_{00}, a_{10}, a_{01}, a_{20}, a_{11}, a_{02}, b_{00}, b_{10}, b_{01}, b_{20}, b_{11}, b_{02})$ be the 12-tuple of the coefficients of systems (2) and denote the corresponding polynomial ring by $\mathbb{R}[a, x, y] = \mathbb{R}[a_{00}, \ldots, b_{02}, x, y]$.

The group $Aff(2, \mathbb{R})$ of the affine transformations on the plane acts on the set **QS** (cf. [32]). For every subgroup $G \subseteq Aff(2, \mathbb{R})$, we have an induced action of G on **QS**. We can identify the set **QS** of systems of the form (2) with a subset of $\mathbb{R}^{12}$ via the map $\mathbf{QS} \longrightarrow \mathbb{R}^{12}$ which associates to each such system the 12-tuple $a = (a_{00}, a_{10}, \ldots, b_{02})$ of its coefficients.

The action of $Aff(2, \mathbb{R})$ on **QS** yields an action of this group on $\mathbb{R}^{12}$. For every $\mathfrak{g} \in Aff(2, \mathbb{R})$ let $r_{\mathfrak{g}} : \mathbb{R}^{12} \longrightarrow \mathbb{R}^{12}$, $r_{\mathfrak{g}}(a) = \tilde{a}$ where $\tilde{a}$ is the 12-tuple of coefficients of the transformed system $\tilde{S}$. It is known that $r_{\mathfrak{g}}$ is linear and that the map $r : Aff(2, \mathbb{R}) \longrightarrow GL(12, \mathbb{R})$ thus obtained is a group homomorphism. For every subgroup G of $Aff(2, \mathbb{R})$, r induces a representation of G onto a subgroup $\mathcal{G}$ of $GL(12, \mathbb{R})$.

Definition 3. *A polynomial $U(a, x, y) \in \mathbb{R}[a, x, y]$ is called a comitant of systems (2) with respect to a subgroup G of $Aff(2, \mathbb{R})$, if there exists $\chi \in \mathbb{Z}$ such that for every $(\mathfrak{g}, a) \in G \times \mathbb{R}^{12}$ and for every $(x, y) \in \mathbb{R}^2$ the following relation holds:*

$$U(r_{\mathfrak{g}}(a), \ \mathfrak{g}(x, y)) \equiv (\det \mathfrak{g})^{-\chi} U(a, x, y),$$

where $\det \mathfrak{g}$ is the determinant of the linear matrix of the transformation $\mathfrak{g} \in Aff(2, \mathbb{R})$. If the polynomial U does not explicitly depend on x and y then it is called invariant. The number $\chi \in \mathbb{Z}$ is called the weight of the comitant $U(a, x, y)$. If $G = GL(2, \mathbb{R})$ (or $G = Aff(2, \mathbb{R})$) then the comitant $U(a, x, y)$ of systems (2) is called GL-comitant (respectively, affine comitant).

Let us consider the polynomials

$$C_i(a, x, y) = yp_i(a, x, y) - xq_i(a, x, y) \in \mathbf{Q}[a, x, y], \ i = 0, 1, 2,$$
$$D_i(a, x, y) = \frac{\partial}{\partial x}p_i(a, x, y) + \frac{\partial}{\partial y}q_i(a, x, y) \in \mathbf{Q}[a, x, y], \ i = 1, 2.$$

As it was shown in [39] the polynomials

$$\left\{ C_0(a, x, y), \quad C_1(a, x, y), \quad C_2(a, x, y), \quad D_1(a), \quad D_2(a, x, y) \right\} \quad (3)$$

of degree one in the coefficients (variable parameters) of systems (2) are *GL*-comitants of these systems.

Let $f, g \in R[a, x, y]$ and we denote

$$(f, g)^{(k)} = \sum_{h=0}^{k} (-1)^h \binom{k}{h} \frac{\partial^k f}{\partial x^{k-h} \partial y^h} \frac{\partial^k g}{\partial x^h \partial y^{k-h}}. \quad (4)$$

$(f, g)^{(k)} \in \mathbb{R}[a, x, y]$ is called the *transvectant of index k* of (f, g) (cf. [20], [24]).

Theorem 1. [42] *Any GL-comitant of systems* (2) *can be constructed from the elements of the set* (3) *by using the operations:* $+, -, \times,$ *and by applying the differential operation* $(f, g)^{(k)}$.

For the definitions of a GL–comitant with some specific properties (i.e., of a T–comitant and a CT–comitant) we refer the reader to the paper [32] (see also [39]). Here we only construct a few invariant polynomials which are responsible for the singularities (finite and infinite) of systems (2).

Consider the differential operator $\mathcal{L} = x \cdot \mathbf{L}_2 - y \cdot \mathbf{L}_1$ acting on $\mathbb{R}[a, x, y]$ constructed in [10], where

$$\mathbf{L}_1 = 2a_{00}\frac{\partial}{\partial a_{10}} + a_{10}\frac{\partial}{\partial a_{20}} + \tfrac{1}{2}a_{01}\frac{\partial}{\partial a_{11}} + 2b_{00}\frac{\partial}{\partial b_{10}} + b_{10}\frac{\partial}{\partial b_{20}} + \tfrac{1}{2}b_{01}\frac{\partial}{\partial b_{11}},$$
$$\mathbf{L}_2 = 2a_{00}\frac{\partial}{\partial a_{01}} + a_{01}\frac{\partial}{\partial a_{02}} + \tfrac{1}{2}a_{10}\frac{\partial}{\partial a_{11}} + 2b_{00}\frac{\partial}{\partial b_{01}} + b_{01}\frac{\partial}{\partial b_{02}} + \tfrac{1}{2}b_{10}\frac{\partial}{\partial b_{11}}.$$

Using this operator and the affine invariant $\mu_0 = \mathrm{Res}\,_x\big(p_2(a, x, y), q_2(a, x, y)\big)/y^4$ we construct the following polynomials

$$\mu_i(a, x, y) = \frac{1}{i!}\mathcal{L}^{(i)}(\mu_0), \ i = 1, .., 4,$$

where $\mathcal{L}^{(i)}(\mu_0) = \mathcal{L}(\mathcal{L}^{(i-1)}(\mu_0))$.

These polynomials are in fact comitants of systems (2) with respect to the group $GL(2, \mathbb{R})$ (see [10]). Their geometrical meaning is revealed in Lemmas 1 and 2 below.

Lemma 1. ([9]) *The total multiplicity of all finite singularities of a quadratic system* (2) *equals* k *if and only if for every* $i \in \{0, 1, \ldots, k - 1\}$ *we have* $\mu_i(a, x, y) = 0$ *in* $\mathbb{R}[x, y]$ *and* $\mu_k(a, x, y) \neq 0$. *Moreover a system* (2) *is degenerate (i.e.,* $\gcd(P, Q) \neq constant$) *if and only if* $\mu_i(a, x, y) = 0$ *in* $\mathbb{R}[x, y]$ *for every* $i = 0, 1, 2, 3, 4$.

Lemma 2. ([10]) *The point* $M_0(0, 0)$ *is a singular point of multiplicity* k ($1 \leq k \leq 4$) *for a quadratic system* (2) *if and only if for every* $i \in \{0, 1, \ldots, k - 1\}$ *we have* $\mu_{4-i}(a, x, y) = 0$ *in* $\mathbb{R}[x, y]$ *and* $\mu_{4-k}(a, x, y) \neq 0$.

Here we apply the invariant polynomials previously constructed (see Remark 1) to a new situation to obtain new results (Main Theorem). In addition, we construct new invariant polynomials, which are responsible for the existence and number of star nodes, arbitrarily located on the plane, as we indicate in the following lemma.

Lemma 3. *A quadratic system* (2) *possesses one star node if and only if one of the following sets of conditions hold:*

$$(i) \ U_1 \neq 0, \ U_2 \neq 0, \ U_3 = Y_1 = 0;$$
$$(ii) \ U_1 = U_4 = U_5 = U_6 = 0, \ \ Y_2 \neq 0;$$

and it possesses two star nodes if and only if

$$(iii) \ \ U_1 = U_4 = U_5 = 0, \ \ U_6 \neq 0, \ \ Y_2 > 0,$$

where

$$U_1(a,x,y) = \widetilde{K} + \widetilde{H}, \quad U_2(a,x,y) = (C_1, \widetilde{H} - \widetilde{K})^{(1)} - 2D_1(\widetilde{K} + \widetilde{H}),$$
$$U_3(a,x,y) = 3\widetilde{D}(D_2^2 - 16\widetilde{K}) + C_2\left[(C_2, \widetilde{D})^{(2)} - 5(D_2, \widetilde{D})^{(1)} + 6\widetilde{F}\right],$$
$$U_4(a,x,y) = 2T_5 + C_1 D_2, \quad U_5(a,x,y) = 3C_1 D_1 + 4T_2 - 2C_0 D_1,$$
$$U_6(a,x,y) = \widetilde{H}, \quad Y_1(a) = \widetilde{A}, \quad Y_2(a,x,y) = 2D_1^2 + 8T_3 - T_4.$$

Here we use the following invariant polynomials constructed through the polynomials appearing in (3):

$$\widetilde{A} = \left(C_1, T_8 - 2T_9 + D_2^2\right)^{(2)} / 144, \quad \widetilde{D} = \left[2C_0(T_8 - 8T_9 - 2D_2^2)\right.$$
$$\left. + C_1(6T_7 - T_6 - (C_1, T_5)^{(1)} + 6D_1(C_1 D_2 - T_5) - 9D_1^2 C_2\right] / 36,$$

$$\widetilde{E} = \left[D_1(2T_9 - T_8) - 3\left(C_1, T_9\right)^{(1)} - D_2(3T_7 + D_1 D_2)\right] / 72,$$

$$\widetilde{F} = \left[6D_1^2(D_2^2 - 4T_9) + 4D_1 D_2(T_6 + 6T_7) + 48C_0\left(D_2, T_9\right)^{(1)} - 9D_2^2 T_4 + 288 D_1 \widetilde{E}\right.$$
$$\left. - 24\left(C_2, \widetilde{D}\right)^{(2)} + 120\left(D_2, \widetilde{D}\right)^{(1)} - 36C_1\left(D_2, T_7\right)^{(1)} + 8D_1\left(D_2, T_5\right)^{(1)}\right] / 144,$$

$$\widetilde{K} = (T_8 + 4T_9 + 4D_2^2)/18, \quad \widetilde{H} = -(8T_9 - T_8 + 2D_2^2)/18.$$

where

$$T_1 = (C_0, C_1)^{(1)}, \quad T_2 = (C_0, C_2)^{(1)}, \quad T_3 = (C_0, D_2)^{(1)},$$
$$T_4 = (C_1, C_1)^{(2)}, \quad T_5 = (C_1, C_2)^{(1)}, \quad T_6 = (C_1, C_2)^{(2)},$$
$$T_7 = (C_1, D_2)^{(1)}, \quad T_8 = (C_2, C_2)^{(2)}, \quad T_9 = (C_2, D_2)^{(1)}.$$

For more information the interested reader could consult [5],[6],[7],[33],[34],[36], [37],[40],[41] for more applications of polynomial invariants and computer calculations; [20],[24],[12] for the theory of algebraic invariants; [14],[1],[29],[26],[43] for quadratic differential systems and singularities.

References

1. Artés, J.C., Llibre, J.: Quadratic Hamiltonian vector fields. J. Differential Equations 107, 80–95 (1994)
2. Artés, J.C., Llibre, J., Schlomiuk, D.: The geometry of quadratic differential systems with a weak focus of second order. International J. Bifurcation and Chaos 16, 3127–3194 (2006)
3. Artés, J.C., Llibre, J., Schlomiuk, D., Vulpe, N.: From topological to geometric equivalence in the classification of singularities at infinity for quadratic vector fields. Rocky Mountain J. Math. (accepted)
4. Artés, J.C., Llibre, J., Schlomiuk, D., Vulpe, N.: Global analysis of infinite singularities of quadratic vector fields. CRM–Report No.3318. Université de Montreal (2011)
5. Artés, J.C., Llibre, J., Vulpe, N.I.: Singular points of quadratic systems: A complete classification in the coefficient space $\mathbb{R}^{12}$. International J. Bifurcation and Chaos 18, 313–362 (2008)

6. Artés, J.C., Llibre, J., Vulpe, N.: Complete geometric invariant study of two classes of quadratic systems. Electron. J. Differential Equations 2012(9), 1–35 (2012)
7. Artés, J.C., Llibre, J., Vulpe, N.: Quadratic systems with an integrable saddle: A complete classification in the coefficient space $\mathbb{R}^{12}$. Nonlinear Analysis 75, 5416–5447 (2012)
8. Baltag, V.A.: Algebraic equations with invariant coefficients in qualitative study of the polynomial homogeneous differential systems. Bull. Acad. Sci. of Moldova. Mathematics 2, 31–46 (2003)
9. Baltag, V.A., Vulpe, N.I.: Affine-invariant conditions for determining the number and multiplicity of singular points of quadratic differential systems. Izv. Akad. Nauk Respub. Moldova Mat. 1, 39–48 (1993)
10. Baltag, V.A., Vulpe, N.I.: Total multiplicity of all finite critical points of the polynomial differential system. Planar nonlinear dynamical systems (Delft, 1995). Differential Equations & Dynam. Systems 5, 455–471 (1997)
11. Bendixson, I.: Sur les courbes définies par des équations différentielles. Acta Math 24, 1–88 (1901)
12. Bularas, D., Calin, I., Timochouk, L., Vulpe, N.: T–comitants of quadratic systems: A study via the translation invariants. Delft University of Technology, Faculty of Technical Mathematics and Informatics, Report No. 96-90 (1996),
ftp://ftp.its.tudelft.nl/publications/tech-reports/
1996/DUT-TWI-96-90.ps.gz
13. Calin, I.: On rational bases of $GL(2, \mathbb{R})$-comitants of planar polynomial systems of differential equations. Bull. Acad. Sci. of Moldova. Mathematics 2, 69–86 (2003)
14. Coppel, W.A.: A survey of quadratic systems. J. Differential Equations 2, 293–304 (1966)
15. Dumortier, F.: Singularities of vector fields on the plane. J. Differential Equations 23, 53–106 (1977)
16. Dumortier, F.: Singularities of Vector Fields. Monografias de Matemática, vol. 32. IMPA, Rio de Janeiro (1978)
17. Dumortier, F., Fiddelaers, P.: Quadratic models for generic local 3-parameter bifurcations on the plane. Trans. Am. Math. Soc. 326, 101–126 (1991)
18. Dumortier, F., Llibre, J., Artés, J.C.: Qualitative Theory of Planar Differential Systems. Universitext. Springer, Berlin (2008)
19. Gonzalez Velasco, E.A.: Generic properties of polynomial vector fields at infinity. Trans. Amer. Math. Soc. 143, 201–222 (1969)
20. Grace, J.H., Young, A.: The algebra of invariants. Stechert, New York (1941)
21. Hilbert, D.: Mathematische Probleme. In: Nachr. Ges. Wiss., Second Internat. Congress Math., Paris, pp. 253–297. Göttingen Math.–Phys., Kl (1900)
22. Jiang, Q., Llibre, J.: Qualitative classification of singular points. Qualitative Theory of Dynamical Systems 6, 87–167 (2005)
23. Llibre, J., Schlomiuk, D.: Geometry of quadratic differential systems with a weak focus of third order. Canad. J. Math. 6, 310–343 (2004)
24. Olver, P.J.: Classical Invariant Theory. London Math. Soc. Student Texts, vol. 44. Cambridge University Press (1999)
25. Nikolaev, I., Vulpe, N.: Topological classification of quadratic systems at infinity. J. London Math. Soc. 2, 473–488 (1997)
26. Pal, J., Schlomiuk, D.: Summing up the dynamics of quadratic Hamiltonian systems with a center. Canad. J. Math. 56, 583–599 (1997)
27. Popa, M.N.: Applications of algebraic methods to differential systems. Piteşi Univers., The Flower Power Edit., Romania (2004)

28. Roussarie, R.: Smoothness property for bifurcation diagrams. Publicacions Matemàtiques 56, 243–268 (1997)
29. Schlomiuk, D.: Algebraic particular integrals, integrability and the problem of the center. Trans. Amer. Math. Soc. 338, 799–841 (1993)
30. Schlomiuk, D.: Basic algebro-geometric concepts in the study of planar polynomial vector fields. Publicacions Mathemàtiques 41, 269–295 (1997)
31. Schlomiuk, D., Pal, J.: On the geometry in the neighborhood of infinity of quadratic differential phase portraits with a weak focus. Qualitative Theory of Dynamical Systems 2, 1–43 (2001)
32. Schlomiuk, D., Vulpe, N.I.: Geometry of quadratic differential systems in the neighborhood of infinity. J. Differential Equations 215, 357–400 (2005)
33. Schlomiuk, D., Vulpe, N.I.: Integrals and phase portraits of planar quadratic differential systems with invariant lines of at least five total multiplicity. Rocky Mountain J. Mathematics 38(6), 1–60 (2008)
34. Schlomiuk, D., Vulpe, N.I.: Integrals and phase portraits of planar quadratic differential systems with invariant lines of total multiplicity four. Bull. Acad. Sci. of Moldova. Mathematics 1, 27–83 (2008)
35. Schlomiuk, D., Vulpe, N.I.: The full study of planar quadratic differential systems possessing a line of singularities at infinity. J. Dynam. Differential Equations 20, 737–775 (2008)
36. Schlomiuk, D., Vulpe, N.I.: Global classification of the planar Lotka–Volterra differential system according to their configurations of invariant straight lines. J. Fixed Point Theory Appl. 8, 177–245 (2010)
37. Schlomiuk, D., Vulpe, N.I.: The global topological classification of the Lotka–Volterra quadratic differential systems. Electron. J. Differential Equations 2012(64), 1–69 (2012)
38. Seidenberg, E.: Reduction of singularities of the differential equation $Ady = Bdx$. Amer. J. Math. 90, 248–269 (1968); Zbl. 159, 333
39. Sibirskii, K.S.: Introduction to the Algebraic Theory of Invariants of Differential Equations, Translated from the Russian. Nonlinear Science: Theory and Applications. Manchester University Press, Manchester (1988)
40. Vulpe, N.: Characterization of the finite weak singularities of quadratic systems via invariant theory. Nonlinear Analysis. Theory, Methods and Applications 74(4), 6553–6582 (2011)
41. Vulpe, N.I.: Affine–invariant conditions for the topological discrimination of quadratic systems with a center. Differential Equations 19, 273–280 (1983)
42. Vulpe, N.I.: Polynomial bases of comitants of differential systems and their applications in qualitative theory, "Ştiinţa", Kishinev (1986) (in Russian)
43. Żołądek, H.: Quadratic systems with center and their perturbations. J. Differential Equations 109, 223–273 (1994)

Singularities of Implicit Differential Equations and Static Bifurcations

Werner M. Seiler

Institut für Mathematik, Universität Kassel, 34132 Kassel, Germany
seiler@mathematik.uni-kassel.de

Abstract. We discuss geometric singularities of implicit ordinary differential equations from the point of view of Vessiot theory. We show that quasi-linear systems admit a special treatment leading to phenomena not present in the general case. These results are then applied to study static bifurcations of parametric ordinary differential equations.

1 Introduction

The Vessiot theory [17] provides a powerful framework for analysing differential equations geometrically. It uses vector fields in contrast to the differential forms of the more familiar Cartan-Kähler theory [2]. Fackerell [3] applied it in the context of symmetry analysis of partial differential equations. In [5, 6], we gave a rigorous proof that Vessiot's construction of solutions works, if and only if one is dealing with an involutive equation.

Vessiot's intention was to provide an alternative proof of the basic existence and uniqueness theorem for general systems of partial differential equations, the Cartan-Kähler Theorem. In our opinion, there are better approaches for this (see e. g. the proof in [14] which mainly follows [9]). However, his results are very useful for the further geometric analysis of involutive equations. In particular, one can easily see that Arnold's treatment of singularities of implicit ordinary differential equations in [1] is in fact based on what we call the Vessiot distribution. This identification provides us with a starting point for extending singularity theory to more general systems. For example, Tuomela [15, 16] used it for differential algebraic equations and we considered in [7] partial differential equations of finite type.

In this work, we first review the theory of geometric singularities of general nonlinear systems of ordinary differential equations which are not underdetermined. Then we discuss the particularities appearing in quasi-linear systems. In contrast to the general case, the Vessiot distribution is here projectable to the base manifold which leads to types of singular behaviour which are not admitted by fully non-linear equations. In particular, solutions can be extended to certain points where strictly speaking the differential equation is not even defined. We recover and clarify here results by Rabier [10–12] with alternative proofs.

Finally, we apply the obtained results to the analysis of static bifurcations in parametric autonomous ordinary differential equations. This is possible due

V.P. Gerdt et al. (Eds.): CASC 2013, LNCS 8136, pp. 355–368, 2013.
© Springer International Publishing Switzerland 2013

to the simple observation that differentiation of the equilibrium condition yields a quasi-linear equation. The projected Vessiot distribution contains now much information about bifurcations: turning points (also called saddle-node bifurcations) are characterised by a vertical distribution, whereas at pitchfork or transcritical bifurcations the distribution vanishes. Furthermore, the bifurcation diagram consists of integral manifolds of the Vessiot distribution.

2   Geometric Theory of Differential Equations

The geometric modelling of differential equations is based on jet bundles [8, 9, 13, 14]. Let $\pi : \mathcal{E} \to \mathcal{T}$ be a fibred manifold. For ordinary differential equations, we may assume that $\dim \mathcal{T} = 1$. For simplicity, we will work in local coordinates, although we will use throughout a "global notation". As coordinate on the base space $\mathcal{T}$ we use t and fibre coordinates in the total space $\mathcal{E}$ will be $\boldsymbol{u} = (u^1, \dots, u^m)$. The first derivative of u^α will be denoted by $\dot{u}^\alpha$; higher derivatives are written in the form $u_k^\alpha = d^k u^\alpha / dt^k$. Adding all derivatives u_k^α with $k \leq q$ (collectively denoted by $\boldsymbol{u}^{(q)}$) defines a coordinate system for the q-th order jet bundle $J_q\pi$. There are natural fibrations $\pi_r^q : J_q\pi \to J_r\pi$ for $r < q$ and $\pi^q : J_q\pi \to \mathcal{T}$. Sections $\sigma : \mathcal{T} \to \mathcal{E}$ of the fibration π correspond to functions $\boldsymbol{u} = \boldsymbol{s}(t)$, as locally they can always be written in the form of a graph $\sigma(t) = \big(t, \boldsymbol{s}(t)\big)$. To such a section σ, we associate its *prolongation* $j_q\sigma : \mathcal{T} \to J_q\pi$, a section of the fibration π^q given by $j_q\sigma(t) = \big(t, \boldsymbol{s}(t), \dot{\boldsymbol{s}}(t), \ddot{\boldsymbol{s}}(t), \dots \big)$.

The geometry of $J_q\pi$ is to a large extent determined by its *contact structure* describing intrinsically the relationship between the different types of coordinates. The *contact distribution* is the smallest distribution $\mathcal{C}_q \subset T(J_q\pi)$ that contains the tangent spaces $T(\mathrm{im}\, j_q\sigma)$ of all prolonged sections and any field in it is a *contact vector field*. In local coordinates, $\mathcal{C}_q$ is generated by one transversal and m vertical fields:

$$C_{\mathrm{trans}}^{(q)} = \partial_t + \sum_{j=0}^{q-1} u_{j+1}^\alpha \partial_{u_j^\alpha} \,, \tag{1a}$$

$$C_\alpha^{(q)} = \partial_{u_q^\alpha} \,, \qquad 1 \leq \alpha \leq m \,. \tag{1b}$$

Proposition 1. *A section $\gamma : \mathcal{T} \to J_q\pi$ is of the form $\gamma = j_q\sigma$ with $\sigma : \mathcal{T} \to \mathcal{E}$, if and only if $T_{\gamma(t)}(\mathrm{im}\,\gamma) \subseteq \mathcal{C}_q|_{\gamma(t)}$ for all points $t \in \mathcal{T}$ where γ is defined.*

The following intrinsic geometric definition of a differential equation generalises the usual one, as it allows for certain types of singular behaviour. It imposes considerably weaker conditions on the restricted projection $\hat{\pi}^q$ which in the standard definition is expected to be a surjective submersion. Note that we do not distinguish between scalar equations and systems. Indeed, when we speak of a differential equation in the sequel, we will always mean a system, if not explicitly stated otherwise.

Definition 2. *An* (ordinary) *differential equation of order q is a submanifold $\mathcal{R}_q \subseteq J_q\pi$ such that the restriction $\hat{\pi}^q$ of the projection $\pi^q : J_q\pi \to \mathcal{T}$ to $\mathcal{R}_q$*

has a dense image. A (strong) solution *is a (local) section* $\sigma : \mathcal{T} \to \mathcal{E}$ *such that* $\operatorname{im} j_q \sigma \subseteq \mathcal{R}_q$.

Locally, a differential equation $\mathcal{R}_q \subseteq J_q \pi$ can be described as the zero set of some smooth functions $\Phi : J_q \pi \to \mathbb{R}$. Differentiating every function yields the *prolonged equation* $\mathcal{R}_{q+1} \subseteq J_{q+1} \pi$ defined by all equations $\Phi(t, \boldsymbol{u}^{(q)}) = 0$ and $D_t \Phi(t, \boldsymbol{u}^{(q+1)}) = 0$ with the formal derivative

$$D_t \Phi = C_{\text{trans}}^{(q)}(\Phi) + \sum_{\alpha=1}^{m} u_{q+1}^{\alpha} C_{\alpha}^{(q)}(\Phi) \ . \tag{2}$$

Iteration of this process gives the higher prolongations $\mathcal{R}_{q+r} \subseteq J_{q+r} \pi$. A subsequent projection leads to the differential equation $\mathcal{R}_q^{(1)} = \pi_q^{q+1}(\mathcal{R}_{q+1}) \subseteq \mathcal{R}_q$ which will be a proper submanifold, if integrability conditions are hidden. $\mathcal{R}_q$ is *formally integrable*, if at any prolongation order $r > 0$ the equality $\mathcal{R}_{q+r}^{(1)} = \mathcal{R}_{q+r}$ holds (see [14] for more details). It is easy to show that (under some regularity assumptions) every consistent ordinary differential equation $\mathcal{R}_q$ leads after a finite number of projection and prolongation cycles to a formally integrable equation $\mathcal{R}_q^{(s)} \subseteq \mathcal{R}_q$. Therefore, without loss of generality, we will always assume in the sequel that we are already dealing with a formally integrable equation.

More precisely, we will study in this work only *square* first-order equations with a local representation

$$\mathcal{R}_1 : \left\{ \ \boldsymbol{\Phi}(t, \boldsymbol{u}^{(1)}) = 0 \right. \tag{3}$$

where $\boldsymbol{\Phi} : J_1 \pi \to \mathbb{R}^m$ (thus we have as many equations as unknown functions) and where we furthermore assume that the symbol matrix, i.e. the Jacobian $\partial \boldsymbol{\Phi} / \partial \dot{\boldsymbol{u}}$, is almost everywhere non-singular. These assumptions are less restrictive as they may appear. Whenever a first-order equation is not underdetermined, its symbol matrix has almost everywhere rank m. Thus locally we may always assume that a general first-order equation splits into an equation of the form considered here plus some purely algebraic equations. Solving the latter ones, we can eliminate some of the dependent variables u^{α} and obtain then a smaller equation of the desired form. At least theoretically this is always possible.

By a classical result on jet bundles, the fibration $\pi_0^1 : J_1 \pi \to \mathcal{E}$ is an affine bundle. A first-order differential equation $\mathcal{R}_1 \subset J_1 \pi$ is *quasi-linear*, if it defines an affine subbundle. For such equations, we assume in the sequel a local representation of the form

$$\mathcal{R}_1 : \left\{ \ A(t, \boldsymbol{u}) \dot{\boldsymbol{u}} = \boldsymbol{r}(t, \boldsymbol{u}) \right. \tag{4}$$

where the $m \times m$ matrix function A is almost everywhere non-singular.

Assumption 3. *In the sequel,* $\mathcal{R}_1 \subseteq J_1 \pi$ *will always be a formally integrable square first-order ordinary differential equation with a local representation of the form (3) or (4), respectively, which is not underdetermined.*

3   The Vessiot Distribution

A key insight of Cartan was to study *infinitesimal solutions* of a differential equation $\mathcal{R}_1 \subseteq J_1\pi$, i. e. to consider at any point $\rho \in \mathcal{R}_1$ those linear subspaces $\mathcal{U}_\rho \subseteq T_\rho\mathcal{R}_1$ which are potentially part of the tangent space of a prolonged solution. We will follow here an approach pioneered by Vessiot [17] which is based on vector fields and dual to the more familiar Cartan-Kähler theory of exterior differential systems (see [4, 6, 14] for modern presentations in the context of the geometric theory). By Proposition 1, the tangent spaces $T_\rho(\operatorname{im} j_1\sigma)$ of prolonged sections at points $\rho \in J_1\pi$ are always subspaces of the contact distribution $\mathcal{C}_1|_\rho$. If the section σ is a solution of $\mathcal{R}_1$, it furthermore satisfies $\operatorname{im} j_1\sigma \subseteq \mathcal{R}_1$ by Definition 2 and hence $T(\operatorname{im} j_1\sigma) \subseteq T\mathcal{R}_1$. These considerations motivate the following construction.

Definition 4. *The* Vessiot distribution *of a first-order ordinary differential equation* $\mathcal{R}_1 \subseteq J_1\pi$ *is the distribution* $\mathcal{V}[\mathcal{R}_1] \subseteq T\mathcal{R}_1$ *defined by*

$$\mathcal{V}[\mathcal{R}_1] = T\mathcal{R}_1 \cap \mathcal{C}_1|_{\mathcal{R}_1} . \tag{5}$$

Computing the Vessiot distribution is straightforward and requires only linear algebra. It follows from Definition 4 that any vector field X contained in $\mathcal{V}[\mathcal{R}_1]$ is a contact field and thus can be written as a linear combination of the basic contact fields (1): $X = aC_{\text{trans}}^{(1)} + \sum_\alpha b^\alpha C_\alpha^{(1)}$. On the other hand, X must be tangent to the manifold $\mathcal{R}_1$. Hence, if $\mathcal{R}_1$ is described by the local system (3), then the field X must satisfy the equations $d\boldsymbol{\Phi}(X) = X(\boldsymbol{\Phi}) = 0$. Evaluation of this condition yields a linear system of equations for the coefficients a, b^α:

$$C_{\text{trans}}^{(1)}(\Phi^\mu)a + \sum_{\alpha=1}^{m} C_\alpha^{(1)}(\Phi^\mu)b^\alpha = 0 , \qquad \mu = 1, \ldots, m . \tag{6}$$

Note that X is vertical with respect to π^1, if and only if the coefficient a vanishes. Concerning our Assumption 3 on the differential equation $\mathcal{R}_1$, we remark that equations of lower order are irrelevant for the Vessiot distribution provided that the equation is indeed formally integrable [14, Prop. 9.5.10].

Determining the Vessiot distribution of $\mathcal{R}_1$ requires essentially the same computations as prolonging it. Indeed, the prolongation $\mathcal{R}_2 \subseteq J_2\pi$ is locally described by the original equations $\Phi^\mu = 0$ together with their prolongations

$$C_{\text{trans}}^{(1)}(\Phi^\mu) + \sum_{\alpha=1}^{m} C_\alpha^{(1)}(\Phi^\mu)\ddot{u}^\alpha = 0 , \quad \mu = 1, \ldots, m . \tag{7}$$

These coincide with (6), if we set $a = 1$ and $b^\alpha = \ddot{u}^\alpha$, i. e. for transversal solutions of (6). One may say that (6) is a "projective" version of (7).

It should be stressed that we allow that the rank of a distribution varies from point to point. In fact, this will be important for certain types of singularities. The following, fairly elementary result is the basis of Vessiot's approach to the existence theory of differential equations. It relates solutions with certain sub-distributions of the Vessiot distribution. We formulate it here only for first-order ordinary differential equations; for the general case see [14, Prop. 9.5.7].

Lemma 5. *If the section $\sigma : \mathcal{T} \to \mathcal{E}$ is a solution of the first-order ordinary differential equation $\mathcal{R}_1 \subseteq J_1\pi$, then the tangent bundle $T(\mathrm{im}\, j_1\sigma)$ is a one-dimensional subdistribution of $\mathcal{V}[\mathcal{R}_1]|_{\mathrm{im}\, j_1\sigma}$ transversal to the fibration π^1. Conversely, if the subdistribution $\mathcal{U} \subseteq \mathcal{V}[\mathcal{R}_1]$ is one-dimensional and transversal, then any integral curve of $\mathcal{U}$ has locally the form $\mathrm{im}\, j_1\sigma$ for a solution σ of $\mathcal{R}_1$.*

Definition 6. *A* generalised solution *of the first-order ordinary differential equation $\mathcal{R}_1 \subseteq J_1\pi$ is an integral curve $\mathcal{N} \subseteq \mathcal{R}_1$ of the Vessiot distribution $\mathcal{V}[\mathcal{R}_1]$. The projection $\pi_0^1(\mathcal{N}) \subseteq \mathcal{E}$ is a* geometric solution.

Note that generalised solutions live in the jet bundle $J_1\pi$ and not in the base manifold $\mathcal{E}$. If the section $\sigma : \mathcal{T} \to \mathcal{E}$ is a classical solution, then the image of its prolongation $j_1\sigma : \mathcal{T} \to J_1\pi$ is a generalised solution. However, not every generalised solution $\mathcal{N}$ projects on a classical one: this will be the case, if and only if $\mathcal{N}$ is everywhere transversal to the fibration $\pi^1 : J_1\pi \to \mathcal{T}$.

4   Geometric Singularities

A *geometric singularity* of a differential equation $\mathcal{R}_1$ is a critical point $\rho \in \mathcal{R}_1$ of the restricted projection $\hat{\pi}_0^1 : \mathcal{R}_1 \to \mathcal{E}$, i.e. a point where the tangent map $T_\rho\hat{\pi}_0^1$ is not surjective. Following the terminology of Arnold [1] in the scalar case, we distinguish three types of points on $\mathcal{R}_1$. Note that this taxonomy makes only sense for formally integrable equations which are not underdetermined.

Definition 7. *Let $\mathcal{R}_1 \subseteq J_1\pi$ be a first-order ordinary differential equation satisfying Assumption 3. A point $\rho \in \mathcal{R}_1$ is* regular, *if $\mathcal{V}_\rho[\mathcal{R}_1]$ is one-dimensional and transversal to π^1, and* regular singular, *if $\mathcal{V}_\rho[\mathcal{R}_1]$ is one-dimensional and vertical. If $\dim \mathcal{V}_\rho[\mathcal{R}_1] > 1$, then ρ is an* irregular singularity.

For an equation of the form (3), we define the $m \times m$ matrix $A(t, \boldsymbol{u}^{(1)})$ and the m-dimensional vector $\boldsymbol{r}(t, \boldsymbol{u}^{(1)})$ by

$$A = \frac{\partial \boldsymbol{\Phi}}{\partial \dot{\boldsymbol{u}}}, \qquad \boldsymbol{d} = \frac{\partial \boldsymbol{\Phi}}{\partial t} + \frac{\partial \boldsymbol{\Phi}}{\partial \boldsymbol{u}} \cdot \dot{\boldsymbol{u}}. \tag{8}$$

In addition to the (symbol) matrix $A(t, \boldsymbol{u}^{(1)})$, we introduce its determinant $\delta(t, \boldsymbol{u}^{(1)}) = \det A(t, \boldsymbol{u}^{(1)})$ and its adjugate $C(t, \boldsymbol{u}^{(1)}) = \mathrm{adj}\, A(t, \boldsymbol{u}^{(1)})$. Because of the well-known identity $\delta \mathbb{1}_m = AC = CA$, we find $\mathrm{im}\, C(\rho) \subseteq \ker A(\rho)$ and $\mathrm{im}\, A(\rho) \subseteq \ker C(\rho)$ for any point $\rho \in J_1\pi$ where $\delta(\rho) = 0$. For later use, we note that if additionally $\dim \ker A(\rho) = 1$ at such a point, then $C(\rho) \neq 0$, as $A(\rho)$ must possess at least one non-vanishing minor. Hence in this case, we even find $\mathrm{im}\, C(\rho) = \ker A(\rho)$ and $\mathrm{im}\, A(\rho) = \ker C(\rho)$.

The following considerations recover results by Rabier [10] from the point of view of Vessiot's theory and provide simpler alternative proofs. They show in particular that geometric singularities are characterised by the vanishing of δ. We therefore call the subset $\mathcal{S}_1 = \{\rho \in \mathcal{R}_1 \mid \delta(\rho) = 0\}$ the *singular locus* of the differential equation $\mathcal{R}_1$.

Theorem 8. *Let $\mathcal{R}_1 \subseteq J_1\pi$ be a first-order ordinary differential equation satisfying Assumption 3. A point $\rho \in \mathcal{R}_1$ is regular, if and only if $\operatorname{rk} A(\rho) = m$ and regular singular, if and only if $\operatorname{rk} A(\rho) = m - 1$ and $\boldsymbol{d}(\rho) \notin \operatorname{im} A(\rho)$. A regular point ρ has an open simply connected neighbourhood $\mathcal{U} \subseteq \mathcal{R}_1$ without any geometric singularity and there exists locally a unique strong solution σ such that $\rho \in \operatorname{im} j_1\sigma \subseteq \mathcal{U}$. A regular singular point ρ has an open simply connected neighbourhood $\mathcal{U} \subseteq \mathcal{R}_1$ without any irregular singularity and there exists locally a unique generalised solution $\mathcal{N}$ such that $\rho \in \mathcal{N} \subseteq \mathcal{U}$. If the neighbourhood $\mathcal{U}$ is chosen sufficiently small, then in both cases the Vessiot distribution is generated in $\mathcal{U}$ by the vector field*

$$X = \delta C_{\text{trans}}^{(1)} - (C\boldsymbol{d})^t \boldsymbol{C}^{(q)} . \tag{9}$$

Proof. The first two assertions follow from the fact that the matrix of the linear system (6) evaluated at a point $\rho \in \mathcal{R}_1$ is $\big(\boldsymbol{d}(\rho) \mid A(\rho)\big)$ and that the Vessiot distribution is one-dimensional, if and only if its rank is m. Since rank is an upper semicontinuous function, the Vessiot distribution will remain one-dimensional in a whole neighbourhood $\mathcal{U}$ of such a point. Hence any such neighbourhood cannot contain an irregular singularity where the rank of the matrix $\big(\boldsymbol{d}(\rho) \mid A(\rho)\big)$ must be less than m. If the point ρ is regular, we can apply the same argument to the matrix $A(\rho)$ and find that $\mathcal{U}$ cannot even contain a regular singularity.

If the neighbourhood $\mathcal{U}$ is chosen simply connected, then the one-dimensional distribution $\mathcal{V}[\mathcal{R}_1]$ can be generated in it by a single vector field without any zero. The explicit generator X of (9) is now obtained by simply multiplying (6) with the adjugate C. Note that the field X does indeed vanish nowhere, as even at a regular singularity where $\delta(\rho) = 0$ we find by the considerations above that $\ker C(\rho) = \operatorname{im} A(\rho)$ and thus $C(\rho)\boldsymbol{d}(\rho) \neq 0$ since $\boldsymbol{d}(\rho) \notin \operatorname{im} A(\rho)$.

The existence of a unique local integral curve $\mathcal{N} \subseteq \mathcal{U}$ of X and thus of a unique generalised solution through ρ follows now by the usual existence and uniqueness theorems for vector fields. If ρ is a regular point, then X (and thus also $\mathcal{N}$) is everywhere on $\mathcal{U}$ transversal to π_0^1 and we can write $\mathcal{N} = \operatorname{im} \gamma$ for some section $\gamma : \mathcal{T} \to J_1\pi$. By Proposition 1, $\gamma = j_1\sigma$ for a section $\sigma : \mathcal{T} \to \mathcal{E}$. Thus in this case ρ even lies on a unique (prolonged) strong solution. $\quad\square$

A different formulation of the existence and uniqeness part of Theorem 8 can be found in [7, Thm. 4.1]. There the standard existence and uniqueness result for ordinary differential equations solved for the derivatives is generalised to arbitrary formally integrable equations without irregular singularities. In particular, it is shown that any strong solution can be extended until its prolongation hits either the boundary of $\mathcal{R}_1$ or a regular singularity.

Note that at a regular singularity $\delta = 0$ and hence the vector field X defined by (9) is indeed vertical. By an extension of the above argument, one can prove the following statement about irregular singularities [7, Thm. 4.2]. Any irregular singularity ρ lies on the boundary of an open simply connected neighbourhood $\mathcal{U} \subseteq \mathcal{R}_1$ without further irregular singularities. By Theorem 8, $\mathcal{V}[\mathcal{R}_1]$ is generated on $\mathcal{U}$ by a vector field X. Then any extension of X to ρ will vanish. The dynamics around an irregular singular point ρ is now to a large extent determined by the eigenvalues of the Jacobian of X at ρ. Usually, there are infinitely many strong (prolonged) solutions beginning or ending at such point.

Example 9. From a geometric point of view, it is straightforward to understand what happens at a regular singularity ρ (cf. also [7]). There are two possibilities.[1] If the sign of the ∂_t-component of X along $\mathcal{N}$ does change at ρ (the generic behaviour), then there exist precisely *two* strong solutions which both either end or start at $\pi_0^1(\rho)$ and which arise through the "folding" of the generalised solution $\mathcal{N}$ during the projection. No prolonged strong solution can go through ρ in this case, as the projection of $\mathcal{N}$ is not a graph at $\pi_0^1(\rho)$. Otherwise, the projection of $\mathcal{N}$ is a strong solution which, however, is only of class $\mathcal{C}^1$, as its second derivative at ρ cannot exist by the above comparison of prolongation and determination of the Vessiot distribution.

As a concrete example, we consider the equation $\dot{u}^3 - u\dot{u} - t = 0$ whose surface $\mathcal{R}_1 \subseteq J_1\pi$ corresponds to the elementary catastrophe known as gather or Whitney pleat. We call this equation the *elliptic gather*. The blue surface in the picture on the right is $\mathcal{R}_1$; the short black lines indicate the direction defined by the Vessiot distribution $\mathcal{V}[\mathcal{R}_1]$ at some points $\rho \in \mathcal{R}_1$. The white curve shows the singular locus. All points on it are regular singular points. The yellow curves depict some generalised solutions determined by numerically integrating the vector field X of Theorem 8 (which has also been computed numerically). Their projection to $\mathcal{E}$ is shown on the red plane.

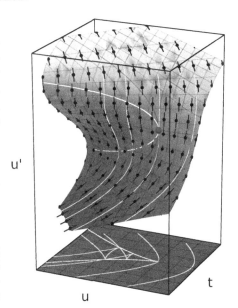

Whenever a generalised solution crosses the singular locus, the projected curve changes its direction and thus ceases to be the graph of a function. An exception is the one generalised solution that goes through the "tip" of the singular locus, as here the Vessiot distribution is tangential to the singular locus. Its projection is still the graph of a function which, however, is at $t = 0$ only once differentiable; the second derivative blows up at this point.

---

[1] We ignore here the degenerate case that the generalised solution $\mathcal{N}$ through ρ is completely vertical, i.e. $\mathcal{N} \subseteq (\pi_0^1)^{-1}\big(\pi_0^1(\rho)\big)$; see, however, the next section.

As the following result shows, generically the Vessiot distribution is transversal to the singular locus at regular singularities and the singular locus is almost everywhere a smooth manifold. We omit a proof of this proposition, as we can simply use the one given by Rabier [10]. He does not interpret the non-degeneracy condition in terms of the Vessiot distribution. But given the expression (9) for the generator X of $\mathcal{V}[\mathcal{R}_1]$, it is easy to see that his results imply the transversality of $\mathcal{V}[\mathcal{R}_1]$ to $\mathcal{S}_1$.

Proposition 10. *Let $\rho \in \mathcal{R}_1$ be a regular singular point. If*

$$\boldsymbol{v}^t \frac{\partial A}{\partial \dot{\boldsymbol{u}}}(\rho)\boldsymbol{v} \notin \operatorname{im} A(\rho) \tag{10}$$

for all non-vanishing vectors $\boldsymbol{v} \in \ker A(\rho)$, then the Vessiot distribution $\mathcal{V}_\rho[\mathcal{R}_1]$ is transversal to the singular locus $\mathcal{S}_1$ and the singular locus $\mathcal{S}_1$ is a smooth manifold in a neighbourhood of ρ.

Remark 11. One should note that we do *not* get here an equivalence between the condition (10) and transversality of the Vessiot distribution—or the vector field X—to the singular locus. The converse may become invalid when the determinant δ is of the form $\delta = \zeta^k$ for some function ζ and an exponent $k > 1$. In this case, transversality cannot be decided by using the differential $d\delta$, as it vanishes everywhere on the singular locus $\mathcal{S}_1$. However, Rabier's proof of Proposition 10 uses $d\delta$ and thus cannot be inverted in such situations.

5   Quasi-linear Equations

We specialise now the results of the previous section to quasi-linear equations of the form (4). As we already indicated with our notations, the matrix A is then indeed the Jacobian with respect to the derivatives. Thus here the only difference is the fact that in the quasi-linear case A does not depend on the derivatives and we will continue to denote its determinant by δ and its adjugate by C. However, the vectors $\boldsymbol{r}$ and $\boldsymbol{d}$ are not related.

Lemma 12. *Let $\mathcal{R}_1 \subseteq J_1\pi$ be a quasi-linear equation satisfying Assumption 3. According to Theorem 8, on a simply connected open subset $\mathcal{U} \subseteq \mathcal{R}_1$ without any irregular singularity the Vessiot distribution $\mathcal{V}[\mathcal{R}_1]$ is generated by the vector field X given by (9). This field is projectable to a vector field $Y = (\pi_0^1)_* X$ defined on $\pi_0^1(\mathcal{U}) \subseteq \mathcal{R}_0^{(1)} \subseteq \mathcal{E}$ by*

$$Y = \delta\partial_t + (C\boldsymbol{r})^t \partial_{\boldsymbol{u}} \,. \tag{11}$$

Proof. Multiplication of (4) with the adjugate C yields the equation $\delta\dot{\boldsymbol{u}} = C\boldsymbol{r}$. Thus on $\mathcal{R}_1$ we may write $X = Y + (C\boldsymbol{d})^t \partial_{\dot{\boldsymbol{u}}}$. As all coefficients of Y depend only on t and $\boldsymbol{u}$, the field X is projectable to $\mathcal{E}$ and $(\pi_0^1)_* X = Y$. $\qquad\square$

The coordinate form (11) shows that the field Y can be locally continued to points outside of the projection $\mathcal{R}_0^{(1)}$. More precisely, it follows from the definition of the adjugate that Y can be defined at any point $\xi \in \mathcal{E}$ where both A and $\boldsymbol{r}$ are defined. This observation allows us to study the singularities of $\mathcal{R}_1$ on $\mathcal{E}$ using the vector field Y instead of working in $J_1\pi$ with the Vessiot distribution $\mathcal{V}[\mathcal{R}_1]$.

Definition 13. *Let $\mathcal{R}_1 \subseteq J_1\pi$ be a quasi-linear differential equation with local representation (4) and $\mathcal{D} \subseteq \mathcal{E}$ the subset where both A and $\mathbf{r}$ are defined. $\xi \in \mathcal{D}$ is a regular point for $\mathcal{R}_1$, if Y_ξ is transversal to the fibration $\pi : \mathcal{E} \to \mathcal{T}$, and an impasse point otherwise. An impasse point is irregular, if $Y_\xi = 0$, and regular otherwise. The set of all impasse points is the impasse hypersurface $\mathcal{S}_0 \subseteq \mathcal{E}$. A geometric solution of the differential equation $\mathcal{R}_1$ is an integral curve $\mathcal{N} \subseteq \mathcal{E}$ of the vector field Y.*

In analogy to Theorem 8, one obtains the following characterisation of the different cases. We stress again that for quasi-linear equation all conditions live in $\mathcal{E}$ and not in $J_1\pi$.

Theorem 14. *Let $\mathcal{R}_1 \subseteq J_1\pi$ be a quasi-linear differential equation satisfying Assumption 3. A point $\xi \in \mathcal{D}$ is regular for $\mathcal{R}_1$, if and only if $\operatorname{rk} A(\xi) = m$. It is a regular impasse point, if and only if $\operatorname{rk} A(\xi) = m - 1$ and $\xi \notin \mathcal{R}_0^{(1)}$.*

Proof. The case of a regular point is obvious. If $\operatorname{rk} A(\xi) < m-1$, then all minors of $A(\xi)$ of size $m-1$ and thus also the adjugate $C(\xi)$ vanish. This fact implies then trivially that $Y_\xi = 0$. Hence there only remains the case $\operatorname{rk} A(\xi) = m - 1$. Recall from above that then $\operatorname{im} A(\xi) = \ker C(\xi)$. Since $\xi \notin \mathcal{R}_0^{(1)}$ is equivalent to $\mathbf{r}(\xi) \notin \operatorname{im} A(\xi)$, we thus find in this case that $Y_\xi \neq 0$, if and only if $\xi \notin \mathcal{R}_0^{(1)}$. $\square$

We compare now these notions with the corresponding ones for general equations introduced in Definition 7. Let $\rho \in \mathcal{R}_1$ be a point on our given differential equation and $\xi = \pi_0^1(\rho)$ its projection to $\mathcal{E}$. If ρ is a regular point, then trivially ξ is regular, too. Indeed, it follows immediately from (9) that in this case X_ρ is transversal to π^1 and hence its projection Y_ξ to π. As discussed above, it follows from [7, Thm. 4.2] that X_ρ vanishes, if ρ is an irregular singularity. Hence in this case ξ is an irregular impasse point.

Implicit in the above proof is the observation that regular singular points of a quasi-linear equation *always* show a degenerate behaviour. Indeed, if ρ is a regular singularity, then the fibre $\mathcal{N} = (\pi_0^1)^{-1}(\xi) \cap \mathcal{R}_1$ is one-dimensional and consists entirely of regular singularities. Furthermore, in this degenerate case $\mathcal{N}$ is the unique generalised solution through ρ and it does not project onto a curve in $\mathcal{E}$ but the single point ξ.

The behaviour normally associated with regular singularities appears for quasi-linear equations at regular impasse points. One possibility is that two strong solutions can be extended so that they either start or end at ξ and together define the geometric solution through ξ. This will happen, if the sign of the ∂_t-component of Y changes at ξ. If the sign remains the same, then we find a unique "strong" solution through ξ which, however, is only $\mathcal{C}^0$ in ξ and thus strictly speaking cannot be considered as a solution. Concrete examples for both cases will be given in the next section.

Again a degenerate situation may arise. Let $\mathcal{R}_1$ be a *semi-linear* equation where the symbol matrix A depends only on the independent variable t. If now $\operatorname{rk} A(t_0) = m - 1$ for some point $t_0 \in \mathcal{T}$, then at all points $\xi \in \pi^{-1}(t_0) \subset \mathcal{E}$ the vector Y_ξ is either vertical or vanishes. Assuming that the latter happens only on some lower-dimensional subset, we find geometric solutions $\mathcal{N}$ lying completely in the fibre $\pi^{-1}(t_0)$. As these project on the single point $t_0 \in \mathcal{T}$, they cannot be interpreted as strong solutions with some singularities.

Example 15. Consider the quasi-linear equation $2u\dot{u} - t = 0$. Its singular locus $\mathcal{S}_1$ is the vertical line $t = u = 0$ which is simultaneously a generalised solution.

Two points on it, $(0, 0, \pm 1)$, are irregular singularities; all other points are
regular singularities. The impasse man-
ifold $\mathcal{S}_0$ given by $u = 0$ contains one irregular impasse point at the origin. An
explicit integration of this equation is
easily possible and the solutions are of
the implicit form $u^2 - t^2/2 = c$ for a
constant $c \in \mathbb{R}$. For $c < 0$, the two
branches of the square root always meet
on $\mathcal{S}_0$ where the solution is not differentiable. Note that on one branch $\dot{u} \to \infty$
whereas on the other branch $\dot{u} \to -\infty$.
For $c = 0$ one obtains the two lines
intersecting at the origin. For $c > 0$
each branch of the square root yields
one strong solution.

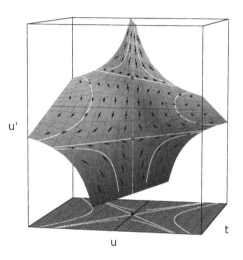

There also exists a special version of Proposition 10 for autonomous quasi-linear equations. In this case the condition (10) must be replaced by

$$v^t \frac{\partial A}{\partial u}(\xi)v \notin \text{im } A(\xi) \tag{12}$$

for all non-vanishing vectors $v \in \ker A(\xi)$ and it ensures that the field Y is transversal to the impasse surface $\mathcal{S}_0$ at the impasse point ξ and that $\mathcal{S}_0$ is a smooth manifold in a neighbourhood of ξ. As already discussed in Remark 11, this condition is only sufficient but not necessary. The quasi-linear equation associated with a hysteresis point (see Example 18 below) represents a concrete counterexample where condition (12) fails at the hysteresis point but nevertheless the vector field Y is there transversal to the smooth manifold $\mathcal{S}_0$.

6   Static Bifurcations

We apply the results of the last section to the analysis of static bifurcations of a parametrised autonomous ordinary differential equation of the standard form $u' = \phi(t, u)$. Opposed to the conventions used so far in this work, t represents now the parameter and we denote the (not explicitly appearing) independent variable by x and hence derivatives with respect to it by u'. We continue to consider t and u as coordinates on a fibred manifold $\pi : \mathcal{E} \to \mathcal{T}$.

For static bifurcations one analyses the dependence of solutions of the algebraic system $\phi(t, \boldsymbol{u}) = 0$ on the parameter t, i.e. how the equilibria change as t varies. The solution set may be considered as a bifurcation diagram. Note that this represents a purely algebraic problem in $\mathcal{E}$. At certain bifurcation values of the parameter t the number of equilibria changes.

Definition 16. *The point $\xi = (t, \boldsymbol{u}) \in \mathcal{E}$ is a* turning point, *if*

$$\phi(\xi) = 0 \quad \wedge \quad \dim \ker \frac{\partial \phi}{\partial \boldsymbol{u}}(\xi) = 1 \quad \wedge \quad \frac{\partial \phi}{\partial t}(\xi) \notin \operatorname{im} \frac{\partial \phi}{\partial \boldsymbol{u}}(\xi) . \tag{13}$$

At a bifurcation point $\xi \in \mathcal{E}$, the third condition is replaced by its converse:

$$\phi(\xi) = 0 \quad \wedge \quad \dim \ker \frac{\partial \phi}{\partial \boldsymbol{u}}(\xi) = 1 \quad \wedge \quad \frac{\partial \phi}{\partial t}(\xi) \in \operatorname{im} \frac{\partial \phi}{\partial \boldsymbol{u}}(\xi) . \tag{14}$$

The rationale behind the above distinction is that at a turning point ξ all solutions of $\phi = 0$ still lie on one smooth curve and the number of solutions only changes because this curve "turns" at ξ. At a bifurcation point several solution curves meet. In the bifurcation literature, much emphasis is put on distinguishing *simple* turning or bifurcation points from higher ones. In particular, the numerical analysis differs for non-simple points. It will turn out that in our approach such a distinction is irrelevant. We try to write the solutions as a function $\boldsymbol{u}(t)$. Differentiating the given algebraic system with respect to the parameter t yields then a square quasi-linear differential equation for this function:

$$\frac{\partial \phi}{\partial \boldsymbol{u}}(t, \boldsymbol{u})\dot{\boldsymbol{u}} + \frac{\partial \phi}{\partial t}(t, \boldsymbol{u}) = 0 . \tag{15}$$

Thus we set $A(t, \boldsymbol{u}) = \frac{\partial \phi}{\partial \boldsymbol{u}}(t, \boldsymbol{u})$ and $\boldsymbol{d}(t, \boldsymbol{u}) = \frac{\partial \phi}{\partial t}(t, \boldsymbol{u})$. The bifurcation diagram consists now of those geometric solutions of (15) on which ϕ vanishes.

Theorem 17. *$\xi \in \mathcal{E}$ is a turning point, if and only if $\phi(\xi) = 0$ and ξ is a regular impasse point of (15). $\xi \in \mathcal{E}$ is a bifurcation point, if and only if $\phi(\xi) = 0$ and ξ is an irregular impasse point of (15) where $\operatorname{rk} A(\xi) = m - 1$. In this case each branch of the bifurcation diagram is tangent to an eigenvector of the Jacobian of the vector field Y at ξ for an eigenvalue with non-vanishing real part.*

Proof. The first two assertions follow immediately from comparing Definition 16 with Theorem 14. For the last assertion we note that an irregular impasse point is, by definition, an equilibrium of the vector field Y and hence it follows from basic dynamical systems theory. □

Example 18. We compare a *simple turning point* or (*saddle node bifurcation*) with a *hysteresis point* (a degenerate turning point). As all our geometric considerations remain invariant under coordinate transformation, we take for simplicity the well-known normal forms of such points: $\phi_1(t, u) = t - u^2$ and $\phi_2 = t - u^3$. The corresponding quasi-linear equations are $2u\dot{u} = 1$ and $3u^2\dot{u} = 1$. Thus in both cases the impasse manifold is given by the equation $u = 0$ and consists entirely of regular impasse points. A straightforward calculation yields for the vector field defined by (11) $Y_1 = 2u\partial_t + \partial_u$ and $Y_2 = 3u^2\partial_t + \partial_u$, respectively. In both cases the origin is the sole turning point in the sense of Definition 16.

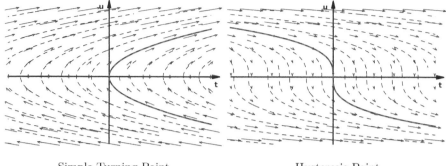

Simple Turning Point   Hysteresis Point

The above pictures show the fields Y_1 and Y_2 and their streamlines.[2] The red curve is the bifurcation diagram. One clearly sees that even in the degenerate case there are no particular numerical problems. The only difference between the two cases is the behaviour of the ∂_t-component of Y: on the left it changes sign when going through the impasse point, on the right it does not. As explained in the previous section, this observation entails that on the left we have two strong solutions starting arbitrarily close to the origin and extendable to the origin, whereas on the right we find one solution going through the origin which is, however, not differentiable there.

Example 19. The normal form of a *pitchfork bifurcation* is $\phi(t, u) = tu - u^3$. The associated quasi-linear equation is then $(t - 3u^2)\dot{u} + u = 0$ with singular locus $\mathcal{S}_1 \subseteq \mathcal{R}_1$ given by the points satisfying in addition $3u^2 = t$. The impasse surface $\mathcal{S}_0 \subseteq \mathcal{E}$ is also described by this equation. The origin is an irregular impasse point; all other points on $\mathcal{S}_0$ are regular. The vector field of (11) is given by $Y = (t - 3u^2)\partial_t - u\partial_u$ and vanishes at the origin as required for an irregular impasse point.

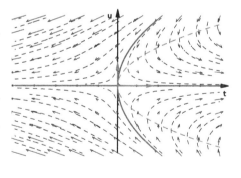

The picture on the right exhibits the vector field Y, its streamlines and the eigenvectors of the Jacobian of Y at the origin $J = \begin{pmatrix} 1 & 0 \\ 0 & -1 \end{pmatrix}$. Obviously, they are tangent to the bifurcation diagram consisting of the invariant manifolds of Y. The picture also nicely shows how the pitchfork bifurcation arises when two turning points moving on the dashed green line collide.

---

[2] Only one of the streamlines satisfies the algebraic equation $\phi(\xi) = 0$ and thus represents the bifurcation diagram we want. The other streamlines can be interpreted as bifurcation diagrams of the perturbed problems $\phi(\xi) = \epsilon$ with a constant $\epsilon \in \mathbb{R}$. Indeed, such a constant disappears when we differentiate in order to obtain our quasi-linear equation. Thus strictly speaking, we simultaneously analyse a whole family of bifurcation problems in our approach.

We remark that from this geometric point of view a pitchfork and a *transcritical bifurcation* are very similar. There arise only minor differences how the invariant manifolds lie relative to the eigenspaces. We omit therefore the details for a transcritical bifurcation.

7   Conclusions

In this work we used Vessiot's vector field based approach to differential equations for an analysis of geometric singularities of ordinary differential equations satisfying certain basic assumptions. We clarified the special role of quasi-linear equations where the geometric analysis can be performed on the basic fibred manifold $\mathcal{E}$ instead of the jet bundle $J_1\pi$. This observation is not surprising, as one can see similar effects also at other places in the theory of differential equations. In [14, Addendum Sect. 9.5] the method of characteristics is reviewed from a geometric point of view. There it also turns out that one may consider the characteristics for linear equations even on $\mathcal{T}$ and for quasi-linear ones also on $\mathcal{E}$. Only for fully non-linear equations one must use the jet bundle $J_1\pi$.

As an application we studied the theory of static bifurcations of autonomous ordinary differential equations. We associated a quasi-linear differential equation and thus a certain vector field Y on the base manifold $\mathcal{E}$ with such a bifurcation problem. Then we showed that the distinction between turning and bifurcation points corresponds to the distinction between regular and irregular impasse points of the associated quasi-linear equation. For equations admitting only static bifurcation points, these results may lead to a simpler way to determine bifurcation diagrams. Instead of using continuation methods one can simply integrate the vector field Y. Furthermore, around a bifurcation point ξ there is no need to search where branches may head, as they can only emerge in the direction of eigenvectors of the Jacobian of Y at ξ.

We did not consider the question of recognising simple turning or bifurcation points. As already noted by Rabier and Rheinboldt [11], simple turning points are characterised by the fact that there also the condition (10) is satisfied. We discussed in Remark 11 that from a geometric point of view this condition and consequently the concept of a simple turning point is not fully satisfactory (or of a more technical nature), since (10) is not equivalent to the transversality of the vector field X to the singular locus $\mathcal{S}_1$ (or the transversality of Y to the impasse hypersurface $\mathcal{S}_0$, respectively). The condition (10) always fails, if the determinant δ provides a degenerate description of $\mathcal{S}_1$ (or $\mathcal{S}_0$).

References

1. Arnold, V.: Geometrical Methods in the Theory of Ordinary Differential Equations, 2nd edn. Grundlehren der mathematischen Wissenschaften, vol. 250. Springer, New York (1988)
2. Bryant, R., Chern, S., Gardner, R., Goldschmidt, H., Griffiths, P.: Exterior Differential Systems. Mathematical Sciences Research Institute Publications, vol. 18. Springer, New York (1991)

3. Fackerell, E.: Isovectors and prolongation structures by Vessiot's vector field formulation of partial differential equations. In: Martini, R. (ed.) Geometric Aspects of the Einstein Equations and Integrable Systems. Lecture Notes in Physics, vol. 239, pp. 303–321. Springer, Berlin (1985)

4. Fesser, D.: On Vessiot's Theory of Partial Differential Equations. Ph.D. thesis, Fachbereich Mathematik, Universität Kassel (2008)

5. Fesser, D., Seiler, W.: Vessiot connections of partial differential equations. In: Calmet, J., Seiler, W., Tucker, R. (eds.) Global Integrability of Field Theories, pp. 111–134. Universitätsverlag Karlsruhe, Karlsruhe (2006)

6. Fesser, D., Seiler, W.: Existence and construction of Vessiot connections. SIGMA 5, 092 (2009)

7. Kant, U., Seiler, W.: Singularities in the geometric theory of differential equations. In: Feng, W., Feng, Z., Grasselli, M., Lu, X., Siegmund, S., Voigt, J. (eds.) Dynamical Systems, Differential Equations and Applications (Proc. 8th AIMS Conference, Dresden 2010, vol. 2, pp. 784–793. AIMS (2012)

8. Lychagin, V.: Homogeneous geometric structures and homogeneous differential equations. In: Lychagin, V. (ed.) The Interplay between Differential Geometry and Differential Equations. Amer. Math. Soc. Transl., vol. 167, pp. 143–164. Amer. Math. Soc., Providence (1995)

9. Pommaret, J.: Systems of Partial Differential Equations and Lie Pseudogroups. Gordon & Breach, London (1978)

10. Rabier, P.: Implicit differential equations near a singular point. J. Math. Anal. Appl. 144, 425–449 (1989)

11. Rabier, P., Rheinboldt, W.: On impasse points of quasilinear differential algebraic equations. J. Math. Anal. Appl. 181, 429–454 (1994)

12. Rabier, P., Rheinboldt, W.: Theoretical and numerical analysis of differential-algebraic equations. In: Ciarlet, P., Lions, J. (eds.) Handbook of Numerical Analysis, vol. VIII, pp. 183–540. North-Holland, Amsterdam (2002)

13. Saunders, D.: The Geometry of Jet Bundles. London Mathematical Society Lecture Notes Series, vol. 142. Cambridge University Press, Cambridge (1989)

14. Seiler, W.: Involution — The Formal Theory of Differential Equations and its Applications in Computer Algebra. Algorithms and Computation in Mathematics, vol. 24. Springer, Berlin (2010)

15. Tuomela, J.: On singular points of quasilinear differential and differential-algebraic equations. BIT 37, 968–977 (1997)

16. Tuomela, J.: On the resolution of singularities of ordinary differential equations. Num. Algo. 19, 247–259 (1998)

17. Vessiot, E.: Sur une théorie nouvelle des problèmes généraux d'intégration. Bull. Soc. Math. Fr. 52, 336–395 (1924)

A Quantum Measurements Model of Hydrogen-Like Atoms in Maple*

L. Sevastianov, A. Zorin, and A. Gorbachev

Joint Institute For Nuclear Research
Peoples' Friendship University of Russia
sevast@sci.pfu.edu.ru, zorin@mx.rudn.ru, alexarus1986@gmail.com

Abstract. Interaction effects between a quantum system and the quantum part of a measurement system leads to the complexification of the initial system's observable operator. Computer algebra methods of Maple help to express operators of the hydrogen-like atom observables in explicit form. We use these operators to solve a spectrum estimation problem using a procedure for computation of the Ritz matrix in Maple.

Keywords: Observable operator of quantum system, Quantum measurement model, Ritz matrix explicit form in Maple, General eigenvalue problem in Maple.

1 Introduction

Description of a quantum-mechanical object usually consists of either theoretical research of a mathematical model or the analysis of experimental observations. Combining these two parts requires specification of the quantum measurement model. Among all quantum measurement models, there are two most popular: the operational model and the quantum estimation model.

Both models can be treated in terms of Hilbert state-space representation and phase-space representation. In his work Wódkiewicz K. [1] suggested phase-space representation of the quantum measurement's operational model. This model, though unfinished by the author, is equal to the model of quantum mechanics with a non-negative quantum distribution function suggested by Kuryshkin V. and well described in detail [2–16].

Ozawa M. [17–20] developed the ideas of Wigner–Araki–Yanaze [21] operational approach to the quantum measurements problem and made them mathematically rigor. In his recent works [22–25], Ozawa proved the equivalence of the operational model of quantum measurements and the quantum estimation model. That makes the Kuryshkin–Wódkiewicz constructive model of quantum measurements the universal model of quantum measurements.

However, explicit form of the observables' operators $O_\rho(A)$ is very complex, even for a simple quantum object. In this work, we introduce dedicated package **QDF** for the computer algebra system Maple. With the help of this package,

---

* The work is partially supported by RFBR grants No. 11-01-00278, 13-01-00595.

V.P. Gerdt et al. (Eds.): CASC 2013, LNCS 8136, pp. 369–380, 2013.
© Springer International Publishing Switzerland 2013

we calculate explicit form of the observables' operators, construct a Ritz matrix, and describe spectral characteristics of hydrogen-like atoms and atoms of alkaline metals in terms of the quantum Kepler problem.

2   Analytical Calculations of the $O_{\{\varphi k\}}(A)$ Operators in Explicit Form

We apply the operational model of quantum measurements to the Kepler problem with Hamilton function $H(q, p) = \frac{p^2}{2} - \frac{2}{|q|}$. Weyl operator of the observable H for the isolated quantum system is $\hat{H} \equiv O_W(H) = -\frac{\hbar^2}{2}\Delta - \frac{2}{|q|}$.

When the process of measurement is applied to the system with a quantum filter in a state $\hat{\rho}_2 = \sum C_k |\psi_k\rangle \langle \psi_k|$, the observable of the measured value $O_{\rho 2}(H)$ can be described by the following expression

$$[O_W(H * W_{\rho 2})u](q) = \int (H * W_{\rho 2})\left(\frac{q + \tilde{q}}{2}, p\right)\exp\left\{\frac{i}{\hbar}p(q - \tilde{q})\right\}u(\tilde{q})\,d\tilde{q} \quad (1)$$

where $W_{\rho 2}(q, p) = \sum C_k W_{\psi k}(q, p)$, and

$$W_{\psi k}(q, p) = \frac{1}{\pi\hbar}\int \bar{\psi}_k(q + \eta)\psi_k(q - \eta)\exp\left\{-\frac{i}{\hbar}2p\eta\right\}d\eta$$

To study measured values of the hydrogen observables (and all alkalines with one outer-shell electron) we use Sturm functions $\Psi_{nlm}(r, \theta, \varphi) = S_{nl}(r)Y_{lm}(\theta, \varphi)$ of the hydrogen atom as basis functions $\{\psi_k(r)\}_{k=1}^\infty$. Here

$$R_{nl}(r) = p(\kappa r)\exp(-\kappa r)$$

and κ is a constant, $p(\kappa r)$ is a κr polynomial (see [12, 28, 29]).

Sturmian functions in Maple

```
Sturm:=proc(np,lp,b) local kk,alpha_nl,N_nl,altLaguerreL; kk:=1/b:
alpha_nl:=kk*np: N_nl:=-sqrt((np-lp-1)!/((np+lp)!)^3/2):
### An alternative notation of the Laguerre polynomial
### instead of the internal(LaguerreL) is used:
altLaguerreL := (n,a,x) -> (-1)^a*n!*LaguerreL(n-a,a,x):
N_nl*exp(-kk*r)*(2*kk*r)^(lp+1)
*simplify(altLaguerreL(np+lp,2*lp+1,2*kk*r)): end proc;
```

The full energy operator $O(H)$ equals to the sum of kinetic energy $O(p^2/2\mu)$ and potential energy $O(-Ze^2/|r|)$ operators of the Coulomb field.

The function $p^2/2\mu$ of the kinetic energy operator does not depend on the coordinates. Therefore, we can reduce the ninefold-iterated integral to the threefold:

$$O\left(\frac{p^2}{2\mu}\right) = \int \frac{(\eta - i\hbar\nabla)^2}{2\mu}\beta_0(\eta)\,d\eta = \sum_j C_j O_j\left(\frac{p^2}{2\mu}\right),$$

where $O_j \left(\frac{p^2}{2\mu} \right) = \int \frac{(\eta - i\hbar \nabla)^2}{2\mu} \left| \tilde{\psi}_j (\eta) \right|^2 d^3\eta$, $\tilde{\psi}_j (\eta) = \frac{1}{(2\pi\hbar)^{3/2}} \int \psi_j (r) \, e^{\frac{i}{\hbar}(p,r)} d^3 r$.

Similar to the kinetic energy operator, the function $-Ze^2 / |r|$ of the potential energy operator does not depend on the momentum:

$$O \left(-\frac{Ze^2}{r} \right) = \int -\frac{Ze^2}{|r + \xi|} \alpha_0 (\xi) \, d\xi = \sum_j C_j O_j \left(-\frac{Ze^2}{r} \right),$$

where $O_j \left(-\frac{Ze^2}{r} \right) = -Ze^2 \int \frac{|\psi_j(\xi)|^2}{|r+\xi|} d^3\xi$.

For evaluation of the potential energy operators $O_j \left(-Ze^2 / |r| \right)$ we use series expansion:

$$\frac{1}{|\vec{r_1} + \vec{r_2}|} = 4\pi \sum_{l=0}^{\infty} \frac{1}{2l+1} \frac{r_-^l}{r_+^{l+1}} \sum_{m=-l}^{l} Y_{lm} (\theta_1, \phi_1) \times Y_{lm}^* (\theta_2, \phi_2) \,,$$

where $r_- = \min (r_1, r_2) \,, \; r_+ = \max (r_1, r_2)$ and

$$\frac{1}{|\vec{r_1} - \vec{r_2}|} = 4\pi \sum_{l=0}^{\infty} \frac{1}{2l+1} \frac{r_-^l}{r_+^{l+1}} \sum_{m=-l}^{l} Y_{lm} (\theta_1, \phi_1) \times Y_{lm}^* (\theta_2, \pi + \phi_2)$$

This leads to the following equation

$$O_{\bar{n} \, \bar{l} \, \bar{m}} \left(-\frac{Ze^2}{|\vec{r}|} \right) = -4\pi \, Z \, e^2 \sum_{l=0}^{\infty} \frac{1}{2l+1} \int \frac{r_-^l}{r_+^{l+1}} R_{\bar{n} \, \bar{l}}^2 (r_2) \sum_{m=-l}^{l} Y_{lm} (\theta_1, \phi_1) \times$$

$$\int Y_{lm}^* (\theta_2, \pi + \phi_2) \left| Y_{\bar{l} \, \bar{m}} (\theta_2, \phi_2) \right|^2 r_2^2 \sin \theta_2 \, dr_2 \, d\theta_2 \, d\phi_2 \quad (2)$$

Products of spherical harmonics can be transformed using the formula

$$Y_{l_1, m_1} (\theta, \phi) \, Y_{l_2, m_2} (\theta, \phi) = \sum_{L,M} \sqrt{\frac{(2l_1 + 1)(2l_2 + 1)}{4\pi (2L + 1)}} C_{l_1 0 l_2 0}^{L0} C_{l_1 m_1 l_2 m_2}^{LM} Y_{LM} (\theta, \phi).$$

To use the orthonormality property of spherical harmonics and extract the finite number of components from the infinite sum (2) we expand the absolute value squared of spherical harmonic $\left| Y_{\bar{l} \, \bar{m}} (\theta_2, \phi_2) \right|^2$ into the following linear combinations of spherical harmonics:

$$|\Upsilon_{00}|^2 = \Upsilon_{00}/\sqrt{4\pi} \,,$$

$$|\Upsilon_{10}|^2 = \Upsilon_{00}/\sqrt{4\pi} + \Upsilon_{20}/\sqrt{5\pi} \,,$$

$$|\Upsilon_{11}|^2 = |\Upsilon_{1-1}|^2 = \Upsilon_{00}/\sqrt{4\pi} - \Upsilon_{20}/\sqrt{20\pi} \,.$$

There is a finite number of terms in the expansion of $\left| Y_{\bar{l} \, \bar{m}} (\theta_2, \phi_2) \right|^2$. Thus, there is a finite number of non-zero terms in the infinite sum (2). Due to orthogonality of spherical harmonics, non-zero terms after the integration of angular variables are given only by l, m combinations. This means that we only need to calculate integrals over r_2. These integrals split in two parts:

$$\int_0^\infty \frac{r_-^l}{r_+^{l+1}} S_{\tilde{n}\,\tilde{l}}^2\,(r_2/_b)\ r_2^2\ dr_2 = \int_0^{|r_1|} \frac{r_2^l}{r_1^{l+1}} S_{\tilde{n}\,\tilde{l}}^2\,(r_2/_b)\ r_2^2\ dr_2 + \int_{|r_1|}^\infty \frac{r_1^l}{r_2^{l+1}} S_{\tilde{n}\,\tilde{l}}^2\,(r_2/_b)\ r_2^2\ dr_2 .$$

Here we present Maple-assisted analytical calculations of the first five potential and kinetic energy operators for the hydrogen-like atom:

$$O_1(-\frac{Ze^2}{r}) = -\frac{Ze^2}{r} + Ze^2 \exp\left\{-\frac{2r}{b_1 r_0}\right\}\left(\frac{1}{r_0} + \frac{1}{br_0}\right),$$

$$O_2(-\frac{Ze^2}{r}) = -\frac{Ze^2}{r} + \frac{Ze^2}{b_2 r_0}\exp\left\{-\frac{r}{b_2 r_0}\right\}\left[\frac{3}{4} + \frac{b_2 r_0}{r} + \frac{1}{4}\frac{r}{b_2 r_0} + \frac{1}{8}\left(\frac{r}{b_2 r_0}\right)^2\right],$$

$$O_3(-\frac{Ze^2}{r}) = -\frac{Ze^2}{r} + \frac{Ze^2}{b_3 r_0}\exp\left\{-\frac{r}{br_0}\right\}\left[\frac{b_3 r_0}{r} + \frac{3}{4} + \frac{1}{4}\frac{r}{b_3 r_0} + \frac{1}{24}\left(\frac{r}{b_3 r_0}\right)^2\right] -$$
$$-\frac{Ze^2}{b_3 r_0}\left(3\cos^2\theta - 1\right)\exp\left\{-\frac{r}{b_3 r_0}\right\} \times$$
$$\times \left[\frac{1}{24}\left(\frac{r}{b_3 r_0}\right)^2 + \frac{1}{4}\frac{r}{b_3 r_0} + 6\left(\frac{b_3 r_0}{r}\right)^3 - \exp\left\{\frac{r}{b_3 r_0}\right\}6\left(\frac{b_3 r_0}{r}\right)^3\right],$$

$$O_4(-\frac{Ze^2}{r}) = -\frac{Ze^2}{r} + \frac{Ze^2}{b_4 r_0}\exp\left\{-\frac{r}{b_4 r_0}\right\}\left[\frac{b_4 r_0}{r} + \frac{3}{4} + \frac{1}{4}\frac{r}{b_4 r_0} + \frac{1}{24}\left(\frac{r}{b_4 r_0}\right)^2\right] -$$
$$-\frac{Ze^2}{b_4 r_0}\left(3\cos^2\theta - 1\right)\exp\left\{-\frac{r}{b_4 r_0}\right\} \times$$
$$\times \left[\frac{1}{48}\left(\frac{r}{b_4 r_0}\right)^2 + \frac{1}{8}\frac{r}{b_4 r_0} + \frac{1}{2} + \frac{3}{2}\frac{b_4 r_0}{r} + 3\left(\frac{b_4 r_0}{r}\right)^2 + 3\left(\frac{b_4 r_0}{r}\right)^3\right] -$$
$$+3\frac{Ze^2}{b_4 r_0}\left(3\cos^2\theta - 1\right)\left(\frac{b_4 r_0}{r}\right)^3,$$

$$O_5(-\frac{Ze^2}{r}) = -\frac{Ze^2}{r} + \frac{Ze^2}{b_5 r_0}\exp\left\{-\frac{r}{b_5 r_0}\right\}\left[\frac{b_5 r_0}{r} + \frac{3}{4} + \frac{1}{4}\frac{r}{b_5 r_0} + \frac{1}{24}\left(\frac{r}{b_5 r_0}\right)^2\right] -$$
$$-\frac{Ze^2}{b_5 r_0}\left(3\cos^2\theta - 1\right)\exp\left\{-\frac{r}{b_5 r_0}\right\} \times$$
$$\times \left[\frac{1}{48}\left(\frac{r}{b_5 r_0}\right)^2 + \frac{1}{8}\frac{r}{b_5 r_0} + \frac{1}{2} + \frac{3}{2}\frac{b_5 r_0}{r} + 3\left(\frac{b_5 r_0}{r}\right)^2 + 3\left(\frac{b_5 r_0}{r}\right)^3\right] +$$
$$+3\frac{Ze^2}{b_5 r_0}\left(3\cos^2\theta - 1\right)\left(\frac{b_5 r_0}{r}\right)^3,$$

$$O_i\left(\frac{\boldsymbol{p}^2}{2\mu}\right) = -\frac{\hbar^2}{2m}\Delta + \frac{\hbar^2}{2mb_i^2}, \quad i = 1, ..., 5.$$

Procedure for calculations of the potential energy operators

```
OH:=proc(np,lp,mp,b)
local l,m,sumout,sumin,intr;
sumout:=0;
for l from 0 to 2*lp do
  sumin:=0;
intr:=int(r2^l/r^(l+1)*subs(r=r2,R(np,lp,b))^2*r2^2,r2=0..r)
+int(r^l/r2^(l+1)*subs(r=r2,R(np,lp,b))^2*r2^2,r2=r..infinity);
  for m from -l to l do
    sumin:=sumin+Y(l,m)*int(int(conjugate((-1)^l*Y(l,m))
    *YY(lp,mp,lp,mp,true)*sin(theta),theta=0..Pi),phi=0..2*Pi);
  od;
  sumout:=sumout+1/(2*l+1)*sumin*intr;
od;
expand(-4*Pi*Z*e^2*simplify(sumout));
end proc;
```

Calculations of the Kinetic Energy Operators

```
for i from 1 by 1 to matrsize
do
  for j from 1 by 1 to matrsize
  do
OHM_cin_sturm[i,j]:=simplify(int(int(int(conf[i]*(diff(r^2*_
diff(f[j],r),r)/r^2+diff(sin(theta)*diff(f[j],theta),theta)/_
r^2/sin(theta)+diff(f[j],phi$2)/r^2/sin(theta)^2)*r^2*sin(theta)_
,phi=0..2*Pi),theta=0..Pi),r=0..infinity ));
  od;
od;
```

If we use the same parameter b for both terms $O_4(-Ze^2/r)$ and $O_5(-Ze^2/r)$ of the potential energy operator, these terms will be equal to each other. Finally we get explicit form of the Hamiltonian, which is due to linearity

$$O\left(H\right) = \sum_{j=1}^{5} C_j \left(O_j \left(\frac{p^2}{2\mu} \right) + O_j(-\frac{Ze^2}{r}) \right),$$

where $\sum_{j=1}^{5} C_j = 1$.

3   Asymptotics and Spectral Properties of the $O_{\{\varphi k\}}(A)$ Operators

To make the discussion on properties of the operators presented above more reasonable we consider asymptotics of the potential energy partial multipole contributions:

$$V_1\left(r \to 0, \theta, \varphi\right) = -\frac{Z}{b_1} + \frac{2Z}{3b_1^3}r^2 - \frac{2Z}{3b_1^4}r^3 + \frac{2Z}{5b_1^5}r^4 + O\left(r^5\right),$$

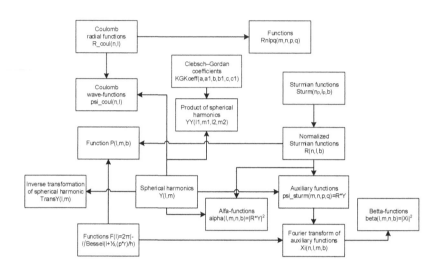

Fig. 1. "SourceFunctions" library

$$V_2\left(r \to 0, \theta, \varphi\right) = -\frac{Z}{4b_2} + \frac{Z}{12b_2^3}r^2 - \frac{Z}{12b_2^4}r^3 + \frac{7Z}{16b_2^5}r^4 + O\left(r^5\right),$$

$$V_3\left(r \to 0, \theta, \varphi\right) = -\frac{Z}{4b_3} - \frac{Z\left(3\cos^2\theta - 1\right)}{120b_3^3}r^2 + \frac{Z\left(10\cos^2\theta - 1\right)}{1120b_3^5}r^4 + O\left(r^5\right),$$

$$V_4\left(r \to 0, \theta, \varphi\right) = -\frac{Z}{4b_4} + \frac{Z\left(3\cos^2\theta - 1\right)}{120b_4^3}r^2 - \frac{Z\left(5\cos^2\theta - 1\right)}{1120b_4^5}r^4 + O\left(r^5\right),$$

and

$$V_1\left(r \to \infty, \theta, \varphi\right) = -\frac{Z}{r} + O\left(\exp\left(-\frac{2r}{b_1}\right)\right),$$

$$V_2\left(r \to \infty, \theta, \varphi\right) = -\frac{Z}{r} + O\left(\exp\left(-\frac{r}{b_2}\right)\right),$$

$$V_3\left(r \to \infty, \theta, \varphi\right) = -\frac{Z}{r} - \frac{6Zb_3^2\left(3\cos^2\theta - 1\right)}{r^3} + O\left(\exp\left(-\frac{r}{b_3}\right)\right),$$

$$V_4\left(r \to \infty, \theta, \varphi\right) = -\frac{Z}{r} - \frac{3Zb_4^2\left(3\cos^2\theta - 1\right)}{r^3} + O\left(\exp\left(-\frac{r}{b_4}\right)\right).$$

The following properties of pseudodifferential operators $O_{\rho 2}\left(H\right)$ were proved for the operators (1) in a number of papers.

Proposition 1. *Operators $O_{\rho 2}(H)$ are bounded from below and essentially self-adjoint operators.*

Proposition 2. *Operators $O_{\rho 2}(H)$ are $\hat{H}$-compact and $C(\hat{\rho}_2)$ are bounded at infinity.*

Proposition 3. *Sturmian functions ψ_k of $\hat{H}$ operator form a complete function system, which is almost orthogonal with respect to the operator $O_{\rho 2}(H)$.*

Proposition 4. *Operator $O_{\rho 2}(H)$ spectrum consists of discrete and essentially continuous parts. The discrete part of the spectrum belongs to the interval $[C(\hat{\rho}_2) - 1, C(\hat{\rho}_2)]$, where $C(\hat{\rho}_2)$ is a constant that depends on the state of the filter $\hat{\rho}_2$, and lies below continuous part of the spectrum. Therefore, minimax principle can be used to find eigenvalues of the operator $O_{\rho 2}(H)$.*

Theorem 1.   *1. Ritz matrix $M_{kl}^N(H, \hat{\rho}_2) = \langle \psi_k, O_{\rho 2}(H)\psi_l \rangle$, $k, l = 1, ..., N$ of the operator $O_{\rho 2}(H)$ with Sturmian functions of the operator $\hat{H}$ as coordinate functions is well conditioned for any dimension N (the number of coordinate functions $\{\psi_k,\ k = 1, ..., N\}$).*
 2. Eigenvalues of the Ritz matrix $M_{kl}^N(H, \hat{\rho}_2)$ form the monotonic sequence λ_j^N converging to the eigenvalues λ_j^∞ of the $O_{\rho 2}(H)$ operator.
 3. Eigenvectors of the Ritz matrix $M_{kl}^N(H, \hat{\rho}_2)$ form the minimizing sequence ψ_j^N, $j = 1, ..., N$ of the Ritz functional for the Rayleigh relationship in the minimax principle and converge to the eigenvectors ψ_j^∞ of the $O_{\rho 2}(H)$ operator.

Eigenvalues' estimations of $O(H)$ were obtained in several works [9, 30, 31]. These estimations can be specified more precisely by applying a specific numerical method to the equation $O(H)\psi = \lambda\psi$. The discrete spectrum of the operator $O(H)$ lies beneath its continuous spectrum [9, 31]. A minimax Rayleigh method, applied to the operator bounded below in essentiality [32], first gives us the eigenvalues of the operator and only after that reaches the lower bound of the continuous spectrum. Numerical implementation of the Rayleigh method is possible only in the case of finite-dimensional approximation of the initial operator $O(H)$ using the Ritz matrix. To construct the Ritz matrix we chose Sturmian functions with the energy parameter E [27] to be coordinate functions as they form an orthonormal basis in the representation space of the operator $O(H)$.

The $N \times N$ Ritz matrix $M_N[c_1, \ldots, c_n; b_1, \ldots, b_n; E](H)$ for the operator $O(H)$ with coordinate functions $S_{nl}(r; E)Y_{lm}(\theta, \varphi)$ was calculated in Maple and depends on the parameter $E \in (-\infty, 0)$, $2n$ parameters $(c_1, \ldots, c_n; b_1, \ldots, b_n)$ and auxiliary functions $\{\varphi_1, \ldots, \varphi_n\}$. The matrix $M_N(H)$ is a hermitian strongly sparse almost orthogonal matrix [11, 33] for all values of the operator. Spectra $\lambda_1^N, \ldots, \lambda_N^N$ of the matrix can be calculated with different algorithms. We chose the Jacobi method as it works fine with complex matrices. Obtained eigenvalues λ_j depend on the parameters $(c_1, \ldots, c_n; b_1, \ldots, b_n; E)$.

4   Constructing Ritz Matrix in Maple

The Hamiltonian $O\left(H\right)$ of the hydrogen-like atom is defined in terms of quantum mechanics with a non-negative quantum distribution function by the following equations

$$D(O(H)) = D(\hat{H}),$$

$$O_{\{\phi_k\}}(H) = \hat{H} + C_{\{\phi_k\}}\hat{I} + V_{\{\phi_k\}}(\boldsymbol{r})$$

for the set of the acceptable functions $\{\phi_k \in L_2(R^3)\}$

Operator $O\left(H\right)$ is self-adjoint and bounded below with a constant $(C-1)$. Value of C depends on a selection of the auxiliary functions.

To calculate the Ritz matrix $O_k^M(H)$ in Maple analytically, where M is the number of dimensions corresponding to the operators $O_k(H)$, first M Sturmian functions $\psi_{nlm}^{E_0}(\boldsymbol{r}) = \tilde{S}_{nl}(k\boldsymbol{r})Y_{lm}(\theta,\phi)$, $k = \sqrt{-E_0}$ were used as basis functions.

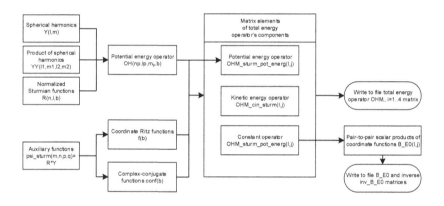

Fig. 2. Matrix coefficients computations

We need to solve a generalized eigenvalue problem $M\boldsymbol{x} = \lambda B\boldsymbol{x}$ because of the non-orthogonal character of the described system. Here M is a Ritz matrix of $O^M(H)$ and B is an inner product of coordinate functions matrix. The Ritz matrix elements are defined by the formula

$$M_{kl}^{(j)} = \int \psi_k^{E_0}(\boldsymbol{r})\left[O_j\left(\frac{\boldsymbol{p}^2}{2\mu}\right) + O_j(-\frac{Z_{eff}e^2}{r})\right]\psi_l^{E_0}(\boldsymbol{r})\,d\boldsymbol{r}$$

Each of the matrices OHM_1, OHM_2, OHM_3, and OHM_4 is calculated using only one of the auxiliary functions. Let us compose a linear combination of these matrices with the arbitrary coefficients $a1$, $a2$, $a3$, and $a4$. The latter, however, are not completely independent since their sum must equal to 1. That's why we can assume $a4 = 1 - a1 - a2 - a3$. Besides, we can substitute different values $b1, b2, b3, b4$ in these four matrices to vary each parameter b independently.

Procedure for calculation of the Hamilton operators spectra

```
spectre:=proc(E0q,b1qq,b2qq,b3qq,b4qq, a1qq,a2qq,a3qq, Z1q, meq,
matrsize)
local Ritz,MM,eigen_vls,L1,Nrgy_spctr;
MM:=matrix(matrsize,matrsize): Ritz:=matrix(matrsize,matrsize):
Ritz:=subs({a1q=a1qq,a2q=a2qq,a3q=a3qq,E0=E0q,Z=Z1q,
me=meq,b1q=b1qq,b2q=b2qq,b3q=b3qq,b4q=b4qq},evalm(matr_E0_)):
MM:=evalm(inv_B_E0&*Ritz): eigen_vls:=eigenvals(MM): L1:=
[seq(eigen_vls[i],i=1..matrsize)]: Nrgy_spctr:=sort(L1); end proc;
```

As a result, the elements of the calculated Ritz matrix depend on $E0$, Z, me, $a1$, $a2$, $a3$, $b1$, $b2$, $b3$, $b4$ (the latter are denoted by $a1q$, $a2q$, $a3q$, $b1q$, $b2q$, $b3q$, $b4q$):

According to the Ritz method, eigenvalues of the Ritz matrix are the spectral values of the observable, i.e., energy.

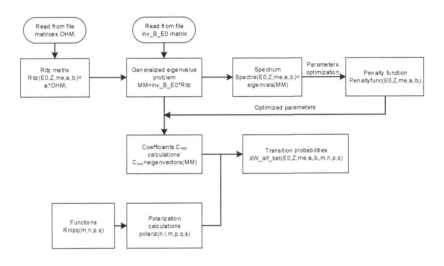

Fig. 3. Transition probabilities computations

5 Conclusion

Spectral properties of the observables' expected values, i.e., operators of the observables obtained by the Weyl (Weyl–Berezin [29, 30]) quantization are well studied for a wide class of quantum systems. Models verification of the studied systems are usually based on the comparison of the calculated values and their characteristics to the corresponding experimental values.

During last several decades, many researches experimentally measured and discussed dependency between the position and characteristics of quantum analyzers and detectors. This dependency decreases the quality of verification process, because theoretically calculated observables in terms of Born postulate do not depend on the properties and characteristics of the measurement instruments.

Generalization of Born postulate in terms of quantum measurement theory solves this contradiction [26]. However, observable operators emerging in this theory seem to be rather complicated and, thus, are studied much less than traditional Weyl operators.

In previous works, we proved the possibility of describing observable operators in terms of Weyl–Kuryshkin quantization rule. Explicit form of these operators is given by the constructive model of Kuryshkin–Wódkiewicz. Model studying bases on the symbolic computations of the Weyl–Kuryshkin operators in explicit form. First part of the paper describes these computations.

Explicit form of the measured observables (Weyl–Kuryshkin operators) allows us to perform comparative analysis with corresponding expected observable values (Weyl–Berezin operators). The analysis leads to the computational investigation of the discrete spectrum of the Weyl–Kuryshkin operators for the hydrogen-like atoms. Main instrument in this research is a Ritz matrix of the corresponding operator in the basis of Sturmian functions for hydrogen atom. Symbolic computation of the Ritz matrix has been described in the second part of this paper.

All symbolic computations were performed in Maple computer program. Dedicated package QDF has been developed to define the Kuryshkin–Wódkiewicz quantization rule, auxiliary and Sturmian functions, products of spherical harmonics, and other necessary functions. However, most of the functions necessary for the study already exists in Maple, makes it easy to construct complex-conjugate functions, solve the generalized eigenvalue problem and scale the Ritz matrix dimension to optimize calculation times.

Carried out symbolic computations open up the opportunities for the further numeric investigations of the observable spectrum. Current numeric studies prove consistency of the Kuryshkin–Wódkiewicz model.

References

1. Wódkiewicz, K.: Operational approach to phase-space measurements in quantum mechanics. Phys. Rev. Lett. 52, 1064 (1984)
2. Kuryshkin, V.V.: On the construction of quantum operators. Izv. VUZov. Phys. 11, 102–106 (1971)
3. Kuryshkin, V.V.: La mécanique quantique avec une fonction nonnegative de distribution dans l'espace des phases. Annales Inst. Henri Poincaré 17, 81–95 (1972)
4. Kuryshkin, V.V.: Une généralisation possible de la mécanique quantique non relativiste. Compt. Rend. Acad. Sc. Paris 274 Série B, 1107–1110 (1972)
5. Kuryshkin, V.V.: L'ossilateur harmonique à une dimension dans la mecanique quantique a fonction de distribution non negative dans l'espace des phases. Compt. Rend. Acad. Sc. Paris 274 Série B, 1163–1165 (1972)
6. Kuryshkin, V.V.: Some problems of quantum mechanics possessing a non-negative phase-space distribution function. Int. J. Theor. Phys. 7, 451–466 (1973)

7. Zorin, A.V., Kuryshkin, V.V., Sevastyanov, L.A.: Description of the spectrum of a hydrogen-like atom. Bull. PFUR. Ser. Phys. 6(1), 62–66 (1998)
8. Zorin, A.V.: Approximate determination of states in quantum mechanics of Kuryshkin. Bull. PFUR, Ser. Physics 12, 81–87 (2004)
9. Zorin, A.V.: The method of study of essential and discrete spectra of the Hamiltonian in quantum mechanics of Kuryshkin. Bull. PFUR, Ser. Appl. and Comp. Math. 3(1), 121–131 (2004)
10. Zorin, A.V., Sevastianov, L.A., Belomestny, G.A., Laveev, E.B.: Quantum systems' modeling by methods of computer algebra. In: Proc. CASC 2004, pp. 497–506. TUM Publ., Munich (2004)
11. Zorin, A.V., Sevastianov, L.A., Belomestny, G.A.: Numerical search for the states with minimal dispersion in quantum mechanics with non–negative quantum distribution function. In: Li, Z., Vulkov, L.G., Waśniewski, J. (eds.) NAA 2004. LNCS, vol. 3401, pp. 613–620. Springer, Heidelberg (2005)
12. Zorin, A.V., Sevastianov, L.A.: Hydrogen-like atom with nonnegative quantum distribution function. Nuclear Phys. 70, 792–799 (2007)
13. Zorin, A.V., Sevastianov, L.A., Tretyakov, N.P.: Computer modeling of hydrogen-like atoms in quantum mechanics with nonnegative distribution function. Programming and Computer Software 33(2), 94–104 (2007)
14. Gorbachev, A.V.: Modeling of Alkaline Metal Spectra using Quantum Mechanics with Nonnegative Quantum Distribution Function. Master's thesis. PFUR (2010) (in Russian)
15. Zorin, A.V., Sevastianov, L.A.: Kuryshkin-Wódkiewicz quantum measurements model. Bull. PFUR. Ser. Math. Inform. Phys. (3), 99–104 (2010)
16. Sevastyanov, L., Zorin, A., Gorbachev, A.: Pseudo-differential operators in an operational model of the quantum measurement of observables. In: Adam, G., Buša, J., Hnatič, M. (eds.) MMCP 2011. LNCS, vol. 7125, pp. 174–181. Springer, Heidelberg (2012)
17. Ozawa, M.: Measurements of nondegenerate discrete observables. Phys. Rev. A 62, 062101(1–13) (2000)
18. Ozawa, M.: Operations, disturbance, and simultaneous measurability. Phys. Rev. A 63, 032109(1–15) (2001)
19. Ozawa, M.: Conservation laws, uncertainty relations, and quantum limits of measurements. Phys. Rev. Lett. 88, 050402(1–4) (2002)
20. Kimura, G., Meister, B.K., Ozawa, M.: Quantum limits of measurements induced by multiplicative conservation laws: Extension of the Wigner–Araki–Yanase theorem. Phys. Rev. A 78, 032106 (2008)
21. Araki, H., Yanase, M.M.: Measurement of quantum mechanical operators. Phys. Rev. 120, 622–626 (1960)
22. Ozawa, M.: Quantum reality and measurement: A quantum logical approach. Foundations of Physics 41, 592–607 (2011), doi:10.1007/s10701-010-9462-y
23. Ozawa, M.: Simultaneous measurability of non-commuting observables and the universal uncertainty principle. In: Proc. 8th Int. Conf. on Quantum Communication, Measurement and Computing, pp. 363–368. NICT Press, Tokyo (2007); Ozawa, M., Kitajima, Y.: Reconstructing Bohr's reply to EPR in algebraic quantum theory. Foundations of Physics 42(4), 475–487 (2012), doi:10.1007/s10701-011-9615-7
24. Ozawa, M.: Mathematical foundations of quantum information: Measurement and foundations. Sugaku 61-2, 113–132 (2009) (in Japanese)
25. Mehta, C.L.: Phase-space formulation of the dynamics of canonical variables. J. Math. Phys. 5(5), 677–686 (1964)

26. Zorin, A.V., Sevastianov, L.A.: Mathematical modeling of quantum mechanics with non-negative QDF. Bull. PFUR. Ser. Phys. 11(2), 81–87 (2003)
27. Rotenberg, M.: Theory and applications of Sturmian functions. In: Bates, D.R., Esterman, I. (eds.) Adv. in Atomic and Molec. Phys., vol. 6, pp. 233–268. Academic Press, New York (1970)
28. Avery, J.: Generalised Sturmians and Atomic Spectra. World Scientific, Singapore (2006)
29. Weyl, H.: Quantenmechanik und Gruppentheorie. Zeitschrift für Physik 46, 1–46 (1927)
30. Sevastianov, L.A., Zorin, A.V.: The method of lower bounds for the eigenvalues of the Hamiltonian differential operator in quantum mechanics of Kuryshkin. Bull. PFUR, Ser. Appl. and Comp. Math. 1(1), 134–144 (2002)
31. Zorin, A.V., Sevastianov, L.A.: Spectral properties of the Hamilton operator in quantum mechanics with non-negative QDF. In: Proc. Second Int. Conf. on Func. Anal. and Diff. Op., pp. 169–170. Fizmatlit-Publ., Moscow (2003)
32. Zorin, A.V., Sevastianov, L.A., Belomestny, G.A.: Analytical calculation of observables' matrices of Hydrogen-like atom in Kuryshkin's Quantum Mechanics. Bull. PFUR, Ser. Appl. and Comp. Math. 3(1), 106–120 (2004)
33. Reed, M., Simon, B.: Methods of Modern Mathematical Physics. Analysis of Operators, vol. IV. Academic Press, New York (1977)

CAS Application to the Construction of the Collocations and Least Residuals Method for the Solution of 3D Navier–Stokes Equations

Vasily P. Shapeev and Evgenii V. Vorozhtsov

Khristianovich Institute of Theoretical and Applied Mechanics,
Russian Academy of Sciences, Novosibirsk 630090, Russia
{shapeev,vorozh}@itam.nsc.ru

Abstract. In the present work, the computer algebra system (CAS) is applied for constructing a new version of the method of collocations and least residuals (CLR) for solving the 3D incompressible Navier–Stokes equations. The CAS is employed at all stages from writing, deriving, and verifying the formulas of the method to their translation into arithmetic operators of the Fortran language. The verification of derived formulas of the method has been done on the test problem solution. Comparisons with the published numerical results of solving the benchmark problem of the 3D driven cavity flow show a high accuracy of the developed method at the level of the most accurate known solutions.

1 Introduction

In the present work, the tools of computer algebra systems are applied for constructing a new version of the method of collocations and least residuals (CLR) for solving the 3D Navier–Stokes (NS) equations. In this version of the method, the solution is sought in the form of the second-degree polynomials in three independent variables. There are here three substantial circumstances. First, the basis of solenoidal polynomials used for solution representation contains 30 elements. Second, the system of the 3D NS equations is in itself a complex nonlinear analytic object. And, third, it is required in the CLR method for improving the properties of the collocation technique that the minimum of the solution residuals be reached on the problem solution.

These circumstances cause the necessity of using the CASs. And the fact that symbolic manipulations with fractional rational functions have been implemented in the leading CASs sufficiently well makes their application for the CLR method development efficient.

In the present work, the CAS is at first applied at the laborious stage of the derivation of the formulas of the method for solving the 3D NS equations with the use of useful functions of CAS *Mathematica*.

The verification of obtained formulas is carried out with the aid of test computations directly in the CAS to identify the algorithmic and other errors.

The CAS is further used for optimizing the analytic solution formulas in terms of arithmetic operations with regard for the fact that they have polynomial form.

V.P. Gerdt et al. (Eds.): CASC 2013, LNCS 8136, pp. 381–392, 2013.
© Springer International Publishing Switzerland 2013

The CAS is applied at the laborious stage of translating the analytic solution formulas, matrix elements, and other expressions in the form of Fortran arithmetic operators. An intermediate result of the Fortran code work is the solution in the form of a polynomial in three variables. It can be differentiated and integrated exactly without using the numerical procedures, which introduce additional errors. Solution recalculation at a passage from one grid to another at the application of the multigrid versions of the CLR method is implemented by a simple change of coordinates also without introducing the extra errors in the constructed solution unlike the smoothing and interpolation procedures, which are applied for these purposes in other methods. The requirement of the error minimization on the solution contributes in itself to the damping of various disturbances arising in the process of the numerical solution of problems. All the needed formulas are obtained at first in CAS and are programmed in Fortran. So that the Fortran code then outputs all final results in the automatic regime.

2   Description of the Modified CLR Method

2.1   Problem Statement

Consider the system of stationary NS equations

$$(\mathbf{V} \cdot \nabla)\mathbf{V} + \nabla p = (1/\mathrm{Re})\varDelta\mathbf{V} - \mathbf{f}, \quad \mathrm{div}\,\mathbf{V} = 0, \quad (x_1, x_2, x_3) \in \varOmega, \qquad (1)$$

which govern the flows of a viscous, non-heat-conducting, incompressible fluid in the cube

$$\varOmega = \{(x_1, x_2, x_3), 0 \le x_i \le X, i = 1, 2, 3\} \qquad (2)$$

with the boundary $\partial\varOmega$, where $X > 0$ is the user-specified length of the edge of the cubic region $\varOmega$, and x_1, x_2, x_3 are the Cartesian spatial coordinates. In equations (1), $\mathbf{V} = \mathbf{V}(x_1, x_2, x_3)$ is the velocity vector having the components $v_1(x_1, x_2, x_3), v_2(x_1, x_2, x_3)$, and $v_3(x_1, x_2, x_3)$ along the axes x_1, x_2, and x_3, respectively; $p = p(x_1, x_2, x_3)$ is the pressure, $\mathbf{f} = (f_1, f_2, f_3)$ is a given vector function. The positive constant Re is the Reynolds number, $\varDelta = \frac{\partial^2}{\partial x_1^2} + \frac{\partial^2}{\partial x_2^2} + \frac{\partial^2}{\partial x_3^2}$, $(\mathbf{V} \cdot \nabla) = v_1\frac{\partial}{\partial x_1} + v_2\frac{\partial}{\partial x_2} + v_3\frac{\partial}{\partial x_3}$.

Four equations (1) are solved under the Dirichlet boundary condition $\mathbf{V}|_{\partial\varOmega} = \mathbf{g}$, where $\mathbf{g} = \mathbf{g}(x_1, x_2, x_3) = (g_1, g_2, g_3)$ is a given vector function.

2.2   Local Coordinates and Basis Functions

In the CLR method, the spatial computational region (2) is discretized by a grid with cubic cells $\varOmega_{i,j,k}$, where i, j, k vary along the axes x_1, x_2, and x_3, respectively. We search for the solution in the form of a piecewise smooth function on this grid. To write the formulas of the CLR method it is convenient to introduce the local coordinates y_1, y_2, y_3 in each cell $\varOmega_{i,j,k}$. The dependence of local coordinates on the global spatial variables x_1, x_2, x_3 is determined by the relations $y_m = (x_m - x_{m,i,j,k})/h$, $m = 1, 2, 3$, where $x_{m,i,j,k}$ is the value of the coordinate x_m at the geometric center of the cell $\Omega_{i,j,k}$, and h is the halved length of the edge of the cubic cell $\Omega_{i,j,k}$. The local coordinates then belong to the interval $y_m \in [-1, 1]$. We now introduce the notations $\mathbf{u}(y_1, y_2, y_3) = \mathbf{V}(hy_1+x_{1,i,j,k}, hy_2+x_{2,i,j,k}, hy_3+x_{3,i,j,k})$, $p(y_1, y_2, y_3) = p(hy_1+x_{1,i,j,k}, hy_2+x_{2,i,j,k}, hy_3+x_{3,i,j,k})$. After the above change of variables the Navier–Stokes equations take the following form:

$$\Delta u_m - \text{Re}h \left(u_1 \frac{\partial u_m}{\partial y_1} + u_2 \frac{\partial u_m}{\partial y_2} + u_3 \frac{\partial u_m}{\partial y_3} + \frac{\partial p}{\partial y_m} \right) = \text{Re} \cdot h^2 f_m; \qquad (3)$$

$$\frac{1}{h} \left(\frac{\partial u_1}{\partial y_1} + \frac{\partial u_2}{\partial y_2} + \frac{\partial u_3}{\partial y_3} \right) = 0, \qquad (4)$$

where $\Delta = \partial^2/\partial y_1^2 + \partial^2/\partial y_2^2 + \partial^2/\partial y_3^2$, $m = 1, 2, 3$.

The basic idea of the method is to use the collocation method in combination with the method of least residuals to obtain numerical solution. We call such a combined method the "collocation and least residuals" (CLR) method. One of the reasons for using this combination was that the application of the least residuals method often improves the properties of the numerical method. In turn, the collocation method is simple in implementation.

We now linearize the Navier–Stokes equations(3) after Newton directly in terms of the functions at the next iteration, which are to be found:

$$\xi \left[\Delta u_m^{s+1} - (\text{Re} \cdot h) \left(u_1^s u_{m,y_1}^{s+1} + u_1^{s+1} u_{m,y_1}^s + u_2^s u_{m,y_2}^{s+1} + u_2^{s+1} u_{m,y_2}^s + u_3^s u_{m,y_3}^{s+1} \right. \right.$$
$$\left. \left. + u_3^{s+1} u_{m,y_3}^s + p_{y_m}^{s+1} \right) \right] = \xi F_m, \quad m = 1, 2, 3, \qquad (5)$$

where s is the iteration number, $s = 0, 1, 2, \ldots$, $F_m = \text{Re}[h^2 f_m - h(u_1^s u_{m,y_1}^s + u_2^s u_{m,y_2}^s + u_3^s u_{m,y_3}^s)]$, $u_{m,y_l} = \partial u_m/\partial y_l$, $p_{y_m} = \partial p/\partial y_m$, $l, m = 1, 2, 3$.

One can see in (5) another difference from the linearization used in [8]: we have introduced a positive user-defined parameter ξ. The optimal choice of this parameter enables the convergence acceleration of the iteration technique. The linearization (5) has the following two advantages over the procedure used in [8]: (i) the right-hand sides F_m have become shorter; (ii) to update the iterative solution one has now no need to add a small increment to the solution from the previous iteration. The both of these features lead to the CPU time savings at the numerical stage of our symbolic-numeric method.

We now present the approximate solution in each cell $\Omega_{i,j,k}$ as a linear combination of the basis vector functions φ_l

$$(u_1^s, u_2^s, u_3^s, p^s)^T = \sum_l b_{i,j,k,l}^s \varphi_l, \qquad (6)$$

where the superscript T denotes the transposition operation. The explicit expressions for the basis functions φ_l may be found in [8]. To approximate the velocity components we use the second-degree polynomials in variables y_1, y_2, y_3, and the first-degree polynomials are used for the approximation of the pressure. The basis functions for the velocity components are solenoidal, that is div $\varphi_l = 0$.

2.3   Derivation of the Overdetermined System from Collocation and Matching Conditions

The number of collocation points and their positions inside each cell $\Omega_{i,j,k}$ are specified by the user, and this can be done in different ways. Let us denote by N_c the number of collocation points inside the cell. We have implemented four variants of the specification of collocation points in our computer code: $N_c = 6$, $N_c = 8$, $N_c = 14$, and $N_c = 27$. For example, at $N_c = 6$, the coordinates of collocation points are as follows: $(\pm\omega, 0, 0)$, $(0, \pm\omega, 0)$, $(0, 0, \pm\omega)$, where ω is the user-specified value in the interval $0 < \omega < 1$.

Substituting the coordinates of collocation points in equations (5) we obtain $3N_c$ equations of collocations:

$$a_{\nu,m}^{(1)} \cdot b_m^{s+1} = f_\nu^s, \quad \nu = 1, \ldots, 3N_c; \ m = 1, \ldots, 30. \tag{7}$$

As in [8] we use on the cell faces the matching conditions ensuring the only piecewise polynomial solution. These conditions of the solution continuity are the linear combinations of the form

$$\begin{aligned}
h\partial(u^+)^n/\partial n + \eta_1(u^+)^n &= h\partial(u^-)^n/\partial n + \eta_1(u^-)^n; \\
h\partial(u^+)^{\tau_1}/\partial n + \eta_2(u^+)^{\tau_1} &= h\partial(u^-)^{\tau_1}/\partial n + \eta_2(u^-)^{\tau_1}; \\
h\partial(u^+)^{\tau_2}/\partial n + \eta_2(u^+)^{\tau_2} &= h\partial(u^-)^{\tau_2}/\partial n + \eta_2(u^-)^{\tau_2}; \\
p^+ &= p^-.
\end{aligned} \tag{8}$$

Here n is the external normal to the cell face, $(\cdot)^n$, $(\cdot)^{\tau_1}$, and $(\cdot)^{\tau_2}$ are the normal and tangential components of the velocity vector with respect to the face between the cells, u^+ and u^- are the limits of the function u at its tending to the cell face from within and from outside the cell; η_1, η_2 are positive parameters, which can affect the conditionality of the obtained system of linear algebraic equations (SLAE) and the convergence rate; $h\frac{\partial}{\partial n} = h\left(n_1\frac{\partial}{\partial x_1} + n_2\frac{\partial}{\partial x_2} + n_3\frac{\partial}{\partial x_3}\right) = n_1\frac{\partial}{\partial y_1} + n_2\frac{\partial}{\partial y_2} + n_3\frac{\partial}{\partial y_3}$. The points at which equations (8) are written are called the matching points. As in the case of collocation points, the specification of the number of matching points for the velocity components and their position on each face may be done in different ways. Let us denote by N_m the number of matching points for the velocity components on the faces of each cell. We have implemented in our Fortran code the cases when $N_m = 6$, $N_m = 12$, and $N_m = 24$. We substitute the coordinates of these points in each of the first three matching conditions in (8) and obtain $3N_m$ matching conditions for velocity components.

If the cell face coincides with the boundary of region Ω, then we use the boundary conditions instead of the matching conditions: $u_m = g_m$, $m = 1, 2, 3$.

The matching conditions for the pressure are specified at six points $(\pm 1, 0, 0)$, $(0, \pm 1, 0)$, $(0, 0, \pm 1)$. To identify the unique solution we specified the pressure at the coordinate origin in the same way as in [8]. Thus, the matching conditions for velocity components and the pressure give in the total $3N_m + 6 + \delta_i^1\delta_j^1\delta_k^1$

linear algebraic equations in each cell (i, j, k), where δ_i^j is the Kronecker symbol, $\delta_i^j = 1$ at $i = j$ and $\delta_i^j = 0$ at $i \neq j$. Let us write these equations in the form

$$a_{\nu,m}^{(2)} \cdot b_m^{s+1} = g_\nu^{s,s+1}, \quad \nu = 1, \ldots, 3N_m + 6 + \delta_i^1 \delta_j^1 \delta_k^1; \ m = 1, \ldots, 30. \quad (9)$$

The right-hand sides $g_\nu^{s,s+1}$ depend both on the quantities b_m^s and on the quantities b_m^{s+1}, which were just computed in some of the neighboring cells.

Let us introduce the matrix $A_{i,j,k}$, which unites the matrices of systems (7) and (9), as well as the column vector of the right-hand sides $\boldsymbol{f}_{i,j,k}^{s,s+1}$. The following SLAE is then solved in each cell:

$$A_{i,j,k} \cdot \boldsymbol{b}^{s+1} = \boldsymbol{f}_{i,j,k}^{s,s+1}, \quad (10)$$

where $\boldsymbol{b}^{s+1} = (b_{i,j,k,1}^{s+1}, \ldots, b_{i,j,k,30}^{s+1})^T$. Equations (9) in this system are similar to the boundary conditions for subdomains in the well-known decomposition method. In the given version, each grid cell is the subdomain. And for solving the problem on the whole, the Schwarz alternating method has been applied here. The matrix $A_{i,j,k}$ contains $3N_c + 3N_m + 6 + \delta_i^1 \delta_j^1 \delta_k^1$ rows and 30 columns. The system (10) is solved with respect to 30 unknowns $b_{i,j,k,l}^{s+1}$, $l = 1, 2, \ldots, 30$, by the QR expansion method. The advantages of this method over the least-squares method are discussed below in Section 4. All these equations were derived on computer in Fortran form by using symbolic computations with *Mathematica*.

At the obtaining of the final form of the formulas for the coefficients of the equations, it is useful to perform the simplifications of the arithmetic expressions of polynomial form to reduce the number of the arithmetic operations needed for their numerical computation. To this end, we employed standard functions of the *Mathematica* system, such as `Simplify` and `FullSimplify` for the simplification of complex symbolic expressions arising at the symbolic stages of the construction of the formulae of the method. Their application enabled a two-three-fold reduction of the length of polynomial expressions.

3   The Multigrid Algorithm

The multigrid methods are now known as very efficient methods for the numerical solution of various mathematical physics problems. The main idea of multigrid is to accelerate the convergence of a basic iterative method by solving the problem on the grids of different resolution. As a result, selective efficient damping of the error harmonics occurs. This constitutes the essence of the efficiency of multigrid algorithms. The convergence of multigrid methods was studied in most detail for the numerical solution of partial differential equations of elliptic type.

The general structure of multigrid algorithms as applied to the system of 3D incompressible Navier–Stokes equations (1) is as follows. Denote the system (3), (4) in local variables y_1, y_2, y_3 as $L\mathbf{u} = \mathbf{f}$. Let $h_1 > h_2 > \cdots > h_N$ ($N \geq 2$) be a sequence of decreasing grid cell sizes. The discretization corresponding to the parameter h_m (level m) is denoted by $L_m\mathbf{u}_m = \mathbf{f}_m$, where L_m is the discretization matrix, $\mathbf{u}_m$ and $\mathbf{f}_m$ are grid functions $(m = 1, \ldots, N)$. We now introduce the restriction operator R, that is the operator for the interpolation from the fine grid to coarse grid. In addition, we also introduce the prolongation operator P for the interpolation from the coarse grid to fine grid.

The CLR method is well suitable for the interpolation of the solution from one grid to another because the solution is presented in the analytic form in each cell of the spatial computing mesh. Let us now describe the implementation of the prolongation and restriction operators within the CLR method.

$1°$. **Prolongation.** Let us illustrate the prolongation algorithm by the example of the velocity component $u_1(y_1, y_2, y_3, b_1, \ldots, b_{30})$. Let $h_1 = h$, where h is the half-step of the coarse grid, and let $h_2 = h_1/2$ be the half-step of a fine grid on which one must find the decomposition of the function u_1 over the basis.
Step 1. Let X_1, X_2, X_3 be the global coordinates of the center of a coarse grid cell. We make the following substitutions in the polynomial expression for u_1:

$$y_l = (x_l - X_l)/h_1, \quad l = 1, 2, 3. \tag{11}$$

As a result we obtain the function

$$U_1(x_1, x_2, x_3, b_1, \ldots, b_{30}) = u_1\left(\frac{x_1 - X_1}{h_1}, \frac{x_2 - X_2}{h_1}, \frac{x_3 - X_3}{h_1}, b_1, \ldots, b_{30}\right). \tag{12}$$

Step 2. Let $(\tilde{X}_1, \tilde{X}_2, \tilde{X}_3)$ be the global coordinates of the center of any of the eight cells of the fine grid, which lie within the coarse grid cell. We make in (12) the substitution $x_l = \tilde{X}_l + \tilde{y}_l \cdot h_2$, $l = 1, 2, 3$. As a result we obtain the function

$$\tilde{U}_1 = \tilde{b}_1 + \tilde{b}_4\tilde{y}_1 + \tilde{b}_{12}\tilde{y}_1^2 + \tilde{b}_{19}\tilde{y}_1^2 + \tilde{b}_7\tilde{y}_2 - 2\tilde{b}_{15}\tilde{y}_1\tilde{y}_2 + \tilde{b}_{20}\tilde{y}_2^2$$
$$+ \tilde{b}_{10}\tilde{y}_3 - 2\tilde{b}_{17}\tilde{y}_1\tilde{y}_3 + \tilde{b}_{23}\tilde{y}_2\tilde{y}_3 + \tilde{b}_{25}\tilde{y}_3^2.$$

The analytic expressions for coefficients $\tilde{b}_1, \ldots, \tilde{b}_{30}$ were found efficiently with the aid of the *Mathematica* function `Coefficient[...]`. To reduce the length of the obtained expressions for the above coefficients we have applied a number of transformation rules as well as the Mathematica function `FullSimplify[...]`. As a result, the length of the final expressions for $\tilde{b}_1, \ldots, \tilde{b}_{30}$ proved to be five times shorter than the length of the original expressions. It turns out that the coordinates X_1, X_2, X_3 and $\tilde{X}_1, \tilde{X}_2, \tilde{X}_3$ enter the $\tilde{b}_l$ $(l = 1, \ldots, 30)$ only in the form of combinations $\delta x_l = (X_l - \tilde{X}_l)/h_1$. For example, $\tilde{b}_4 = (h_2/h_1) \cdot (b_4 - 2b_{12}\delta x_1 - 2b_{19}\delta x_1 + 2b_{15}\delta x_2 + 2b_{17}\delta x_3)$. In accordance with (11) the quantity $-\delta x_l = (\tilde{X}_l - X_l)/h_1$ has the meaning of the local coordinate on the coarse grid of the coordinate $\tilde{X}_l$ of the center of the fine grid cell.

$2°$. **Restriction.** The purpose of the restriction algorithm is to project the approximate numerical solution from the fine grid onto the coarse grid. To this end, we use eight cells of the fine grid, from which we compose a single cell of the coarse grid. In view of the availability of the analytic solution obtained on the fine grid, we have used instead of the averaging operation, which is applied for obtaining the initial approximation on coarse grid at a passage from the fine grid to coarse grid, a more accurate procedure of constructing the necessary initial condition by recalculating the local coordinates on fine grid at first to global coordinates and then to the local coordinates on coarse grid.

The work formulas both for the prolongation and restriction operators were calculated in symbolic form and then were stored in external files in the Fortran form. After that these Fortran operators were included in our Fortran code.

4   Convergence Acceleration Algorithm Based on Krylov's Subspaces

To accelerate the convergence of the iterations used for the approximate solution construction we have implemented here a new variant [3,7] of the well-known method [4,9,5,6,3,7], which is based on Krylov's subspaces [4]. Different variants of this method are described in [5]. Let us present the formulas of the new variant by using, for convenience of explanations, the iteration process

$$X^{n+1} = \mathrm{T}X^n + f, \qquad n = 0, 1, \ldots \tag{13}$$

of solving the SLAE $\mathrm{A}X = d$, where A and T are the square real matrices, f and d are the vectors of right-hand sides, X^n is the approximation to the solution at the iteration with number n. Let the iteration process (13) converge, and the system $\mathrm{A}X = d$ is equivalent to the system

$$X = \mathrm{T}X + f. \tag{14}$$

Substituting the value X^n in system (14) and accounting for (13), we have a residual of equation (14) at the nth iteration $r^n = \mathrm{T}X^n + f - X^n = X^{n+1} - X^n$. Using the last relationship it is not difficult to show that the relation $Z^{n+1} = \mathrm{T}Z^n$ takes place for the solution error at the nth iteration $Z^n = X - X^n$ and similarly for the residual $\mathrm{T}r^n = r^{n+1}$. Accounting for the last relations, one can also show after simple transformations that the following equation is valid for any n:

$$(\mathrm{T}^{-1} - E)Z^{n+1} = r^n. \tag{15}$$

In the method under consideration, we search at the $k + 1$th iteration for the $Y^{k+1} = \sum_{i=1}^{k} \alpha_i r^i$ as an approximate value of the error Z^{k+1}. The requirement that the approximate error value satisfies equation (15) at $n = k$ implies the following SLAE for the sought α_i:

$$\left(r^1 - r^0 \right) \alpha_1 + \ldots + \left(r^k - r^{k-1} \right) \alpha_k = -r^k. \tag{16}$$

Its solution α_i, $i = 1, 2, \ldots, k$ enables the obtaining of an approximation to the solution of SLAE (14), which is obtained at the $(k + 1)$th iteration, assuming $X \approx X^{*k+1} = X^{k+1} + Y^{k+1}$. This technique accelerates considerably the convergence of the iterations of the solution of SLAE (14). Its application does not depend on the way of writing the iteration process of the solution of the given

SLAE $A\boldsymbol{X} = \boldsymbol{d}$. Let the SLAE be obtained by the linearization of the original nonlinear problem. If the iterations in nonlinearity are combined with the iterations of solving such a SLAE, and at a single combined iteration, the coefficients of nonlinear equations of the original problem vary in their modulus much less than $\|\boldsymbol{X}^{k+1} - \boldsymbol{X}^{k}\|$, then the presented technique may accelerate substantially the process of solving the original nonlinear problem. Numerous numerical experiments conducted for the application of this algorithm at the solution of the Navier–Stokes equations showed that depending on the Re number and the value of k, more than a five-fold reduction of the CPU time was reached. At Re=1000, the acceleration of the iterative process increased when k was increased up to 10. The application of the given algorithm for the correction of the kth iteration with $k > 20$ did not produce a significant further acceleration of the process in comparison with the cases of $k \leq 10$.

At the convergence of iterations, the residual vector $\boldsymbol{r}^{n}$ decreases, and system (16) becomes worse and worse conditioned. In order to avoid a catastrophic drop of the accuracy of corrections computation additional techniques are necessary in the region of small residuals. The normalization of the columns of the matrix of system (16) is done at first to avoid the arithmetic operations on the numbers close to the machine zero. The substitution of unknowns is carried out: $\beta_i = \alpha_i \| \boldsymbol{r}^{i} - \boldsymbol{r}^{i-1}\|_2$, $i = 1, \ldots, k$. As a result, SLAE (16) takes the form

$$B\boldsymbol{\beta} = \left(\boldsymbol{B}_1 \right) \beta_1 + \ldots + \left(\boldsymbol{B}_k \right) \beta_k = -\boldsymbol{r}^{k}, \tag{17}$$

where $\boldsymbol{B}_i = (\boldsymbol{r}^{i} - \boldsymbol{r}^{i-1})/\| \boldsymbol{r}^{i} - \boldsymbol{r}^{i-1}\|_2$ are the columns of matrix B, $i = 1,\ldots,k$. System (16) was solved in [6] by the method of least squares (MLS). The MLS is known to yield the pseudo-solution of the overdetermined system, on which the minimum of the residual is reached among all its pseudo-solutions. In the given case, this is the solution of system $B^T B \boldsymbol{\beta} = -B^T \boldsymbol{r}^{k}$, where the superscript T by the matrix denotes its transposition. But the shortcoming of the MLS is that the conditionality of matrix $B^T B$ may be much worse than the conditionality of matrix B. Here, as in [3], (17) is solved by the QR-decomposition with column pivoting at the construction of matrix Q. And the solution of system (17) reduces to the solution of system $R\boldsymbol{\beta} = -Q^T \boldsymbol{r}^{k}$. The conditionality of matrix R is equal to the conditionality of matrix B. It is not difficult to show that at some choice of the sequence of equations in system (17), the pseudo-solutions obtained by the MLS method and by the QR decomposition coincide in the absence of rounding errors at their computation. This constitutes a significant advantage of the given realization of the algorithm for acceleration of iterations, which makes the computation of corrections after Krylov more stable in the region of small residuals. We omit here for the sake of brevity the presentation of one more technique for increasing the stability of the construction of the Krylov's subspace, which was also applied in [3].

The refined vector of the $k + 1$th approximation $\boldsymbol{X}^{*k+1}$ is used as the initial approximation for further continuation of the sequence of iterations.

One of the advantages of the considered method for acceleration of iterations is that it can easily be applied to already programmed iteration processes. To

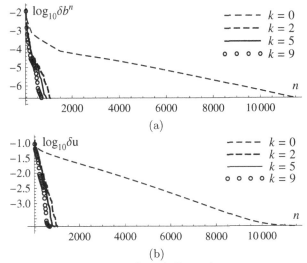

Fig. 1. The influence of the quantity k in (16) on the convergence rate of the CLR method: (a) the logarithm of error δb^n; (b) the logarithm of the error $\delta \mathbf{u}$

this end, it is sufficient to incorporate into the existing computer code a small procedure for computing the correction Y^{k+1} in accordance with the above.

5   Numerical Results

5.1   Test with Exact Analytic Solution

Let us take the exact solution of the Navier–Stokes equations (1) in the cubic region (2) from [8]. The region (2) was discretized by a uniform grid of cubic cells. The half-size h of the cell edge was equal to $h = X/(2M)$, where M is the number of cells along each coordinate direction.

Since the present method uses the linearization of the Navier–Stokes equations, the iterations in nonlinearity are necessary. The zero initial guess for the quantities u_i and p was used.

To determine the absolute numerical errors of the method on a specific uniform grid with half-step h we have computed the same root mean square errors $\delta \mathbf{u}(h)$ and $\delta p(h)$ as in [8]. Let us further use the same formulas for convergence orders ν_u and ν_p as in [8]. Let us denote the value of the coefficient $b_{i,j,k,l}$ in (6) at the nth iteration by $b_{i,j,k,l}^n$, $n = 0, 1, \ldots$. The following condition was used for termination of the iterations in nonlinearity: $\delta b^{n+1} < \varepsilon$, where

$$\delta b^{n+1} = \max_{i,j,k} \left(\max_{1 \leq l \leq 30} \left| b_{i,j,k,l}^{n+1} - b_{i,j,k,l}^n \right| \right) \tag{18}$$

and ε is a small positive user-specified number, $\varepsilon < h^2$.

Figure 1 illustrates the influence of quantity k in (16) on the convergence rate of the CLR method. These computations were done on the grid of $20 \times 20 \times 20$

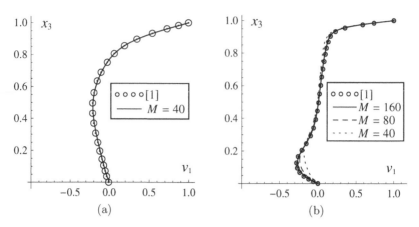

Fig. 2. Profiles of the velocity component v_1 on the central line $x_1 = x_2 = 0.5$ for Re $= 100$ (a) and Re $= 1000$ (b)

cells; $\eta_1 = \eta_2 = 2$ in (8), $\xi = 0.01$ in (5). The satisfaction of inequality $\delta b^n < 2 \cdot 10^{-7}$ was the criterion for the finalization of computation. The case $k = 0$ corresponds to the computation without using the Krylov's algorithm. It is seen that the convergence rate of the numerical solution by the CLR method increases with increasing number of residuals k employed in the Krylov's method. The number of iterations needed for satisfying the inequality $\delta b^n < 2 \cdot 10^{-7}$ was less than in the case when the Krylov's algorithm was not used by the factors of 11, 13, and 17, respectively, at $k = 2, k = 5$, and $k = 9$.

The numerical results presented in Tables 1 and 2 were obtained by using the value $\omega = 1/2$ at the specification of collocation points. It can be seen from Table 1 that in the case of the Reynolds number Re $= 100$, the convergence order ν_u lies in the interval $2.10 \leq \nu_u \leq 3.05$, that is it is somewhat higher than the second order. But with the increasing Reynolds number the convergence order ν_u decreases, see Table 2.

Table 1. The errors $\delta \mathbf{u}, \delta p$ and the convergence orders ν_u, ν_p on a sequence of grids, Re $= 100$

M	$\delta \mathbf{u}$	δp	ν_u	ν_p
10	$0.364 \cdot 10^{-3}$	$0.585 \cdot 10^{-2}$		
20	$0.852 \cdot 10^{-4}$	$0.232 \cdot 10^{-2}$	2.10	1.33
30	$0.247 \cdot 10^{-4}$	$0.131 \cdot 10^{-2}$	3.05	1.41

Table 2. The errors $\delta \mathbf{u}, \delta p$ and the convergence orders ν_u, ν_p on a sequence of grids, Re $= 1000$

M	$\delta \mathbf{u}$	δp	ν_u	ν_p
10	$0.462 \cdot 10^{-3}$	$0.341 \cdot 10^{-2}$		
20	$0.189 \cdot 10^{-3}$	$0.240 \cdot 10^{-2}$	1.29	0.51
30	$0.112 \cdot 10^{-3}$	$0.155 \cdot 10^{-2}$	1.29	1.08

It is to be noted that at the application of the CLR method for solving any problems, it is important that the equations in the overdetermined system, which play the equal role in the approximate solution, have equal weight coefficients.

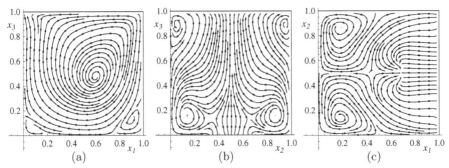

Fig. 3. Pseudo streamlines in different sections of the cubic cavity at Re = 1000: (a) section $x_2 = 0.5$; (b) section $x_1 = 0.5$; (c) section $x_3 = 0.5$

5.2   Flow in the Lid-driven Cavity

Consider the flow of a viscous incompressible fluid in a cubic cavity whose lid moves in the given direction at a constant speed. The computational region is the cube (2). The coordinate origin lies in one of the cube corners, and the Ox_3 axis is directed upwards. The cube upper face moves in dimensionless coordinates at the unit velocity in the positive direction of the Ox_1 axis. The remaining faces of cube (2) are at rest. The no-slip conditions are specified on all cube faces: $v_1 = 1$, $v_2 = v_3 = 0$, if $x_3 = X$, and $v_m = 0$, $m = 1, 2, 3$, at the remaining boundary faces.

Some results of the numerical computations of the problem under consideration were presented in [8] for the Reynolds number Re = 100. Figures 2 and 3 show some results for the Reynolds numbers Re = 100 and Re = 1000.

The high-accuracy numerical solutions of the lid-driven cubic cavity problem, which were obtained in [1], are until now the only benchmark solutions of the given problem. The computations were done in [1] with the aid of a method representing the extension of the method of [2] for the case of three spatial variables. The flow fields are singular along the edges between the moving wall and stationary walls. These singularities deteriorate the convergence of the numerical method and the accuracy of the final solution. Therefore, a superposition of stationary local asymptotic solutions was applied in [1] for all singular edges and the remaining smooth solution. Figure 2 presents, for comparison, also the solutions obtained in [1]. It can be seen that the numerical solution by the CLR method agrees very well with the result from [1] despite the fact that we have not used in our computation the asymptotic solutions near singular edges as this was done in [1,3]. Figure 2, (b) shows the convergence dynamics with the increasing number of the grid cells M in each spatial direction.

We have used the so-called pseudo streamlines for a clear graphic representation of the 3D computed results. Let us give the definition of this concept. Let us project the velocity vector of each fluid particle lying in the chosen plane onto this plane. We obtain a planar vector field. We call the lines, which are tangent to the vectors of this field, the pseudo streamlines. Figure 3 presents the patterns of pseudo streamlines in different sections, and only in section $x_2 = 0.5$, the pseudo streamlines coincide with the true streamlines. The pseudo streamlines, as the streamlines, show clearly the flow structure. The pseudo streamlines shown in Fig. 3 were obtained on the grid of $80 \times 80 \times 80$ cells with the aid of the *Mathematica* function `ListStreamPlot[...]`.

6 Conclusions

The computer algebra system *Mathematica* has been applied for constructing a new version of the method of collocations and least residuals (CLR) for solving the 3D Navier–Stokes equations. A large amount of symbolic computations, which arose in the work, was done efficiently with *Mathematica*. It is very important that the application of CAS has facilitated greatly this work, reduced at all its stages the probability of errors usually introduced by the mathematician-numerist at the development of a new algorithm.

The verification of the method accuracy by solving the well-known benchmark problem of the lid-driven cubic cavity flow and comparison with the most accurate published solutions of this problem [1], which were obtained by other researchers, have shown a high accuracy of the constructed method. This has confirmed additionally the efficiency and benefit of using the CASs for constructing new analytic-numerical methods.

References

1. Albensoeder, S., Kuhlmann, H.C.: Accurate three-dimensional lid-driven cavity flow. J. Comp. Phys. 206, 536–558 (2005)
2. Botella, O., Peyret, R.: Benchmark spectral results on the lid-driven cavity flow. Comput. Fluids 27, 421–433 (1998)
3. Isaev, V.I., Shapeev, V.P.: High-accuracy versions of the collocations and least squares method for the numerical solution of the Navier–Stokes equations. Computat. Math. and Math. Phys. 50, 1670–1681 (2010)
4. Krylov, A.N.: On the numerical solution of the equation, which determines in technological questions the frequencies of small oscillations of material systems. Izv. AN SSSR, Otd. Matem. i Estestv. Nauk. (4), 491–539 (1931) (in Russian)
5. Saad, Y.: Numerical Methods for Large Eigenvalue Problems. Manchester University Press, Manchester (1991)
6. Semin, L.G., Sleptsov, A.G., Shapeev, V.P.: Collocation and least -squares method for Stokes equations. Computat. Technologies 1(2), 90–98 (1996) (in Russian)
7. Shapeev, V.P., Isaev, V.I., Idimeshev, S.V.: The collocations and least squares method: application to numerical solution of the Navier-Stokes equations. In: CD-ROM Proc. 6th ECCOMAS. Vienna Univ. of Tech (September 2012) ISBN: 978-3-9502481-9-7
8. Shapeev, V.P., Vorozhtsov, E.V.: Symbolic-numeric implementation of the method of collocations and least squares for 3D Navier–Stokes equations. In: Gerdt, V.P., Koepf, W., Mayr, E.W., Vorozhtsov, E.V. (eds.) CASC 2012. LNCS, vol. 7442, pp. 321–333. Springer, Heidelberg (2012)
9. Sleptsov, A.G.: On convergence acceleration of linear iterations II. Modelirovanie v Mekhanike 3(5), 118–125 (1989) (in Russian)

Construction of Classes
of Irreducible Bivariate Polynomials

Doru Ştefănescu

University of Bucharest, Romania
stef@rms.unibuc.ro

Abstract. We describe a method for constructing classes of bivariate polynomials which are irreducible over algebraically closed fields of characteristic zero. The constructions make use of some factorization conditions and apply to classes of polynomials that includes the generalized difference polynomials.

Keywords: Irreducible polynomials, Polynomial factorization, Algebraic algorithms.

Introduction

We consider bivariate polynomials polynomials over an algebraically closed field k of characteristic zero. It is known that the ring $k[X, Y]$ of these polynomials is a unique factorization domain. The irreducible elements in $k[X, Y]$ are the irreducible polynomials, and they play the key role in polynomial factorization.

There exist several results concerning the construction of bivariate irreducible polynomials, see [1], [5], [6], [7]. They apply for polynomials for which the leading coefficient of a variable is a nonzero constant, namely

$$F(X, Y) = cY^n + \sum_{i=1}^{n} P_i(X)Y^{n-i},\qquad(1)$$

where $c \in k \setminus \{0\}$, $\in \mathbb{N}^*$, $P_i(X) \in k[X]$.

We remind that such a polynomial is called a *generalized difference polynomial* if

$$\deg(P_i) < i\,\frac{\deg(P_n)}{n}\quad\text{for all}\quad i,\ 1 \le i \le n-1.$$

We consider the degree-index

$$p_Y(F) = \max\left\{\frac{\deg(P_i)}{i}\,;\, 1 \le i \le n\right\}$$

considered by Panaitopol–Ştefănescu [6]. It was proved that for particular values of $p_Y(F)$, the polynomial $F(X, Y)$ is irreducible in $k[X, Y]$, see, for example [1], [2], [3], [5], [6]. They key tool for constructing irreducible polynomials using the degree index is the consideration of the Newton polygon of a product of two polynomials, see [6]. In fact:

V.P. Gerdt et al. (Eds.): CASC 2013, LNCS 8136, pp. 393–400, 2013.
© Springer International Publishing Switzerland 2013

Proposition 1 (Panaitopol–Ştefănescu, 1990). *If $F = F_1 F_2$ is factorization in $k[X, Y]$ and $p_Y(F) = \deg(P_n)/n$, we have*

$$p_Y(F) = p_Y(F_1) = p_Y(F_2).$$

The previous result can be restated for univariate polynomials with coefficients in a valued field, see, for example [4].

In this paper, we give a method for the construction of bivariate irreducible polynomials of the form (1) for which the degree index is not equal to $\deg(P_n)/n$. Such polynomials are called *quasi–difference polynomials* (cf. [3]). More precisely, we will give factorization conditions in function of the difference between the degree index $p_Y(F)$ and $\deg(P_n)/n$.

Factorization Conditions

From now on, we consider a family of polynomials $F \in k[X, Y]$ which contains the generalized difference polynomials.

Theorem 1. *Let*

$$F(X, Y) = cY^n + \sum_{i=1}^{n} P_i(X)Y^{n-i} \in k[X, Y], c \in k \setminus \{0\}$$

for which there exists $s \in \{1, 2, ..., n\}$ such that the following conditions are satisfied:

(a) $\dfrac{deg P_i}{i} \leq \dfrac{deg P_s}{s}$, *for all $i \in \{1, 2, ..., n\}$.*

(b) $(deg P_s, s) = 1$.

(c) $\dfrac{deg P_s}{s} - \dfrac{deg P_n}{n} \leq \dfrac{1}{sn}$.

Then $F(X, Y)$ is irreducible in $k[X, Y]$ or has a factor whose degree with respect to Y is a multiple of s.

Proof: Let us suppose that there exists a nontrivial factorization $F = F_1 F_2$ of the polynomial F. We put $m = \deg(P_n)$ and $a = \deg(P_s)$. By hypothesis (a) we have

$$P_Y(F) = \frac{a}{s}.$$

On the other hand, by condition (c),

$$an - sm \leq 1.$$

If $an - sm = 0$ we have

$$P_Y(P) = \frac{m}{n}$$

and by Proposition 1 we have

$$p_Y(F) = p_Y(F_1) = p_Y(F_2). \tag{2}$$

We have $an = sm$ and, by hypotheses, $(a, s) = 1$, so s should divide $n = \deg_Y(F)$. On the other hand, by (2)

$$\frac{a}{s} = p_Y(F_1) = \frac{m_1}{n_1},$$

where $n_1 = \deg(F_1)$, $m_1 = \deg_X F(X, 0)$. Therefore,

$$an_1 = sm_1.$$

But $(a, s) = 1$, so s should divide $n_1 = \deg_Y(F_1)$.

We consider now the case $an - sm = 1$.

By Theorem 1 from [6], we know that $p_Y(F) = \max\{p_Y(F_1), p_Y(F_2)\}$
We observe that

$$\frac{m_1}{n_1} \leq p_Y(F_1) \leq p_Y(F) = \frac{a}{s},$$

which gives

$$an_1 - sm_1 \geq 0. \tag{3}$$

We put $n_2 = \deg_Y(F_2)$ and $m_2 = \deg_X F_2(X, 0)$ and we observe that

$$\frac{m_2}{n_2} = \frac{m - m_1}{n - n_1} \leq p_Y(F_2) \leq p_Y(F) = \frac{a}{s}.$$

We successively obtain

$$
\begin{aligned}
s(m - m_1) &\leq a(n - n_1), \\
sm - sm_1 &\leq an - an_1, \\
(an - sm) + (sm_1 - an_1) &\geq 0, \\
1 + (sm_1 - an_1) &\geq 0, \\
an_1 - sm_1 &\leq 1.
\end{aligned}
$$

Therefore, using (3), we have $an_1 - sm_1 \in \{0, 1\}$.

In the case $an_1 - sm_1 = 0$, because a and s are coprime, it follows that s divides $n_1 = \deg_Y(F_1)$, hence the conclusion.

If $an_1 - sm_1 = 1$ we successively obtain

$$
\begin{aligned}
an_1 - sm_1 &= 1, \\
a(n - n_2) - s(m - m_2) &= 1, \\
(an - sm) + (sm_2 - an_2) &= 1, \\
1 + (sm_2 - an_2) &= 1, \\
sm_2 - an_2 &= 0.
\end{aligned}
$$

From $sm_2 = an_2$ and the assumption that a and s are coprime, it follows that s divides $n_2 = \deg_Y(F_2)$.

Corollary 1. *If $s \in \{1, n-1\}$ and F has no linear factors with respect to Y, the polynomial F is irreducible in $k[X,Y]$.*

Proof. By Theorem 1, if F is not irreducible it must have a divisor of degree s with respect to Y. So $F = F_1 F_2$, where one of the polynomials F_1 or F_2 has the degree 1 with respect to Y. But F has no linear factors with respect to Y. $\square$

Corollary 2. *If $n > 3$ and $s > n/2$ the polynomial F is irreducible or has a divisor of degree s with respect to Y.*

Proof. By Theorem 1 the polynomial F is irreducible or has a divisor G of degree ds. In the second case we have

$$n > ds > d \cdot \frac{n}{2}.$$

It follows that $d < 2$, so $d = 1$. $\square$

Proposition 2. *Let $F(X,Y) = Y^n + \sum\limits_{i=1}^{n} P_i(X)Y^{n-i} \in k[X,Y]$ and suppose that there exists $s \in \{1, 2, ..., n\}$ such that $(degP_s, s) = 1$, $\dfrac{degP_i}{i} \leq \dfrac{degP_s}{s}$ for all $i \in \{1, 2, ..., n\}$ and*

$$\frac{degP_s}{s} - \frac{degP_n}{n} = \frac{u}{sn}, \quad where \quad u \in \{2, 3\}.$$

Then one of the following statements is satisfied:

1. The polynomial $F(X,Y)$ is irreducible in $k[X,Y]$.

2. The polynomial F has a divisor whose degree with respect to Y is a multiple of s.

3. The polynomial F factors in a product of two polynomials such that the difference of their degrees with respect to Y is a multiple of s.

4. The polynomial F factors in a product of two polynomials such that the difference between the double of the degree of one of them and the degree of the other with respect to Y is a multiple of s.

Proof. With the notation from Theorem 1 we have

$$as - sm = 2 \quad or \quad 3.$$

The case $as - sm = 2$.

This gives

$$a(n_1 + n_2) - s(m_1 + m_2) = 2,$$

i.e.,

$$(an_1 - sm_1) + (an_2 - sm_2) = 2.$$

If $an_1 - sm_1 = 0$ or $an_1 - sm_1 = 2$ we have the conclusions from Theorem 1, i.e., in this case the polynomial $F(X,Y)$ is irreducible in $k[X,Y]$ or has a divisor of degree with respect to Y which is a multiple of s.

If
$$an_1 - sm_1 = 1,$$
$$an_2 - sm_2 = 1$$

we consider the solution of the Diophantine equation $ax - sy = 1$. If (x_0, y_0) is a solution, we have

$$\begin{cases} n_1 = x_0 + t_1, \; n_2 = x_0 + t_2 s, \\ n_2 = y_0 + t_1, \; m_2 = y_0 + t_2, \end{cases}$$

where $t_1, t_2 \in \mathbb{Z}$.

We obtain
$$n_1 - n_2 = (t_1 - t_2)s,$$
$$m_1 - m_2 = (t_1 - t_2)a.$$

It follows that the difference of the degrees with respect to Y of the two divisors is a multiple of s.

The case $as - sm = 3$.

It follows that
$$a(n_1 + n_2) - s(m_1 + m_2) = 3,$$

that is
$$(an_1 - sm_1) + (an_2 - sm_2) = 3.$$

We have the following possibilities:

$$\begin{aligned} an_1 - sm_1 &= 0 \text{ and } an_2 - sm_2 = 3, \\ an_1 - sm_1 &= 1 \text{ and } an_2 - sm_2 = 2, \\ an_1 - sm_1 &= 2 \text{ and } an_2 - sm_2 = 1, \\ an_1 - sm_1 &= 3 \text{ and } an_2 - sm_2 = 0. \end{aligned} \tag{4}$$

It is sufficient to examinate the first two cases.

If $an_1 - sm_1 = 0$ and $an_2 - sm_2 = 3$ we have $an_1 = sm_1$, so s divides n_1, and we are in case 2 of the conclusions.

Suppose that $an_1 - sm_1 = 1$ and $an_2 - sm_2 = 2$. Substracting these relations we obtain
$$a(n_2 - n_1) + s(m_2 - m_1) = 1$$

and substracting from this the relation $an_1 - sm_1$ we finally have

$$a(n_2 - 2n_1) - s(m_2 - 2m_1) = 0. \tag{5}$$

Relation (5) proves that s divides $n_2 - 2n_1$, so we are in case 4 from the conclusions. □

Remark 1. Note that if $u = 2$ we have the conclusions 1, 2 or 3, while if $u = 3$ one of the statements 1, 2 or 4 is satisfied.

Applications

We use the previous results for studying factorization properties of some families of polynomials and the construct of classes of irreducible polynomials.

Example 1. Corollary 1 produces families of irreducible polynomials in $k[X, Y]$. It is sufficient to apply the following steps:

- Fix $n \geq 4$ and $s = n - 1$.
- Fix the natural numbers $a_1, a_2, \ldots, a_{n-2}$ and a_n.
- Compute

$$M = \max \left\{ \frac{a_i}{i} ; 2 \leq i \leq n, i \neq s \right\}.$$

- Compute $a = a_s \in \mathbb{N}^*$ such that

$$\frac{a}{n-1} > M \quad \text{and} \quad (a, n - 1) = 1.$$

- Compute polynomials P_i such that $\deg(P_i) = a_i$ for all $i \in \{1, 2, \ldots, n\}$.
- Check if the polynomial $F(X, Y) = Y^n + \sum_{i=1}^{n} P_i(X) Y^{n-i}$ has linear factors with respect to Y.

If $F(X, Y)$ has no linear divisors with respect to Y conclude that it is irreducible in $k[X, Y]$.

Example 2. We consider

$$F(X, Y) = Y^n + p(X) Y^2 + q(X),$$

where $p, q \in k[X]$, $n \in \mathbb{N}$, $n \geq 4$, and 3 does not divide n.
Note that in this case $m = \deg(q)$.

We suppose that $\deg(p)$ and $n - 2$ are coprime and that

$$\frac{\deg(p)}{n-2} > \frac{\deg(q)}{n}.$$

and we can apply Theorem 1 or Proposition 2 provided we have

$$\frac{a}{s} - \frac{m}{n} = \frac{\deg(p)}{n-2} - \frac{\deg(q)}{n} \leq \frac{3}{(n-2)n}.$$

Particular Case:

We consider $\deg(p) = n - 1$ and $\deg(q) = n + 1$. Then we have

$$\frac{a}{s} - \frac{m}{n} = \frac{n(n-1) - (n-2)(n+1)}{(n-2)n} = \frac{2}{(n-2)n}.$$

The hypotheses of Proposition 2 are fulfilled. We have $a = n - 1$ and $s = n - 2$. Indeed, $n - 1$ and $n - 2$ are coprime and

$$s = n - 2 \geq \frac{n}{2}.$$

If we are in case 2, let G be a nontrivial divisor. Then $\deg_Y(G) = k(n-2)$, with $k \geq 1$. It follows that $k = 1$, so $\deg_Y(G) = n-2$. We deduce that the other divisor of F has the Y-degree equal to 2, so F has a quadratic factor with respect to Y.

If we are in case 3, let $F = GH$ be a nontrivial factorization in $k[X,Y]$. Since $|\deg_Y(G) - \deg_Y(H)| = k(n-2)$ we have $|\deg_Y(G) - \deg_Y(H)| = n-2$. Let us suppose that $\deg_Y(G) \geq \deg_Y(H)$. We have $\deg_Y(G) - \deg_Y(H) = n-2$, hence $\deg_Y(G) = \deg_Y(H) + n - 2 \geq n - 1$.

Because $\deg_Y(H) \geq 1$ we have $\deg_Y(G) = n-1$ and $\deg_Y(H) = 1$, therefore, one of the divisors of F is linear with respect to Y.

Therefore, if $\deg(p) = n-1$ and $\deg(q) = n+1$ the polynomial $F(X,Y) = Y^n + p(X)Y^2 + q(X)$ is irreducible or has a factor of degree 1 or 2 with respect to Y.

Remark 2. If, in the previous case, the polynomial $q(X)$ is square free, then $F(X,Y)$ is irreducible or has a quadratic factor with respect to Y. Indeed, if there is a linear factor $Y - r(x)$ then $r^n + pr^2 + q = 0$, so r^2 would divide q.

Example 3. The polynomial $F(X,Y) = Y^n + X^2Y^2 + X^3$ is irreducible in $\mathbb{Z}[X,Y]$ for all $n \in \mathbb{N}^*$, n is not divisible by 3.

If $n \geq 7$ we have

$$\frac{m}{n} = \frac{3}{n} > \frac{2}{n-2} = \frac{a}{s},$$

so $p_Y(F) = 3/n$ and F is a generalized difference polynomial. By hypotheses n is not a multiple of 3, by Corollary 3 from [6], the polynomial F is irreducible.

For $n < 7$ we have to check the irreducibility for $n \in \{1, 2, 4, 5\}$. In each case, the polynomial is irreducible.

Example 4. We consider

$$F(X,Y) = Y^n + p(X)Y^3 + q(X)Y^2 + r(X), \quad \text{where} \quad p, q, r \in k[X], n \geq 5.$$

In this case, $m = \deg(r)$.

We suppose that

$$\frac{\deg(q)}{n-2} > \frac{\deg(r)}{n} = \frac{m}{n}.$$

We consider

$$\deg(p) = n-4, \quad \deg(q) = n-1, \quad \deg(r) = n+1$$

the previous conditions are satisfied. We note that we have

$$\frac{a}{s} - \frac{m}{n} = \frac{3}{sn},$$

so we can use Proposition 2.

If a factor has the degree multiple of $s = n-2$, then it has degree $n-2$. So the other factor is quadratic or the square of a linear factor.

If we are in case 4 from the conclusions, let G, H be two factors such that $\deg(G) - 3\deg(H)$ be a multiple of $s = n - 2$. This gives information on the divisors in particular cases.

In the case $n = 5$, for example, we have $\deg(G) = 3\deg(H) + 3t$ with $t \in \mathbb{N}$, so $\deg(G)$ is a multiple of 3. Therefore, $\deg(G) = 3$, and the other factor is quadratic or the square of a linear factor.

References

1. Angermüller, G.: A generalization of Ehrenfeucht's irreducibility criterion. J. Number Theory 36, 80–84 (1990)
2. Ayad, M.: Sur les polynômes f(X,Y) tels que K[f] est intégralement fermé dans K[X,Y]. Acta Arith. 105, 9–28 (2002)
3. Bhatia, S., Khanduja, S.K.: Difference polynomials and their generalizations. Mathematika 48, 293–299 (2001)
4. Bishnoi, A., Khanduja, S.K., Sudesh, K.: Some extensions and applications of the Eisenstein irreducibility criterion. Developments in Mathematics 18, 189–197 (2010)
5. Cohen, S.D., Movahhedi, A., Salinier, A.: Factorization over local fields and the irreducibility of generalized difference polynomials. Mathematika 47, 173–196 (2000)
6. Panaitopol, L., Ştefănescu, D.: On the generalized difference polynomials. Pacific J. Math. 143, 341–348 (1990)
7. Rubel, L.A., Schinzel, A., Tverberg, H.: On difference polynomials and hereditary irreducible polynomials. J. Number Theory 12, 230–235 (1980)

Algebraic Attacks Using IP-Solvers*

Ehsan Ullah

Universität Passau, Passau, Germany
ehsanmath@gmail.com

Abstract. The main task for carrying out a successful algebraic attack on a cipher (or for examining the security of a cipher) is to solve a multivariate polynomial system over a finite field. We study recent suggestions of using IP-solvers for this task. After formulating the solution of a system of polynomial equations as a mixed integer linear programming problem, we apply state-of-the-art IP-solvers, such as CPLEX [12], inside our algebraic techniques. In particular, we highlight a new technique and develop several strategies for converting the polynomial system to a set of linear equalities and inequalities. We also generalize the approach in [7,14]. Finally, the efficiency of these techniques is examined using standard cryptographic examples such as Small Scale AES, and CTC.

Keywords: algebraic attacks, IP-solver, system solving, conversion technique.

1 Introduction

Cryptosystems or encryption schemes are important building blocks in cryptographic protocols that play an essential role in software engineering to ensure IT-security. It is well-known that any encryption map between finite dimensional vector spaces over a finite field is polynomial (see page 330, [6]). Thus, it is natural to represent the task of breaking a cryptosystem by the problem of solving a multivariate polynomial system of equations over a finite field, especially over $\mathbb{F}_2$. This type of attacks is known as *algebraic attacks* and is studied in *algebraic cryptanalysis*. Therefore, in this paper we study techniques that can be used in the context of polynomial systems derived from algebraic attacks to examine the security of different ciphers.

There are several techniques, for an overview see [14], which solve systems of multivariate polynomial equations over finite fields. It is well-known that the problem of solving a system of multivariate polynomial equations, even over a finite field, is **NP**-hard. On the other hand, the Mixed Integer Linear Programming (MILP) problem (a problem from discrete optimization) is also **NP**-hard. Inspired by the possibility that solution of either one of them could be used for solving the other, since all **NP**-complete problems are polynomially equivalent, we study recent suggestions of transferring the problem of solving a system of polynomial equations into a mixed integer linear programming problem. Until

---

* This work was completed with the support of an HEC-DAAD Scholarship.

V.P. Gerdt et al. (Eds.): CASC 2013, LNCS 8136, pp. 401–411, 2013.
© Springer International Publishing Switzerland 2013

now, the use of IP-solvers in polynomial system solving over finite fields is very new and limited.

Some methods for representing polynomials over $\mathbb{F}_2$ as polynomials over $\mathbb{R}$ (resp. $\mathbb{Z}$) can be found in the literature, but they have not been used for our purpose. In [6], an overview of possible representations is listed. Later, this study was extended slightly in [17], but the main idea behind the representation methods was basically unaltered. Very recently, some techniques have been proposed for transferring the problem of solving a system of polynomial equations over $\mathbb{F}_2$ into a MILP problem. In [14], M. Kreuzer provided a conversion algorithm based on converting polynomial equations over $\mathbb{F}_2$ into polynomial equations over $\mathbb{Z}$. In [7], J. Borghoff et. al provided another conversion technique based on converting polynomial equations over $\mathbb{F}_2$ into polynomial equations over $\mathbb{R}$. J. Borghoff et. al studied their method for systems of polynomial equations coming from stream cipher *Bivium*, but an algorithm for general systems of polynomial equations is missing. The commercial solver package CPLEX [12] was then used to solve the resulting MILP problem, which corresponds to recovering the internal state of Bivium. In the case of stream cipher Bivium A, solving this MILP problem using CPLEX takes less than 4.5 hours, which is faster than Raddums approach (about a day) [19], but much slower than using MiniSAT (21 seconds) [8].

Recently, solving a system of polynomial equations over $\mathbb{F}_2$ by converting it to a set of propositional logic clauses achieved a lot of success. The first study of efficient methods for converting boolean polynomial systems to CNF clauses was presented in [5]. Later this study was extended slightly in [3,9] and [20] but the procedure was basically unaltered. The latest effort is due to P. Jovanovic and M. Kreuzer [13]. They examined different conversion strategies, i.e. different ways to convert the polynomial system into a satisfiability problem.

Our first contribution, presented in Section 2 of this paper, is to highlight a new conversion technique based on propositional logic and pseudo-boolean optimization. In particular, first we convert a system of polynomial equations to a set of propositional logic clauses, according to proposals in [13], and then exploit the connection between propositional clauses and 0-1 inequalities to model the polynomial system as a MILP problem. This enables us to export several strategies from propositional logic for modeling a MILP problem. Therefore, the new conversion technique also has the ability to exploit several strategies for formulating more economical MILP models.

The connection between propositional clauses and linear 0-1 inequalities not only provides a new conversion technique, but also provides strategies to generalize the methods proposed by M. Kreuzer [14] and J. Borghoff et. al [7]. This is the topic of Section 3. Such strategies can be used to achieve more economical constraints while replacing nonlinear terms with new 0-1 variables. This leads us towards the development of new hybrid conversion techniques which seem to outperform their standard versions proposed by M. Kreuzer [14] and J. Borghoff et al. [7].

In Section 4 we report on some experiments and timings using the author's implementation of techniques and strategies in C++ which are also available online in package `glpk` of the computer algebra system ApCoCoA [2]. The experiments, on the Courtois Toy Cipher (CTC), and small scale AES, show that our new polynomial conversion technique and strategies perform better than the known techniques. Moreover, we show that a suitably chosen conversion strategy can save a substantial amount of newly introduced variables and inequalities in the MILP problem. The benefit of a suitably chosen conversion technique and strategy is then a significant speed-up of the IP-solvers which are applied to these MILP problems. We shall also see that, for sparse systems, our techniques are highly competitive to state-of-the-art solving techniques coming from Gröbner basis theory. Our new additions, make IP techniques perform better than Magma implementations in certain cases.

The implementations of the techniques are made available by the author in package `glpk` of ApCoCoA and the polynomial systems used for experiments can be download from `http://apcocoa.org/polynomialsystems/`. This paper is based on a part of the authors Ph.D. thesis. We adhere to the notation and terminology introduced in the books [15,16] unless mentioned otherwise.

2   Converting Boolean Polynomials to Inequalities

Let $\mathbb{F}_2$ be the finite field with two elements and let $f_1, \ldots, f_m \in P = Fqt[x_1, \ldots, x_n]$ be non-zero polynomials. Let each f_i be a squarefree polynomial (boolean polynomial), i.e. all terms in the support of f_i will be squarefree, but this is not an essential hypothesis. We are interested in finding $\mathbb{F}_2$-rational solutions of the following system of polynomial equations.

$$f_1(x_1, \ldots, x_n) = 0$$
$$\vdots$$
$$f_m(x_1, \ldots, x_n) = 0$$

This task can be restated as follows: Find a tuple $(a_1, \ldots, a_n) \in \{0, 1\}^n$ such that

$$F_1(a_1, \ldots, a_n) \equiv 0 \pmod{2}$$
$$\vdots \tag{1}$$
$$F_m(a_1, \ldots, a_n) \equiv 0 \pmod{2}$$

where $F_i \in \mathbb{Z}[X_1, \ldots, X_n]$ is a canonical representative of f_i. So we are looking for an integer solution $(a_1, \ldots, a_n)$ of the system (1) which satisfies $0 \leq a_i \leq 1$. The idea is to formulate these congruences (1) as a system of linear equalities and inequalities over $\mathbb{Z}$ and solve it using an IP-solver. There could be many ways to reach such a formulation. In the following we highlight a new technique for this purpose.

In what follows, let $X = \{X_1, \ldots, X_n\}$ be a set of boolean variables (atomic formulas), and let $\widehat{X}$ be the set of all (propositional) logical formulas that can be constructed from them, i.e. all formulas involving the operations $\neg$, $\wedge$, and $\vee$. The conversion procedure for converting boolean polynomial systems to CNF clauses as suggested in [5] consists of the following steps.

(1) Linearize the system by introducing a new indeterminate for each term in the support of one of the polynomials.
(2) Having written a polynomial as a sum of indeterminates, introduce new indeterminates to cut it after a certain number of terms. (This number is called the *cutting number*.)
(3) Convert the reduced sums into their logical equivalents using a XOR-CNF conversion.

The topic discussed in [13], is to examine different *conversion strategies*, i.e. different ways to convert the polynomial system into a satisfiability problem. The crucial point is that the linearization phase (1) usually produces too many new indeterminates. The goal is to substitute not single terms, but term combinations, in order to save indeterminates and clauses in the CNF output. In particular, author's introduce *standard strategy (SS)*, *linear partner strategy (LPS)*, *double partner strategy (DPS)*, *quadratic partner substitution (QPS)*, and *cubic partner substitution (CPS)*. By combining the choice of a substitution strategy with the other steps of the conversion procedure, we can benefit from these conversion strategies while converting boolean polynomial systems to CNF clauses.

Surprisingly, it turns out that there are deep connections between propositional logic, and pseudo-boolean optimization (see [11], Chapter 5). In particular, there is a connection, as given by the following lemma, between propositional clauses and 0-1 inequalities which can be used to model the solution of polynomial system over $\mathbb{F}_2$ (boolean polynomial system) as a MILP problem. This enables us to use the above conversion technique and strategies to model a MILP problem.

Lemma 1. *Let* $C = \{X_1 \vee \cdots \vee X_r \vee \neg Y_1 \vee \cdots \vee \neg Y_s \mid 1 \leq r, s \leq n\}$ *be a set of clauses. Then the set* C *is satisfiable if and only if the system of clausal inequalities* $I_c = \{X_1 + \cdots + X_r - Y_1 - \cdots - Y_s \geq 1 - s \mid 1 \leq r, s \leq n\}$ *together with the bounds* $0 \leq X_i, Y_j \leq 1$ *for all* $i, j \in \{1, \ldots, n\}$, *has an integer solution.*

Proof. Let $c \in C$ be a clause. If $c = X_1 \vee \cdots \vee X_r$ then by the definition of satisfiability at least one of the X_i is true. In other words at least one of the X_i is 1. This gives us the clausal inequality $X_1 + \cdots + X_r \geq 1$ together with the bounds $0 \leq X_i \leq 1$. If $c = \neg Y_1 \vee \cdots \vee \neg Y_s$ then by the definition of satisfiability at least one of the Y_j is false. In other words at least one of the $1 - Y_j$ is 1. This gives us the clausal inequality $(1 - Y_1) + \cdots + (1 - Y_s) \geq 1$ together with the bounds $0 \leq Y_j \leq 1$. If $c = X_1 \vee \cdots \vee X_r \vee \neg Y_1 \vee \cdots \vee \neg Y_s$ then it follows from the first two cases that $X_1 + \cdots + X_r + (1 - Y_1) + \cdots + (1 - Y_s) \geq 1$ is the corresponding clausal inequality together with the bounds $0 \leq X_i, Y_j \leq 1$. Therefore, the clause

$$c = X_1 \vee \cdots \vee X_r \vee \neg Y_1 \vee \cdots \vee \neg Y_s$$

can be translated into a clausal inequality

$$X_1 + \cdots + X_r + (1 - Y_1) + \cdots + (1 - Y_s) \geq 1$$

$$\text{or} \quad X_1 + \cdots + X_r - Y_1 - \cdots - Y_s \geq 1 - s$$

and the clause set C is satisfiable if and only if the corresponding system of clausal inequalities I_c together with the bounds $0 \leq X_i, Y_j \leq 1$ has an integer solution. Therefore, reasoning in propositional logic can be seen as a special case of reasoning with linear inequalities in integer variables.

Now its time to explicitly describe our first conversion algorithm using several strategies which we implemented and used for the applications and timings.

Proposition 1. (Logical Polynomial Conversion (LPC))
Let $f_1, \ldots, f_m \in \mathbb{F}_2[x_1, \ldots, x_n]$ be a system of polynomials which has at least one zero in $\mathbb{F}_2^n$. Let $\ell \geq 3$ be the desired cutting number. Then the following instructions define an algorithm which computes a tuple $(a_1, \ldots, a_n) \in \{0,1\}^n$ whose residue class in $\mathbb{F}_2^n$ represent a zero of the 0-dimensional radical ideal $I = \langle f_1, \ldots, f_m, x_1^2 + x_1, \ldots, x_n^2 + x_n \rangle$.

1) Let $G = \emptyset$. Perform the following steps 2)–5) for $i = 1, \ldots, m$.
2) Repeat the following step 3) until no polynomial g can be found anymore.
3) Find a subset of $\operatorname{Supp}(f_i)$ which defines a polynomial g of the type required by the chosen conversion strategy. Introduce a new indeterminate y_j, replace f_i by $f_i - g + y_j$, and append $g + y_j$ to G.
4) Perform the following step 5) until $\#\operatorname{Supp}(f_i) \leq \ell$. Then append f_i to G.
5) If $\#\operatorname{Supp}(f_i) > \ell$ then introduce a new indeterminate y_j, let g be the sum of the first $\ell - 1$ terms of f_i, replace f_i by $f_i - g + y_j$, and append $g + y_j$ to G.
6) For each polynomial in G, compute a logical representation in CNF and form the set of all clauses C of all these logical representations.
7) For each clause $c \in C$ form a clausal inequality I_c.
8) For all $\alpha \in \{1, \ldots, n\}$, let $I_\alpha: X_\alpha \leq 1$ and for each j let $I_j: Y_j \leq 1$.
9) Choose a linear polynomial $L \in \mathbb{Q}[X_i, Y_j]$ and use an IP-solver to find the tuple of natural numbers (a_i, b_j) which solves the system of equations and inequalities $\{I_c, I_j, I_\alpha\}$ and minimizes C.
10) Return $(a_1, \ldots, a_n)$ and stop.

Proof. Consider the conversion procedure in the introduction of this section. It can be seen easily that steps 2)−3) correspond to the linearization part (1) of the procedure. Steps 4)−5) are an explicit description of the cutting part (2) of the procedure. Step 6) is based on [13], Lemma 2, [13], Lemma 3, [13], Proposition 6 or [13], Proposition 8 for the conversion of polynomials $g + y_j$ from step 3), and on the standard XOR-CNF conversion for the linear polynomials from steps 4)−5). Finally, step 7) follows from Lemma 1. This concludes our claim.

Remark 1. Assume that we are in the setting of the algorithm in Proposition 1. A natural question could be to ask about the nature of the clausal inequalities in step 7). As claimed by Lemma 1, the variables Y_j are continuous in the interval $[0, 1]$. Since the initial variables X_α are forced to be binary, the variables Y_j take on integer values automatically. The good news is the continuity of these variables because the difficulty of solving a mixed-integer program depends more on the number of integer variables than on the number of continues variables. Another nice property of these conversion strategies is the possibility to reduce the number of new variables and inequalities.

3   New Hybrid Conversion Techniques

The strategies described in Section 2 can be used to achieve more economical constraints while replacing nonlinear terms with new 0-1 variables. Actually, in [7], J. Borghoff et. al suggested two conversion methods. The first conversion method, which is based on converting polynomial equations over $\mathbb{F}_2$ into polynomial equations over $\mathbb{Z}$, is the same as given by M. Kreuzer in [14] and we called it *Integer Polynomial Conversion (IPC)*. The second conversion method, which is based on converting polynomial equations over $\mathbb{F}_2$ into polynomial equations over $\mathbb{R}$, will be called as *Real Polynomial Conversion (RPC)*. Essentially, the idea behind the IPC conversion is that the problem of solving the polynomial equation system $f_1 = \cdots = f_m = 0$ can be restated as finding a tuple for congruence system 1. In the following, we combine the choice of a substitution strategy with the other steps of the IPC and spell out the generalized version which we implemented and used for the applications and timings in Section 4. In the following we also refer some times the boolean variables $X_1, \ldots, X_n$ as integer indeterminates which may assume value 0 or 1.

Proposition 2. (Integer Polynomial Conversion (IPC))
Let $f_1, \ldots, f_m \in P = \mathbb{F}_2[x_1, \ldots, x_n]$. Then the following instructions define an algorithm which computes a tuple $(a_1, \ldots, a_n) \in \{0, 1\}^n$ whose residue class in $\mathbb{F}_2^n$ represent a zero of the 0-dimensional radical ideal $I = \langle f_1, \ldots, f_m, x_1^2 + x_1, \ldots, x_n^2 + x_n \rangle$ of P.

1) *Reduce $f_1, \ldots, f_m$ modulo the field equations, i.e. make their support square-free. Let $G = \emptyset$.*
2) *Repeat the following step 3) until no polynomial g can be found anymore.*
3) *Find a subset of $\mathrm{Supp}(f_i)$ which defines a polynomial g of the type required by the chosen conversion strategy. Introduce a new indeterminate x_{n+j}, replace f_i by $f_i - g + x_{n+j}$, and append $g + x_{n+j}$ to G.*
4) *For each polynomial in G, compute a logical representation in CNF and form the set of all clauses C of all these logical representations.*
5) *For each clause $c \in C$ form a clausal inequality I_c.*
6) *For $i = 1, \ldots, m$, let S_i be the set of new indeterminates x_{n+j} in f_i, and let $s_i = \#\mathrm{Supp}(f_i)$.*
7) *For $i = 1, \ldots, m$, introduce a new integer indeterminate K_i and write down the linear inequality $I_i : K_i \leq \lfloor s_i/2 \rfloor$.*
8) *For $i = 1, \ldots, m$, write $f_i = \sum_j x_{n+j} + \ell_i$ where the sum extends over all j such that $x_{n+j} \in S_i$ and where $\ell_i \in P_{\leq 1}$. Form the equation $F_i : \sum_j X_{n+j} + L_i - 2K_i = 0$, where $L_i \in \mathbb{Z}[X_1, \ldots, X_n]_{\leq 1}$.*
9) *For all $\alpha \in \{1, \ldots, n\}$, let $I'_\alpha : X_\alpha \leq 1$.*

10) *Choose a linear polynomial* $L \in \mathbb{Z}[X_\alpha, X_{n+j}, K_i]$ *and use an IP-solver to find the tuple of natural numbers* (a_α, a_{n+j}, c_i) *which solves the system of equations and inequalities* $\{I_i, F_i, I_c, I'_\alpha\}$ *and minimizes* L.

11) *Return* $(a_1, \ldots, a_n)$ *and stop.*

Proof. Note that steps 2)−3) linearize the polynomials f_i by introducing new indeterminates x_{n+j}. It is easy to see that step 4) is based on [13], Lemma 2, [13], Lemma 3, [13], Proposition 6 or [13], Proposition 8 for the polynomials $g + x_{n+j}$ from step 3). Step 5) follows from Lemma 1. In step 8) the polynomials f_i are linear polynomials in the indeterminates x_α and x_{n+j}. Next it follows from F_i that $F_i(a_1, \ldots, a_n) = 2K_i$ is an even number, and I_i is nothing but the trivial bound for K_i implied by the size of the support of f_i.

For $\alpha = 1, \ldots, n$, we are looking for natural numbers a_α for which I'_α holds, so we have $a_\alpha \in \{0, 1\}$. Moreover, we have $a_{n+j} \in \{0, 1\}$ by I'_α and steps 2) − 5). In this way the solutions of the IP problem correspond uniquely to the tuples $(a_1, \ldots, a_n) \in \{0, 1\}^n$ which satisfy the above reformulation of the given polynomial system. The claim follows easily from these observations.

Note that the indeterminates X_α take on binary values. The clausal inequalities in step 5) of the algorithm keep the variables X_{n+j} continuous. Finally, the variables K_i will take on integer values in the interval $[0, \lfloor s_i/2 \rfloor]$. If we use the standard strategy, the algorithm coincides with the method given by M. Kreuzer and J. Borghoff et al. Furthermore, it is straightforward to see that the above strategies can also be used in the settings of the RPC conversion.

4   Applications and Timings

In this section we report on experimental results with the new conversion method and strategies. We compare the LPC and IPC conversions using the SS, LPS, DPS, and QPS strategies. Furthermore, we compare some of the timings we obtained to the straightforward Gröbner basis technique. For the CTC and AES cryptosystem, we used the ApCoCoA implementations by J. Limbeck [18]. The output of the conversion algorithms are files in the format which is used by CPLEX [12]. The LPC conversion generally used the cutting number 4 unless mentioned otherwise. All timings are obtained on a computer with a 2.1 GHz AMD Opteron 6172 processor having 48 cores and 64GB RAM. For each timing, we run CPLEX in parallel on two cores.

The conversion algorithms are implemented in C++. All timings are in seconds, unless mentioned otherwise. The timings for the conversion algorithms were ignored, since they do not contribute to the complexity of solving and take very little time. While modeling a MILP problem, we choose the objective function as the sum over all the initial variables and *maximization* as optimization direction. Finally, we note that all timings can be reproduced, if we do not permute the set of input linear equalities and inequalities. The reason for this is that IP-solvers also implement randomized algorithms which rely heavily on heuristic methods.

4.1 The Courtois Toy Cipher CTC

Given the CTC cryptosystem and a plaintext-ciphertext pair, we construct an overdetermined algebraic system of equations in terms of the indeterminates representing key bits and certain intermediate quantities (see [57]). Then the task is to solve the system for the key bits. The size of the system depends mainly on two parameters: the number B of simultaneous S-boxes and the number N of encryption rounds used. We denote a particular instance of CTC by $\mathrm{CTC}_{B,N}$. The S-boxes were modelled using first 7 equations out of 14. In the following table we collect the number of additional variables ($\#v$), equalities ($\#e$), inequalities ($\#i$), and CPLEX timings we obtain using LPC and IPC conversions along with different strategies.

Table 1. LPC and IPC conversion for CTC

		LPC				IPC			
		$\mathrm{CTC}_{4,4}$	$\mathrm{CTC}_{5,5}$	$\mathrm{CTC}_{6,6}$	$\mathrm{CTC}_{7,7}$	$\mathrm{CTC}_{4,4}$	$\mathrm{CTC}_{5,5}$	$\mathrm{CTC}_{6,6}$	$\mathrm{CTC}_{7,7}$
SS	$\#v$	225	326	505	687	396	630	900	1218
	$\#i$	2257	3191	5137	6969	920	1460	2088	2828
	$\#e$	0	0	0	0	268	430	612	826
	time	45	1145	53915	40260	49	1517	46910	>15h
LPS	$\#v$	209	351	469	638	396	630	900	1218
	$\#i$	2001	3591	4561	6185	920	1460	2088	2828
	$\#e$	0	0	0	0	268	430	612	826
	time	8	827	30726	208352	13	1286	39604	>15h
DPS	$\#v$	161	251	361	491	396	630	900	1218
	$\#i$	1713	2741	3913	5303	920	1460	2088	2828
	$\#e$	0	0	0	0	268	430	612	826
	time	21	436	3417	53915	14	318	3555	10415

Table 1 shows that by combining our techniques with other conversion techniques, significantly larger $\mathrm{CTC}_{B,N}$ examples can be solved.

4.2 Small Scale AES

In [10], C. Cid et. al defined small scale variants of the AES. These variants inherit the design features of the AES and provide a suitable framework for comparing different cryptanalytic methods. Without going into details, let us recall the arguments of possible configurations of the small scale AES cryptosystem, denoted by $AES(n, r, c, e)$, presented in [10]. We denote by

- $n \in \{1, \dots, 10\}$ the number of (encryption) rounds,
- r the number of rows in the rectangular arrangement of the input,
- c the number of columns in the rectangular arrangement of the input,
- e the size (in bits) of a word.

The word size e describes the field $\mathbb{F}_{2^e}$ over which the equations are defined. For instance, $e = 4$ corresponds to $\mathbb{F}_{16}$ and $e = 8$ to $\mathbb{F}_{256}$. For more details and a way to express this cipher as a multivariate equation system over $\mathbb{F}_2$ we refer to [10,13]. Note that the polynomial systems arising from an algebraic attack on small scale variants of AES [10] are naturally defined over a finite extension field $\mathbb{F}_{2^e}$ of $\mathbb{F}_2$. To convert a polynomial system over $\mathbb{F}_{2^e}$ to a polynomial system over $\mathbb{F}_2$, we use the method given in [13], Section 3. Small scale AES yields linear polynomials and homogeneous polynomials of degree 2 only. Thus SS and QPS are the only conversion strategies suitable for small scale AES.

Table 2. Sizes of small scale AES MILP problems

	LPC		IPC	
	$(\#v_{SS}, \#i_{SS})$	$(\#v_{DPS},$ $\#i_{DPS})$	$(\#v_{SS}, \quad \#i_{SS},$ $\#e_{SS})$	$(\#v_{DPS}, \#i_{DPS},$ $\#e_{DPS})$
AES(6,1,1,4)	(1692,1216)	(1652,11489)	(1568,3904,800)	(1568,4480,800)
AES(8,1,1,4)	(2263,16233)	(1943,15337)	(2080,5184,1056)	(2080,5952,1056)
AES(9,1,1,4)	(2540,18223)	(2180,17215)	(2336,5824,1184)	(2336,6688,1056)
AES(10,1,1,4)	(2829,20279)	(2429,19159)	(2592,6464,1312)	(2592,7424,1312)
AES(4,2,1,4)	(2778,20499)	(2458,19603)	(1212,5248,1088)	(1212,6016,1088)
AES(1,2,2,4)	(969,7595)	(849,7259)	(960,2304,576)	(960,2592,576)
AES(2,2,2,4)	(2336,17981)	(2096,17309)	(1792,4352,1024)	(1792,4928,1024)
AES(1,1,1,8)	(3859,28389)	(3491,26613)	(1664,4352,640)	(1664,7088,640)

In Table 2 we list the number of additional variables ($\#v$), equalities ($\#e$), and inequalities ($\#i$), each technique and strategy produces. Furthermore, we provide some timings for CPLEX in Table 3. The timings also depend on the chosen plaintext-ciphertext pairs. Above we used one plaintext-ciphertext pair in all examples.

For sparse systems the running times of the IP techniques are better than the running times of a Gröbner basis technique, even with individually tailored Gröbner basis methods, such as the ones reported in [1]. Furthermore, when we

Table 3. Timings for small scale AES MILP problems

	LPC		IPC	
	SS	DPS	SS	DPS
AES(6,1,1,4)	67	119	44	6
AES(8,1,1,4)	1975	3908	1986	226
AES(9,1,1,4)	527	26406	417	236
AES(10,1,1,4)	14298	6994	2655	1982
AES(4,2,1,4)	7416	1377	789	3147
AES(1,2,2,4)	51	42	13	20
AES(2,2,2,4)	21735	19970	7830	81014
AES(1,1,1,8)	56815	42354	4684	9323

compare these timings to the Gröbner basis approach in [18], we see that IP-solvers are highly competitive to state-of-the-art solving techniques coming from

Gröbner basis theory. In certain cases, this technique outperforms the Magma implementations in terms of both time and memory consumption. This can be easily seen by comparing Table 3 with the timings given in [18].

Finally, we note that the timings seem to depend on the cutting number, optimization direction, objective function and chosen restrictions on variables in a rather subtle and unpredictable way. We consider CTC(6,6) and AES(6,1,1,4) to present the impact of the LPC conversion algorithm using different strategies and cutting numbers. Table 4 lists the number of additional variables ($\#v$) and inequalities ($\#i$) each strategy produces during the conversion of CTC(6,6) and AES(6,1,1,4) to a set of clausal inequalities. The symbol ∞ indicates that the respective strategy is ineffective due to structure properties of the system or do not appear to provide substantial improvements over the standard strategy.

Table 4. LPC conversion with different cutting numbers

		CTC(6,6) using half Sbox				AES(6,1,1,4)			
		3	4	5	6	3	4	5	6
SS	$\#v$	1021	505	397	361	2845	1692	1254	1076
	$\#i$	5509	5137	5497	6793	12621	12164	14307	18237
LP	$\#v$	841	469	361	289	∞	∞	∞	∞
	$\#i$	4789	4561	4849	5713	∞	∞	∞	∞
DLP	$\#v$	769	361	325	325	∞	∞	∞	∞
	$\#i$	4573	4201	4345	4777	∞	∞	∞	∞
QP	$\#v$	∞	∞	∞	∞	2701	1596	1206	1028
	$\#i$	∞	∞	∞	∞	12333	11873	13827	1781

Clearly, the use of IP-solvers in polynomial system solving techniques opens up a wealth of new possibilities.

Acknowledgment. First, the author is thankful to IBM for providing their commercial software CPLEX for academic research. The author is also thankful to M. Kreuzer (Universität Passau, Passau, Germany) for deep and valuable discussions on the subjects of this paper. The author is indebted to Jan Limbeck for providing his implementations of various cryptosystems in ApCoCoA [18].

References

1. Albrecht, M.: Algebraic attacks on the Courtois Toy Cipher. Diploma thesis, Universität Bremen (2006)
2. The ApCoCoA Team.: ApCoCoA: Approximate Computations in Commutative Algebra, http://www.apcocoa.org
3. Bard, G.V.: Algebraic cryptanalysis. Springer (2009)
4. Bard, G.V.: On the rapid solution of systems of polynomial equations over lowde-gree extension fields of GF(2) via SAT-solvers. In: 8th Central European Conf. on Cryptography (2008)

5. Bard, G.V., Courtois, N.C.: Efficient methods for conversion and solution of sparse systems of low-degree multivariate polynomials over GF(2) via SAT-Solvers. Cryptology ePrint Archive 2007(24) (2007)
6. Becker, T., Weispfenning, V.: Gröbner Bases: A Computational Approach to Commutative Algebra. Springer, New York (1993)
7. Borghoff, J., Knudsen, L.R., Stolpe, M.: Bivium as a Mixed-Integer linear programming problem. In: Parker, M.G. (ed.) Cryptography and Coding 2009. LNCS, vol. 5921, pp. 133–152. Springer, Heidelberg (2009)
8. Cameron, M., Chris Charnes, J.P.: An algebraic analysis of Trivium ciphers based on the Boolean Satisfiability problem, Cryptology ePrint Archive, Report 2007/129 (2007), http://eprint.iacr.org/
9. Chen, B.: Strategies on algebraic attacks using SAT solvers. In: 9th Int. Conf. for Young Computer Scientists. IEEE Press (2008)
10. Cid, C., Murphy, S., Robshaw, M.: Small scale variants of the AES. In: Gilbert, H., Handschuh, H. (eds.) FSE 2005. LNCS, vol. 3557, pp. 145–162. Springer, Heidelberg (2005)
11. Hooker, J.: Logic-based methods for optimization: combining optimization and constraint satisfaction. John Wiley and Sons, Canada (2000)
12. ILOG CPLEX, http://www.ilog.com/products/cplex/
13. Jovanovic, P., Kreuzer, M.: Algebraic attacks using AST-Solvers. Groups - Complexity - Cryptology 2, 247–259 (2010)
14. Kreuzer, M.: Algebraic attacks galore! Groups - Complexity - Cryptology 1, 231–259 (2009)
15. Kreuzer, M., Robbiano, L.: Commputational commutative algebra 1. Springer, Heidelberg (2000)
16. Kreuzer, M., Robbiano, L.: Commputational commutative algebra 2. Springer, Heidelberg (2005)
17. Lamberger, M., Nad, T., Rijmen, V.: Numerical solvers in cryptanalysis. J. Math. Cryptology 3, 249–263 (2009)
18. Limbeck, J.: Implementation und optimierung algebraischer angriffe. Diploma thesis, Universität Passau (2008)
19. Raddum, H.: Cryptanalytic results on Trivium. eSTREAM report 2006/039 (2006), http://www.ecrypt.eu.org/stream/triviump3.html
20. Soos, M., Nohl, K., Castelluccia, C.: Extending SAT solvers to cryptographic problems. In: Kullmann, O. (ed.) SAT 2009. LNCS, vol. 5584, pp. 244–257. Springer, Heidelberg (2009)

Stationary Points for the Family of Fermat–Torricelli–Coulomb-Like Potential Functions

Alexei Yu. Uteshev and Marina V. Yashina

Faculty of Applied Mathematics, St. Petersburg State University
Universitetskij pr. 35, Petrodvorets, 198504, St. Petersburg, Russia
{alexeiuteshev,marina.yashina}@gmail.com

Abstract. Given the points $\{P_j\}_{j=1}^K \subset \mathbb{R}^n$ and the positive numbers $\{m_j\}_{j=1}^K$ we investigate the set of stationary points of the function $F(P) = \sum_{j=1}^K m_j |PP_j|^L$ for different values of the exponent $L \in \mathbb{R}$.

Keywords: Multivariate irrational functions, Stationary points, Fermat–Torricelli problem, Coulomb potential

1 Introduction

Given the coordinates of K points $\{P_j\}_{j=1}^K \subset \mathbb{R}^n$, find the coordinates of the stationary point $P_* \in \mathbb{R}^n$ for the function

$$F(P) = \sum_{j=1}^K m_j |PP_j|^L \ . \tag{1}$$

Here $\{m_j\}_{j=1}^K$ are assumed to be real positive numbers, the exponent $L \in \mathbb{R}$ is nonzero while $|\cdot|$ stands for the euclidean distance.

The stated problem in its particular case $L = 2$ has a well-known solution: the point

$$P_* = \frac{\sum_{j=1}^K m_j P_j}{\sum_{j=1}^K m_j}$$

provides the global minimal value for the function $F(P)$. If $\{m_j\}_{j=1}^K$ are interpreted as masses, then P_* is just the center of mass (barycenter).

For the case $L = 1, n \geq 2$, the problem is known as the generalized Fermat–Torricelli (or the Fermat–Weber) problem [2], [5]; this problem is the origin of the branch of Operation Research known as Facility Location or Location Analysis [3], [4].

The case $L = -1, n \in \{2, 3\}$ corresponds to the Electrostatics problem of finding the equilibrium position for the charged particle of the positive charge being subjected to the Coulomb repulsing forces created by the system of point charges $\{m_j\}_{j=1}^K$ fixed at the positions $\{P_j\}_{j=1}^K$. One may also interpret this problem in

V.P. Gerdt et al. (Eds.): CASC 2013, LNCS 8136, pp. 412–426, 2013.
© Springer International Publishing Switzerland 2013

the language of Mechanics with $\{m_j\}_{j=1}^{K}$ treated again as masses of stationary particles and with the potential of the induced gravitational field equal to $(-F)$. Potential functions with other values of the exponent L appear in Atomic Physics problems. For instance, the van der Waals interaction is most often modelled using the Lennard–Jones potential $\sum_{j=1}^{K}(m_j |PP_j|^{-12} - \tilde{m}_j |PP_j|^{-6})$.

The aim of the present paper is to analyze the set of stationary points of (1) for different values of the exponent L. We will mainly treat the case $K = n + 1$, and our approach will be based on analytical computations.

2   Direct Problem

We first treat the planar case, i.e., let $n = 2$. Let the points $\{P_j = (x_j, y_j)\}_{j=1}^{3}$ be noncollinear. The generalized Fermat–Torricelli problem possesses an analytical solution:

Theorem 1. *Denote by* $\alpha_1, \alpha_2, \alpha_3$ *the corner angles of the triangle* $P_1 P_2 P_3$. *If the conditions*

$$\begin{cases} m_1^2 < m_2^2 + m_3^2 + 2\,m_2 m_3 \cos\alpha_1, \\ m_2^2 < m_1^2 + m_3^2 + 2\,m_1 m_3 \cos\alpha_2, \\ m_3^2 < m_1^2 + m_2^2 + 2\,m_1 m_2 \cos\alpha_3 \end{cases} \tag{2}$$

are satisfied then there exists a unique stationary point $P_* = (x_*, y_*) \in \mathbb{R}^2$ *for the function*

$$F(P) = m_1|PP_1| + m_2|PP_2| + m_3|PP_3| \tag{3}$$

lying inside the triangle $P_1 P_2 P_3$. *Its coordinates are as follows:*

$$P_* = \frac{K_1 K_2 K_3}{4|S|\sigma d}\left(\frac{P_1}{K_1} + \frac{P_2}{K_2} + \frac{P_3}{K_3}\right), \tag{4}$$

with

$$F(P_*) = \min_{P \in \triangle P_1 P_2 P_3} F(P) = \sqrt{d} \ .$$

Here

$$d = \frac{1}{2\sigma}(m_1^2 K_1 + m_2^2 K_2 + m_3^2 K_3) \tag{5}$$

$$= 2|S|\sigma + \frac{1}{2}\left[m_1^2(r_{12}^2 + r_{13}^2 - r_{23}^2) + m_2^2(r_{23}^2 + r_{12}^2 - r_{13}^2) + m_3^2(r_{13}^2 + r_{23}^2 - r_{12}^2)\right] \ . \tag{6}$$

and

$$r_{j\ell} = \sqrt{(x_j - x_\ell)^2 + (y_j - y_\ell)^2} = |P_j P_\ell| \quad for \ \{j, \ell\} \subset \{1, 2, 3\} \ ;$$

$$S = x_1 y_2 + x_2 y_3 + x_3 y_1 - x_1 y_3 - x_3 y_2 - x_2 y_1 \ ; \tag{7}$$

$$\sigma = \frac{1}{2}\sqrt{-m_1^4 - m_2^4 - m_3^4 + 2\,m_1^2 m_2^2 + 2\,m_1^2 m_3^2 + 2\,m_2^2 m_3^2}\ ; \tag{8}$$

and

$$\begin{cases} K_1 = (r_{12}^2 + r_{13}^2 - r_{23}^2)\sigma + (m_2^2 + m_3^2 - m_1^2)S, \\ K_2 = (r_{23}^2 + r_{12}^2 - r_{13}^2)\sigma + (m_1^2 + m_3^2 - m_2^2)S, \\ K_3 = (r_{13}^2 + r_{23}^2 - r_{12}^2)\sigma + (m_1^2 + m_2^2 - m_3^2)S. \end{cases} \tag{9}$$

If any of the conditions (2) is violated, then $F(P)$ attains its minimal value at the corresponding vertex of the triangle.

For the proof of (and geometrical sense of the constants appeared in) this theorem we refer to [7]. This result states that at least for the case $L = +1, n = 2, K = 3$, the problem of stationary point localization of potential function (1) can be solved in radicals[1]. This is not true for the case $L = +1, n = 2$ and $K \geq 5$ points *in general position*: even the equal weighted ($m_1 = \cdots = m_K = 1$) Fermat–Torricelli problem does not allow the computational algorithm with arithmetic operations and extraction of roots [1].

We failed to find publications where the problem of stationary point localization for Coulomb potential

$$F(P) = \frac{m_1}{|PP_1|} + \frac{m_2}{|PP_2|} + \frac{m_3}{|PP_3|} \tag{10}$$

is treated in the ideology of analytical computations. The lack in such a treatment can probably be explained by the following reasoning.

Example 1. Let $P_1 = (1,1), P_2 = (5,1), P_3 = (2,6)$. Analyse the behavior of the set of stationary points of the function (10) lying inside the triangle $P_1 P_2 P_3$ for $m_1 = 1$ and m_2, m_3 treated as parameters.

Solution. Stationary points of the function (10) are given by the system of equations

$$-\frac{\partial F}{\partial x} = \frac{x-1}{\sqrt{(x-1)^2 + (y-1)^2}^3} + \frac{m_2(x-5)}{\sqrt{(x-5)^2 + (y-1)^2}^3}$$
$$+ \frac{m_3(x-2)}{\sqrt{(x-2)^2 + (y-6)^2}^3} = 0,$$

$$-\frac{\partial F}{\partial y} = \frac{y-1}{\sqrt{(x-1)^2 + (y-1)^2}^3} + \frac{m_2(y-1)}{\sqrt{(x-5)^2 + (y-1)^2}^3}$$
$$+ \frac{m_3(y-6)}{\sqrt{(x-2)^2 + (y-6)^2}^3} = 0.$$

---

[1] More formally, in *extended radicals*, since the solution formulae include modulus taking operation.

Denote by A_1, A_2, and A_3 the summands in any of the above equations. In order to eliminate radicals, one may utilize the following squaring procedure:

$$A_1 + A_2 + A_3 = 0 \quad \Rightarrow \quad (A_1 + A_2)^2 = A_3^2 \quad \Rightarrow \quad (2\,A_1 A_2)^2 = (A_3^2 - A_1^2 - A_2^2)^2\,.$$

The resulting system can be reduced to an algebraic one

$$F_1(x, y, m_2, m_3) = 0, \quad F_2(x, y, m_2, m_3) = 0$$

where F_1 and F_2 are polynomials of the degree 28 with respect to the variables x and y and with the coefficients of the orders up to 10^{19}. Finding all the real solutions of this system with the aid of elimination of variable procedure (resultant or the Gröbner basis computation) is a hardly feasible task.

In the next section, the stated problem will be solved via a more effective approach. For now we mention only the fact that, in comparison with the generalized Fermat–Torricelli problem, solution to the present one is not unique. The set of stationary points of the Coulomb potential (10) inside the triangle $P_1 P_2 P_3$ contains from 2 to 4 points depending on the values of the parameters m_2 and m_3. $\qquad\square$

3   Inverse Problem

Given the point $P_* \in \mathbb{R}^n$, we wish to find the values for the weights $\{m_j\}_{j=1}^{n+1}$ with the aim for the corresponding objective function $\sum_{j=1}^{n+1} m_j |PP_j|^L$ to posses a minimum point precisely at P_*. Although this problem looks like equivalent in complexity to the direct one, it admits a surprisingly simple solution.

Theorem 2. *Let the points $\{P_j = (x_{j1}, \dots, x_{jn})\}_{j=1}^{n+1}$ be counted in such a manner that the condition*

$$V = \begin{vmatrix} 1 & 1 & \dots & 1 \\ x_{11} & x_{21} & \dots & x_{n+1,1} \\ \vdots & \vdots & & \vdots \\ x_{1n} & x_{2n} & \dots & x_{n+1,n} \end{vmatrix} > 0 \qquad (11)$$

is satisfied. Denote by V_j the determinant obtained on replacing the j-th column of (11) by the column[2] $[1, x_{*1}, \dots, x_{*n}]^\top$. Then for the choice

$$\{m_j^* = |P_* P_j|^{2-L} V_j\}_{j=1}^{n+1} \qquad (12)$$

the function

$$F_*(P) = \sum_{j=1}^{n+1} m_j^* |PP_j|^L \qquad (13)$$

---

[2] Here $^\top$ denotes transposition.

has its stationary point at $P_* = (x_{*1}, \ldots, x_{*n})$. If P_* lies inside the simplex $P_1 P_2 \ldots P_{n+1}$, then the values (12) are all positive, and

$$F_*(P_*) = \sum_{j=1}^{n+1} V_j |P_* P_j|^2 \qquad (14)$$

$$= - \begin{vmatrix} 1 & 1 & \ldots & 1 & 1 \\ x_{11} & x_{21} & \cdots & x_{n+1,1} & x_{*1} \\ \vdots & \vdots & & \vdots & \vdots \\ x_{1n} & x_{2n} & \cdots & x_{n+1,n} & x_{*n} \\ \sum_{\ell=1}^{n} x_{1\ell}^2 & \sum_{\ell=1}^{n} x_{2\ell}^2 & \cdots & \sum_{\ell=1}^{n} x_{n+1,\ell}^2 & \sum_{\ell=1}^{n} x_{*\ell}^2 \end{vmatrix}. \qquad (15)$$

Proof. One has:

$$\frac{\partial F_*}{\partial x_1} = L \sum_{j=1}^{n+1} m_j^* |P P_j|^{L-1} \frac{(x_1 - x_{j1})}{|P P_j|} \quad \text{and} \quad \left. \frac{\partial F_*}{\partial x_1} \right|_{P=P_*} \overset{(12)}{=} L \sum_{j=1}^{n+1} (x_{*1} - x_{j1}) V_j \,.$$

The sum in the right-hand side can be represented in the determinantal form as

$$(-1)^n \begin{vmatrix} 1 & 1 & 1 & \ldots & 1 \\ x_{*1} & x_{11} & x_{21} & \cdots & x_{n+1,1} \\ \vdots & \vdots & \vdots & & \vdots \\ x_{*n} & x_{1n} & x_{2n} & \cdots & x_{n+1,n} \\ 0 & (x_{*1} - x_{11}) & (x_{*1} - x_{21}) & \cdots & (x_{*1} - x_{n+1,1}) \end{vmatrix}.$$

Add the second row of this determinant to the last one:

$$= (-1)^n \begin{vmatrix} 1 & 1 & 1 & \ldots & 1 \\ x_{*1} & x_{11} & x_{21} & \cdots & x_{n+1,1} \\ \vdots & \vdots & \vdots & & \vdots \\ x_{*n} & x_{1n} & x_{2n} & \cdots & x_{n+1,n} \\ x_{*1} & x_{*1} & x_{*1} & \cdots & x_{*1} \end{vmatrix}.$$

The first row is now proportional to the last one; therefore, the determinant equals just zero. Similar arguments lead one to the conclusion $\{\partial F_*/\partial x_j = 0\}_{j=2}^{n}$ for $P = P_*$.

For the idea of the proof of the second statement of the theorem, we refer to [7] where the case $L = 1, n = 2$ is treated. $\qquad \square$

Remark 1. For the particular case $n = 2$, we will use different notation for the values[3] (12). Set

$$S_1(x, y) = \begin{vmatrix} 1 & 1 & 1 \\ x & x_2 & x_3 \\ y & y_2 & y_3 \end{vmatrix}, \quad S_2(x, y) = \begin{vmatrix} 1 & 1 & 1 \\ x_1 & x & x_3 \\ y_1 & y & y_3 \end{vmatrix}, \quad S_3(x, y) = \begin{vmatrix} 1 & 1 & 1 \\ x_1 & x_2 & x \\ y_1 & y_2 & y \end{vmatrix}. \qquad (16)$$

---

[3] One of the reasons for this is to keep the coherence with the notation of the paper [7].

The following equality is easily verified:

$$S_1(x, y) + S_2(x, y) + S_3(x, y) \equiv \begin{vmatrix} 1 & 1 & 1 \\ x_1 & x_2 & x_3 \\ y_1 & y_2 & y_3 \end{vmatrix} \overset{(7)}{=} S \ . \tag{17}$$

For the choice

$$\left\{ m_j^* = |P_* P_j|^{2-L} S_j(x_*, y_*) \right\}_{j=1}^3 \tag{18}$$

the critical value of the corresponding function

$$F_*(P) = \sum_{j=1}^3 m_j^* |P P_j|^L \tag{19}$$

is given by

$$F_*(P_*) = \sum_{j=1}^3 S_j(x_*, y_*) |P_* P_j|^2 = \begin{vmatrix} 1 & 1 & 1 & 1 \\ x_* & x_1 & x_2 & x_3 \\ y_* & y_1 & y_2 & y_3 \\ x_*^2 + y_*^2 & x_1^2 + y_1^2 & x_2^2 + y_2^2 & x_3^2 + y_3^2 \end{vmatrix} \ . \tag{20}$$

Remark 2. Solution of the inverse problem is determined up to a common positive multiplier: if $(m_1^*, \ldots, m_{n+1}^*)$ is a solution, then $(\tau m_1^*, \ldots, \tau m_{n+1}^*)$ is also a solution for any value of $\tau > 0$.

The last remark gives one an opportunity to obtain an approach for solving the problem of finding the stationary point set for the function $F(P) = \sum_{j=1}^{n+1} m_j |P P_j|^L$ which is an alternative to the traditional one based on treating the system of equations

$$\partial F / \partial x_1 = 0, \ldots, \partial F / \partial x_n = 0 \ . \tag{21}$$

Let us find this set via the inversion of the solution for the inverse problem. We generate a new system from the ratio

$$m_1 : \ldots : m_{n+1} = \left(|P P_1|^{2-L} V_1 \right) : \ldots : \left(|P P_{n+1}|^{2-L} V_{n+1} \right) \ . \tag{22}$$

Theorem 3. *Any solution of the system (21), not coinciding with $\{P_j\}_{j=1}^{n+1}$, is a solution of the system (22).*

Proof. We outline the idea of the proof treating the case $n = 2$, i.e., we consider system (21) in the form

$$\begin{cases} m_1 |P P_1|^{L-2}(x - x_1) + m_2 |P P_2|^{L-2}(x - x_2) + m_3 |P P_3|^{L-2}(x - x_3) = 0, \\ m_1 |P P_1|^{L-2}(y - y_1) + m_2 |P P_2|^{L-2}(y - y_2) + m_3 |P P_3|^{L-2}(y - y_3) = 0. \end{cases}$$

Let it posses a solution $P = (x_0, y_0) \in \mathbb{C}^2$. Then, for these values of x and y, let us treat this system as a linear one with respect to m_1, m_2, and m_3. Since this system possesses a nontrivial solution, then any triple of these solution values (m_1, m_2, m_3) should satisfy the generalized Cramer's rule:

$$m_1 : m_2 : m_3$$

$$= \begin{vmatrix} |PP_2|^{L-2}(x - x_2) & |PP_3|^{L-2}(x - x_3) \\ |PP_2|^{L-2}(y - y_2) & |PP_3|^{L-2}(y - y_3) \end{vmatrix}$$

$$: - \begin{vmatrix} |PP_1|^{L-2}(x - x_1) & |PP_3|^{L-2}(x - x_3) \\ |PP_1|^{L-2}(y - y_1) & |PP_3|^{L-2}(y - y_3) \end{vmatrix}$$

$$: \begin{vmatrix} |PP_1|^{L-2}(x - x_1) & |PP_2|^{L-2}(x - x_2) \\ |PP_1|^{L-2}(y - y_1) & |PP_2|^{L-2}(y - y_2) \end{vmatrix}$$

$$= |PP_1|^{L-2}|PP_2|^{L-2}|PP_3|^{L-2}$$

$$\times \left(|PP_1|^{2-L} S_1(x, y) : |PP_2|^{2-L} S_2(x, y) : |PP_3|^{2-L} S_3(x, y) \right)$$

with $\{S_j(x, y)\}_{j=1}^3$ defined by (16). This completes the proof.   □

It can easily be verified that for the *even* values of the exponent $L \geq 2$, the system (22) does not have any advantage over (21), since the degrees of the algebraic equations generated by (22) equal to $L - 1$. However, for the *odd* values of L, manipulations with the system (22) look more promising compared to (21). We back up this statement with the treatment of the case $n = 2$. For this case, system (22) gives rise to the following one

$$\left[\frac{m_2 S_1}{m_1 S_2} \right]^2 = \left[\frac{|PP_2|}{|PP_1|} \right]^{2(2-L)} , \quad \left[\frac{m_2 S_3}{m_3 S_2} \right]^2 = \left[\frac{|PP_2|}{|PP_3|} \right]^{2(2-L)} . \qquad (23)$$

Solution to Example 1. For this example, the system (23) reduces to

$$\tilde{F}_1(x, y, m_2, m_3) = 0, \quad \tilde{F}_2(x, y, m_2, m_3) = 0 , \qquad (24)$$

where

$$\tilde{F}_1(x, y, m_2, m_3) = (5x + 3y - 28)^2(x^2 + y^2 - 2x - 2y + 2)^3 m_2^2$$
$$- (5x - y - 4)^2(x^2 + y^2 - 10x - 2y + 26)^3, \qquad (25)$$
$$\tilde{F}_2(x, y, m_2, m_3) = (4y - 4)^2(x^2 + y^2 - 4x - 12y + 40)^3 m_2^2$$
$$- m_3^2(5x - y - 4)^2(x^2 + y^2 - 10x - 2y + 26)^3 \qquad (26)$$

and the degree of $\tilde{F}_1$ and $\tilde{F}_2$ with respect to the variables x and y equals 8. This time, in comparison with the direct squaring algorithm outlined in Section 2, it is realistic to eliminate any variable from the obtained algebraic system. For instance, the resultant of these polynomials treated with respect to x

$$\mathcal{Y}(y, m_2, m_3) = \mathcal{R}_x(\tilde{F}_1, \tilde{F}_2) \qquad (27)$$

is[4] the polynomial of the degree 34 in y . For any specialization of parameters m_2 and m_3, it is possible to find the exact number of real zeros and to localize them

---

[4] On excluding an extraneous factor.

in the ideology of symbolic computations (e.g., via the Sturm series construction). For instance, there are two stationary points $(3.631035, 1.747577)$ and $(1.495580, 2.820345)$ for the case $m_2 = 0.5, m_3 = 3$ and four stationary points $(2.096880, 2.735988)$, $(2.667764, 2.666618)$, $(2.723424, 1.427458)$ and $(2.996571, 2.818614)$ for the case $m_2 = 1.8, m_3 = 3$ (inside the triangle).

An extra advantage of the proposed approach consists in possibility of evaluating the influence of parameters m_2 and m_3 variation on the geometry of the stationary point set. Indeed, if we eliminate the parameter m_2 from system (24), then we obtain the algebraic equation in the form $\Psi(x, y, m_3) = 0$ which defines, for every particular value of m_3, the implicitly represented curve containing the stationary points of the function (10) for all the values of m_2. However, in our example, it would be more wiser to eliminate first the parameter m_3. Since the polynomial (25) does not depend on m_3, the first equation of system (24) gives the answer. Branches of the curve $\tilde{F}_1 = 0$ corresponding to some values of m_2 are displayed in Fig. 1 (only the ovals lying inside the triangle).

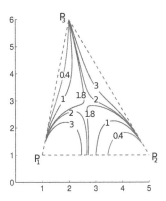

Fig. 1. Bifurcation diagram

The figure drops a hint at the existence of the bifurcation value for m_2 lying in the interval $[1.8, 2.0]$. To evaluate it, one needs to eliminate the variables x and y from the system

$$\tilde{F}_1 = 0, \ \partial \tilde{F}_1/\partial x = 0, \ \partial \tilde{F}_1/\partial y = 0 \ .$$

This results in the equation

$$84500000 \, m_2^8 - 1177878637 \, m_2^6 + 3414199390 \, m_2^4 - 1540302833 \, m_2^2 + 144500000 = 0$$

which indeed has a zero $m_2 \approx 1.81315$.

More topical is the problem of establishing the bifurcation diagram in the parameter (m_2, m_3)-plane. Bifurcation values for these parameters can be found from the condition of changing the number of real solutions for the system (24). Hence, the bifurcation values correspond to the case when the multiple zero for polynomial (27) appears. This condition is equivalent to vanishing of the discriminant

$$\mathcal{D}_y(\mathcal{Y}) = \mathcal{R}_y(\mathcal{Y}, \mathcal{Y}'_y) .$$

This is a huge expression, which[5] gives rise to the *discriminant curve* of the order 48 in the (m_2, m_3)-plane. As yet, we have failed to establish its geometry. Every point in this curve corresponds to the values of m_2 and m_3 such that the number of stationary points of the function F lying inside the triangle $P_1 P_2 P_3$ equals exactly 3. For instance, the bifurcation pair $m_2 \approx 1.842860, m_3 \approx 4.157140$ provides the stationary points

$$(2.691693, 1.930238); \; (1.821563, 2.558877); \; (3.374990, 2.739157)$$

with the first point being a degenerate one. □

The proof of Theorem 3 drops a hint on how it is possible to extend the approach outlined in the present section for investigation of stationary points of potential function (1) to the case when $K > n + 1$. Consider, for instance, the case $n = 2, K = 4$; let L be an odd number. One can first resolve the system

$$\sum_{j=1}^{4} m_j |PP_j|^{L-2}(x - x_j) = 0, \; \sum_{j=1}^{4} m_j |PP_j|^{L-2}(y - y_j) = 0 \qquad (28)$$

with respect to, e.g., m_1 and m_2:

$$\begin{cases} m_1 |PP_1|^{L-2} S_3 = m_3 |PP_3|^{L-2} S_1 + m_4 |PP_4|^{L-2} S_4, \\ m_2 |PP_2|^{L-2} S_3 = m_3 |PP_3|^{L-2} S_2 + m_4 |PP_4|^{L-2} S_5, \end{cases} \qquad (29)$$

Here S_1, S_2, and S_3 are defined by (16) while

$$S_4(x, y) = \begin{vmatrix} 1 & 1 & 1 \\ x & x_2 & x_4 \\ y & y_2 & y_4 \end{vmatrix}, \; S_5(x, y) = \begin{vmatrix} 1 & 1 & 1 \\ x_1 & x & x_4 \\ y_1 & y & y_4 \end{vmatrix} .$$

Next, the squaring procedure can be applied for both equations of the system (29). For the Coulomb case $L = -1$, this procedure results in the algebraic equations of degree 12 in x and y. This result should be treated as an essential simplification in comparison with the squaring algorithm applied directly to system (28): the latter generates algebraic equations of degree 24.

4 Stability

Our aim now is to establish the conditions under which the stationary point P_* is in fact the minimum point for (13). For the case of Coulomb potential, this task has the meaning that we wish to guarantee *the stability property* for the charged particle (of the positive charge) placed at P_*. We start with the case $n = 2$.

---

[5] On excluding an extraneous factor.

Theorem 4. *Let the points $\{P_j = (x_j, y_j)\}_{j=1}^3$ be noncollinear and counted counterclockwise, $\{S_j(x, y)\}_{j=1}^3$ be defined by (16) and S by (7). Denote*

$$\Phi(x, y) = \frac{S_1(x, y) S_2(x, y) S_3(x, y)}{|PP_1|^2 |PP_2|^2 |PP_3|^2} \begin{vmatrix} 1 & 1 & 1 & 1 \\ x & x_1 & x_2 & x_3 \\ y & y_1 & y_2 & y_3 \\ x^2 + y^2 & x_1^2 + y_1^2 & x_2^2 + y_2^2 & x_3^2 + y_3^2 \end{vmatrix} . \quad (30)$$

Stationary point $P_* = (x_*, y_*)$ is a minimum point for the function (19) if

1. *$L \geq 1$ or*
2. *$L < 1$ and the condition*

$$\Phi(x_*, y_*) > \frac{1 - L}{(L - 2)^2} S^2 . \quad (31)$$

   is valid.

Proof. In the following proof, it will be assumed that $L \notin \{0, 2\}$. Construct the Hessian matrix for F_* at the point P_*. First compute the second-order partial derivatives[6]

$$\partial^2 F_* / \partial x^2 = L \sum m_j^* \left((L - 2) |PP_j|^{L-4} (x - x_j)^2 + |PP_j|^{L-2} \right)$$

and

$$\left. \frac{\partial^2 F_*}{\partial x^2} \right|_{P=P_*} \overset{(18)}{=} L(L - 2) \sum S_j \frac{(x_* - x_j)^2}{|P_* P_j|^2} + L \sum S_j$$

$$\overset{(17)}{=} L(L - 2) \sum S_j \frac{(x_* - x_j)^2}{|P_* P_j|^2} + LS .$$

Similar expression can be obtained for the value $\partial^2 F_* / \partial y^2$ at $P = P_*$ while

$$\left. \frac{\partial^2 F_*}{\partial x \partial y} \right|_{P=P_*} = L(L - 2) \sum S_j \frac{(x_* - x_j)(y_* - y_j)}{|P_* P_j|^2} .$$

Thus, the Hessian matrix is as follows:

$$H(x_*, y_*) = \left. \begin{bmatrix} \partial^2 F_* / \partial x^2 & \partial^2 F_* / \partial x \partial y \\ \partial^2 F_* / \partial x \partial y & \partial^2 F_* / \partial y^2 \end{bmatrix} \right|_{P=P_*}$$

$$= \begin{bmatrix} LS + L(L - 2) \sum S_j \frac{(x_* - x_j)^2}{|P_* P_j|^2} & L(L - 2) \sum S_j \frac{(x_* - x_j)(y_* - y_j)}{|P_* P_j|^2} \\ L(L - 2) \sum S_j \frac{(x_* - x_j)(y_* - y_j)}{|P_* P_j|^2} & LS + L(L - 2) \sum S_j \frac{(y_* - y_j)^2}{|P_* P_j|^2} \end{bmatrix} .$$

---

[6] In the rest of the proof, the sum is always taken for $j \in \{1, 2, 3\}$, and S_j stands for $S_j(x_*, y_*)$.

Compute its characteristic polynomial:

$$\det\left(H(x_*, y_*) - \lambda \begin{bmatrix} 1 & 0 \\ 0 & 1 \end{bmatrix} \right).$$

Substitution

$$\tilde{\lambda} = \frac{\lambda}{L(L-2)} - \frac{S}{L-2} \tag{32}$$

yields:

$$L^2(L-2)^2 \det \begin{bmatrix} \sum S_j \frac{(x_*-x_j)^2}{|P_*P_j|^2} - \tilde{\lambda} & \sum S_j \frac{(x_*-x_j)(y_*-x_j)}{|P_*P_j|^2} \\ \sum S_j \frac{(x_*-x_j)(y_*-x_j)}{|P_*P_j|^2} & \sum S_j \frac{(y_*-y_j)^2}{|P_*P_j|^2} - \tilde{\lambda} \end{bmatrix}$$

(in what follows we neglect the factor $L^2(L-2)^2$)

$$= \tilde{\lambda}^2 - \left(\sum S_j \frac{(x_*-x_j)^2}{|P_*P_j|^2} + \sum S_j \frac{(y_*-y_j)^2}{|P_*P_j|^2} \right) \tilde{\lambda}$$

$$+ \left(\sum S_j \frac{(x_*-x_j)^2}{|P_*P_j|^2} \right)\left(\sum S_j \frac{(y_*-y_j)^2}{|P_*P_j|^2} \right) - \left(\sum S_j \frac{(x_*-x_j)(y_*-x_j)}{|P_*P_j|^2} \right)^2$$

$$\overset{(17)}{=} S\tilde{\lambda}^2 - S\tilde{\lambda}$$

$$+ S_1 S_2 \frac{[(y_*-y_1)(x_*-x_2) - (x_*-x_1)(y_*-y_2)]^2}{|P_*P_1|^2 |P_*P_2|^2}$$

$$+ S_1 S_3 \frac{[(y_*-y_1)(x_*-x_3) - (x_*-x_1)(y_*-y_3)]^2}{|P_*P_1|^2 |P_*P_3|^2}$$

$$+ S_2 S_3 \frac{[(y_*-y_2)(x_*-x_3) - (x_*-x_2)(y_*-y_3)]^2}{|P_*P_2|^2 |P_*P_3|^2}$$

$$= \tilde{\lambda}^2 - S\tilde{\lambda} + \frac{S_1 S_2 S_3^2}{|P_*P_1|^2 |P_*P_2|^2} + \frac{S_1 S_2^2 S_3}{|P_*P_1|^2 |P_*P_3|^2} + \frac{S_1^2 S_2 S_3}{|P_*P_2|^2 |P_*P_3|^2}$$

$$= \tilde{\lambda}^2 - S\tilde{\lambda} + \frac{S_1 S_2 S_3}{|P_*P_1|^2 |P_*P_2|^2 |P_*P_3|^2} \left(S_1 |P_*P_1|^2 + S_2 |P_*P_2|^2 + S_3 |P_*P_3|^2 \right)$$

$$\overset{(20)}{=} \tilde{\lambda}^2 - S\tilde{\lambda} + \frac{S_1 S_2 S_3 F_*(P_*)}{|P_*P_1|^2 |P_*P_2|^2 |P_*P_3|^2}$$

and now we come back to the variable λ and recall the neglected factor $L^2(L-2)^2$:

$$= \lambda^2 - L^2 S\lambda + L^2(L-2)^2 \left(\frac{L-1}{(L-2)^2} S^2 + \frac{S_1 S_2 S_3 F_*(P_*)}{|P_*P_1|^2 |P_*P_2|^2 |P_*P_3|^2} \right).$$

Hessian matrix $H(x_*, y_*)$ is positive definite iff its characteristic polynomial has both its (real!) roots positive. Due to Descartes rule of signs and the formula (20), the latter is equivalent to the statement of the theorem. □

In the language of the charged particles, the last theorem claims that for the case $L \geq 1$, any point inside the triangle $P_1 P_2 P_3$ can be made the minimum point for the function $F(P)$ by means of an appropriate choice of the charges m_1, m_2, and m_3. This will not be true for the case $L < 1$: one can stabilize the charge placed at the point $P_* \in \triangle P_1 P_2 P_3$ only if it belongs to the domain surrounded by the branch of the curve

$$\Phi(x,y) = \frac{1-L}{(L-2)^2}S^2 \tag{33}$$

where $\Phi(x,y)$ is defined by (30). We will further refer to this domain as *the stability domain* (corresponding to the value of L).

Example 2. Let $P_1 = (1,1), P_2 = (5,1), P_3 = (2,6)$. Construct the stability domains for different values of L.

Solution. Here $S = 5$ and

$$\Phi(x,y) = \frac{(28 - 5\,x - 3\,y)(5\,x - y - 4)(y-1)(-52 + 30\,x + 32\,y - 5\,x^2 - 5\,y^2)}{((x-1)^2 + (y-1)^2)((x-5)^2 + (y-1)^2)((x-2)^2 + (y-6)^2)}.$$

Therefore, curve (33) is an algebraic one of order 6. Its level curves for some values of L are displayed in Fig. 2 (only the ovals lying inside the triangle).

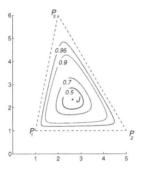

Fig. 2. Stability domains

We restrict ourselves to the case $L \in [0,1]$ since the curve corresponding to a negative value of L coincides with the curve corresponding to the positive value $1 + 1/(L-1) \in [0,1]$. Thus, the stability domain for the Coulomb case $L = -1$ coincides with the domain corresponding to $L = 1/2$. □

Hypothesis. There exists a unique point common for all the stability domains: the point J with the coordinates

$$x_{J} = \frac{x_1|P_2P_3| + x_2|P_1P_3| + x_3|P_1P_2|}{|P_2P_3| + |P_1P_3| + |P_1P_2|}, \quad y_{J} = \frac{y_1|P_2P_3| + y_2|P_1P_3| + y_3|P_1P_2|}{|P_2P_3| + |P_1P_3| + |P_1P_2|}$$

is a stationary point for the function $\Phi(x, y)$. It coincides with the center of the circle inscribed into the triangle $P_1 P_2 P_3$ (the *incenter* of the triangle). For the previous example, one gets

$$J = \left(\frac{\sqrt{34} + 5\sqrt{26} + 8}{\sqrt{34} + \sqrt{26} + 4}, \frac{\sqrt{34} + \sqrt{26} + 24}{\sqrt{34} + \sqrt{26} + 4} \right) \approx (2.634034, 2.339587) .$$

We now consider the case $n = 3$, i.e.,

$$F_*(P) = \sum_{j=1}^{4} m_j^* |PP_j|^L . \tag{34}$$

with $\{m_j^*\}_{j=1}^{4}$ defined by (12).

Theorem 5. *Let the points $\{P_j = (x_j, y_j, z_j)\}_{j=1}^{4}$ satisfy the assumptions of Theorem 2 and P_* be chosen inside the tetrahedron $P_1 P_2 P_3 P_4$. Stationary point $P_* = (x_*, y_*, z_*)$*

1. *is not a minimum point for the function (34) if $L \leq -1$;*
2. *is a minimum point for the function (34) if $L \geq 2$.*

Proof. The idea of the proof is similar to that one of Theorem 4. Let us compute the characteristic polynomial for the Hessian matrix $H(x_*, y_*, z_*)$ of the function (34) at $P = P_*$:

$$\det(H - \lambda I) .$$

Here I stands for the identity matrix of the third order. Substitution similar to (32)

$$\tilde{\lambda} = \frac{V}{L - 2} - \frac{\lambda}{L(L - 2)}$$

allows one to represent this polynomial in the form

$$L^3 (L - 2)^3 \left(\tilde{\lambda}^3 + t_1 \tilde{\lambda}^2 + t_2 \tilde{\lambda} + t_3 \right) .$$

Expressions for its coefficients can be obtained in a similar manner to their counterparts from the planar case[7]

$$t_1 = V;$$

$$t_2 = \sum_{1 \leq j < k \leq 4} V_j V_k \frac{\det(M_{jk} \cdot M_{jk}^\top)}{|P_* P_j|^2 |P_* P_k|^2}$$

where

$$M_{jk} = \begin{bmatrix} x_* - x_j & y_* - y_j & z_* - z_j \\ x_* - x_k & y_* - y_k & z_* - z_k \end{bmatrix};$$

---

[7] Although this time the intermediate computations are much more longer and, in the mercy to the reader, we skip them.

and
$$t_3 = \frac{V_1 V_2 V_3 V_4 F_*(P_*)}{|P_*P_1|^2 |P_*P_2|^2 |P_*P_3|^2 |P_*P_4|^2} \ .$$

Note that the values t_1, t_2, t_3 are nonnegative for all the permissible by the conditions of the theorem values of parameters. Coming back to the variable λ, one gets:

$$= -\lambda^3 + L(L+1)V\lambda^2 - L^2 \left[(2L-1)V^2 + t_2(L-2)^2\right]\lambda$$

$$+ L^3 \left[(L-1)V^3 + t_2 V(L-2)^2 + t_3(L-2)^3\right] \ .$$

For $L \leq -1$, the coefficient of λ^2 is not positive, therefore, due to Descartes rule of signs, at least one of the eigenvalues of the Hessian matrix should be non-positive. From this follows the first statement of the theorem. On the contrary, for $L \geq 2$, we get three variations in sign in the sequence of the coefficients of the characteristic polynomial. This means that all the eigenvalues of $H(x_*, y_*, z_*)$ are positive. Wherefrom follows the second statement of the theorem.   □

For the Coulomb case $L = -1$, the last theorem states that, in comparison to the planar case, in $\mathbb{R}^3$ it is impossible to stabilize the charged particle via an appropriate choice of charges $\{m_j\}_{j=1}^4$ fixed at the given positions $\{P_j\}_{j=1}^4$. This negative result might be treated as a particular case of the so-called *Earnshaw's theorem* which states that a collection of point charges in $\mathbb{R}^3$ cannot be maintained in a stable stationary equilibrium configuration solely by the electrostatic interaction of the charges [6].

5   Conclusions

Analytical approach for the investigation of the set of stationary points for the general potential function $F(P) = \sum_{j=1}^K m_j |PP_j|^L$ in $\mathbb{R}^n$ is developed. This approach is based on the replacement of system (21) by an algebraic one more simple for dealing with. This gives one an opportunity of studying the influence of parameters $\{m_j\}_{j=1}^K$ on the geometry of the stationary point set. For further investigation, there remains the problem of discriminant curve geometry mentioned in Section 3 and also the stability problem in $\mathbb{R}^3$ for the cases not covered by the statement of Theorem 5.

Acknowledgments. The authors are grateful to the anonymous referees for valuable suggestions that helped to improve the quality of the paper and to Professor Sergei Andrianov for his patient consulting on complex matters of the modern physics. This work was supported by the St.Petersburg State University research grant # **9.38.674.2013**.

References

1. Bajaj, C.: The algebraic degree of geometric optimization problems. Discr. Comput. Geom. 3, 177–191 (1988)
2. Courant, R., Robbins, H.: What is Mathematics? Oxford University Press, London (1941)
3. Drezner, Z., Hamacher, H.W. (eds.): Facility Location: Applications and Theory. Springer, Heidelberg (2004)
4. Eiselt, H.A., Marianov, V. (eds.): Foundations of Location Analysis. Springer, Heidelberg (2011)
5. Fermat–Torricelli problem. Encyclopedia of Mathematics (2012),
   http://www.encyclopediaofmath.org/
   index.php?title=Fermat-Torricelli_problem&oldid=22419
6. Tamm, I.: Fundamentals of the Theory of Electricity. Mir Publishers, Moscow (1979)
7. Uteshev, A.Y.: Analytical solution for the generalized Fermat–Torricelli problem. ArXiv. 1208.3324.; Amer. Math. Monthly (to appear)

Symbolic-Numerical Algorithm for Generating Cluster Eigenfunctions: Tunneling of Clusters through Repulsive Barriers

Sergue Vinitsky[1], Alexander Gusev[1], Ochbadrakh Chuluunbaatar[1],
Vitaly Rostovtsev[1], Luong Le Hai[1,2],
Vladimir Derbov[3], and Pavel Krassovitskiy[4]

[1] Joint Institute for Nuclear Research, Dubna, Moscow Region, Russia
vinitsky@theor.jinr.ru
[2] Belgorod State University, Belgorod, Russia
[3] Saratov State University, Saratov, Russia
[4] Institute of Nuclear Physics, Almaty, Kazakhstan

Abstract. A model for quantum tunnelling of a cluster comprising A identical particles, coupled by oscillator-type potential, through short-range repulsive potential barriers is introduced for the first time in the new symmetrized-coordinate representation and studied within the s-wave approximation. The symbolic-numerical algorithms for calculating the effective potentials of the close-coupling equations in terms of the cluster wave functions and the energy of the barrier quasistationary states are formulated and implemented using the Maple computer algebra system. The effect of quantum transparency, manifesting itself in nonmonotonic resonance-type dependence of the transmission coefficient upon the energy of the particles, the number of the particles $A = 2, 3, 4$, and their symmetry type, is analyzed. It is shown that the resonance behavior of the total transmission coefficient is due to the existence of barrier quasistationary states imbedded in the continuum.

1 Introduction

During a decade, the mechanism of quantum penetration of two bound particles through repulsive barriers [1] attracts attention from both theoretical and experimental viewpoints in relation with such problems as near-surface quantum diffusion of molecules [2–4], fragmentation in producing very neutron-rich light nuclei [5, 6], and heavy ion collisions through multidimensional barriers [7–14]. Within the general formulation of the scattering problem for ions having different masses, a benchmark model with long-range potentials was proposed in Refs. [15–17]. The generalization of the two-particle model over a quantum system of A identical particles is of great importance for the appropriate description of molecular and heavy-ion collisions. *The aim of this paper is to present the convenient formulation of the problem stated above and the calculation methods, algorithms, and programs for solving this problem.*

V.P. Gerdt et al. (Eds.): CASC 2013, LNCS 8136, pp. 427–442, 2013.
© Springer International Publishing Switzerland 2013

We consider a *new method* for the description of the penetration of A identical quantum particles, coupled by short-range oscillator-like interaction, through a repulsive potential barrier. We assume that the spin part of the wave function is known, so that only the spatial part of the wave function is to be considered, which may be symmetric or antisymmetric with respect to a permutation of A identical particles. The initial problem is reduced to the penetration of a composite system with the internal degrees of freedom, describing an $(A-1) \times d$-dimensional oscillator, and the external degrees of freedom describing the center-of-mass motion of A particles in d-dimensional Euclidian space. For simplicity, we restrict our consideration to the so-called s-wave approximation [1] corresponding to one-dimensional Euclidean space $(d = 1)$.

We seek for the solution in the form of Galerkin expansion in terms of cluster functions in the *new symmetrized coordinate representation* (SCR) [18] with unknown coefficients having the form of matrix functions of the center-of-mass variable. As a result, the problem is reduced to a boundary-value problem for a system of ordinary second-order differential equations with respect to the center-of-mass variable. Conventional asymptotic boundary conditions involving unknown amplitudes of reflected and transmitted waves are imposed on the desired matrix solution. Solving the problem was implemented as a complex of the *symbolic-numeric algorithms and programs* in CAS MAPLE and FORTRAN environment. The results of calculations are analyzed with particular emphasis on the effect of quantum transparency that manifests itself as nonmonotonic energy dependence of the transmission coefficient due to resonance tunnelling of the bound particles in S (A) states through the repulsive potential barriers.

The paper is organized as follows. In Section 2, we present the problem statement in symmetrized coordinates. In Section 3, we introduce the SCR of the cluster functions of the considered problem and the asymptotic boundary conditions involving unknown amplitudes of reflected and transmitted waves. In Section 4, we formulate the boundary-value problem for the close-coupling equations in the Galerkin form using the SCR. In Section 5, we analyze the results of numerical experiment on the resonance transmission of a few coupled identical particles in S(A) states, whose energies coincide with the resonance eigenenergies of the barrier quasi-stationary states embedded in the continuum. In Conclusion, we sum up the results and discuss briefly the perspectives of application of the developed approach.

2   Problem Statement

We consider a system of A identical quantum particles having the mass m and a set of the Cartesian coordinates $x_i \in \mathbf{R}^d$ in d-dimensional Euclidian space, considered as vector $\tilde{\mathbf{x}} = (\tilde{x}_1, ..., \tilde{x}_A) \in \mathbf{R}^{A \times d}$ in $A \times d$-dimensional configuration space. The particles are coupled by the pair potentials $\tilde{V}^{pair}(\tilde{x}_{ij})$ depending upon the relative coordinates, $\tilde{x}_{ij} = \tilde{x}_i - \tilde{x}_j$, similar to a harmonic oscillator potential $\tilde{V}^{hosc}(\tilde{x}_{ij}) = \frac{m\omega^2}{2}(\tilde{x}_{ij})^2$ with the frequency ω. The resulting clusters are subject to the influence of the potentials $\tilde{V}(\tilde{x}_i)$ describing the external field of a target. The appropriate Schrödinger equation takes the form

$$\left[-\frac{\hbar^2}{2m}\sum_{i=1}^{A}\frac{\partial^2}{\partial\tilde{x}_i^2}+\sum_{i,j=1;i<j}^{A}\tilde{V}^{pair}(\tilde{x}_{ij})+\sum_{i=1}^{A}\tilde{V}(\tilde{x}_i)-\tilde{E}\right]\tilde{\Psi}(\tilde{\mathbf{x}})=0,$$

where $\tilde{E}$ is the total energy of the system of A particles, and $\tilde{P}^2=2m\tilde{E}/\hbar^2$, $\tilde{P}$ is the total momentum of the system, and $\hbar$ is Planck constant. Using the oscillator units $x_{osc}=\sqrt{\hbar/(m\omega\sqrt{A})}$, $p_{osc}=\sqrt{(m\omega\sqrt{A})/\hbar}=x_{osc}^{-1}$, and $E_{osc}=\hbar\omega\sqrt{A}/2$ to introduce the dimensionless coordinates $x_i=\tilde{x}_i/x_{osc}$, $x_{ij}=\tilde{x}_{ij}/x_{osc}=x_i-x_j$, $E=\tilde{E}/E_{osc}=P^2$, $P=\tilde{P}/p_{osc}=\tilde{P}x_{osc}$, $V^{pair}(x_{ij})=\tilde{V}^{pair}(x_{ij}x_{osc})/E_{osc}$, $V^{hosc}(x_{ij})=\tilde{V}^{hosc}(x_{ij}x_{osc})/E_{osc}=\frac{1}{A}(x_{ij})^2$ and $V(x_i)=\tilde{V}(x_ix_{osc})/E_{osc}$, one can rewrite the above equation in the form

$$\left[-\sum_{i=1}^{A}\frac{\partial^2}{\partial x_i^2}+\sum_{i,j=1;i<j}^{A}\frac{1}{A}(x_{ij})^2+\sum_{i,j=1;i<j}^{A}U^{pair}(x_{ij})+\sum_{i=1}^{A}V(x_i)-E\right]\Psi(\mathbf{x})=0,\,(1)$$

where $U^{pair}(x_{ij})=V^{pair}(x_{ij})-V^{hosc}(x_{ij})$, i.e., if $V^{pair}(x_{ij})=V^{hosc}(x_{ij})$, then $U^{pair}(x_{ij})=0$.

The problem of tunnelling of a cluster of A identical particles in the symmetrized coordinates $(\xi_0,\boldsymbol{\xi})$, where $\boldsymbol{\xi}=\{\xi_1,...,\xi_{A-1}\}$:

$$\xi_0=\frac{1}{\sqrt{A}}\left(\sum_{t=1}^{A}x_t\right),\ \xi_s=\frac{1}{\sqrt{A}}\left(x_1+\sum_{t=2}^{A}a_0x_t+\sqrt{A}x_{s+1}\right),\ s=1,...,A-1,(2)$$

in terms of total potential $U(\xi_0,\boldsymbol{\xi})=V(\xi_0,\boldsymbol{\xi})+U^{eff}(\xi_0,\boldsymbol{\xi})$ reads as [18]

$$\left[-\frac{\partial^2}{\partial\xi_0^2}+\sum_{i=1}^{A-1}\left(-\frac{\partial^2}{\partial\xi_i^2}+(\xi_i)^2\right)+U(\xi_0,\boldsymbol{\xi})-E\right]\Psi(\xi_0,\boldsymbol{\xi})=0,\qquad(3)$$

$$U^{eff}(\xi_0,\boldsymbol{\xi})=\sum_{i,j=1;i<j}^{A}U^{pair}(x_{ij}(\boldsymbol{\xi})),\quad V(\xi_0,\boldsymbol{\xi})=\sum_{i=1}^{A}V(x_i(\xi_0,\boldsymbol{\xi})),$$

which is invariant under permutations $\xi_i\leftrightarrow\xi_j$ at $i,j=1,...,A-1$, i.e., the invariance of Eq. (1) under permutations $x_i\leftrightarrow x_j$ at $i,j=1,...,A$ survives the transformation.

3   Cluster Functions and Asymptotic Boundary Conditions

For simplicity we restrict our consideration to the so-called s-wave approximation [1], i.e., one-dimensional Euclidian space $(d=1)$. Cluster functions $\tilde{\Phi}_j(\xi_0,\boldsymbol{\xi})$, where $\boldsymbol{\xi} = \{\xi_1, ..., \xi_{A-1}\}$, corresponding to the threshold energies $\tilde{\epsilon}_j(\xi_0)$ dependent on ξ_0 as a parameter, are solutions of the parametric eigenvalue problem

$$\left(-\frac{\partial^2}{\partial\boldsymbol{\xi}^2} + \boldsymbol{\xi}^2 + U(\xi_0, \boldsymbol{\xi}) - \tilde{\epsilon}_j(\xi_0)\right)\tilde{\Phi}_j(\xi_0, \boldsymbol{\xi}) = 0, \int_{-\infty}^{+\infty}\tilde{\Phi}_i(\xi_0, \boldsymbol{\xi})\tilde{\Phi}_j(\xi_0, \boldsymbol{\xi})d^{A-1}\boldsymbol{\xi} = \delta_{ij}, (4)$$

where $U(\xi_0, \boldsymbol{\xi}) = V(\xi_0, \boldsymbol{\xi}) + U^{eff}(\xi_0, \boldsymbol{\xi})$ is the total potential that enters Eq. (3). The effective potential $U^{eff}(\xi_0, \boldsymbol{\xi})$ can be approximated also by the deformed Wood–Saxon potential in the single-particle oscillator approximation [9]. We seek for the cluster functions $\Phi_i(\xi_0, \boldsymbol{\xi})$ in the form of an expansion over the eigenfunctions $\Phi_{j'}^{S(A)}(\boldsymbol{\xi})$, symmetric (S) or antisymmetric (A) with respect to a permutation of the initial A Cartesian coordinates of A identical particles. These functions correspond to eigenenergies $E_i^{S(A)}$ of the $(A-1)$-dimensional oscillator, generated by the algorithm SCR [18], with unknown coefficients $\tilde{\alpha}_{j'}^{(i)}(\xi_0)$:

$$\tilde{\Phi}_i(\xi_0, \boldsymbol{\xi}) = \sum_{j'=1}^{j'_{max}}\tilde{\alpha}_{j'}^{(i)}(\xi_0)\Phi_{j'}^{S(A)}(\boldsymbol{\xi}). \tag{5}$$

Thus, the eigenvalue problem (4) is reduced to a linearized version of the Hartree–Fock algebraic eigenvalue problem

$$\sum_{j'=1}^{j'_{max}}\left(\delta_{ij'}E_i^{S(A)} + U_{ij'}(\xi_0) - \delta_{ij'}\tilde{\epsilon}_i(\xi_0)\right)\tilde{\alpha}_{j'}^{(i)}(\xi_0) = 0, \quad \sum_{j'=1}^{j'_{max}}\tilde{\alpha}_{j'}^{(i')}(\xi_0)\tilde{\alpha}_{j'}^{(i)}(\xi_0) = \delta_{ii'}, (6)$$

where the potentials $U_{ij'}^{pair}$ and $V_{ij'}(\xi_0)$ are expressed in terms of the integrals

$$U_{ij'}^{pair} = \int d^{A-1}\boldsymbol{\xi}\Phi_i^{S(A)}(\boldsymbol{\xi})U^{eff}(\boldsymbol{\xi})\Phi_{j'}^{S(A)}(\boldsymbol{\xi}), \tag{7}$$

$$V_{ij'}(\xi_0) = \int d^{A-1}\boldsymbol{\xi}\Phi_i^{S(A)}(\boldsymbol{\xi})\left(\sum_{k=1}^{A}V(x_k(\xi_0, \boldsymbol{\xi}))\right)\Phi_{j'}^{S(A)}(\boldsymbol{\xi}). \tag{8}$$

The parametric *algorithm SCR*, i.e., *algorithm PSCR*, for solving the above parametric eigenvalue problem was implemented by means of subroutines [19, 20], or in the single-particle approximation by means of the subroutine [9] in CAS MAPLE and FORTRAN environment.

(G) If $U_{ij'}(\xi_0) = U_{ij'}^{pair}$ are independent on ξ_0, then $\tilde{\epsilon}_i(\xi_0) = \tilde{\epsilon}_i$ and $\tilde{\alpha}_{j'}^{(i)}(\xi_0) = \tilde{\alpha}_{j'}^{(i)}$ are also independent of ξ_0, and (5) reduces to $\tilde{\Phi}_i(\boldsymbol{\xi}) = \sum_{j'=1}^{j'_{max}}\tilde{\alpha}_{j'}^{(i)}\Phi_{j'}^{S(A)}(\boldsymbol{\xi})$.

(O) If $V^{pair}(x_{ij}) = V^{hosc}(x_{ij})$ and $U_{ij'}^{pair} = 0$, then $\tilde{\epsilon}_i = E_i^{S(A)}$ and $\tilde{\alpha}_{j'}^{(i)} = \delta_{ij'}$.

For the short-range barrier potentials $V(\xi_0, x_i(\boldsymbol{\xi}))$ in terms of the asymptotic cluster functions $\tilde{\Phi}_j(\boldsymbol{\xi}) \to \tilde{\Phi}_j(\xi_0, \boldsymbol{\xi})$ at $|\xi_0| \to \infty$ the asymptotic boundary conditions for the solution $\Psi(\xi_0, \boldsymbol{\xi}) = \{\Psi_{i_o}(\xi_0, \boldsymbol{\xi})\}_{i_o=1}^{N_o}$ in the asymptotic region $|\boldsymbol{\xi}|/|\xi_0| \ll 1$ have the form [16]

$$\overset{\leftrightarrow}{\Psi_{i_o}}(\xi_0 \to \pm\infty, \boldsymbol{\xi}) \to \tilde{\Phi}_{i_o}(\boldsymbol{\xi})\frac{\exp(\mp\imath(p_{i_o}\xi_0))}{\sqrt{p_{i_o}}} + \sum_{j=1}^{N_o}\tilde{\Phi}_j(\boldsymbol{\xi})\frac{\exp(\pm\imath(p_j\xi_0))}{\sqrt{p_j}}\overset{\leftarrow}{R_{ji_o}}(E),$$

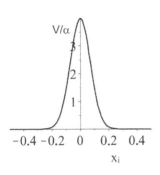 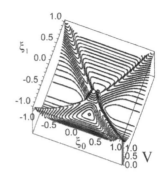

Fig. 1. The Gaussian-type potential (16) at $\sigma = 0.1$ (in oscillator units) and the corresponding 2D barrier potential at $\alpha = 1/10$, $\sigma = 0.1$

$$\Psi_{i_o}^{\overleftarrow{\rightarrow}}(\xi_0 \to \mp\infty, \boldsymbol{\xi}) \to \sum_{j=1}^{N_o} \tilde{\Phi}_j(\boldsymbol{\xi}) \frac{\exp\left(\mp\imath\,(p_j\xi_0)\right)}{\sqrt{p_j}} T_{ji_o}^{\overleftarrow{\rightarrow}}(E), \tag{9}$$

$$\Psi_{i_o}^{\overleftarrow{\rightarrow}}(\xi_0, |\boldsymbol{\xi}| \to \infty) \to 0.$$

Here $v = \leftarrow, \rightarrow$ indicates the initial direction of the particle motion along the ξ_0 axis, N_o is the number of open channels at the fixed energy E and momentum $p_{i_o}^2 = E - E_{i_o} > 0$ of cluster; $R_{ji_o}^{\leftarrow} = R_{ji_o}^{\leftarrow}(E)$, $R_{ji_o}^{\rightarrow} = R_{ji_o}^{\rightarrow}(E)$ and $T_{ji_o}^{\leftarrow} = T_{ji_o}^{\leftarrow}(E)$, $T_{ji_o}^{\rightarrow} = T_{ji_o}^{\rightarrow}(E)$ are the unknown amplitudes of the reflected and transmitted waves. We can rewrite Eqs. (9) in the matrix form $\boldsymbol{\Psi} = \tilde{\boldsymbol{\Phi}}^T \boldsymbol{F}$ describing the incident wave and the outgoing waves at $\xi_0^+ \to +\infty$ and $\xi_0^- \to -\infty$ as

$$\begin{pmatrix} \boldsymbol{F}_\rightarrow(\xi_0^+) & \boldsymbol{F}_\leftarrow(\xi_0^+) \\ \boldsymbol{F}_\rightarrow(\xi_0^-) & \boldsymbol{F}_\leftarrow(\xi_0^-) \end{pmatrix} = \begin{pmatrix} \mathbf{0} & \mathbf{X}^{(-)}(\xi_0^+) \\ \mathbf{X}^{(+)}(\xi_0^-) & \mathbf{0} \end{pmatrix} + \begin{pmatrix} \mathbf{0} & \mathbf{X}^{(+)}(\xi_0^+) \\ \mathbf{X}^{(-)}(\xi_0^-) & \mathbf{0} \end{pmatrix} \mathbf{S}. \tag{10}$$

Here the unitary and symmetric scattering matrix $\mathbf{S}$

$$\mathbf{S} = \begin{pmatrix} \mathbf{R}_\rightarrow & \mathbf{T}_\leftarrow \\ \mathbf{T}_\rightarrow & \mathbf{R}_\leftarrow \end{pmatrix}, \quad \mathbf{S}^\dagger\mathbf{S} = \mathbf{S}\mathbf{S}^\dagger = \mathbf{I}, \tag{11}$$

where $\mathbf{S}^\dagger$ is the conjugate transpose of $\mathbf{S}$. It is composed of the matrices, whose elements are reflection and transmission amplitudes that enter Eqs. (9) and possess the following properties[16, 17]:

$$\mathbf{T}_\rightarrow^\dagger\mathbf{T}_\rightarrow + \mathbf{R}_\rightarrow^\dagger\mathbf{R}_\rightarrow = \mathbf{I}_{oo} = \mathbf{T}_\leftarrow^\dagger\mathbf{T}_\leftarrow + \mathbf{R}_\leftarrow^\dagger\mathbf{R}_\leftarrow,$$
$$\mathbf{T}_\rightarrow^\dagger\mathbf{R}_\leftarrow + \mathbf{R}_\rightarrow^\dagger\mathbf{T}_\leftarrow = \mathbf{0} = \mathbf{R}_\leftarrow^\dagger\mathbf{T}_\rightarrow + \mathbf{T}_\leftarrow^\dagger\mathbf{R}_\rightarrow, \tag{12}$$
$$\mathbf{T}_\rightarrow^T = \mathbf{T}_\leftarrow, \quad \mathbf{R}_\rightarrow^T = \mathbf{R}_\rightarrow, \quad \mathbf{R}_\leftarrow^T = \mathbf{R}_\leftarrow.$$

4   Close-Coupling Equations in the SCR

We seek for the solution of problem (3) in the symmetrized coordinates in the form of Galerkin (G) expansion over the asymptotic cluster functions $\tilde{\Phi}_j(\boldsymbol{\xi})$

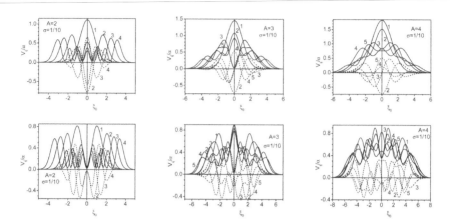

Fig. 2. Diagonal V_{jj} (solid lines) and nondiagonal V_{j1}, (dashed lines) effective potentials for $A = 2$, $A = 3$ and $A = 4$ of the S- (upper panels) and A- (lower panels) of the particles at $\sigma = 1/10$

corresponding to the eigenvalues $\tilde{\epsilon}_i$, which are also independent of ξ_0, from (6) under the (G) condition, with unknown coefficient functions $\chi_{ji_o}(\xi_0)$:

$$\Psi_{i_o}(\xi_0, \boldsymbol{\xi}) = \sum_{j=1}^{j_{\max}} \tilde{\Phi}_j(\boldsymbol{\xi})\chi_{ji_o}(\xi_0), \ \chi_{ji_o}(\xi_0) = \int d^{A-1}\boldsymbol{\xi}\tilde{\Phi}_j(\boldsymbol{\xi})\Psi_{i_o}(\xi_0, \boldsymbol{\xi}). \quad (13)$$

The set of close-coupling Galerkin equations in the symmetrized coordinates has the form

$$\left[-\frac{d^2}{d\xi_0^2} + \tilde{\epsilon}_i - E\right]\chi_{ii_o}(\xi_0) + \sum_{j=1}^{j_{\max}} \tilde{V}_{ij}(\xi_0)\chi_{ji_o}(\xi_0) = 0, \quad (14)$$

where the effective potentials $\tilde{V}_{ij}(\xi_0)$ are calculated using the set of eigenvectors $\tilde{\alpha}_{j'}^{(i)}$ of the noparametric algebraic problem (6) under the above condition (G): $U_{ij'}(\xi_0) = U_{ij'}^{pair} \neq 0$,

$$\tilde{V}_{ij}(\xi_0) = \sum_{j'=1}^{j'_{\max}} \sum_{j''=1}^{j'_{\max}} \tilde{\alpha}_{j'}^{(i)} V_{j'j''}(\xi_0)\tilde{\alpha}_{j''}^{(j)}, \quad (15)$$

and the integrals $V_{ij'}(\xi_0)$ are defined in (8) and calculated in CAS MAPLE. In the examples considered below, we put $U_{ij'}(\xi_0) = U_{ij'}^{pair} = 0$ in (6), then we have the (O) condition: $\tilde{\epsilon}_i = E_i^{S(A)}$, $\tilde{\alpha}_{j'}^{(i)} = \delta_{ij'}$ and $\tilde{V}_{ij}(\xi_0) = V_{ij}(\xi_0)$. The repulsive barrier is chosen to have the Gaussian shape

$$V(x_i) = \frac{\alpha}{\sqrt{2\pi\sigma}}\exp(-\frac{x_i^2}{\sigma^2}). \quad (16)$$

Table 1. Resonance values of the energy E_S (E_A) for S (A) states for $A = 2, 3, 4$ ($\sigma = 1/10$, $\alpha = 20$) with approximate eigenvalues E_i^D, for the first ten states $i = 1, ..., 10$, calculated using the truncated oscillator basis (D) till $j_{max} = 136, 816, 1820$ at $A = 2, 3, 4$. The asterisk labels two overlapping peaks of transmission probability

i	1	2	3	4	5	6	7	8	9	10
					$A = 2$					
E_S	5.72	9.06	9.48	12.46	12.57	13.46	15.74	15.78	16.65	17.41
E_A	5.71	9.06	9.48	12.45	12.57	13.45	15.76*	15.76*	16.66	17.40
E_i^D	5.76	9.12	9.53	12.52	12.64	13.52	15.81	15.84	16.73	17.47
					$A = 3$					
E_S	8.18	11.11		12.60	13.93		14.84	15.79		16.67
	8.31	11.23			14.00		14.88			16.73
E_A			11.55			14.46			16.18	
			11.61			14.56			16.25	
E_i^D	8.19	11.09	11.52	12.51	13.86	14.42	14.74	15.67	16.11	16.53
					$A = 4$					
E_S	10.12	11.89	12.71	14.86	15.19	15.41	15.86	16.37	17.54	17.76
E_i^{D31}	10.03		12.60	14.71	15.04			16.18	17.34	17.56
E_i^{D22}		11.76				15.21	15.64			

Figure 1 illustrates the Gaussian potential and the corresponding barrier potentials in the symmetrized coordinates at $A = 2$. This potential has the oscillator-type shape, and two barriers are crossing at the right angle. In the case $A \geq 3$, the hyperplanes of barriers are crossing at the right angle, too.

The effective potentials $V_{ij}(\xi_0)$ calculated using the *algorithm SCR* [18] and *algorithm DC* (see Section 5), are shown in Fig. 2. In comparison with the symmetric basis, for antisymmetric one the increase of the numbers i and/or j results in stronger oscillation of the effective potentials V_{ij} and weaker decrease of them to zero at $\xi_0 \to \infty$. At $A = 2$, all effective potentials are even functions, and at $A \geq 3$, some effective potentials are odd functions.

Thus, the scattering problem (3) with the asymptotic boundary conditions (9) is reduced to the boundary-value problem for the set of close-coupling equations in the Galerkin form (14) under the boundary conditions at $d = 1$, $\xi_0 = \xi_{min}$ and $\xi_0 = \xi_{max}$:

$$\frac{dF(\xi_0)}{d\xi_0}\bigg|_{\xi_0=\xi_{min}} = \mathcal{R}(\xi_{min})F(\xi_{min}), \quad \frac{dF(\xi_0)}{d\xi_0}\bigg|_{\xi_0=\xi_{max}} = \mathcal{R}(\xi_{max})F(\xi_{max}), \quad (17)$$

where $\mathcal{R}(\xi)$ is an unknown $j_{max} \times j_{max}$ matrix function, $F(\xi_0) = \{\chi_{i_o}(\xi_0)\}_{i_o=1}^{N_o} = \{\{\chi_{ji_o}(\xi_0)\}_{j=1}^{j_{max}}\}_{i_o=1}^{N_o}$ is the required $j_{max} \times N_o$ matrix solution, and N_o is the number of open channels, $N_o = \max_{2E \geq \check{e}_j} j \leq j_{max}$, calculated using the third version of KANTBP 3.0 program [21, 22], implemented in CAS MAPLE and FORTRAN environment and described in [16, 17].

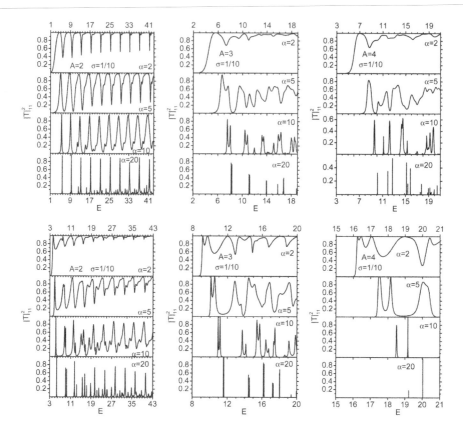

Fig. 3. The total transmission probability $|T|^2_{11}$ vs energy E (in oscillator units) for the system of $A = 2, 3, 4$ S- (upper panels) and A- (lower panels) particles coupled by the oscillator potential and being initially in the ground cluster state penetrating through the repulsive Gaussian-type potential barriers (16) with $\sigma = 0.1$ and $\alpha = 2, 5, 10, 20$

5   Resonance Transmission of a Few Coupled Particles

In the (O) case, i.e., $V^{pair}(x_{ij}) = V^{hosc}(x_{ij})$, the solution of the scattering problem described above yields the reflection and transmission amplitudes $R_{ji_o}(E)$ and $T_{ji_o}(E)$ that enter the asymptotic boundary conditions (9) as unknowns. $|R_{ji_o}(E)|^2$ ($|T_{ji_o}(E)|^2$) is the probability of a transition to the state described by the reflected (transmitted) wave and, hence, will be referred as the reflection (transmission) coefficient. Note that $|R_{ji_o}(E)|^2 + |T_{ji_o}(E)|^2 = 1$.

In Figs. 3 and 4, we show the energy dependence of the total transmission probability $|T|^2_{ii} = \sum_{j=1}^{N_o} |T_{ji}(E)|^2$. This is the probability of a transition from a chosen state i into any of N_o states found from Eq. (13) by solving the boundary-value problem in the Galerkin form, (14) and (17), using the KANTBP 3.0 program [21, 22] on the finite-element grid $\Omega_\xi\{-\xi_0^{\max}, \xi_0^{\max}\}$ with N_{elem} fourth-order Lagrange elements between the nodes. For S-solutions at $A = 2, 3, 4$

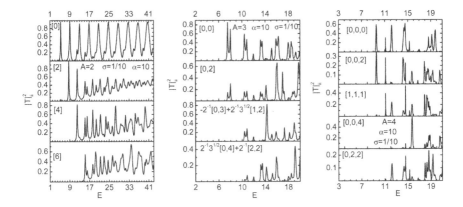

Fig. 4. The total transmission probability $|T|_{ii}^2$ vs the energy E (in oscillator units) for the system of $A = 2, 3, 4$ particles, coupled by the oscillator potential and being initially in the ground and excited S-states, penetrating through the repulsive Gaussian-type potential barriers (16) with $\sigma = 0.1$ and $\alpha = 10$. We use the notation of the S-states, $[i_1, ..., i_{A-1}] = 1/\sqrt{N_\beta} \sum_{i'_1, ..., i'_{A-1}} \prod \bar{\Phi}_{i'_k}(\xi_k)$, with summation over all (N_β) multiset permutations of $i_1, ..., i_{A-1}$ of $A - 1$-dimensional oscillator functions [18]

the following parameters were used: $j_{\max} = 13, 21, 39$, $\xi_0^{\max} = 9.3, 10.5, 12.8$, $N_{\text{elem}} = 664, 800, 976$, while for A-solutions we used $j_{\max} = 13, 16, 15$, $\xi_0^{\max} = 9.3, 10.5, 12.2$, $N_{\text{elem}} = 664, 800, 976$ that yield an accuracy of the solutions of an order of the fourth significant figures.

Figure 3 demonstrates non-monotonic behavior of the total transmission probability versus the energy, and the observed resonances are manifestations of the quantum transparency effect. With the barrier height increasing, the peaks become narrower, and their positions shift to higher energies. The multiplet structure of the peaks in the symmetric case is similar to that in the antisymmetric case. For three particles, the major peaks are double, while for two and four particles, they are single. For $A = 2$ and $\alpha = 10, 20$, one can observe the additional multiplets of small peaks.

Figure 4 illustrates the energy dependence of the total transmission probabilities from the exited states. As the energy of the initial excited state increases, the transmission peaks demonstrate a shift towards higher energies, the set of peak positions keeping approximately the same as for the transitions from the ground state and the peaks just replacing each other, like it was observed in the model calculations [12]. For example, for $A = 3$, the position of the third peak for transitions from the first two states ($E = 10.4167$ and $E = 10.4156$) coincides with the position of the first peak for the transitions from the second two states ($E = 10.4197$ and $E = 10.4298$).

Calculation of Energy Position of the Barrier Quasistationary States. In the considered case, the potential barrier $V(x_i)$ is narrow, and $V^{pair}(x_{ij}) = V^{hosc}(x_{ij})$, so that we solve Eq. (1) in the Cartesian coordinates $x_1, ..., x_A$ in one

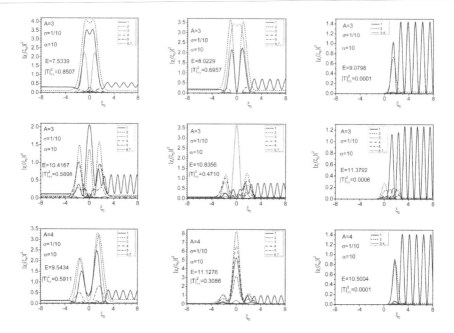

Fig. 5. The probability densities $|\chi_i(\xi_0)|^2$ for the coefficient functions of the decomposition (13), representing the incident wave function of the ground S-state of the particles at the values of the collision energy E corresponding to individual maxima and minima of the transmission coefficient in Fig. 3. The parameters of the Gaussian barrier are $\alpha = 10$ and $\sigma = 0.1$

of the $2^A - 2$ subdomains, defined as $p_i x_i > 0$, $p_i = \pm 1$, under the Dirichlet conditions (DC): $\Psi(x_1, ..., x_A)|_{\cup_{i=1}^{A}\{x_i=0\}} = 0$ at the internal boundaries $\cup_{i=1}^{A}\{x_i = 0\}$. Here the value $p_i = \pm 1$ indicates the location of the ith particle at the right or left side of the barrier, respectively. Thus, in the DC procedure we seek for the solution in the form of a Galerkin expansion over the orthogonal truncated oscillator basis, $\Psi_i^D(\mathbf{x}) = \sum_{j=1}^{j_{max}} \bar{\Phi}_j(\mathbf{x})\Psi_{ji}^D$ composed of A-dimensional harmonic oscillator functions $\bar{\Phi}_j(\mathbf{x})$, odd in each of the Cartesian coordinates $x_1, ..., x_A$ in accordance with the above DCs, with unknown coefficients Ψ_{ji}^D. As a result, we arrive at the algebraic eigenvalue problem $D\Psi^D = \Psi^D E^D$ with a dense real-symmetric $j_{max} \times j_{max}$ matrix. So, in the DC procedure we seek for an approximate solution in one of the potential wells, i.e., we neglect the tunnelling through the barriers between wells. Therefore, we cannot observe the splitting inherent in exact eigenvalues corresponding to S and A eigenstates, differing in permutation symmetry. However, we can explain the mechanism of their appearance and give their classification, which is important, too. This *algorithm DC* was implemented in CAS MAPLE and FORTRAN environment.

Remark. The DC procedure is similar to solving Eq. (3) in the symmetrized coordinates $\xi_0, \boldsymbol{\xi}$ related to the Cartesian ones by Eq. (2), implemented the following two steps:

(i) we approximate the narrow barriers by impenetrable walls $x_k(\xi_0, \boldsymbol{\xi}) = 0$;
(ii) we superpose these mutually perpendicular walls with the coordinate hyperplanes using rotations.

Actually, the two approaches yield the same boundary-value problem formulated in different coordinates (1), (3).

The algorithm DC:

Input:
A is the number of identical particles;
x_k, $k = 1, ..., A$ are the Cartesian coordinates of the identical particles;
$p_k = \pm 1$ indicates the location of the kth particle ;
j_{max} is the number of the eigenfunctions of A-dimensional harmonic oscillator;
Output:
$D = \{D_{j'j}\}$ is the $j_{max} \times j_{max}$ matrix ;
E_i^D and Ψ_{ji}^D are the real-value eigenenergies and eigenvectors;
Local:
$\Phi_j = \sqrt{2^A} \prod_{k=1}^{A} \bar{\Phi}_{i_k}(x_k)$;
$I(i_k', i_k) = \int_0^\infty \bar{\Phi}_{i_k'}(x) \bar{\Phi}_{i_k}(x) dx = \frac{2^{(i_k'+i_k)/2} {}_2F_1(i_k', i_k; (2-i_k'-i_k)/2; 1/2)}{\Gamma((2-i_k'-i_k)/2)\sqrt{i_k'! i_k!}}$;
$\Gamma(*)$ is the gamma-function, ${}_2F_1(*, *; *; *)$ is the hypergeometric function;
1: $Eq := (-\Delta + \sum(p_k x_k - p_{k'} x_{k'})/2A)$;
2: $Eq := \sqrt{A/(A-1)}(Eq, \Delta \to \Delta/(A/(A-1)), x_k \to x_k \sqrt[4]{A/(A-1)}$;
3: $Eq := Eq, p_k^2 \to 1, \Delta = \sum_k (x_k^2 - (2n_k + 1))$;
4: $Eq := Eq \prod \bar{\Phi}_{i_k}(x_k)$;
5: $Eq := x_k = (\sqrt{i_k + 1}\bar{\Phi}_{i_k+1}(x_k) + \sqrt{i_k}\bar{\Phi}_{i_k-1}(x_k))/(\sqrt{2}\bar{\Phi}_{i_k}(x_k))$;
6: **for** $j, j' = 1, ..., j_{max}$ **do**
    $D_{j'j} := \Phi_{i_k}(x_k) \to I(i_k', i_k)$;
    end for
7: $D\Psi_{ji}^D = \Psi_{ji}^D E_i^D \to E_i^D$ and Ψ_{ji}^D;

In Table 1, we present the resonance values of the energy E_S (E_A) calculated by solving the boundary-value problem (14) and (17), using the KANTBP 3.0 program, for S (A) states at $A = 2, 3, 4$ $\sigma = 1/10$, $\alpha = 20$ that correspond to the maxima of transmission coefficients $|T|_{ii}^2$ in Fig. 3 up to values of energy $E < 18$ and corresponding resonance values of the energy E_D calculated by means of the algorithm DC. One can see that the accepted approximation of the narrow barrier with impermeable walls using in the algorithm DC provides the appropriate approximations E_i^D of the above high accuracy results E_S (E_A) with the error smaller than 2%. Below we give a comparison and qualitative analysis of the obtained results.

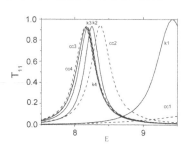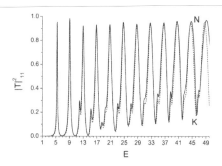

Fig. 6. a. The comparison of convergence rate of Galerkin (cc*) and Kantorovich (k*) close-coupling expansions in calculations of transmission coefficient $|T|^2_{11}$ for the S-states, $A = 2$ at $\alpha = 10$, $\sigma = 0.1$, like epure of the first peak from Fig. 3. b. The comparison of Galerkin and Kantorovich methods (G=K) with Finite-Difference Numerov method (N)

For two particles, $A = 2$ (see Fig. 1), there are two symmetric potential wells. In each of them both symmetric and asymmetric wave functions are constructed. Since the potential barrier separating the wells is sufficiently high, the appropriate energies are closely spaced, so that each level describes the states of both S and A type. The lower energy levels form a sequence "singlet-doublet-triplet, etc.", which is seen in Fig. 3. The resonance transmission energies for a pair of particles in S states are lower than that for a pair of those in A states. This is due to the fact that in the vicinity of the collision point, the wave function is zero. When $A = 3$ there are six similar wells, three of them at each side of the plane $\xi_0 = 0$. The symmetry with respect to the plane $\xi_0 = 0$ explains the presence of doublets. The presence of states with definite symmetry is associated with the fact that the axis ξ_0 is a third-order symmetry axis. However, in contrast to the case $A = 2$, one can obtain either S or A combinations of states. For example, the first four solutions of the problem, in one of the wells (e.g., the one restricted with the pair-collision planes "13" and "23") possess the dominant components $2\sqrt{2}\bar{\Phi}_1(x_1)\bar{\Phi}_1(x_2)\bar{\Phi}_1(x_3)$, $2(\bar{\Phi}_1(x_1)\bar{\Phi}_3(x_2) + \bar{\Phi}_3(x_1)\bar{\Phi}_1(x_2))\bar{\Phi}_1(x_3)$, $2(\bar{\Phi}_1(x_1)\bar{\Phi}_3(x_2) - \bar{\Phi}_3(x_1)\bar{\Phi}_1(x_2))\bar{\Phi}_1(x_3)$, $2\sqrt{2}\bar{\Phi}_1(x_1)\bar{\Phi}_1(x_2)\bar{\Phi}_3(x_3)$. Note that the first, second, and fourth of these functions are symmetric with respect to the permutation $x_1 \leftrightarrow x_2$, while the third one is antisymmetric. Hence, in all six wells using the first four solutions one can obtain six S and two A states.

When $A = 4$ there are 14 wells. Six wells at the center correspond to the case when two particles are located at one side of the barrier and the rest two at the other side. The corresponding eigenenergy is denoted E_i^{D22}. The rest eight wells correspond to the case when one particle is located at one side of the barrier and the rest three at the other side. The corresponding eigenenergy is denoted E_i^{D31}. For these states, doublets must be observed, similar to the case of three particles. However, the separation between the energy levels is much smaller, because the 4-well groups are strongly separated by two barriers, instead of only one barrier in the case $A = 3$.

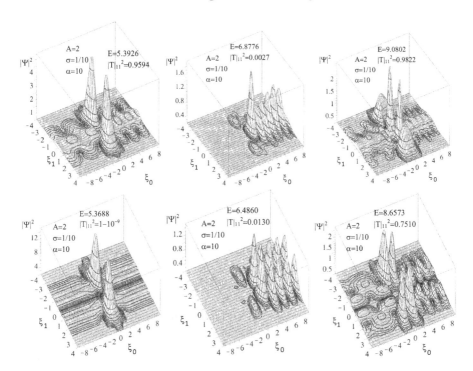

Fig. 7. The profiles of probability densities $|\Psi(\xi_0, \xi_1)|^2$ for the S- (upper panel) and A- (lower panel) states of $A = 2$ particles, revealing resonance transmission and total reflection at resonance energies, shown in Figs. 3

The necessary condition for the quasi-stationary state being symmetric (antisymmetric) is that the wave functions must be symmetric (antisymmetric) with respect to those coordinates x_i and x_j, for which $p_i = p_j$.

The effect of quantum transparency is caused by the existence of barrier quasistationary states imbedded in the continuum. Fig. 5 shows that in the case of resonance transmission, the wave functions depending on the center-of-mass variable ξ_0 are localized in the vicinity of the potential barrier center ($\xi_0 = 0$).

For the energy values corresponding to some of the transmission coefficient peaks in Fig. 3 at $\alpha = 10$ within the effective range of barrier potential action, the wave functions demonstrate considerable increase (from two to ten times) of the probability density in comparison with the incident unit flux. This is a fingerprint of quasistationary states, which is not a quantitative definition, but a clear evidence in favor of their presence in the system[23]. In the case of total reflection, the wave functions are localized at the barrier side, on which the wave is incident, and decrease to zero within the effective range of the barrier action.

Note that the explicit explanation of the quantum transparency effect is achieved in the framework of Kantorovich close-coupling equations because of the multi-barrier potential structure of the effective potential, appearing explicitly even in the diagonal or adiabatic approximation, in particular, in the S case for $A = 2$ [1, 16]. Nevertheless, in Galerkin close-coupling equations, the multi-barrier potential structure of the effective potential is observed explicitly in the A case (see Fig. 2).

As an example, Fig. 6a, which is an epure of Fig. 3, shows the comparison of convergence rates of Galerkin (13) and Kantorovich close-coupling expansions in calculations of transmission coefficient $|T|^2_{11}$ for S wave functions, $A = 2$ at $\alpha = 10$, $\sigma = 0.1$. One can see that the diagonal approximation of the Kantorovich method provides better approximations of the positions of the transmission coefficient $|T|^2_{11}$ resonance peaks. With the increasing number of basis functions, i.e., the number $j_{\max}$ of close-coupling equations with respect to the center-of-mass coordinates in Galerkin (14) and Kantorovich form, respectively, the convergence rates are similar and confirm the results obtained by solving the problem by means of the Finite-Difference Numerov method in 2D domain [1], see Fig. 6 b. This is true for the considered short-range potentials (16), while for long-range potentials of the Coulomb type, the Kantorovich method can be more efficient [16].

Figure 7 shows the profiles of $|\boldsymbol{\Psi}|^2 \equiv |\boldsymbol{\Psi}^{(-)}_{Em\rightarrow}|^2$ for the S and A total wave functions of the continuous spectrum in the (ξ_0, ξ_1) plane with $A = 2$, $\alpha = 10$, $\sigma = 1/10$ at the resonance energies of the first and the second maximum and the first minimum of the transmission coefficient demonstrating *resonance transmission* and *total reflection*, respectively. It is seen that in the case of resonance transmission, the redistribution of energy from the center-mass degree of freedom to the internal (transverse) ones takes place, i.e., the transverse oscillator undergoes a transition from the ground state to the excited state, while in the total reflection, the redistribution of energy is extremely small, and the transverse oscillator returns to infinity in the same state.

6   Conclusion

We considered a model cluster of A identical particles bound by the oscillator-type potential that undergo quantum tunnelling through the short-range repulsive barrier potentials. The model was formulated in the new representation, which we referred as the Symmetrized Coordinate Representation (SCR, see forthcoming paper [18]), that implies construction of symmetric (asymmetric) combinations of oscillator wave functions in new coordinates. The approach was implemented as a complex of the symbolic-numeric algorithms and programs.

For clarity, a system of several identical particles was considered in one-dimensional Euclidian space ($d = 1$). We calculated only the spatial part of the wave function, symmetric or antisymmetric under permutation of A identical particles. If necessary, the spin part of the wave function can be introduced using the conventional procedure for more rigorous calculation.

We analyzed the effect of quantum transparency, i.e., the resonance tunnelling of several bound particles through repulsive potential barriers. We demonstrated that this effect is due to the existence of sub-barrier quasistationary states imbedded in the continuum. For the considered type of symmetric Gaussian barrier potential, the energies of the S and A quasistationary states are slightly different because of the similarity of the multiplet structure of oscillator energy levels at a fixed number of particles. This fact explains a similar behavior of transmission coefficients for S and A states shifted by threshold energies. The multiplet structure of these states is varied with increasing the number of particles, e.g., for three particles, the major peaks are double, while for two and four particles, they are single. Our calculations have also shown that with increasing the energy of the initial excited state of few-body clusters, the transmission peaks demonstrate a shift towards higher energies, the set of peak positions keeping approximately the same as for the transitions from the ground state and the peaks just skipping from one position to another.

The proposed approach can be adapted and applied to tetrahedral-symmetric nuclei, quantum diffusion of molecules and micro-clusters through surfaces, and fragmentation mechanism in producing very neutron-rich light nuclei. In connection with the intense search for superheavy nuclei, a particularly significant application of the proposed approach is the mathematically correct analysis of mechanisms of sub-barrier fusion of heavy nuclei and the study of fusion rate enhancement by means of resonance tunnelling.

The authors thank Professors V.P. Gerdt, A. Góźdź, and F.M. Penkov for collaboration. The work was supported by grants 13-602-02 JINR, 11-01-00523 and 13-01-00668 RFBR, 0602/GF MES RK and the Bogoliubov-Infeld program.

References

1. Pen'kov, F.M.: Quantum transmittance of barriers for composite particles. JETP 91, 698–705 (2000)
2. Pijper, E., Fasolino, A.: Quantum surface diffusion of vibrationally excited molecular dimers. J. Chem. Phys. 126, 014708-1–014708-10 (2007)
3. Bondar, D.I., Liu, W.-K., Ivanov, M.Y.: Enhancement and suppression of tunneling by controlling symmetries of a potential barrier. Phys. Rev. A 82, 052112-1–052112-9 (2010)
4. Shegelski, M.R.A., Pittman, J., Vogt, R., Schaan, B.: Time-dependent trapping of a molecule. European Phys. J. Plus 127, 17-1–17-13 (2012)
5. Ershov, S.N., Danilin, B.V.: Breakup of two-neutron halo nuclei. Phys. Part. Nucl. 39, 1622–1720 (2008)
6. Nesterov, A.V., Arickx, F., Broeckhove, J., Vasilevsky, V.S.: Three-cluster description of properties of light nuclei with neutron and proton access within the algebraic version of the resonating group method. Phys. Part. Nucl. 41, 1337–1426 (2010)
7. Hofmann, H.: Quantummechanical treatment of the penetration through a two-dimensional fission barrier. Nucl. Phys. A 224, 116–139 (1974)
8. Krappe, H.J., Möhring, K., Nemes, M.C., Rossner, H.: On the interpretation of heavy-ion sub-barrier fusion data. Z. Phys. A 314, 23–31 (1983)
9. Cwiok, S., Dudek, J., Nazarewicz, W., Skalski, J., Werner, T.: Single-particle energies, wave functions, quadrupole moments and g-factors in an axially deformed Woods-Saxon potential with applications to the two-centre-type nuclear problems. Comput. Phys. Communications 46, 379–399 (1987)

10. Hagino, K., Rowley, N., Kruppa, A.T.: A program for coupled-channel calculations with all order couplings for heavy-ion fusion reactions. Comput. Phys. Commun. 123, 143–152 (1999)
11. Zagrebaev, V.I., Samarin, V.V.: Near-barrier fusion of heavy nuclei: coupling of channels. Phys. Atom. Nucl. 67, 1462–1477 (2004)
12. Ahsan, N., Volya, A.: Quantum tunneling and scattering of a composite object reexamined. Phys. Rev. C 82, 064607-1–064607-19 (2010)
13. Shotter, A.C., Shotter, M.D.: Quantum mechanical tunneling of composite particle systems: Linkage to sub-barrier nuclear reactions. Phys. Rev. C 83, 054621-1–054621-11 (2011)
14. Shilov, V.M.: Sub-barrier fusion of intermediate and heavy nuclear systems. arXiv:1012.3683 [nucl-th] Phys. Atom. Nucl. 75, 485–490 (2012)
15. Chuluunbaatar, O., Gusev, A.A., Derbov, V.L., Krassovitskiy, P.M., Vinitsky, S.I.: Channeling problem for charged particles produced by confining environment. Phys. Atom. Nucl. 72, 768–778 (2009)
16. Gusev, A.A., Vinitsky, S.I., Chuluunbaatar, O., Gerdt, V.P., Rostovtsev, V.A.: Symbolic-numerical algorithms to solve the quantum tunneling problem for a coupled pair of ions. In: Gerdt, V.P., Koepf, W., Mayr, E.W., Vorozhtsov, E.V. (eds.) CASC 2011. LNCS, vol. 6885, pp. 175–191. Springer, Heidelberg (2011)
17. Gusev, A.A., Chuluunbaatar, O., Vinitsky, S.I.: Computational scheme for calculating reflection and transmission matrices, and corresponding wave functions of multichannel scattering problems. In: Uvarova, L.A. (ed.) Proc. Second International Conference "The Modeling of Non-linear Processes and Systems", Yanus, Moscow, pp. 978–975 (2011)
18. Gusev, A., Vinitsky, S., Chuluunbaatar, O., Rostovtsev, V., Hai, L., Derbov, V., Góźdź, A., Klimov, E.: Symbolic-numerical algorithm for generating cluster eigenfunctions: identical particles with pair oscillator interactions. In: Gerdt, V.P., Koepf, W., Mayr, E.W., Vorozhtsov, E.V. (eds.) CASC 2013. LNCS, vol. 8136, pp. 155–168. Springer, Heidelberg (2013)
19. Vinitsky, S.I., Gerdt, V.P., Gusev, A.A., Kaschiev, M.S., Rostovtsev, V.A., Samoilov, V.N., Tupikova, T.V., Chuluunbaatar, O.: A symbolic-numerical algorithm for the computation of matrix elements in the parametric eigenvalue problem. Programming and Computer Software 33, 105–116 (2007)
20. Bunge, C.F.: Fast eigensolver for dense real-symmetric matrices. Comput. Phys. Communications 138, 92–100 (2001)
21. Chuluunbaatar, O., Gusev, A.A., Vinitsky, S.I., Abrashkevich, A.G.: KANTBP 2.0: New version of a program for computing energy levels, reaction matrix and radial wave functions in the coupled-channel hyperspherical adiabatic approach. Comput. Phys. Commun. 179, 685–693 (2008)
22. Chuluunbaatar, O., Gusev, A.A., Vinitsky, S.I., Abrashkevich, A.G.: KANTBP 3.0 - New version of a program for computing energy levels, reflection and transmission matrices, and corresponding wave functions in the coupled-channel adiabatic approach, Program library "JINRLIB", http://wwwinfo.jinr.ru/programs/jinrlib/kantbp/indexe.html
23. de Carvalho, C.A.A., Nussenzweig, H.M.: Time delay. Phys. Rept. 364, 83–174 (2002)

Author Index